Agricultural Unmanned Systems: Empowering Agriculture with Automation

Agricultural Unmanned Systems: Empowering Agriculture with Automation

Editor

Shubo Wang

MDPI • Basel • Beijing • Wuhan • Barcelona • Belgrade • Novi Sad • Cluj • Manchester

Editor
Shubo Wang
School of Automation
Qingdao University
Qingdao
China

Editorial Office
MDPI AG
Grosspeteranlage 5
4052 Basel, Switzerland

This is a reprint of articles from the Special Issue published online in the open access journal *Agronomy* (ISSN 2073-4395) (available at: https://www.mdpi.com/journal/agronomy/special_issues/QMUH6L819W).

For citation purposes, cite each article independently as indicated on the article page online and as indicated below:

Lastname, A.A.; Lastname, B.B. Article Title. *Journal Name* **Year**, *Volume Number*, Page Range.

ISBN 978-3-7258-1620-0 (Hbk)
ISBN 978-3-7258-1619-4 (PDF)
doi.org/10.3390/books978-3-7258-1619-4

Cover image courtesy of Shubo Wang

Contents

About the Editor

Shubo Wang

Shubo Wang is a doctoral and master's supervisor and an outstanding young talent at Qingdao University. He obtained a Ph.D. in Mechanical Engineering from China Agricultural University in 2021, conducted postdoctoral research at China Agricultural University from August 2021 to July 2023, and has been teaching at the School of Automation, Qingdao University since August 2023. His main research areas are autonomous navigation, guidance, and control technologies for unmanned systems, as well as research on the collaborative perception network and collaborative operation system of "air ground integration" fine operation information. He has published more than 30 SCI/EI papers as the first or corresponding author, participated in multiple national key research and development programs and national natural science foundation general projects, and led projects such as the China Computer Society and Baidu Apollo autonomous driving project.

agronomy

Editorial

Agricultural Unmanned Systems: Empowering Agriculture with Automation

Mengke Zhang and Shubo Wang *

School of Automation, Institute of Intelligent Unmanned System, Qingdao University, Qingdao 266071, China; zhangmengke@qdu.edu.cn
* Correspondence: shubowang@qdu.edu.cn

Automation is crucial for the advancement of modern agriculture. It plays a significant role in enhancing production efficiency and output, reducing labor costs, addressing natural disasters, and boosting sustainability. Automation utilizes big data and artificial intelligence to monitor agricultural production. It introduces new farming models that adapt to the challenges of scalability and environmental changes, achieving precise and efficient agricultural development.

In the field of smart agriculture, the emergence of unmanned systems represents a significant evolutionary breakthrough. Currently, smart agricultural unmanned systems encompass four spatial dimensions: sky (including navigation, remote sensing, meteorological, and communication satellites) [1,2], air (comprising plant protection drones, remote sensing mapping drones, long-duration solar drones, airships, and biomimetic flying robots) [3,4], land (featuring unmanned farming/harvesting machinery, biomass energy systems, soil improvement biomimetic robots, and unmanned livestock robots) [5,6], and water (including unmanned underwater vehicles, underwater operation robots, and unmanned aquaculture systems) [7,8]. These developments promise a bright future. This Special Issue, titled "Agricultural Unmanned Systems: Empowering Agriculture with Automation", focuses on sharing knowledge related to integrated and precise operational agriculture systems in the sky, air, land, and water. It explores intelligent sensing and control technologies in smart agricultural unmanned systems to advance the progress of unmanned agriculture. Establishing global demonstration sites is essential. These sites support the revolutionary advancement of smart agricultural machinery in automated, intelligent, unmanned, and cluster operations.

Citation: Zhang, M.; Wang, S. Agricultural Unmanned Systems: Empowering Agriculture with Automation. *Agronomy* **2024**, *14*, 1203. https://doi.org/10.3390/agronomy14061203

Received: 23 May 2024
Accepted: 30 May 2024
Published: 2 June 2024

The following five studies explore the application of intelligent algorithms in precision agriculture. To effectively cover the canopy area of tall spindle-shaped apple trees, Wang et al. [9] developed an improved lightweight transfer learning model for citrus pest detection. They utilized networks such as ResNet50, InceptionV3, VGG16, and MobileNetV3, in conjunction with a pre-trained single-shot multibox detector (SSD). This system can classify and rapidly detect pests in citrus orchards, and can be integrated into mobile devices for quick testing and pest counting. It assists farm managers in assessing pest damage and making informed pesticide decisions in orchard management. Peng et al. [10] proposed a new method that utilizes an optimized neural network to quickly identify crop water and nitrogen content. They specifically improved a traditional backpropagation neural network by incorporating particle swarm optimization (PSO). This improved network, with a dual hidden layer structure, enhances prediction accuracy. The enhanced PSO-BPNN model demonstrates a 9.87% increase in accuracy compared to conventional BPNN models. This advancement establishes a strong foundation for precision irrigation and fertilization in modern agriculture. It offers the potential to greatly improve resource management and crop yields. Xiao et al. [11] investigated the present status of target detection and recognition technologies for fruit- and vegetable-harvesting robots, with a focus on digital image processing and traditional machine learning methods. They assessed how these technologies affect the robots' accuracy, speed, and robustness, identifying current challenges and future developments to enhance robotic harvesting through improved computer vision technologies. Xiao et al. [12] provided a comprehensive overview of the advancements in fruit detection and automated harvesting using deep learning, particularly through Convolutional Neural Networks (CNNs), from 2018 to the present. They detailed the challenges faced, proposed solutions, and future research directions aimed at enhancing the accuracy, speed, and robustness of visual detection systems for fruit while reducing overall complexity and costs. This work serves as a reference for future research in the field of deep learning-based automatic fruit harvesting detection and recognition. Ji et al. [13] investigated the progress of the "eye-brain-hand" harvesting system in smart agriculture. This system integrates sensor technology, machine vision algorithms, and intelligent control to simulate human functions for automated and precise fruit and vegetable picking. It explores the development of robotic arms, visual recognition, and decision systems, emphasizing technologies such as image processing and deep learning. The review also evaluates the system's application across various crops and environments. It emphasizes future challenges in algorithm optimization and mechanical device reliability.

In the field of agricultural intelligent robotics technology and applications, Xiong et al. [14] proposed an optimized design method for an efficient dual-mechanical-arm harvesting system. For the typical spatial distribution of fruits in dwarf dense plants, a pair of vertically synchronized, three-degree-range Cartesian coordinate dual mechanical arms was designed. Through the development of a multi-objective optimization model and evaluation using the CRITIC-TOPSIS method, simulation analysis determined the optimal configuration to maximize harvesting efficiency, advancing robotic fruit-picking technology. Zhang et al. [15] implemented a mechanized picking method in trellised pear orchards by designing an integrated picker–placer end effector. They utilized the YOLOv5s object detection algorithm and a depth camera for precise fruit localization. By introducing a simulated annealing algorithm to optimize the picking order and proposing a task allocation method, the system was experimentally verified to improve picking efficiency by 30%. This study provides important references for the further development of robotic picking technologies. Shang et al. [16] addressed the issue of collisions with obstacles in unmanned agricultural machinery by proposing an obstacle detection algorithm based on two-dimensional LiDAR. This method utilizes differences between LiDAR data frames to determine collision incidents; employs preprocessing, median filtering, and DBSCAN to detect obstacles; and computes collision timing following the 6σ principle. Utilizing this algorithm, a pre-collision system was designed, integrated into agricultural navigation software, and tested on a harvester, achieving high accuracy and recall. This system enables emergency stops when farm machinery encounters obstacles during automated operation. It lays the groundwork for unmanned driving in more complex scenarios. Zhang et al. [17] investigated the application of mobile robots in agriculture, specifically emphasizing full-coverage path planning for orchard lawnmowers. They proposed a simplified motion model designed for orchard environments and enhanced the A* algorithm to optimize the lawnmower's navigation paths, reducing unnecessary turns during traversal. These improvements were validated through MATLAB 2020b simulations and field tests, demonstrating the method's effectiveness in enhancing navigation efficiency and task allocation in agricultural settings. Existing methods for controlling drive wheel slip in prototype machines have limitations, which result in suboptimal cultivation and finishing operations. To address this issue, Luo et al. [18] proposed a slip rate control method that is adjusted by both wheel speed and tillage depth. This method was validated using a New Holland T1404 power shift tractor. This method controls slip within an optimal range while ensuring maximum operational quality (tillage depth).

Additionally, the following three papers discuss and research improvements in automated agricultural equipment. Jiang et al. [19] discussed issues related to significant droplet loss, pesticide wastage, and environmental pollution caused by improper spray parameters.

They conducted a two-factor, five-level experiment focusing on power gradient and foliar area volume density (FAVD) to analyze the impact of these factors and the position of sampling points (considering horizontal distance, forward distance, and height) on droplet coverage. This research improves sprayer efficiency and establishes a foundation for future studies on precision spraying. It contributes to more sustainable agricultural practices. Liu et al. [20] developed an innovative baler feed rate detection model by utilizing power monitoring of the pickup platform. They utilized advanced signal processing techniques to mitigate the impact of machine vibration and precisely detect the feed rate of the baler. The model analyzes the dynamic characteristics of the pickup platform and utilizes frequency domain filtering to eliminate noise signals. This approach effectively correlates the power output of the pickup platform with the feed rate. Field experiments have confirmed the model's high accuracy and stability. This significantly improves the precision of feed rate measurements. The model meets the requirements of baler feed rate monitoring in field operations. Hui et al. [21] addressed low throughput and uneven treatment in plasma seed equipment. They designed a dielectric barrier discharge vibrating homogeneous material plasma seed treatment device. They systematically analyzed the structure and working principles of the vibrating homogeneous material equipment and established a mathematical model of seed force. Utilizing EDEM 2021 discrete element simulation software and a three-factor, three-level orthogonal experiment for empirical testing, the study demonstrated promising improvements in seed vitality, germination, and growth. This marks a significant advancement in seed treatment technology.

This Special Issue extends its deepest gratitude to all contributors. The papers included represent a broad range of research in the field of smart agriculture. However, there are still gaps related to 'sky', 'air', and 'water'. In the future, we anticipate more innovative research in these areas, especially focusing on how technological advancements can address current challenges and advance the holistic development of smart agriculture.

Conflicts of Interest: The authors declare no conflicts of interest.

References

1. Martos, V.; Ahmad, A.; Cartujo, P.; Ordoñez, J. Ensuring agricultural sustainability through remote sensing in the era of agriculture 5.0. *Appl. Sci.* **2021**, *11*, 5911. [CrossRef]
2. Sharifi, A.; Mahdipour, H.; Moradi, E.; Tariq, A. Agricultural field extraction with deep learning algorithm and satellite imagery. *J. Indian Soc. Remote Sens.* **2022**, *50*, 417–423. [CrossRef]
3. Rejeb, A.; Abdollahi, A.; Rejeb, K.; Treiblmaier, H. Drones in agriculture: A review and bibliometric analysis. *Comput. Electron. Agric.* **2022**, *198*, 107017. [CrossRef]
4. Abbas, A.; Zhang, Z.; Zheng, H.; Alami, M.M.; Alrefaei, A.F.; Abbas, Q.; Naqvi, S.A.H.; Rao, M.J.; Mosa, W.F.A.; Abbas, Q.; et al. Drones in plant disease assessment, efficient monitoring, and detection: A way forward to smart agriculture. *Agronomy* **2023**, *13*, 1524. [CrossRef]
5. Wu, H.; Wang, S. Design and optimization of intelligent orchard frost prevention machine under low-carbon emission reduction. *J. Clean. Prod.* **2023**, *433*, 139808. [CrossRef]
6. Cheng, C.; Fu, J.; Su, H.; Ren, L. Recent advancements in agriculture robots: Benefits and challenges. *Machines* **2023**, *11*, 48. [CrossRef]
7. Biazi, V.; Marques, C. Industry 4.0-based smart systems in aquaculture: A comprehensive review. *Aquac. Eng.* **2023**, *103*, 102360. [CrossRef]
8. Wu, Y.; Duan, Y.; Wei, Y.; An, D.; Liu, J. Application of intelligent and unmanned equipment in aquaculture: A review. *Comput. Electron. Agric.* **2022**, *199*, 107201. [CrossRef]
9. Wang, L.; Shi, W.; Tang, Y.; Liu, Z.; He, X.; Xiao, H.; Yang, Y. Transfer Learning-Based Lightweight SSD Model for Detection of Pests in Citrus. *Agronomy* **2023**, *13*, 1710. [CrossRef]
10. Peng, Y.; He, M.; Zheng, Z.; He, Y. Enhanced Neural Network for Rapid Identification of Crop Water and Nitrogen Content Using Multispectral Imaging. *Agronomy* **2023**, *13*, 2464. [CrossRef]
11. Xiao, F.; Wang, H.; Li, Y.; Cao, Y.; Lv, X.; Xu, G. Object detection and recognition techniques based on digital image processing and traditional machine learning for fruit and vegetable harvesting robots: An overview and review. *Agronomy* **2023**, *13*, 639. [CrossRef]
12. Xiao, F.; Wang, H.; Xu, Y.; Zhang, R. Fruit detection and recognition based on deep learning for automatic harvesting: An overview and review. *Agronomy* **2023**, *13*, 1625. [CrossRef]

13. Ji, W.; Huang, X.; Wang, S.; He, X. A Comprehensive Review of the Research of the "Eye–Brain–Hand" Harvesting System in Smart Agriculture. *Agronomy* **2023**, *13*, 2237. [CrossRef]
14. Xiong, Z.; Feng, Q.; Li, T.; Xie, F.; Liu, C.; Liu, L.; Guo, X.; Zhao, C. Dual-manipulator optimal design for apple robotic harvesting. *Agronomy* **2022**, *12*, 3128. [CrossRef]
15. Zhang, H.; Li, X.; Wang, L.; Liu, D.; Wang, S. Construction and Optimization of a Collaborative Harvesting System for Multiple Robotic Arms and an End-Picker in a Trellised Pear Orchard Environment. *Agronomy* **2023**, *14*, 80. [CrossRef]
16. Qu, J.; Zhang, Z.; Qin, Z.; Guo, K.; Li, D. Design and Test of Obstacle Detection and Harvester Pre-Collision System Based on 2D Lidar. *Agronomy* **2023**, *13*, 388. [CrossRef]
17. Zhang, M.; Li, X.; Wang, L.; Jin, L.; Wang, S. A Path Planning System for Orchard Mower Based on Improved A* Algorithm. *Agronomy* **2024**, *14*, 391. [CrossRef]
18. Luo, C.; Wen, C.; Meng, Z.; Liu, H.; Li, G.; Fu, W.; Zhao, C. Research on the slip rate control of a power shift tractor based on wheel speed and tillage depth adjustment. *Agronomy* **2023**, *13*, 281. [CrossRef]
19. Jiang, S.; Li, W.; Yang, S.; Zheng, Y.; Tan, Y.; Xu, J. Factors Affecting Droplet Loss behind Canopies with Air-Assisted Sprayers Used for Fruit Trees. *Agronomy* **2023**, *13*, 375. [CrossRef]
20. Liu, H.; Gao, N.; Meng, Z.; Zhang, A.; Wen, C.; Li, H.; Zhang, J. Construction and Test of Baler Feed Rate Detection Model Based on Power Monitoring. *Agronomy* **2023**, *13*, 425. [CrossRef]
21. Hui, Y.; Huang, C.; Liao, Y.; Wang, D.; You, Y.; Bai, X. The Medium-Blocking Discharge Vibration-Uniform Material Plasma Seed Treatment Device Based on EDEM. *Agronomy* **2023**, *13*, 2055. [CrossRef]

agronomy

MDPI

Review

A Comprehensive Review of the Research of the "Eye–Brain–Hand" Harvesting System in Smart Agriculture

Wanteng Ji [1,2,3,†], Xianhao Huang [1,2,3,†], Shubo Wang [1,2,4,*] and Xiongkui He [1,2,4,*]

1 Centre for Chemicals Application Technology, China Agricultural University, Beijing 100193, China; jwt17863978031@163.com (W.J.); xianhao@cau.edu.cn (X.H.)
2 College of Agricultural Unmanned System, China Agricultural University, Beijing 100193, China
3 College of Engineering, China Agricultural University, Beijing100083, China
4 College of Science, China Agricultural University, Beijing 100193, China
* Correspondence: shubowang@qdu.edu.cn (S.W.); xiongkui@cau.edu.cn (X.H.)
† These authors contributed equally to this work.

Abstract: Smart agricultural harvesting robots' vision recognition, control decision, and mechanical hand modules all resemble the human eye, brain, and hand, respectively. To enable automatic and precise picking of target fruits and vegetables, the system makes use of cutting-edge sensor technology, machine vision algorithms, and intelligent control and decision methods. This paper provides a comprehensive review of international research advancements in the "eye–brain–hand" harvesting systems within the context of smart agriculture, encompassing aspects of mechanical hand devices, visual recognition systems, and intelligent decision systems. Then, the key technologies used in the current research are reviewed, including image processing, object detection and tracking, machine learning, deep learning, etc. In addition, this paper explores the application of the system to different crops and environmental conditions and analyzes its advantages and challenges. Finally, the challenges and prospects for the research on picking robots in the future are presented, including further optimization of the algorithm and improvement of flexibility and reliability of mechanical devices. To sum up, the "eye–brain–hand" picking system in intelligent agriculture has great potential to improve the efficiency and quality of crop picking and reduce labor pressure, and it is expected to be widely used in agricultural production.

Keywords: smart agriculture; visual recognition; decision control; end-effector; harvesting robots; research review

Citation: Ji, W.; Huang, X.; Wang, S.; He, X. A Comprehensive Review of the Research of the "Eye–Brain–Hand" Harvesting System in Smart Agriculture. *Agronomy* **2023**, *13*, 2237. https://doi.org/10.3390/agronomy13092237

Academic Editor: Baohua Zhang

Received: 21 July 2023
Revised: 17 August 2023
Accepted: 23 August 2023
Published: 26 August 2023

1. Introduction

In agricultural production, harvest is one of the most important links. It directly relates to the quality of product harvest. However, traditional manual picking methods have problems such as high cost, low efficiency, and labor shortage, which seriously restrict the development of agricultural production. Therefore, it is urgent and necessary to study the technology of fruit and vegetable-picking robots, which is an important way to solve this problem. The visual recognition system, decision control system, and end-effector system are the key contents of the research on harvesting robot technology, and their technological maturity directly affects the harvesting effect of the robot. Intelligent harvesting system refers to a system that utilizes modern technology to achieve autonomous harvesting by the steps of recognition, decision, control, and grasping during the agricultural harvesting process. The intelligent picking system has the characteristics of efficiency, precision, and reliability, which can greatly improve agricultural production efficiency, reduce the labor burden of fruit farmers, and improve the quality and safety of agricultural production [1–8]. The literature distribution of intelligent harvesting systems for different crops is shown in Figure 1.

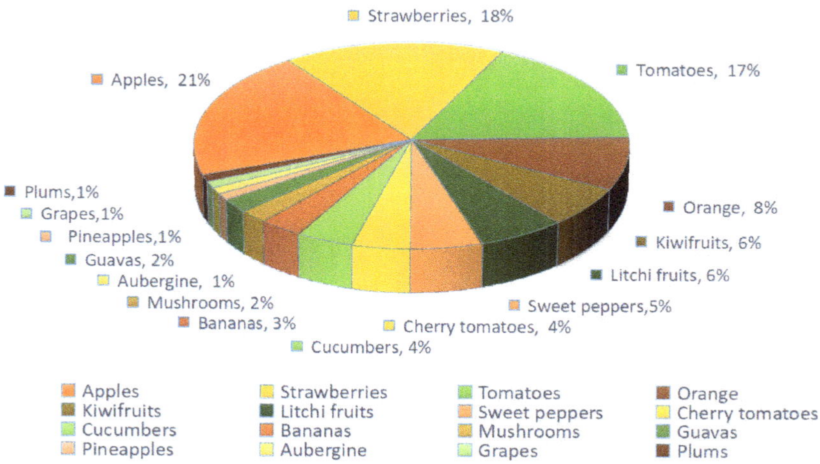

Figure 1. Distribution of the referenced literature per crop type.

The unmanned agricultural harvesting system mainly consists of three parts: a recognition module; a decision control module; and an end-effector module [9–11]. Therefore, this review will start from the recognition, decision control, and end-effector gripper of unmanned agricultural harvesting systems and analyze the current situation of intelligent harvesting systems. Among them, the recognition module is a key part of the system to obtain crop information, which mainly uses technologies such as machine vision and lidar. By using RGB [12,13], depth cameras, various sensors, or other devices, information about picked objects is obtained based on target features, feature fusion, or deep learning, including position, shape, size, maturity, surrounding environment, etc. [14–20]. To enhance the precision and accuracy of image segmentation, various image preprocessing techniques have been utilized, facilitating superior decision and control in subsequent stages [21–23]. The decision and control module serves as a crucial component in autonomous decision and control during the harvesting process. During the decision and control process, a broad range of network models and optimization algorithms [24–26], including Support Vector Machine (SVM) [27,28], decision trees, deep learning, genetic algorithms, and particle swarm optimization, are extensively employed [9,29–33]. These models and algorithms provide intelligent decision control adapted to different harvesting situations, leading to optimal harvesting outcomes. The end-effector module, a key hardware component responsible for automated fruit picking, influences the harvesting results directly through its design type, harvesting method, and size dimensions. Researchers categorize the end-effector grippers into four major types: negative pressure adsorption end-effectors; shearing-style end-effectors; negative pressure adsorption end-effectors; cavity retrieval end-effectors; and flexible grasping end-effectors. [10,34–45]. The selection of an appropriate end-effector gripper is determined by the physical characteristics of the fruits, such as their types, sizes, and the hardness of their peels. These three modules, encompassing vision, manipulation, and decision capabilities, work synergistically within the integrated eye–brain–hand system to accomplish the harvesting task.

In the context of the continuous advancement of artificial intelligence and robotics, unmanned systems for agricultural harvesting are gaining increasing attention as a novel method of crop collection. Within these agricultural harvesting robots, the perception system, decision-control system, and end-effectors play pivotal roles, serving as the essential components for achieving automated harvesting. Harvesting robots equipped with eye–brain–hand-integrated systems are not only characterized by high precision, efficiency, and reliability, but they are also adaptable to various harvesting environments, demonstrating a broad range of prospective applications.

Studying agricultural harvesting robots is not only an important measure to adapt to agricultural modernization and market demand but also a crucial step in solving labor shortages and improving the quality of agricultural products. Through continuous innovation and improvement of technology, agricultural production can be automated, refined, and intelligent, making positive contributions to the sustainable development of global agriculture. This article summarizes the important research progress of over 120 agricultural harvesting robots in the past 6 years. This research focuses on the "eye–brain–hand" of robots and comprehensively analyzes the key role of robots in the agricultural product-picking process. This article delves into various technological breakthroughs, including high-precision visual perception technology, intelligent path-planning and decision-making methods, and advanced end-effector design. By summarizing the research results of these aspects, this review provides insight into the field of agricultural picking robots and provides important guidance and enlightenment for future research and application. These contributions will have a profound impact on promoting the improvement of agricultural production efficiency, solving labor problems, and ensuring the quality of agricultural products. This article mainly discusses the three main parts of agricultural harvesting robots, namely "eye, brain, and hand", based on the "structure–activity relationship method". In these three parts, more detailed descriptions are provided according to the "time sequence method" and the "current situation countermeasure method". In the fifth part, the challenges and prospects in the field of harvesting robots are summarized according to the "current situation countermeasure method". The outline of this article is shown in Figure 2. In this paper, Section 2 introduces common target perception hardware systems, perception methods, and image preprocessing techniques. Hardware systems can be categorized into active vision, passive vision, and applications combining various sensors. The perception methods are primarily based on three aspects: target features; feature fusion; and deep learning. Section 3 primarily elaborates on decision strategies and control methods, encompassing regional division and task allocation, active and passive obstacle avoidance strategies, path planning based on various technologies, and numerous control methods. Section 4 presents the various end-effector mechanisms and evaluation metrics for harvesting robots. Section 5 presents challenges and prospects for agricultural harvesting robots. Finally, a summary of this article is provided in Section 6.

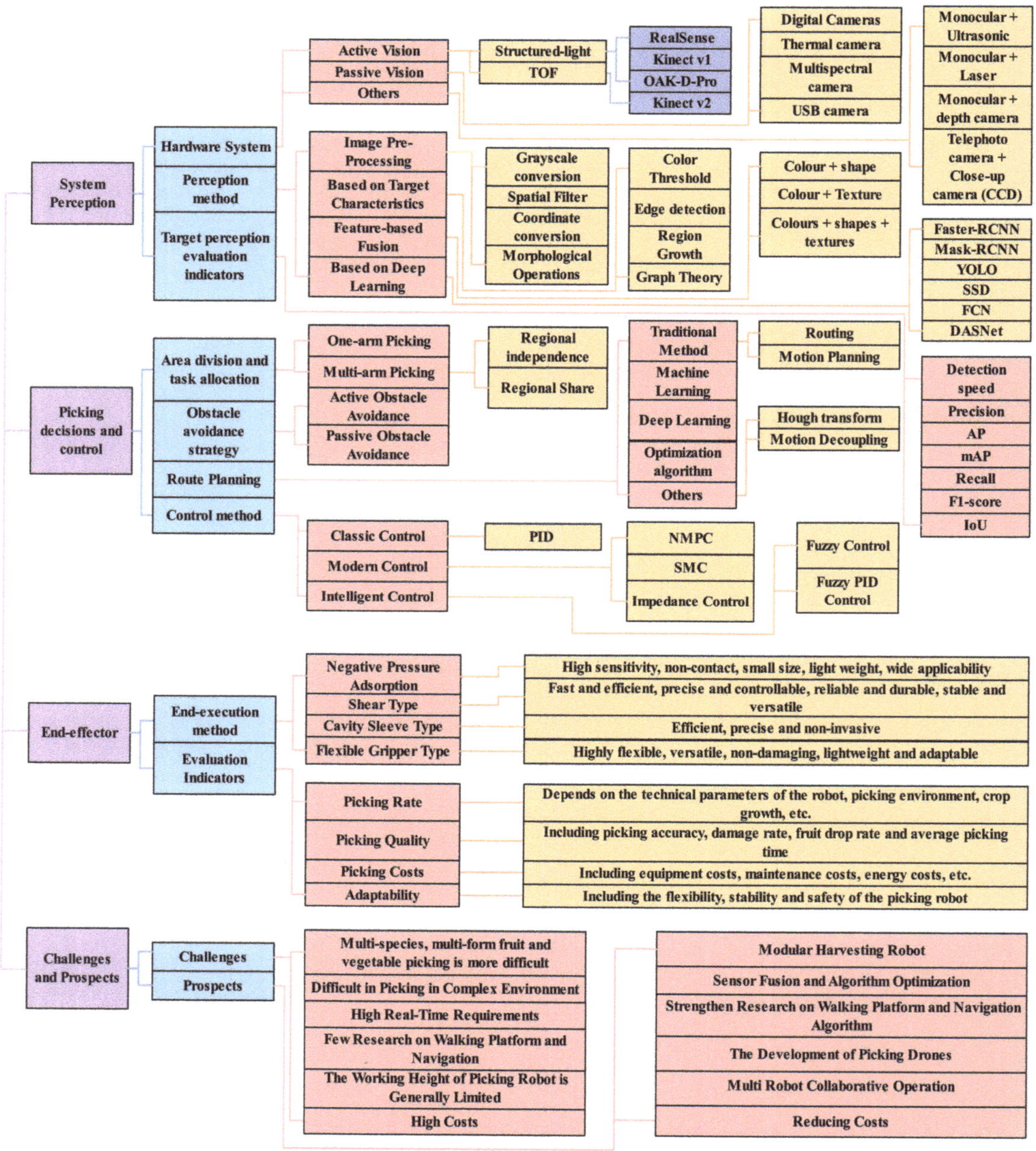

Figure 2. Outline of the article.

2. Intelligent Harvesting "Eye" System

2.1. Perception Hardware System

Considering the intelligent point of view of unmanned system, it is necessary to see accurately in order to achieve better expected results. Therefore, the system perception as the first part of the picking system has been widely paid attention to both at home and abroad. At present, the perception methods of agricultural picking robots mainly include binocular vision, LiDAR, and the combination of monocular cameras and other sensors. Due to cost constraints, the LiDAR-based perception method is less applied, and the other two perception methods are mainly described here.

2.1.1. Object Perception Based on Binocular Vision

Binocular vision measurement is similar to the stereo perception of the human eye; it uses two cameras to image the object to be measured from different angles, based on the stereoscopic parallax of the corresponding points in the two images, combined with the principle of triangulation to realize the acquisition of 3D information of the object. Binocular vision measurement techniques can be divided into two types based on whether the light source is actively projected during the detection process, active vision and passive vision, both of which are often used in hand–eye harvesting. The classification, characteristics, and examples of active and passive visual cameras are shown in Table 1.

(a) Active vision

At present, the active vision technology based on Structured light is popular in the market for binocular cameras, which has the advantages of strong anti-interference ability, a wide application environment, and a mature technical scheme. Therefore, active visual cameras—especially those from the RealSense series from Intel, the Kinect series from Microsoft, and the OAK-D-Pro from OpenAI—are more frequently employed in the actual harvesting of fruits and vegetables. The guava harvesting robot designed by Lin et al. [46] used a Kinect V2 camera as a visual sensor, which consists of an RGB color camera, an infrared camera, and an IR light source. The IR light source actively projects the near-infrared spectrum, which forms random reflection spots when illuminated on an object. These spots will be received by infrared cameras to read the depth information of the object. Based on this Time of Flight (TOF) principle, it obtains depth images of targets between 0.5 and 4.5 m. In addition, the camera is inexpensive and stable. Due to its many advantages, the Kinect-V2 depth camera was also equipped by Ning et al. [47] on a pepper-picking robot to realize pepper picking in a greenhouse planting environment. In order to determine the location of the kiwifruit in an RGB image, extract coordinates, and locate it, Mu et al. [48] also used this camera as their machine vision equipment in the construction of a kiwifruit harvesting robot. Different from the above research, Zhang et al. [49] equipped the cherry tomato picking robot with Intel's RealSense D415 depth camera and installed it on the side of the base of the mechanical arm to ensure that it is not blocked during tomato harvesting. Similarly, Yu et al. [50] used the DF810-HD depth camera to provide visual support to the picking robot during tomato picking. Considering the height of the greenhouse and the growth of tomato plants, and in order to meet the demand for real-time control, the camera is placed on a liftable platform and fixed, which effectively increases the sensing range while reducing the calculation amount of the visual system. For the harvesting of short, densely planted apples in a greenhouse environment, Li et al. [51] chose to combine a RealSense D455 depth camera with a multi-Cartesian mechanical arm and used a task planning algorithm to achieve accurate identification and effective harvesting of short, densely planted apples. Similarly, Kang et al. [52] used the RealSense D-435 depth camera as the hardware support of the visual system of the apple-picking robot and applied the depth neural network DAS-Net to identify the fruit. The experiment showed that the visual system has high precision in fruit detection and segmentation. However, the setting of a single camera is difficult to adapt to the complex and changeable working environment. In order to adapt to an unstructured harvesting environment, Sarabu et al. [53] designed a double-arm harvesting robot consisting of a grab arm and a search arm for the same apple harvesting task, and each arm was equipped with an RGB-D depth camera (Hand-Eye). The camera on the grab arm is used to locate the picking apple in the field of view. The camera on the search arm can detect the target outside the dead zone of the grab arm camera and quickly plan a clear and suitable picking path in combination with relevant algorithms. Moreover, multi-camera and multi-view method can solve the problem of overlapping and blocking fruit detection to a certain extent. Gong et al. [54] provided us with a new idea when designing a greenhouse tomato-picking robot, which is to improve the image segmentation accuracy in a complex environment through multi-source fusion images. RGB, depth, and infrared images are acquired by the Kinect v2 camera, fused to obtain RGB-D-I images, and target segmentation accuracy is improved by 7.6% in combination with the extended Mask-RCNN network. Although active vision cameras can be used in a wide range of scenarios, they are, in some cases, influenced by environmental factors. To achieve better sensing in the night-time environment, Fu et al. [55] equipped four 850 Lumens LED lights on their apple-picking robot platform, and the Kinect V2 camera was installed in the center of the four LED lights, which provided a bright and stable night-time working environment for the picking platform. Although the presence of an active light source can enhance the depth camera's perception in low-light conditions, it remains challenging to maintain optimum and consistent perception in situations with strong light or varying light intensity. To solve this problem, Xiong [56] applied a U-shaped straw-picking robot to a structured greenhouse-picking environment, with two independent picking systems on both sides of the arched structure and a real D435 depth camera for the visual system. The design of a U-shaped full shielding frame structure can greatly reduce or even eliminate the impact of ambient light changes on the quality of acquired images and greatly improve the detection and positioning accuracy of the visual system without any specific correction algorithms.

(b) Passive Vision

Passive vision technology uses a pure RGB camera and binocular parallax principle to detect and locate the target. Compared with active vision, passive vision is widely used in scenes with bright vision and wide space due to its simple structure and low power consumption. Although color is the most intuitive feature to distinguish the target fruit from the background, color-based methods of identification are susceptible to factors such as varying light. To this end, we can reduce the impact of light by means of relevant algorithms. Lv et al. [57] used Sony's Cyber-shot color camera to provide RGB images for their apple-picking robot and then corrected the images of apples affected by external light with an adaptive Gamma algorithm, thus greatly improving the image segmentation accuracy. In addition, the shape-based recognition method is not easily affected by changing lighting. Wang et al. [58] reduced the minimum relative error of Apple, ranging up to 0.96%, by combining a linear fusion detection algorithm and AD-Census matching algorithm based on CMOS binocular camera. However, the apple-picking robot designed by Yu et al. [59] is based on a binocular camera combined with color threshold and edge detection for target identification, and the success rate of apple identification is up to 82.5%. Similarly, Yang et al. [44] also used an RGB camera to identify the color and texture of the target in the Hangzhou chrysanthemum-picking robot. After eliminating the noise with the bilateral filter, the color and texture characteristics of the image were extracted through the RGB value and the grayscale paragenesis matrix of the image and then input into the Least Squares Support Vector Machine model (LS-SVM). The segmentation time of the trained model for the Hangzhou chrysanthemum was as low as 0.7 s. In addition, during tomato fruit picking, Zhou et al. [60] used a variable baseline USB binocular camera (HNY-CV-002), combined with the identification method of circular Hough transform and RGB color space, to achieve efficient picking of the target tomato. In addition, Jin et al. [61] applied depth learning technology to the binocular camera to identify the target fruit in order to better realize the perception of tomato fruit and achieved good results. Similar techniques, such as those discussed above for active vision, are also used for passive binocular vision in order to optimize the recognition impact. Ye et al. [62] installed the Micro-vision MV-VD120SC industrial camera on the end-effector of the litchi-picking robot and planned auxiliary target pickup points for the robot. After the end-effector arrived at the auxiliary target pickup point from the initial point, it would perform environmental perception again and plan the motion between the target pickup points. This strategy can avoid interference between the end-effector and obstacles around the target as much as possible while compensating for visual errors and improving positioning accuracy.

Table 1. Characteristics of Active and Passive Vision Technology and Representative Camera Examples.

Active/Passive Vision	Type	Advantages	Disadvantages	Representative Cameras	
Active Vision	Structured-light	More mature and easier to miniaturize Low power consumption Can be used at night High accuracy and resolution within a certain range	Easily disturbed by ambient light The accuracy deteriorates as the detection distance increases	RealSenseD435i	
				Kinect v1	
				OAK-D-Pro	
	TOF	Long detection distance Less interference by ambient light	High requirements for equipment High power consumption Low edge accuracy Lower frame rate and resolution	Kinect v2	
				PMD CamCube 3.0	
Passive Vision	-	Low hardware requirements and low cost Suitable for both indoor and outdoor use	Very sensitive to ambient light Not for monotonous scenes that lack texture Calculations are more complex	Digital Cameras	
				Thermal camera	
				Multispectral camera	

2.1.2. Target Perception Based on Multi-Sensor Combination

Due to the low accuracy and poor fault tolerance of a single sensor, the perception strategy based on multi-source sensor fusion is widely used in the fields of autonomous driving and industrial robots. Similarly, this technology has also been borrowed by agricultural robot experts for fruit's 3D perception. At present, there are several combinations of multi-source sensors used for fruit perception: monocular + ultrasound, monocular + laser, and monocular + depth camera. The relevant descriptions of different types of multi-sensor combination perception are shown in Table 2.

Oktarina et al. [63] used a combination of a Pi camera placed on a robotic arm and an ultrasonic sensor HC-SR04 to achieve recognition and positioning of red and green tomatoes. Lower-cost network cameras with appropriate resolution offer color image information of the target, and ultrasonic sensors offer depth information of the target fruits. A new visual unit was created by Feng et al. [64] for his cherry tomato-picking robot. A monocular camera and a laser sensor are both mounted on the manipulator arm. The target tomato is sensed and recognized using an RGB camera, and its distance from the vision system is calculated using a laser sensor. Using a suitable combination of the two sensors, corresponding algorithms, and a shear end actuator, harvesting success can be increased by 83%. In contrast to the examples above, Sepulveda et al. [65] created a dual-arm eggplant harvesting robot with a visual system made up of two monocular cameras, specifically, the SR4000 depth measurement capture camera and the Prosilica GC2450C color camera. The former provides high-resolution color images, while the latter provides depth information of images. In addition, in order to identify and locate strawberries more accurately, Feng et al. [66] creatively developed a vision system combining far and near. After the far-sighted unit acquires a larger field-of-view image, the robotic arm, carrying the end-effector and the close-range camera, approaches one by one from the left side to the right side to sense and pick the ripe fruits again.

Table 2. Comparison of different forms of multi-sensor combination perception.

Applied Crops	Perception	Sensors	Characteristic	Structure	Effect	Ref.
Red tomato, Green tomato	Monocular RGB +Ultrasonic	Pi camera (mobile) + ultrasonic sensor HC-SR04 (mobile)	Simple method, low cost, and adaptable to the limited computing resources of microcontrollers		Picking time Red: 4.932 s Green: 5.276 s	[63]
Cherry tomato	Monocular RGB + Laser	FL3-U type RGB camera + LY-LDS-61 type laser sensor	Simple structure, accurate distance measurement, avoiding obstacle obstruction	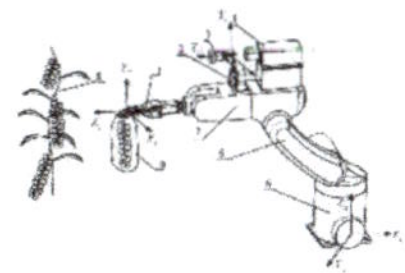	Harvesting success rate: 83%	[64,65]
Eggplant	Monocular RGB + depth camera (TOF)	Prosilica GC2450C RGB camera + Mesa SwissRanger SR4000 depth measurement capture camera	High precision and sensitivity		Sensing time: 0.81 s	[65]
Strawberry	Far-field monocular RGB + Close up monocular RGB	8 mm lens 1280 × 976 pixels 1/3 inch CCD telephoto camera (fixed) + 5 mm lens 640 × 480 pixels 1/2 inch CCD close-up camera (mobile)	Global and local image information can be obtained simultaneously and possible occlusions can be avoided, but it takes longer.	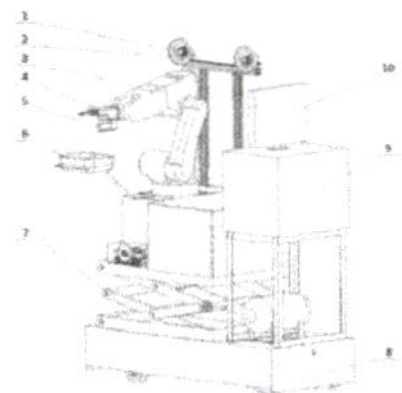	Harvesting success rate: 84% Average harvest time: 10.7 s	[66]

2.2. Target Perception Methods

2.2.1. Image Preprocessing Methods

The fruit and vegetable-picking robot must assess the image of the target fruit based on the target features or the trained neural network model after acquiring the original image through the system perception hardware. However, in order to eliminate background noise, recover real information, enhance detectability, and simplify data, the original image is generally preprocessed first. Commonly used preprocessing methods mainly include grayscale transformation (contrast enhancement, contrast compression, gamma correction, etc.), spatial filtering (Gaussian filtering, mean filtering, median filtering, edge detection, etc.), coordinate transformation (translation, mirroring, rotating), morphological operations (erosion, dilation, open operations, closed operations), and so on.

During image recognition, Xiong et al. [67] first converted RGB images into HSV color saturation images and then judged whether the adaptive color threshold was reached and determined which strawberries could be picked. Moreover, Feng et al. [64] improved the image quality and region delineation accuracy by using the R-G color model to enhance the color features of the images acquired from the camera and then determining the candidate regions of ripe tomato bunches based on the column pixel grey scale statistics. In addition, in order to ensure the stability of obtaining basic information about the target, besides the addition of an external light source, as mentioned earlier, such problems can also be solved by using lighting balance algorithms to preprocess the original image. Zhuang et al. [68] proposed an iterative Retinex algorithm based on the weighted intensity of the fruit region in RGB color images, which can adaptively improve images with poor light distribution, and more than 97% of pixels within the litchi region were correctly segmented after light compensation. In order to enhance the segmentation accuracy and success rate of oil palm fruit under complex backgrounds, different illumination, and different fruit maturity, Huang et al. [69] transformed RGB color space into Lab color space and then obtained a region of interest (ROI) containing oil palm fruit by using Otsu algorithm and morphological operation. After the ROI image in the color space is converted to a Grayscale and smoothed by a Gaussian filter, the target at the edge of the image can be clearly detected. These image preprocessing methods are not only commonly used in traditional image analysis techniques based on target features but also widely used in deep learning techniques. These methods are generally used to reasonably augment a limited dataset by stretching, scaling, rotating, panning, and contrast adjustment as a way to achieve data augmentation and to improve the accuracy and robustness of the neural network model. In the greenhouse of the Guangdong Academy of Agricultural Sciences, Ning et al. [47,70] designed a sweet pepper-picking robot that collected 400 images of 9882 sweet peppers from multiple angles in a variety of weather conditions with a depth camera. In order to provide the YOLO-V4-CBAM model with a sufficient training set and to improve the model detection accuracy, the training set was augmented with data using exposure, blurring, mirroring, and rotation, and 1500 images were obtained, totaling 33,780 sweet peppers. In terms of noise reduction, Mao et al. [70], in order to overcome the interference of complex backgrounds, such as soil, hay, and irrigation pipeline in the cucumber image, the original image of the cucumber was processed under G component to filter out the objects with large color difference in the background. The image is then smoothed using a 3×3 median filter and segmented using the Otsu algorithm to obtain a preliminary denoised background image. After that, MSER (Maximum Stable Extreme Region) is used to further eliminate leaf noise, which enables deep learning to extract cucumber features from complex backgrounds more easily.

2.2.2. Perception Methods Based on Target Features

The traditional techniques for image segmentation based on target features are mainly color thresholding-based, edge detection-based, region growing-based, and graph theory-based. Empirical thresholds and adaptive thresholds are the two types of threshold segmentation technologies that are utilized in real applications. Empirical thresholds are more frequently used and can be adjusted to meet production needs. In the automation process, an adaptive threshold is utilized more frequently, and adaptive algorithms typically have to select an adaptive threshold. To achieve real-time detection of strawberries, Xiong et al. [71] used a simple color thresholding algorithm based on RGB channels with faster processing speed to detect strawberries. At the same time, in order to remove the noisy pixels and fill the holes, morphological opening and closing operations are performed on the original RGB image based on erosion and expansion of the binary image and objects that are too far or too close to the robot are removed by depth filtering of the depth image. Feng et al. [64] used the R-G color model to enhance the difference between the target fruit and the background by analyzing the color features of the images captured by the RGB camera and selecting the candidate region of the saint fruit from the R-G image based on the gray statistics of the column pixels. Finally, the fruits were recognized using the CogPMAlignTool in the Cognex Vision Pro image processing class library. In litchi picking, the localization of the picking point has always been an important part of orchard operations by picking robots, but the localization accuracy of the picking point is easily affected by unstructured growing environments, such as light intensity variations. In order to eliminate the effect of illumination variations, Zhuang et al. [68] improved the illuminance distribution of weakly illuminated images by employing an adaptive iterative Retinex algorithm while keeping the illuminance distribution of well-illuminated images unchanged. The stem is segmented, and noise is filtered using the histogram of intensity distribution after the litchi region has been divided up into RGB color space. Finally, the location of the picking point was determined based on the connection and positional relationship between the segmented litchi and the stem. Although the segmentation method based on color thresholding has been widely used in image segmentation, its shortcomings are also obvious. This method is only applicable to targets whose colors differ significantly from the image background and whose ripe fruits have a relatively single color, and it fails in the face of fruits and vegetables whose ripe fruits are similar in color to the surrounding environment or have multiple colors. In order to achieve effective segmentation of oil palm fruits with various shapes and colors, Septiarini et al. [69] used an edge detection method widely used for fruit segmentation—Canny detection. In order to reduce the noise interference and improve the image quality, Gaussian smoothing is used to connect the small discontinuities in the image before Canny detection. Then, morphological extension, filling, and reconstruction were carried out, and two morphological operations—opening and closing—were used to correct misclassification. A comparison of image segmentation methods based on different target features is shown in Table 3.

Table 3. Comparison of image segmentation methods based on different target features.

Splitting Technology	Applied Crops	Description	Advantages	Disadvantages	Applicable Environment	Examples	Ref.
Color Threshold	Strawberry Cherry Tomato Litchi	One or several thresholds to classify the grayscale histogram, grayscale values in the same category belong to the same object	Most commonly used, simple, fast and efficient calculation	Cannot effectively segment targets with little difference in grayscale values and overlap, more sensitive to noise	Applicable when the difference between image background and target features is obvious	Otsu K-means clustering Maximum entropy method	[64,68,71]
Edge detection	Oil palm fruit	Different images have different grayscale, and there are generally distinct edges at the boundary, so use this feature for image segmentation	Faster retrieval and better detection of different image edges	More sensitive to noise, conflicts between noise immunity and detection accuracy	Applicable when low noise, large difference in edge features between different regions	Canny Sobel Robert Prewitt Laplaeian	[69]
Regional Growth	Eggplant Kiwifruit Chili Guava	Divide the image into different segmentation regions according to the similarity criterion	It has better area characteristics and overcomes the disadvantage of continuous segmentation area that exists in other methods	Prone to over-segmentation	Applicable when a more definite structural division of the area is required	Meyer Watershed Method, Adams Seed Area Growth Method, Gonzalez Regional Split Merge Method	[46,65,72]
Graph Theory	-	The essence is to remove specific edges and divide the graph into several subgraphs to achieve segmentation.	Suitable for a wide range of target shapes	Longer operation time	-	Graph Cuts Grab Cuts Random Walk	-

2.2.3. Feature Fusion-Based Perception Methods

The previous section summarized some methods for object detection based on single features, such as color, texture, and shape, and listed some application examples. However, the above methods often do not perform well enough when encountering complex working environments, and in order to solve this problem, researchers have proposed a detection method based on multi-target feature fusion. This method can integrate the sensing advantages between different features and effectively improve the detection accuracy and robustness of the target-sensing system under complex working conditions.

Kiwifruit image recognition is a bit difficult because of the interference of occlusion and overlapping; in order to solve this problem, many scholars use a feature fusion-based approach for effective perception. Liu et al. [72] also proposed a more complete set of methods for this purpose. In the image processing stage, after converting the RGB color space to HSV, frequency domain filtering and homomorphic filtering techniques are used to eliminate a large amount of noise in the original image and to improve the contrast. Then, the images of kiwifruit were segmented in three stages by combining the Otsu algorithm, the regional growth method, and a dynamic fast identification algorithm. Yang et al. [44] proposed an image segmentation algorithm based on LS-SVM for the visual detection and location of Hangzhou white chrysanthemum. The color and texture features in the RGB space are input to the LS-SVM model after being de-noised by a bilateral filter. The experiment showed that the trained model could effectively separate images of Hangzhou white chrysanthemums from the front, back, and shadow illumination, with an accuracy of more than 90% and a segmentation time of only 0.7 s.

Generally, the above method can only identify and detect one kind of fruit, but it is not applicable to other kinds of fruit and vegetables. Is there any algorithm that can detect multiple types of fruits and vegetables at the same time? To solve this problem, Lin et al. [46] proposed a novel detection method. This technique uses an SVM classifier based on angle, color, and shape characteristics to detect spherical or cylindrical fruit that is common in natural environments. It integrates a clustering algorithm based on region growth and a three-dimensional shape detection algorithm based on m-estimated sample conformance (MSAC). The experiment demonstrates that, for pepper, eggplant, and guava, the algorithm's detection accuracy is 0.866, 0.888, and 0.866, respectively, and that the average detection time for a single fruit is 1.41 s, 4.07 s, and 4.70 s. Similarly, Sepulveda et al. [65] proposed an image segmentation algorithm composed of support vector machine (SVM), watershed transformation, and point cloud extraction for eggplant picking under complex working conditions. The supervised training of the Cubic SVM support vector machine is carried out according to the color characteristics of different scene elements. The trained classifier can identify and segment eggplants in most cases, and the watershed transformation can effectively segment eggplant images in the overlapping state. The above methods improve the perception of the mature target fruit by means of feature fusion but do not detect and analyze the quality of the fruit. If inefficient picking can be prevented by recognizing the rotten and damaged fruits, the picking quality of the picking robot can be enhanced to some amount. In this respect, Kurpaska et al. [73] conducted some research and proposed a method to detect and judge the quality of strawberries based on texture, color, and contour shape analysis. This method uses the analysis of geological samples based on color and texture analysis to detect and analyze the quality of strawberries. The experiment showed that the comprehensive detection method can effectively distinguish different quality strawberries.

2.2.4. Perception Methods Based on Deep Learning

Deep learning is a new research direction in the field of machine learning, first proposed by Hinton et al. in 2006. The neural network in deep learning can be divided into three layers: the input layer; the hidden layer; and the output layer. After the input layer obtains the input image information, it passes the information to the hidden layer for feature extraction, and finally, the hidden layer outputs the model results. The working principle of the neural network presented in fully connected form is shown in Figure 3. Compared with the traditional shallow learning structures such as support vector machine (SVM) and artificial neural network (ANN), deep learning can extract the hidden features in the image and automatically learn to obtain the hierarchical feature representation (as shown in Figure 4a), which is more conducive to the classification or feature visualization. As the amount of training data increases, the advantages of deep learning models become more and more obvious (as shown in Figure 4b). In addition, deep learning has the flexibility to choose the number of network layers according to the designer's needs. Based on the above advantages, in recent years, deep learning has been widely used in the target detection of fruits, vegetables, and other crops with good results.

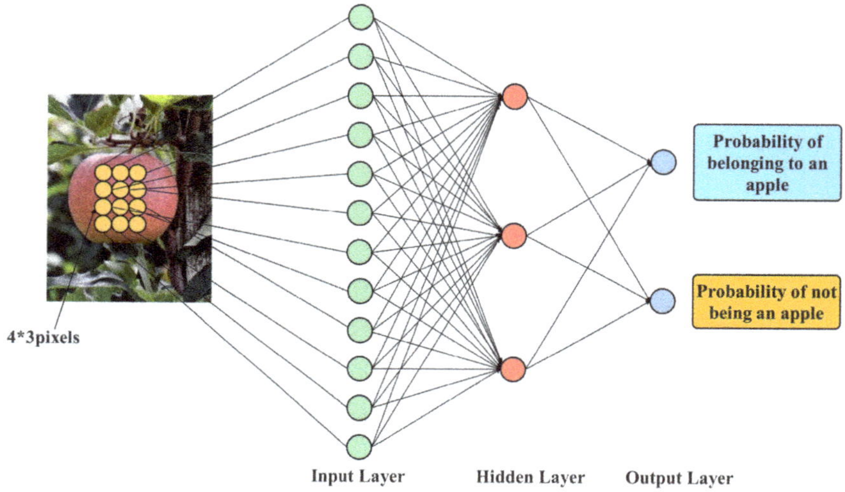

Figure 3. Working diagram of neural network in fully connected form.

Unlike traditional manual feature-based detection algorithms (VJ, HOG, DPM), there are many detection algorithms in deep learning, which can be roughly divided into one-stage detection algorithms and two-stage detection algorithms according to the detection stage (as shown in Figure 5). Among them, two-stage detection algorithms mainly include RCNN, Faster R-CNN, Mask R-CNN, and so on. This type of algorithm performs target detection by first generating a pre-selected box that may contain the object to be detected (Proposal box) and then completing the identification and localization of the target after further detection based on the characteristics of the object. This kind of algorithm was quickly developed in the early stages of the application of deep learning technology because it has high detection precision and accuracy, but it also has the drawback of slower detection speed and is time-consuming. Unlike two-stage detection algorithms, single-stage detection algorithms, such as YOLO and SSD, do not require a region candidate network (PRN) and can directly extract features in the network to predict object classification and location, which is characterized by a one-step process and faster detection speed. The Yolo series of algorithms can reach 200 fps, much higher than the 5 fps of the two-stage algorithm Mask R-CNN, which is especially suitable for mobile platforms, but its detection accuracy is a bit poorer than that of algorithms such as Faster-RCNN. Table 4 compares and analyzes various network models used by different researchers.

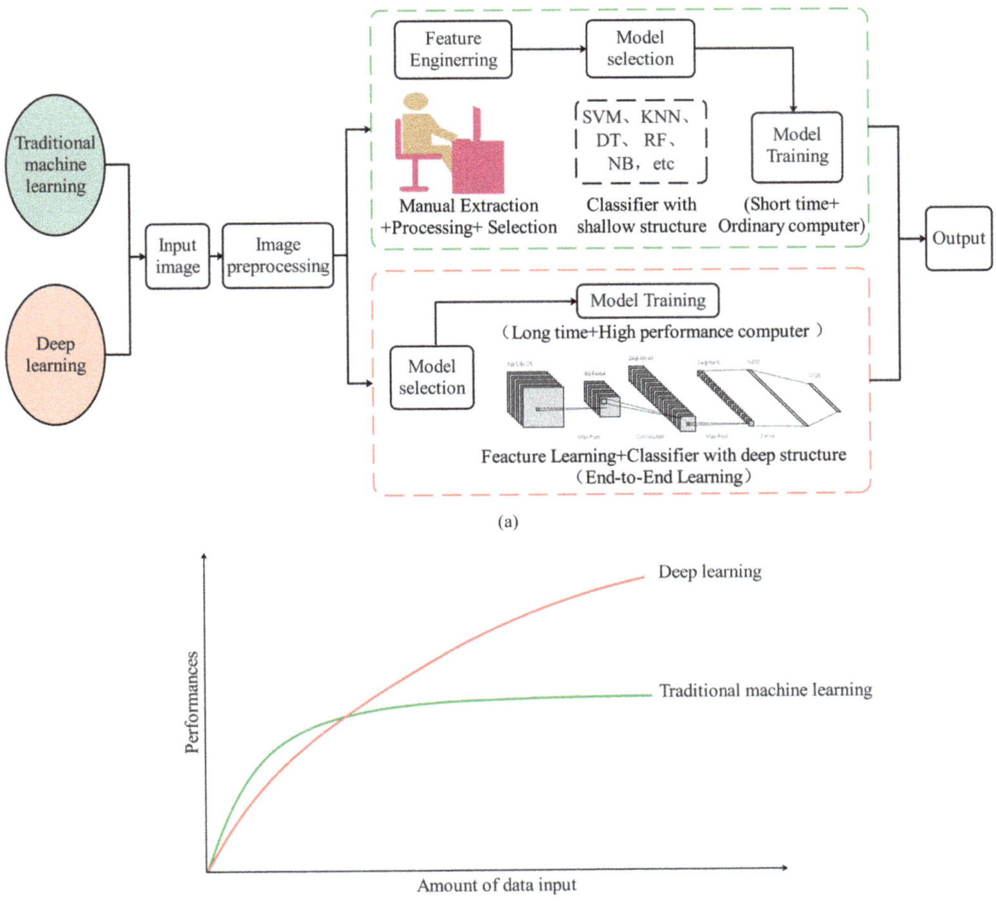

Figure 4. Comparison of Process and Performance between Traditional Machine Learning and Deep Learning in Image Processing: (**a**) Comparison of Image Processing Processes between Traditional Machine Learning and Deep Learning; (**b**) The relationship between the performance of traditional machine learning and deep learning and the amount of data input.

(a) Faster R-CNN

Faster R-CNN is a more classical two-stage target detection network, a detection algorithm that was proposed in 2015 after R-CNN and Fast RCNN. Architecturally, it consists of two main networks, Fast R-CNN and RPN (Regional Proposal Network). Compared with the previous two, Faster -RCNN integrates feature extraction, proposal extraction, bounding box, regression, and classification in a single network, which significantly improves the detection speed and greatly improves the comprehensive performance. Mu et al. [48] used Faster R-CNN for kiwifruit recognition, where color and depth images acquired from a Kinectv2 camera were fed into a convolutional neural network, and the neural network was used to detect and extract the coordinates of kiwifruit. The picking robot applying this network model showed an extremely high picking success rate of 94.2% in an orchard test containing 240 samples, with an average picking time of 4–5 s. Similarly, Fu et al. [55] selected two network structures (ZFNet and VGG16) based on Faster-RCNN for apple picking and used the network of two structures to detect the Original-RGB and Foreground-RGB images acquired from Kinectv2. The experimental results showed that the VGG16 network has the highest average detection accuracy (AP) of 0.893 for Foreground-RGB images.

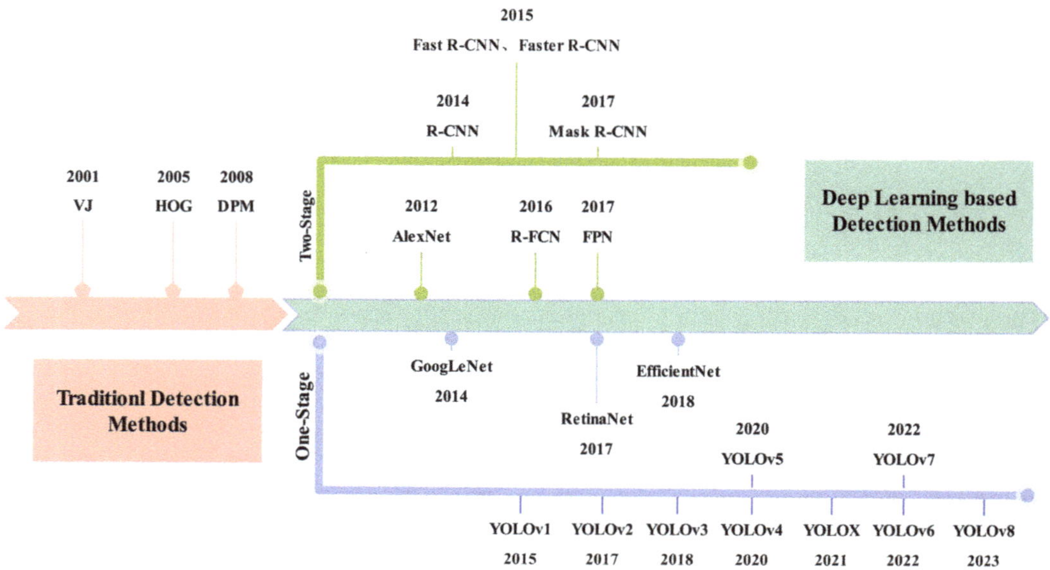

Figure 5. Target detection technology timeline.

(b) Mask-RCNN

Mask-RCNN is another classical deep learning network after Faster R-CNN in the two-stage network, which is based on Faster R-CNN with a fully convolutional Mask Prediction Branch added to the Head layer. The ROI Pooling is improved, and ROI Align is proposed, which solves the problem of twice region mismatch caused by rounding in ROI Pooling in Faster R-CNN. Different from Faster R-CNN, which uses VGG as the skeleton network, Master R-CNN uses ResNet50 or ResNet101 as the skeleton network. Combined with the FCN network structure, four modes can be formed, namely, ResNet50, ResNet101, ResNet50 + FPN, and ResNet101 + FPN. The ROI generation method, the selection of RPs, and the selection of RPs to be projected onto the feature map will be different for different combinations, and the size of the feature maps into the Head layer will also be different so that the researchers can choose flexibly according to their needs. Compared with Faster R-CNN, Mask-RCNN is able to simultaneously achieve target detection, target classification, and pixel-level target segmentation by combining object detection and semantic segmentation. Yu et al. [74] used Mask-RCNN as the detection network of the vision module in order to improve the target detection performance of a strawberry-picking robot and chose Resnet50 as the skeleton network, which was combined with a feature pyramid network (FPN) architecture for feature acquisition of target strawberries. The target detection experiments showed that the average detection accuracy (AP) of the trained model is 95.78%, which is particularly effective for strawberry detection under complex growth states such as changing light intensity, overlap, and occlusion. Similarly, in order to better detect tomatoes that are in an overlapping state with smooth texture and uniform color, Gong et al. [54] used Mask-RCNN, which has a better performance in dealing with overlapping targets, as the basic network and used RGB-D-I fused images as the training set. The test results showed that the target segmentation accuracy is improved by 7.6% over the RGB-based mask R-CNN using the extended Mask R-CNN model trained with fused images. In addition, in order to solve the problem of fruit recognition and localization under different occlusion states, Yang et al. [75] purposely proposed a citrus fruit and branch recognition model based on Mask R-CNN. While constructing a training dataset including multiple complex conditions, a segmentation labeling method is proposed for irregular branches. The experiments showed that the average detection accuracies of the trained model for fruits and branches were 88.15% and 96.27%, respectively, and the average measurement errors for citrus transverse, longitudinal, and branch diameters were 2.52 mm, 2.29 mm, and 1.17 mm, respectively.

(c) YOLO

YOLO series networks belong to one-stage representative networks. Unlike Faster R-CNN and Mask-RCNN, YOLO does not have an RPN network structure and combines object classification and object localization (bounding box) into a regression problem in the detection process. Different from R-CNN's "Look twice" (candidate box extraction and classification), YOLO only needs to Look Once, so the detection speed of the YOLO network is much faster than the two-stage network of the R-CNN series.

In view of the advantages of YOLO series networks and in order to detect banana fruits quickly and accurately in a complex orchard environment, Fu et al. [76] proposed a banana fruit detection method based on YOLOv4. Experimental results showed that the detection rate of the algorithm was 99.29%, the average detection time was 0.171 s, and the AP value was 0.9995. Similarly, in order to meet the identification and positioning requirements of litchi fruits and stems in the nighttime environment, Liang et al. [77] proposed a litchi fruit detection method based on YOLOv3. Under high, medium, and low brightness conditions, the mean Average Precision (mAP) of the model for fruit detection is 96.43%, and the average detection time is 0.026 s. For the segmentation of stem, the accuracy is 95.54%, and the average segmentation time is 0.071 s. In order to verify whether different classification patterns will affect the detection effect of the kiwi detection model, Suo et al. [78] collected and classified 1160 kiwi images according to picking strategies and occlusion conditions and input them into two network models of YOLOv4 and YOLOv3 for training and testing. The experimental results showed that labeling and classifying the data set as detailed as possible can effectively improve the detection accuracy of the network model. Ning et al. [47] used YOLO-V4-CBAM based on YOLO-V4 to identify and locate sweet peppers in dense planting environments so as to improve the recognition and positioning accuracy of sweet pepper-picking robots for multi-target fruits in complex planting environments. Experimental results showed that the F1-score of the proposed method for sweet pepper in a dense planting environment is 91.84%, which is 9.14% higher than that of YOLO-V4, and the positioning accuracy is 89.55%. On the basis of previous research, Xiong et al. [56] combined YOLOv4, Deep SORT, and color threshold to develop a faster and more accurate vision system for strawberry real-time detection, tracking, and positioning. Field experiments showed that the picking success rate of strawberry picking robots using the new system is 62.4%, which is 36.8% higher than before. Similarly, Yu et al. [79] designed a fruit pose estimator called R-YOLO for their new strawberry ridge-harvesting robot. This model is based on YOLOv3 and uses the lightweight network Mobilenet-V1 as the backbone network for feature extraction, which improves the running speed of the model. Tests showed that the model has an average recognition rate of 94.43%, and the processing speed of a single image is 3.6 times faster than YOLOV3. Xu et al. [80] proposed a green mango detection model Light-YOLOv3 based on YOLOv3 for picking green mangoes under different lighting and occlusion environments. This model uses a lightweight unit based on the green mango's color, texture, and shape features to replace the Resnet unit in YOLOv3 and combines the MSCA (Multiscale context aggregation) module to concatenate and predict multi-layer features, which effectively improves the detection effect of green mangoes. Similarly, in order to solve the problem of tomato detection in complex scenes and adapt to embedded devices, Xu et al. [81] proposed a fast detection method based on YOLOv3-tiny. The new model uses improved depth-wise separable convolution and residual structure to replace the standard convolutional network, which increases the depth of the network and greatly reduces the number of Flops. Experiments showed that the f1-score of the new model is 12% higher than that of YOLOv3-tiny, and the detection speed reaches 25 frames per second. In addition, in order to solve the problem of information loss and insufficient semantic feature extraction of small targets in the process of network transmission of Yolov3, Chen et al. [82] proposed an improved Yolov3 cherry tomato detection algorithm YOLOv3-DPN based on DPNs. The improved algorithm can extract richer semantic features of small targets and reduce the information loss in the propagation process. It is worth mentioning that many groups have performed a lot of research on how to better detect objects, but very little attention has been paid to the problem of phasing objects. To this end, Wang et al. [83] proposed a multi-stage strawberry fruit detection method based on Detailed Semantic Enhancement (DSE-YOLO) on the basis of YOLOv3. This model includes the DSE module, EBCE, and DEMSE loss functions, which solve the problem of foreground class imbalance of the original model and can distinguish different stages of fruits with higher accuracy while better detecting small fruits. Aiming at the problems of low accuracy and poor robustness in traditional green pepper detection methods, Li et al. [84] proposed an improved green pepper object detection algorithm based on Yolov4_tiny. The algorithm was based on the backbone network in the classical object detection model and introduced adaptive feature fusion and feature attention mechanism. It improves the recognition accuracy of green pepper small targets and ensures classification accuracy. Similarly, according to the characteristics of the small shape and dense growth of plums, Wang et al. [85] proposed an improved version of YOLOv4 lightweight model based on YOLOv4. This model uses Darknet53 generation MobilenetV3 on the backbone network and uses Depthwise Separable Convolution (DSC) to replace the standard convolution so as to lightweight the model. At the same time, the 152×152 feature layer is introduced to improve the target extraction ability in the dense state. Experiments showed that the model has a higher Mean Average Precision (mAP) than YOLOv4, YOLOv4-Tiny, and MobileNet-SSD. The size is 77.85% smaller than YOLOv4, and the detection speed is 112% faster than YOLOv4. At present, most apple detection algorithms cannot distinguish apples occluded by branches from apples occluded by other apples, which is highly likely to cause damage to the target apple, the robotic arm, and the end-effector during the picking process. In order to solve this problem, Yan et al. [86] proposed an apple detection algorithm based on improved YOLOv5s. Experimental results showed that the algorithm can effectively distinguish between pickable and non-pickable apples. Compared with the classical model, the proposed method effectively improves the mAP while compressing the model size, and the average detection time of a single image is only 0.015 s, which can meet the needs of real-time detection.

(d) SSD

SSD is also a one-stage network; unlike the YOLO series, the SSD network has different scales and aspect ratios of Prior boxes, which allows for the use of different sizes of feature maps for the detection of targets of various scales. Qian et al. [87] proposed an SSD-based method for accurate and real-time mushroom detection and location and optimized the backbone network in the original SSD model to improve the real-time detection performance in the embedded device. The model performs well in tests, with an F1 score of 0.951 and an average localization error of 2.43 mm for mushrooms.

(e) FCN

FCN is the pioneering work of deep learning for semantic segmentation. Compared with CNN, FCN replaces the fully connected layer with a convolutional layer and solves the problem of smaller image size due to convolution and pooling by using up-sampling to recover the image size. FCN does not include a full convolution network with a full connection layer, but it can adapt to target input of any size. Its convolution layer can refine the output results as much as possible, and FCN combined with the jump structure of different depth layer results can also ensure robustness and accuracy. In order to achieve collision-free automatic picking of guava, Lin et al. [12] used a Full Convolutional Network (FCN) for the segmentation of guava color images, and the experimental results showed that the average accuracy of the FCN model for the fruit class is 0.893, and the IOU is 0.806, which indicates that the model is able to be able to segment the guava fruits very well. Unlike Lin et al., in order to improve the accuracy and increase the efficiency of the vision system of the picking robot, Liu et al. [88] combined deep learning algorithms with machine vision and proposed a novel detection algorithm R-FCN combining region-proposed network (RPN) and full convolutional neural network (FCN). The algorithm utilizes FCN to convolve the input image to achieve pixel-level feature extraction and uses RPN to generate multiple candidate frames on the feature map after the convolution operation to effectively separate the foreground and background of the image. In the identification test of apples and oranges, the detection accuracy of the algorithm reaches 97.66% and 96.50%, and the identification accuracy of large fruit bananas reaches 82.30%.

In addition to the above common network models, Li et al. [89] proposed a semantic segmentation method based on Deeplabv3 to segment the fruit, branches, and background in RGB images in order to adapt to the complex growth environment of litchi and detect and locate the fruit branches of multiple litchi clusters. The experiment showed that the extraction accuracy of the test set is 83.33%, and the mean intersection over union (MIOU) is 79.46%, which has a good segmentation effect. Similarly, during the picking process of litchi, in order to better detect the branches and avoid them from damaging the picking robot, Peng et al. [90] used the DeepLabV3+ semantic segmentation model based on the Xception_65 feature extraction network for target detection of litchi. The experimental results showed that the model has an MIoU of 0.765, which is an improvement of 0.144 over the original DeepLabV3+ model, as well as a stronger robustness. Likewise, for fruit and branch segmentation of apples, Kang et al. [52] used the DASNet network model. The f1 score and IoU of the model for fruit detection and segmentation accuracy were 0.871 and 0.862, respectively, according to the test results in the lab and in the orchard setting, indicating that the model was able to precisely and successfully detect and segment orchard apples.

Table 4. Comparison of different object detection network models based on DL.

	Model Based on	Applied Crops	Data			Evaluation Indicators		Feature	Ref
			Total	Training Sets	Testing Sets	Detection Speed	Others		
One-Stage	YOLOV3	Litchi	545	-	-	26 ms	mAP: 96.43%	The detection speed is faster than Faster RCNN and SSD, enabling real-time detection	[77]
	Light-YOLOv3	Green Mane	500	-	-	192 fps (5.21 ms)	FLOPs: 10.12 BN Volume: 44 MB F1-score: 97.7%	The problem of insufficient location and semantic information in YOLOv3 prediction feature maps is solved, and the operation speed is improved by 5 times	[80]
	YOLOv3-tiny	Tomato	-	5500	-	25 fps (40 ms)	F1-score: 91.92%	Adapts to detection in complex environments and to embedded devices	[81]
	Yolov3-DPN	Virgin fruit	1825	1460	365	58 ms	Precision Light changes: 93.54% Fruit shading: 94.59% F1-score: 94.18%	Richer semantic features of small targets can be extracted and information loss in the propagation process can be reduced	[82]
	R-YOLO	Strawberry	2000	1900	100	56 ms	Precision: 94.43% Recall: 93.46%	Detection speed is 3.6 times faster than YOLOv3, with good real-time performance	[79]
	DSE-YOLO	Strawberry	21,921	14,614	7307	18.2 fps (55 ms)	mAP: 86.58% F1-score: 81.59%	Better detection of small fruits and more accurate differentiation of different stages of fruits	[83]
	YOLOv4	Kiwifruit	1160	928	232	25.5 ms	mAP: 91.9%.	More detailed classification of the dataset can improve the detection of YOLOv4	[78]
	YOLOV4-CBAM	Sweet Pepper	-	-	100	-	Positioning accuracy: 89.55%. F1-score: 91.84%	Compared to YOLO-V4, YOLO-V4- cbam has a higher F1 score	[47]
	Deep sort-YOLOv4	Strawberry	-	-	-	-	Cluster picking success rate: 62.4%	The cluster selecting success rate increased by 36.8% from the previous rate to 62.4%.	[56]
	YOLOv4	Banana	1164	835	120 Validation set (Vs): 209	171 ms	Detection rate: 99.29% AP: 0.9995	-	[76]
	Improved-YOLOv4	Plum	1890	1512	378	42.55 fps (23.5 ms)	mAP: 88.56%	77.85% size compression and 112% faster detection than YOLOv4	[85]
	Improved-Yolov4_tiny	Green Pepper	1500	1355	145	89 fps (11.24 ms)	AP: 95.11%, Precision: 96.91% Recall: 93.85%	It can ensure real-time production and can effectively improve the detection of difficult samples of green pepper.	[84]
	Improved-YOLOv5s	Apple	1214	1014	200	66.7 fps (15 ms)	Recall: 91.48% Precision: 83.83% mAP: 86.75% F1-score: 87.49%	It can effectively identify apples that are obscured by leaves and branches	[86]
	SSD	Mushroom	4300	4000	300		F1-score: 0.951	-	[87]

Table 4. *Cont.*

	Model Based on	Applied Crops	Data			Evaluation Indicators		Feature	Ref
			Total	Training Sets	Testing Sets	Detection Speed	Others		
Two-Stage	Faster-RCNN	Apple	800	560	120 Vs: 120	181 ms	AP: 0.893	The VGG16 foreground- rgb image has an AP of up to 0.893, allowing for almost real-time monitoring	[55]
	Mask-RCNN	Tomato	-	-	500	456 ms	Iou: 0.916	The segmentation accuracy is effectively improved by the model trained based on RGB-D-I fused images	[54]
	Mask-RCNN	Strawberry	2000	1900	100	8 fps (125 ms)	MIou: 89.85% AP: 95.78% Recall: 95.41%	-	[74]
	Mask-RCNN	Citrus	-	1000	-	-	MAP Fruits: 88.15% Branches: 96.27%	It can effectively detect citrus and tree branches at the same time, and can plan pick-up paths and perform reasonable obstacle avoidance.	[75]
	FCN	Guava	437	350	87	565 ms	Mean Accuracy0.893 IOU: 0.806	-	[12]
	R-FCN	Apple Orange Banana	160,000	80,000	40,000 Vs: 40,000		Accuracy Apple: 97.66% Orange: 96.50% Banana: 82.30%	Better robustness in real-world engineering	[88]
	Deeplabv3	Litchi	-	-	90	464 ms	Precision: 83.33% IOU: 79.46%	-	[89]
	DeepLabV3+	Litchi	65,625	50,000	15,625	-	MIoU: 0.765	MIoU improves 0.144 over the original DeepLabV3+ model, while having stronger robustness and higher detection accuracy	[90]
	DASNet	-	1277	567	560 Vs: 150	477 ms	Precision: 0.88 F1-score: 0.871 Recall: 0.868 IoU: 0.862	-	[52]

3. Intelligent Harvesting "Brain" System

The picking decision and control of fruit and vegetable picking robots are key to ensuring the normal work and efficient picking of the robots. On the one hand, the design of the picking strategy needs to carry out a picking feasibility analysis according to the characteristics of the target fruits, maturity degree, growth environment, and other factors, and combine the hardware facilities of the picking platform, such as robotic arm, end-effector, sensor, to formulate a reasonable picking route and picking mode. On the other hand, picking control needs to realize the accurate positioning of the robot and accurate control of the motion of the manipulator, avoid damage to fruits and vegetables, ensure the picking efficiency and speed, and combine the actual scene and picking strategy for real-time adjustment and optimization. Therefore, reasonable picking strategy and accurate picking control are necessary conditions to ensure the efficient and stable operation of fruit and vegetable picking robots and are also one of the key technologies to realize agricultural production automation.

The decision of picking time and the location of the target fruit are mainly completed by the visual perception system, which is described in detail in Section 2. This section mainly focuses on region division and task allocation, obstacle avoidance strategies, path planning, and control methods.

3.1. *Spatial Partitioning and Task Allocation*

Based on the number of different robotic arms, we divide the region division and task allocation strategy into single-arm harvesting and multi-arm harvesting. The different strategies adopted by the researchers are shown in Table 5.

3.1.1. Single Mechanical Arm Harvesting

Single robotic arm picking is a common picking mode at present. It has high picking flexibility, strong picking consistency and stability, and can be used with different end-effectors to complete the picking of various fruits, vegetables, and flowers, which can better adapt to diverse agricultural picking needs. In terms of the division of the working area of a single robotic arm, Zhang et al. [49] divided the picking space into several vertical bar subspaces according to the growth characteristics of tomatoes and screened out invalid subspaces by calculating whether there was enough space volume between adjacent branch obstacles to carry the string of tomatoes with claws. This method can effectively solve the problem of a difficult return journey caused by volume increase after successful harvesting.

3.1.2. Multi-Mechanical Arm Harvesting

Compared with single-arm picking, multi-arm cooperative picking can effectively shorten the picking time, improve the picking efficiency, and better adapt to the complex and changing unstructured picking environment, more suitable for different types, different shapes, and different sizes of crop picking, with stronger adaptability and flexibility. Divided by working area, multi-robot cooperative picking can be categorized into two picking strategies: regional independent and regional shared.

Regional independence: The region-independent strategy means that each robotic arm is individually responsible for a completely independent picking region, that there is no cross overlap between the sub-regions, and each robotic arm is only responsible for picking the target fruits in each sub-region, and this kind of task allocation can avoid the collision interference between multiple arms, and the requirements for the control system are also relatively low. Xiong et al. [67] used a low-cost dual Cartesian robotic arm in their preliminary study of strawberry picking, in which the two robotic arms had completely independent working partitions during picking, and each sub-area was divided into left and right half-areas, and the two arms started picking from the left half-area or the right half-area at the same time according to the density of the target strawberries, which ensured that there was a sufficient safety distance between the arms and avoided possible collisions. In the next study, Xiong et al. [91] put a new type of U-shaped arch-picking robot into the application; two three-degree-of-freedom robotic arms with non-contact fixtures were installed on both sides of the arch, which were responsible for picking strawberries in the left and right regions, respectively, which was a more complete independent region division, completely solving the problem of collision between multiple arms and at the same time, could effectively reduce the complexity of the control system.

Table 5. Space planning and task allocation of harvesting robots.

Applied Crops	Classification Type	Mechanical Arm	Feature	Sketch Map	Ref.
Tomato	-	Single	Sieve out invalid subspace, solve the problem of difficult return, improve work efficiency	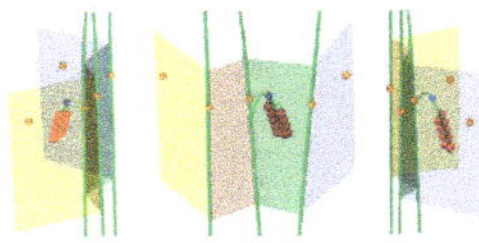	[49]
Strawberry	Regional independence	Double	Sufficient safety distance to avoid collision of two robotic arms		[67]
Strawberry		Double	The picking area is completely independent, completely solving the collision problem		[91]
Apple	Regional sharing	Quadra	Effectively solving the problem of double picking and missed picking		[51]
Eggplant		Double	Avoiding collisions between robotic arms, effectively shortening picking time	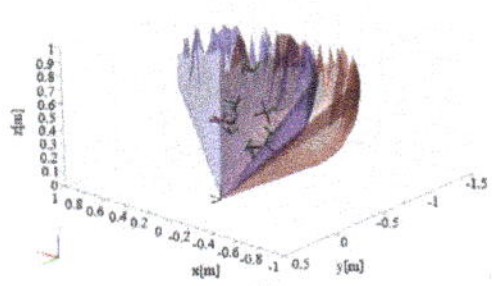	[65]

Regional sharing: The region-shared strategy means that multiple robotic arms are jointly responsible for a large picking area, which is divided into several small sub-areas, with shared overlap between the sub-areas; each robotic arm works independently in the sub-area it is responsible for, and neighboring robotic arms collaborate with each other in the shared area. Under this strategy, each robotic arm collaborates with each other, which can effectively avoid the occurrence of repeated picking and missed picking.

In a dwarfed and densely planted environment, Li et al. [51] used a four-armed robot for the collaborative picking of target apples and planned work partitions for each robotic arm. In addition to the exclusive picking area of the four robotic arms, there are four overlapping picking areas between each neighboring robotic arm. However, to reduce the amount of computation and control difficulty, at most one robotic arm is allowed to enter the overlapping picking area at the same time, and the whole picking task is categorized as an asynchronous overlapped multiple traveling salesman problem, which can effectively shorten the traversal time. In contrast, to solve the eggplant picking problem in an occluded environment, Sepulveda et al. [65] designed a dual-arm cooperative picking robot. This picking platform can not only simultaneously pick target fruits within the respective working range of the two arms but also pick occluded targets by cooperative operation in the shared area of the two arms. Experiments showed that its average harvesting success rate is as high as 91.67%.

3.2. Obstacle Avoidance Strategies

3.2.1. Passive Obstacle Avoidance Strategies

Passive obstacle avoidance is the most common and widely used obstacle avoidance strategy. It mainly refers to taking some passive measures to avoid collision or conflict when planning the path, considering the obstacles that the robot or unmanned aerial vehicle may encounter when performing the task. It is mainly realized in the path-planning stage by modeling the surrounding environment and adding real-time obstacle avoidance factors to obtain a smooth route without collision.

Considering the obstacle avoidance problem after tomato bunch picking, Zhang et al. [49] proposed a real-time motion path-planning algorithm (OPS) based on spatial segmentation. This method can plan an effective picking subspace for the robotic arm in advance based on the position information of the environment and tomato bunches and to avoid exploring the path in the invalid subspace. In addition, the OPS algorithm can adjust the end attitude of the robotic arm in real time, according to the relative position between the obstacle and the robotic arm, to realize obstacle avoidance. Experiments showed that the picking time of a single bunch of tomatoes by this method is 12.51 s, and the picking success rate is close to 100%.

3.2.2. Active Obstacle Avoidance Strategies

In the actual picking process, especially in complex unstructured environments, dense foliage or compact fruit distribution will make the passive obstacle avoidance "bypassing" strategy fail and then will need some more complex active strategies to solve this problem. Unlike passive obstacle avoidance, active obstacle avoidance can be used to "push away" obstacles through a series of complex sequential movements or multi-arm coordination, which is more suitable for target picking under dense shade. To solve the problem of eggplant picking in an occluded environment, Sepulveda et al. [65] used a strategy of pushing away obstacles with one arm and picking with the other arm in their dual-arm picking robot. Experiments showed that the robot had a high success rate of 81.25% in pushing away from obstacles, which is an effective active obstacle avoidance strategy. In addition, regarding obstacle avoidance techniques for strawberry picking in structured growing environments, Xiong et al. [56,67,91] performed extensive research. In their previous work, to determine the number and location of obstacles around the target, Xiong et al. [67] set up a simpler region of interest (ROI) around the target strawberries. This region divides the obstacles into two layers, top and bottom, with six sub-parts in each layer, which is combined with a simple linear operation to push away the possible obstacles at the top and bottom of the target (as shown in Figure 6a). However, for long-stalked strawberries such as "Murano", a single linear push would be ineffective when there are multiple neighboring obstacles around the target, so Xiong et al. [91] added zigzag pushes in the upward and horizontal directions to the original linear push strategy. In addition, a handheld drag operation (in-hand drag) that can avoid accidentally swallowing the upper obstacle is proposed, and a more complex four-layer structure ROI is set around the target (as shown in Figure 6b), which can better solve the obstacle avoidance problem in complex environments. However, it is not reasonable to measure the presence and number of obstacles by the sub-blocks with point cloud information in the region of interest; for this reason, the research team redefined the layout of the ROI area [56] and used the push–drag maneuvers to accurately separate obstacles based on their exact location (as shown in Figure 6c). In addition, to obtain the information of the obstacles after dragging in time, the middle and top layers use continuous "look and move" for real-time sensing and determine a new round of push–drag operation. The experiments showed that under the premise of constant picking speed, the cluster picking success rate of the improved method reaches 62.4%, which is 36.8% higher than the previous one.

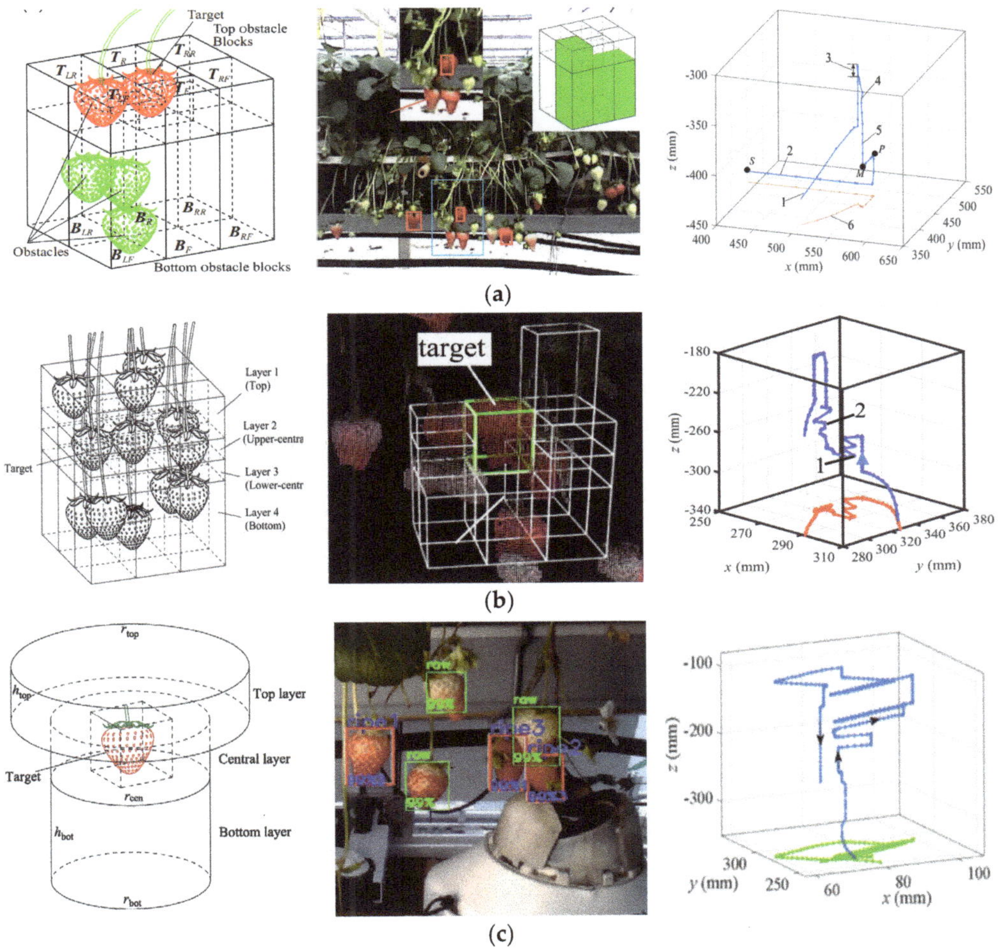

Figure 6. Three Different Active Obstacle Avoidance Strategies: (**a**) a simple linear obstacle avoidance strategy (Ref. [67], 2020, Xiong, Y.); (**b**) a zigzag obstacle avoidance strategy (Ref. [91], 2020, Xiong, Y.); (**c**) obstacle avoidance strategy with new ROI regions and continuous "look and move" (Ref. [56], 2021, Xiong, Y.).

3.3. *Path-Planning Techniques*

3.3.1. Classic Path-Planning Algorithms

The task area needs to be modeled before the path is planned out, and the various obstacle information in the task area is obtained through modeling, and the optimal path for the whole area is planned on this basis. Classical path-planning algorithms include global path-planning algorithms and local path-planning algorithms, and common global planning algorithms include Dijkstra's algorithm, A* algorithm, RRT algorithm, and so on. Since global path planning needs to consider more factors, such as obstacles, work area size, time, etc., it is time-consuming and not easy to cope with the dynamically changing environment. So Sarabu et al. [53] adopted an improved RRT-based algorithm, RRT-Connect, for apple-picking path planning in complex environments. Preliminary experiments showed that this method achieves good results without complex optimization. Moreover, in the apple-picking process, Kang et al. [52] used the Octrees algorithm to preprocess and model the surrounding environment and searched the optimal path through eight subspaces for picking. Compared with other methods, Octrees is more advantageous in terms of storage efficiency.

3.3.2. Machine Learning-Based Path-Planning Algorithms

Although dynamic path planning in classical path planning can achieve real-time adjustment of the picking path according to the surrounding environment, it can only avoid individual obstacles and cannot achieve global optimization, while machine learning-based path planning can effectively solve this drawback. The machine learning-based path-planning method applies shallow neural networks or decision trees and other models to path planning; this method requires a large amount of data input and training through the learning of historical data and can be based on the prediction of the results to determine the next action, for complex environments and tasks can achieve good results.

In order to push away the occluded fruits and reach the specified location to pick the target strawberry successfully in a dense planting environment, Mghames et al. [92] proposed a path-planning algorithm known as Interactive Probabilistic Motion Proto-Principle (I-ProMP) and experimentally verified the starting validity, which is well-suited to be used for solving the problem of obstacle avoidance and path planning in the three-dimensional space, and the computation time is very short, which is about 100 ms.

3.3.3. Deep Learning-Based Path-Planning Algorithms

Through deeper neural networks, deep learning-based path planning can learn and forecast the best routes, which is capable of handling more complex environments and tasks, as well as being adaptive and efficient compared to machine learning. Common fruit and vegetable recognition and path-planning methods generally suffer from poor recognition robustness and difficulty in generating collision-free picking paths in dense and complex environments; for this reason, Ning et al. [47] proposed an algorithm for sweet pepper recognition and picking sequence planning—AYDY algorithm, which combines the improved YOLOV4 detection algorithm, an improved DPC algorithm with an anti-collision picking sequence method that introduces a winner-takes-all strategy. The experimental results showed that the AYDY algorithm can effectively shorten the traversal path and picking time and enhance robustness, and the collision-free harvesting success rate is as high as 90.04% compared with the traditional sequential and random traversal algorithms. Similarly, Yang et al. [75] utilized an integrated system developed based on a Masked Region Convolutional Neural Network (Mask R-CNN) and Branch Segment Merging algorithm, which can efficiently plan reasonable collision-free harvesting paths for the harvesting robots while detecting citrus and tree branches.

3.3.4. Optimization Algorithm-Based Path-Planning Strategies

The path-planning method based on an optimization algorithm is a kind of optimization algorithm to find an optimal path that satisfies the constraints under the conditions of a given starting point and end point. This method usually transforms the path-planning problem into an optimization problem and obtains the optimal path by finding the optimal solution in multiple aspects, such as time, energy consumption, and pulsation.

Sepulveda et al. [65] adopted a stochastic trajectory optimization algorithm (STOMP) to deal with the path-planning problem in the picking process in the design of the two-armed eggplant picking platform, which generates an optimal picking path based on the workspace, the position of the fruits, and the configuration of the arms and determines the sequence of motions required to grasp and separate the eggplants. This algorithm performs a global search while avoiding the problem that traditional algorithms need to traverse the entire search space by random sampling, which greatly reduces the computational complexity. Similarly, to improve the operational efficiency of a multi-mechanical arm of a collaborative picking robot in a dwarf-dense planting environment, Li et al. [51] generalized the multi-mechanical arm-picking problem with overlapped domains into an asynchronous overlapped multiple traveling salesman problem and solved it optimally, based on genetic algorithm. Experiments showed that the task-planning method based on a genetic algorithm reduces the job traversal time dramatically relative to the random traversal method and the sequential planning method and effectively improves the operational efficiency under the premise of ensuring that each robotic arm does not conflict. In addition, to address the drawbacks of traditional path-planning algorithms that are time-consuming, as well as to solve the problem of low picking success rate caused by the collision between robotic arms and branches in unstructured environments, Ye et al. [62] obtained collision-free picking poses during litchi picking by an improved adaptive weight particle swarm optimization (APSO) algorithm and used an optimization algorithm based on the Bi-RRT algorithm (AtBi-RRT) to quickly determine the appropriate collision-free picking path. Simulation results showed that the average computation time of the At Bi-RRT algorithm is 3.71 s shorter than the TRRT algorithm.

In the past decades, many teams have focused on visual perception and path planning but neglected the research on motion planning; yet, stable motion planning is crucial for the realization of efficient and lossless picking. In order to achieve stable, efficient, and lossless harvesting of apples, Cao et al. [93] proposed an improved multi-objective particle swarm optimization algorithm (GMOPSO). The algorithm combines the methods of variation operator, annealing factor, and feedback mechanism to optimize the motion trajectory in terms of time, energy consumption, and pulsation so as to accelerate the convergence speed while satisfying the stable motion and avoiding the local optimal solution, and finally realize the optimal trajectory of the robotic arm. Tests showed that the picking platform optimized by the GMOPSO algorithm can effectively achieve stable, efficient, and lossless picking, and its average picking time is 25.5 s, with a success rate of 96.67%.

In addition to the above four common path-planning methods, there are some other path-planning and trajectory optimization methods. For example, in the greenhouse cucumber picking process, Chen et al. [94] used an improved prediction point Hough transform to quickly and accurately fit the path of a cucumber picking robot to obtain a smoother and easier-to-handle path. Aiming at the shortcomings of the traditional Hough transform in terms of large traversal angle range, wide intersection detection range, and being time-consuming, this method makes relevant improvements in three aspects: traversal angle range; intersection detection range; and fitting accuracy. Experiments showed that this method is more time-saving than the traditional Hough transform while having higher accuracy and better robustness. For another example, Colucci et al. [95] proposed a simplified motion planning algorithm based on motion decoupling for precise agricultural applications, which can simplify the complex motion planning problem into a series of simple sub-problems, significantly reduce the computational cost and, thus, improve the efficiency and accuracy of motion planning.

3.4. Control Methods

In this section, we divide control methods into two categories for description: traditional control methods (classical and modern control); and intelligent control methods. The comparison of the advantages and disadvantages of specific control strategies is shown in Table 6.

Table 6. Comparison of the characteristics of different control methods.

Control Type	Control Method	Applied Crops	Mechanical Arm	Advantages	Disadvantages	Ref.
Classic Control	PID	Strawberry	Single	Simple to implement Easy to adapt Fast response time Good stability	Sensitive to noise Difficult to adjust parameters Unable to handle non-linear systems Unable to handle time-varying systems	[71]
		Eggplant	Double			[65]
Modern Control	NMPC	-	Double	Wide applicability Robustness Optimizable for multiple objectives Can handle constraints	Large calculation volume Difficult to adjust parameters High impact of model error Poor stability	[96]
	Impedance Control	Apple	Single	Wide adaptability High robustness High control accuracy Flexible interaction possible	Large calculation volume Difficult parameter adjustment High requirements for sensors Not very stable	[97]
	SMC	Famous Tea	Single	Robust Rapid response	High-frequency oscillation Complexity of nonlinear design	[56]
Intelligent Control	Fuzzy Control	Wolfberry	-	Robustness Wide adaptability Adjustable control effects Flexible knowledge representation	Large calculation volume Difficult parameter adjustment Unstable control effect	[33]
	Fuzzy PID Control	Wool	-	Robust Flexible fuzzy rules Easy operation	Computationally complex Poor interpretability Difficulty in choosing parameters	[98]

3.4.1. Classical and Modern Control Methods

Traditional control methods refer to control methods based on mathematical models and control theories, specifically PID control, state feedback control, optimal control, and so on. These methods are normally based on accurate mathematical models through modeling and analysis of the system and designing controllers to achieve stable control of the system. They are characterized by good stability and controllability.

Among them, PID control is a common classical control method that can calculate the error of the robot by measuring parameters such as position, speed, and acceleration of the robot and adjusting the control parameters of the robot according to the error so that it can better control the motion trajectory and posture. Sepulveda et al. [65] used a PID controller to receive the trajectory points generated by the STOMP planning algorithm containing information and provided motion execution commands to the picking robot. This information includes the positions, velocities, and accelerations of all the joints of both arms, as well as the start point of the next trajectory path. In addition, Xiong et al. [71] used PID control in an earlier study of a strawberry-picking robot and combined it with information obtained from a vision system to move the arm to the optimal cutting position.

In addition, nonlinear model predictive control (NMPC) belongs to a kind of optimal control in modern control. It is an advanced control strategy that uses a nonlinear model to describe the system and predict the future behavior of the system and computes a series of control inputs so that the system achieves optimal control under certain constraints. NMPC is usually a better choice in some control problems that require higher accuracy. In order to better solve the picking problem of a multi-arm robot in an orchard, FléCher et al. [96] proposed a VPC strategy combining NMPC and IBVS (Image-Based Visual Servoing), by which different end-effectors are controlled to approach the specified target fruits. Simulation experiments showed that the control strategy can enable the multi-arm robot to perform multiple tasks effectively in a shared space.

Impedance control, which also belongs to modern control, is a control method based on the relationship between force and position, which can realize the control of force and position of the robots when interacting with the environment. In the apple harvesting process, to reduce the damage to apples by the picking robots, Ji et al. [97] proposed an adaptive impedance control method based on impedance control, which can adaptively adjust the impedance parameters to adapt to different environments and tasks, so that the end-effector can grasp apples quickly, stably, and with a low overshoot even when the environmental stiffness and position are not clear.

Sliding mode control (SMC) is a nonlinear control technique; its main idea is to introduce a specific switching function on a sliding surface so that the system state slides rapidly on this sliding surface and remains on it, which helps realize the robust control and anti-disturbance ability of the system. On the automatic picking platform of famous and high-quality tea, Zhou et al. [56] designed and optimized the control strategy of robotic arm picking based on sliding mode control, which effectively suppressed the vibration phenomenon of the sliding mode surface during rapid convergence. In the testing process, it also showed a high picking success rate and integrity rate.

3.4.2. Intelligent Control Methods

Intelligent control methods are control methods based on artificial intelligence techniques, such as fuzzy control, neural network control, and genetic algorithm control. These methods are usually based on a data-driven approach, where the controller is designed to realize the intelligent control of the system by learning and analyzing the data of the system. Compared with traditional control methods, intelligent control methods are characterized by high adaptability and good robustness.

Fuzzy control is a control method based on fuzzy logic, which realizes control by establishing a fuzzy rule base and a fuzzy inference mechanism and is able to deal with the fuzzy or uncertainty problems of the system with strong adaptability and robustness. Therefore, to solve the problem of accurate navigation in the unstructured Goji berries environment, Ma et al. [33] used the fuzzy control method to control the navigation of a Goji berry-picking robots, and the experimental results showed that this method could effectively reduce the influence of environmental variables on the picking platform and improve the robustness of the control system. In order to better study the influence of various factors on the effect of cotton picking, Wang et al. [98] developed a cotton-picking measurement and control system, which is a fuzzy PID control system that integrates classical PID control with fuzzy control. This system can realize the continuous adjustability of cotton-picking speed, conveyor belt speed, and fan speed, and its research results can provide support for the optimization of the picking mechanism.

4. Intelligent Picking "Hand" System

Harvesting robots typically employ end-effectors to accomplish the task of harvesting. The end-effector serves as a crucial component of harvesting robots responsible for executing specific tasks, for instance, tasks such as picking, transporting, or assembling. In the context of harvesting robots, the principal function of the end-effectors is to facilitate the efficacious picking of crops while concurrently preserving the integrity of the plants. In terms of end-effectors in harvesting robots, commonly employed types include negative-pressure adsorption, shearing-style, cavity-retrieval, and flexible grasping mechanisms. This section will provide an in-depth discussion of these diverse types of end-effectors.

4.1. End-Effector Modes of Operation

The modes of operation for end-effectors in agricultural harvesting robots typically encompass four methods: negative-pressure adsorption; shearing; cavity retrieval; and flexible grasping. Negative-pressure adsorption end-effectors, as depicted in Figure 7a, principally utilize the principle of negative pressure adsorption to adhere the crops onto the robot's end-effector, after which they are harvested via the robotic arm or other components. Shearing end-effectors, as illustrated in Figure 7b, predominantly employ a clamping method akin to scissors, severing the crops from their branches or stems. Cavity retrieval end-effectors, as demonstrated in Figure 7c, function by extending the cavity retrieval device into the crop, using the robotic arm and leveraging the air pressure difference to secure the crop within the cavity, followed by its extraction. Flexible grasping end-effectors, as portrayed in Figure 7d, leverage the properties of flexible materials, enabling the robotic arm to drive the grasper in securing the crop, thereby accomplishing the harvesting task. Its advantage lies in its suitability for fruits and vegetables of various shapes and sizes, with the capability of adopting different grasper shapes and sizes for different crops. The following sections will provide detailed insights into the research developments and applications of these four distinct types of end-effectors.

(**a**) (**b**) (**c**) (**d**)

Figure 7. Four types of end-effectors for harvesting robots: (**a**) negative pressure adsorption end-effectors (ref. [99], 2018, Ramin Shamshiri et al.); (**b**) shearing-style end-effectors (ref. [63], 2020, Oktarina et al.); (**c**) cavity retrieval end-effectors (ref. [100], 2021, Navas et al.); (**d**) Flexible grasping end-effectors (ref. [59], 2021, Yu et al.).

4.1.1. Negative-Pressure Adsorption End-Effectors

Negative-pressure adsorption end-effectors in agricultural harvesting robots represent a type of end-effector that utilizes negative-pressure adsorption forces during the picking of fruits and vegetables. This end-effector is typically comprised of a suction cup and a negative-pressure system. Such technology necessitates varying designs according to different crop shapes and sizes to ensure sufficient contact with the crop surface and generate ample adsorption force for secure harvesting. Relative to traditional mechanical claws and arms, this end-effector offers superior flexibility and precision, better accommodating crops of diverse shapes and sizes while simultaneously minimizing crop damage.

Over the past few years, significant advancements have been realized in the technology of negative-pressure adsorption end-effectors. Presently, this technology has found applicability in the harvesting of fruits and vegetables with relatively regular shapes, such as apples and tomatoes. A team led by Wang et al. [10] investigated a gripper composed of a flexible silicone funnel, as illustrated in Table 7 (a), which employs vacuum suction for apple harvesting. Through multiple prototype testing, the team designed an optimal funnel shape, considering parameters such as edge thickness, funnel angle, and size while striking a balance between flexibility and robustness. Experimental outcomes revealed that even prolonged exposure of the apple to relatively low vacuum levels did not inflict any damage. Pertaining to the pneumatic harvesting of apples, there are also techniques such as those exemplified in Table 7 (b), a vacuum mechanism robot for apple picking from Abundant Robotics (Hayward, CA, USA). It is a single-suction-cup end-effector capable of autonomous recognition and location of apples, accomplishing harvesting tasks through negative pressure adsorption. This robotic system demonstrated low damage rates and high harvesting precision in experimental trials [99]. As depicted in Table 7 (c), the harvesting gripper design consists of three components: adsorption, clamping, and twisting for the fruit. Upon the gripper's movement to the targeted location, it encapsulates the entire fruit within a sleeve. The rapid inflation of an airbag tightly clamps onto the surface of the fruit. Subsequently, the rotation of the sleeve enables the disengagement of the tomato fruit from its stem, thus culminating in the successful harvesting of the fruit. Nevertheless, challenges still exist for the negative-pressure adsorption end-effectors when it comes to harvesting crops with irregular shapes and soft textures. Variations in surface texture, size, and shape of such fruits and vegetables can affect the performance of the adsorption force. Relevant studies have indicated that negative-pressure adsorption end-effectors can be utilized not only for the harvesting of firm-textured and relatively regular crops, such as apples and tomatoes but also for delicate flowers, such as Hangzhou white chrysanthemums. As demonstrated in Table 7 (d), Yang et al. [44] specifically designed a unique end-effector with a special structure to avoid damage during the harvesting process of Hangzhou white chrysanthemums. This end-effector is equipped with an airbag device at its tip, allowing for the clamping of the chrysanthemum flowers through the inflation of the airbag. The utilization of a segmentation algorithm in conjunction with the end-effector featuring the airbag device effectively ensures the successful harvesting of Hangzhou white chrysanthemums.

In conclusion, considerable progress has been achieved both in research and practical application of the vacuum adsorption end-effector technology in agricultural harvesting robots. Beyond its application in agricultural harvesting, the vacuum adsorption end-effector has potential uses in other sectors, such as part handling and assembly in manufacturing industries. Ultimately, the vacuum adsorption end-effectors in agricultural harvesting robots will continue to evolve and improve in terms of automation, intelligence, multifunctionality, sustainability, and industrial promotion, thereby fostering significant transformation and progress in agricultural production. Therefore, further research and improvements can lead to broader applications and commercialization.

4.1.2. Shearing-Style End-Effectors

Shear-style end-effectors in agricultural harvesting robots are a prevalent type of end-effector, primarily utilized to sever the peduncles of fruits, thereby accomplishing the harvesting task. The following is a detailed overview and current development status of shear-style end-effectors in agricultural harvesting robots.

Table 7. Comparison of different styles of negative pressure adsorption end-effectors.

Figure		Applied Crops	Advantages	Improvements	Gripper Size	Recognition Accuracy	Picking Success Rate	Picking Time	Ref.
	a	Apple	Minimized damage rate	picking speed, and accuracy	Diameter: 10.5 cm	-	-	10.3 s	[10]
	b	Apple	Minimized damage rate, Less costly	Picking rate	-	91%	85%	12 s	[99]
	c	Tomato	Simple structure, Less costly	Positioning speed, picking speed	Diameter: 9 cm	-	-	9.6 s	[10]
	d	Hangzhou white chrysanthemum	High recognition rate	Picking efficiency and accuracy	Diameter: 3 cm	90	80%	12.5 s	[44]

Xiong et al. [71] have dedicated their research to the development of a strawberry harvesting robot. After years of research and successive iterations, A novel strawberry harvesting robot has been developed in this study. This robot is comprised of a newly designed gripper mounted on an industrial arm, which, in turn, is mounted on a mobile base along with an RGB-D camera. The novel cable-driven gripper can open fingers to "swallow" a target. Since it is designed to target the fruit and not the stem, it only requires the fruit location for picking. Moreover, equipped with internal sensors, the gripper can sense and correct positional errors and is robust to the localization errors introduced by the vision module. Another important feature of the gripper is the internal container that is used to collect berries during picking. Since the manipulator does not need to go back and forth between each berry and a separate box, picking time is reduced significantly. The vision system uses color thresholding combined with a screening of the object area and the depth range to select ripe and reachable strawberries, which is fast for processing. These components are integrated into a complete system whose performance is analyzed, starting with the four main failure cases of the vision system: undetected, duplicate detections, inaccurate localization, and segmentation failure. The integration enables the robot to harvest continuously by moving the platform with a joystick. Field experiments show that the average cycle time of continuous single strawberry picking is 7.5 s and 10.6 s when including all procedures. This strawberry-harvesting robot can be considered the most advanced and intelligent strawberry-harvesting robot currently available in the agricultural machinery field.

In the context of harvesting cluster fruits, such as litchi, Ye et al. [62] from the South China Agricultural University developed a harvesting machine consisting primarily of an end-effector equipped with a terminal gripper and a rotating blade disc. The robot, during its harvesting operation, uses a collision-free motion planning algorithm, rendering the harvesting process safer and more convenient. Similarly, for the harvesting of cluster fruits, such as cherry tomatoes, Feng et al. [64] developed an end-effector akin to a pair of scissors. As illustrated in Table 8 (c), it is designed based on the mechanical characteristics of the stem, with dual cutting blades used for severing the stem. The handle, fixed to the cutting blades, can close or open to grasp or release the stem, allowing for reliable cutting and handling of the fruit and facilitating its separation from the plant. This design has enhanced the precision of the harvesting end-effector, providing superior stability during the harvesting process.

In the case of tomato harvesting, Oktarina et al. [63] from Indonesia have designed a tomato-harvesting robot, as shown in Table 8 (d). This robot features a simple yet vivid structure, with a scissor-style end-effector that is sharp and flexible. The harvesting process is facilitated through the drive of a servo motor. In a similar context to tomato harvesting, Jin et al. [61] developed an intelligent tomato harvesting robot system based on multimodal deep feature analysis. The end-effector of this system comprises two mechanical fingers and a three-degree-of-freedom mechanical cutter, utilizing digital servos for the rotation of the joints of the mechanical arm and cutter. This setup enhances the precision requirements of the mechanical arm and cutter, effectively addressing issues of labor shortages and high costs encountered in the tomato harvesting process.

For grape harvesting, Liu et al. [101] designed a harvesting hand, as shown in Table 8 (e), which represents a single-degree-of-freedom grasp-and-cut integrated end-effector. The opening and closing of the two fingers are driven by a helical and symmetrical oscillating linkage mechanism; one fingertip is composed of a floating clamp and blade. Upon contact with the pedicel, the compressive force closes the floating clamp around the pedicel while the blade continues to close, thus completing the cut. This type of end-effector enables low-vibration and rapid operation to minimize fruit drop.

Table 8. Comparison of different types of shear end-effectors.

Figure	Number	Applied Crops	Advantages	Improvements	Gripper Size	Recognition Accuracy	Picking Success Rate	Picking Time	Ref.
	a	Strawberry	Accurately separate obstacles	Targeting accuracy, picking rate	Maximum clamping diameter 60 mm, open diameter 45 mm	-	96.8%	10.6 s	[71]
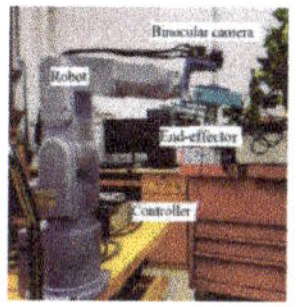	b	Litchi	Non-destructive picking	Picking rate	-	-	-	-	[62]
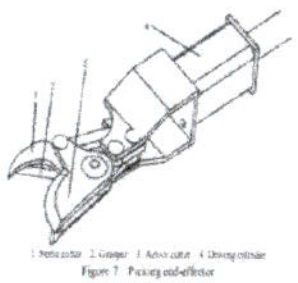	c	Cherry Tomatoes	Stable clamping, low fruit falling rate	Picking success rate	-	-	83%	8 s	[64]
	d	Tomatoes	Fast picking rate	Recognition accuracy	-	-	-	9.676 s	[63]
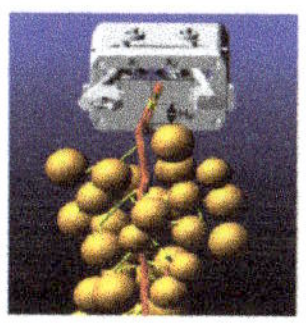	e	Grape	Small size, Flexible	Robustness	Gripping shaft length of 30 mm	-	-	9.6 s	[101]

For sweet pepper harvesting, Ning et al. [47] developed a shear-style end-effector with a Robotic 2F-85 gripper as the terminal execution component of the robot harvesting system, providing a clamping force of 20–235 N and a payload capacity of 5 kg. The harvesting of sweet peppers is not confined to shear-style techniques; the following sections will also introduce flexible grabbing-style end-effectors, among others.

In summary, shear-style end-effectors have been extensively implemented in agricultural harvesting robots and have attracted growing attention from researchers. With continual technological development and innovation, shear-style end-effectors are anticipated to play a more significant role in future agricultural harvesting robots, offering more efficient and reliable solutions for automated agricultural production.

4.1.3. Cavity Retrieval End-Effectors

The cavity retrieval end-effector is another commonly used end-effector in agricultural harvesting robots. It accomplishes the harvesting task through mechanical clamping and encasing the crops. Its structure includes an external casing and an internal cavity, wherein the gas pressure within the cavity is controlled by the casing to either grip or release the crop. During the harvesting process, the cavity-insertion style end-effector needs to be adjusted according to the weight and size of the fruit to ensure that it is securely fixed within the cavity, thus preventing it from falling or being damaged during the harvesting process.

Cavity retrieval end-effectors are generally used for fruits with harder textures and regular shapes, such as pineapples and apples. To realize automated pineapple harvesting, Du et al. [45] designed a pineapple-harvesting gripper, as shown in Table 9 (a). It consists of a gripping mechanism and a cutting mechanism that can sequentially and cleanly sever the pineapple stem, thus minimizing damage to the stem. For the cavity-insertion style harvesting of apples, Wei et al. [102] developed a spherical double-finger structure gripper, as shown in Table 9 (c), which can effectively reduce fruit damage rates. Taking into account the shape characteristics of apples, Miao et al. [103] designed an end-effector, as shown in Table 9 (d), the advantage of which is that it does not damage the fruit during harvesting. The cavity-insertion style is not only used for harvesting pineapples and apples but is also suitable for harvesting softer fruits such as strawberries. The strawberry-harvesting robot developed by Xiong et al. [67] opens the cavity during harvesting, then "swallows" the fruit, and the blade severs the fruit stem, thus completing a cycle of strawberry harvesting.

Table 9. Comparison of different types of cavity extraction end-effectors.

Figure	Number	Applied Crops	Advantages	Improvements	Gripper Size	Recognition Accuracy	Picking Success Rate	Picking Time	Ref.
	a	pineapple	Minimized damage rate, High picking success rate	Picking rate	Cylindrical radius: 100 mm Blade diameter: 130 mm	95%	80%	14.9 S	[45]
	b	mulberry	Accurate separation of obstacles	Positioning accuracy	Maximum clamping diameter: 40 mm, open diameter: 25 mm	-	-	10.6 s	[100]
	c	Apple	Minimized damage rate	Picking rate and accuracy	-	-	-	7.81 s	[102]
	d	Apple	Minimized damage rate	Picking rate	Maximum diameter: 14 cm	91%	82%	9.8 s	[103]

4.1.4. Flexible Gripping End-Effectors

Flexible harvesting end-effectors refer to terminal robotic arms that can mimic the actions of human fingers and palms, characterized by their flexibility, malleability, and ease of operation. The purpose of these effectors is to simulate human organs, such as fingers and palms, enabling precise picking and handling operations for objects of various shapes and sizes. They find wide-ranging applications in fields such as agriculture, manufacturing, and healthcare.

Apples represent one of the most commonly encountered fruit types within the broad spectrum of agricultural produce. Substantial scientific research and developmental efforts have been directed toward enhancing flexible apple-picking methodologies. For instance, Liu et al. [104] have devised a flexible gripper, as demonstrated in Figure 8a, which consists of two curved, flexible fingers. This apparatus has been extensively refined and optimized, enabling the harvesting not only of apples but also other fruits, such as pomegranates and grapefruits. In order to further minimize apple damage during the harvesting process, Pi et al. [105] were inspired by the physical properties of octopus tentacles to study and develop a biomimetic three-fingered flexible gripper, depicted in Figure 8b. Figure 8c illustrates the pneumatic pinch structure of an end-effector developed by Hohimer et al. [106]. This tool is capable of performing apple-picking tasks with significant flexibility and high precision.

Figure 8. Flexible gripping end-effector: (**a**–**l**) (ref. [104], 2018, Liu et al.; ref. [105], 2021, Pi et al.; ref. [106], 2019, Hohimer et al.; ref. [86], 2021, Yan et al.; ref. [59], 2021, Yu et al.; ref. [65], 2020, Sepulveda et al.; ref. [107], 2020, Vu et al.; ref. [50], 2022, Yu et al.; ref. [108], 2021, Chen et al.; ref. [109], 2019, Yung et al.; ref. [110], 2021, Zhang et al.; ref. [111], 2021, Habegger et al.; ref. [112], 2018, Xiong et al.); (**m**,**n**) (https://www.tevel-tech.com/ (accessed on 27 June 2023)).

Furthering the field of flexible robotic technology, Yan et al. [86] developed a flexible gripper mounted on a six-axis robotic arm for apple harvesting, as shown in Figure 8d, which exhibits a high degree of flexibility. From an ergonomics perspective and based on the structural characteristics of the human body, Yu et al. [59] have engineered a three-fingered gripper made from flexible materials, as illustrated in Figure 8e. This design takes advantage of the widely recognized fin effect, where a fin bends toward the direction of applied pressure and reverts to its original state once the pressure is relieved. Such a claw-like structure is conducive to protecting apples from damage, thus achieving the function of damage-free harvesting.

For the flexible harvesting of tomatoes, researchers around the globe have devoted considerable effort toward the cause. As depicted in Figure 8f, Sepulveda et al. [65] have developed an end-effector for harvesting that resembles the human hand, capable of swiftly and accurately picking tomatoes. Vu et al. [107]., on the other hand, have designed a four-fingered mechanical hand module equipped with an internal vacuum system, as shown in Figure 8g. This innovative vacuum claw module enhances the reliability and safety of fruit-picking operations. Yu et al. [50] have developed a flexible claw endowed with a thin film pressure sensor constructed of rubber via injection molding. This three-fingered device, driven by 42 stepping motors, delivers precise grip capabilities. As displayed in Figure 8h, Chen et al. [108] have designed an end-effector for a tomato-harvesting robot based on pneumatic damage-free clamping. This tool effectively reduces the damage rate during picking, thus achieving damage-free harvesting. Figure 8i shows an end-effector designed by Yung et al. [109] for harvesting tomato seedlings. This flexible harvesting tool boasts rapid collection speeds, further enhancing the efficiency of tomato harvesting.

Flexible grippers have also been employed in the harvesting of other fruits. For instance, as shown in Figure 8j, Zhang et al. [110] have investigated a robotic end-effector equipped with adaptive grasping and tactile sensors. This end-effector, using flexible fingers and integrated force and bend sensors, can measure the distribution of contact forces on the contact surface and the deformation of the fingers, enabling adaptive grasping of various spherical fruits. In Figure 8k, Habegger et al. [111] have designed a flexible end-effector specifically for sweet pepper harvesting composed of four fin-ray grippers. This mechanism ensures that no damage is inflicted upon the peppers during the harvesting process. As depicted in Figure 8l, a strawberry-harvesting robot developed by Preter et al. [112] features an end-effector consisting of two flexible two-fingered structures resembling a palm, greatly reducing damage to strawberries during the picking process. Figure 8m illustrates a tomato-harvesting robot whose end-effector is made of flexible materials, enabling damage-free harvesting. In Figure 8n, the Israel-based company Tevel Aerobotics has invented a fruit-picking drone that employs a simple and convenient end-effector structure suitable for picking a variety of fruits, including apples, nectarines, and plums.

In general, flexible end-effectors for harvesting exhibit high precision, strong flexibility, and easy operability, serving as vital tools for enhancing the efficiency of mechanized picking and handling tasks. The literature review above demonstrates that flexible gripper-style end-effectors have gained extensive application within the realm of agricultural harvesting robots, showing good adaptability and efficiency across diverse crop harvesting tasks.

End-effectors of agricultural harvesting robots are crucial components of agricultural robotic systems, with their performance and functionality directly influencing harvesting efficiency and quality. By leveraging ergonomic principles, these end-effectors can be optimized in terms of design and control to achieve more efficient, precise, and safe harvesting operations. The School of Mechanical and Electronic Engineering at Northwest A&F University has conducted extensive research and experimentation in this area, yielding a series of experimental conclusions [59,113–116].

Regarding flexible materials, they are essential components of flexible grippers. Common flexible materials currently used include elastomers, silicone, polyurethane, and airbags. These materials possess excellent flexibility and adaptability, enabling them to grasp objects of various shapes and sizes. When designing flexible grippers, it is necessary to consider the material's strength, durability, and elasticity to meet the requirements of harvesting operations.

Additionally, the grasping angle and gripper size are significant factors influencing the performance of end-effectors in agricultural harvesting robots. The grasping angle refers to the angle between the end-effector and the object during grasping. Different fruits and crops have varying horticultural characteristics, thus requiring different optimal grasping angles. Hence, it is crucial to determine the optimal grasping angle through research and analysis of crop characteristics. The gripper size should be designed based on the size and shape of different crops to ensure the end-effector can adapt to grasping fruits and crops of varying sizes and shapes. Harvesting modes also play a significant role in the performance of end-effectors in agricultural harvesting robots. Common harvesting modes include rotation, stretching, and combined rotation and stretching. Different harvesting modes are suitable for different fruits and crops. For example, rotational harvesting is suitable for smaller crops, while stretching is appropriate for larger fruits.

In summary, the design and control of end-effectors in agricultural harvesting robots necessitate considering multiple factors, including flexible materials, grasping angles, gripper sizes, and harvesting modes. In the future, advancements in materials, sensors, and control technologies can further enhance the performance and intelligence level of end-effectors in agricultural harvesting robots to meet the harvesting requirements of various fruits and crops, thus promoting the development and application of agricultural robotics technology.

4.2. Overview of Harvesting Effect Evaluation Indicators

Evaluating the harvesting performance of agricultural robots is of utmost importance as it directly reflects the efficiency and quality of harvesting, thereby influencing the profitability of agricultural production. This section provides an overview of common metrics used for evaluating harvesting performance (as shown in Table 10).

Recognition Rate: The recognition rate refers to the speed at which agricultural robots can identify fruits or vegetables during harvesting operations. Specifically, it measures the ratio between the number of images processed and recognized by the robot during visual recognition and the corresponding processing time. Generally, a higher recognition rate enables the robot to complete harvesting tasks more quickly, thereby improving harvesting efficiency. Indicators for the recognition rate include the number of images recognized per second, the number of items recognized per second, and the amount of data processed per second.

Harvesting Rate: The harvesting rate is a vital metric for assessing the efficiency of harvesting robots. It is closely related to the technical parameters of the harvesting robot, the complexity of the harvesting site, and the growth conditions of the crops.

Harvesting Quality: Harvesting quality is another important metric for evaluating the harvesting performance of robots. It encompasses indicators such as harvesting accuracy, damage rate, fruit drop rate, and average harvesting time. Harvesting accuracy refers to the consistency of the size, shape, color, and ripeness of the harvested fruits with predetermined standards, while the damage rate reflects the level of fruit damage during the harvesting process.

Harvesting Cost: Harvesting cost is one of the indicators used to measure the economic viability of harvesting robots. It includes factors such as equipment acquisition costs, maintenance and upkeep costs, and energy consumption costs.

Adaptability: Adaptability is a crucial metric for assessing the ability of harvesting robots to adapt to various crops and different harvesting environments. This includes aspects such as the flexibility, stability, and safety of the harvesting robot.

In conclusion, the evaluation metrics for harvesting performance serve as key indicators for assessing harvesting robots. In practical applications, it is necessary to select appropriate evaluation metrics based on specific harvesting tasks and requirements, thereby enabling a scientific and rational assessment and optimization process.

Table 10. Comparison of end-effector indicators for different fruits.

Fruit	Gripper Category	Recognition Rate	Recognition Accuracy	Average Picking Time	Picking Success Rate	Ref.	Year
		-	82.5%	14.6 s	72%	[59]	2021
	flexible grasping	0.012 s	-	-	100%	[105]	2021
		-	-	25.5 s	96.67%	[93]	2021
Apples		-	-	7.3 s	67%	[106]	2019
	shearing-style	0.015 s	-	-	-	[86]	2021
	-	0.181 s	89%	-	-	[55]	2020
	-	0.235 s	87.1%	7 s	-	[52]	2020
		-	92.8%	-	73.04%	[54]	2021
	shearing-style	-	-	9.676 s	-	[63]	2019
		0.021 s	94%	-	100%	[117]	2020
Tomatoes	-	0.096 s	-	-	91.9%	[118]	2018
		-	91.92%	-	-	[81]	2020
	flexible grasping	0.016 s	-	8 s	-	[113]	2021
	-	-	98%	-	-	[110]	2021
		-	89%	-	-	[119]	2020
	cavity retrieval	0.136 s	-	6.1 s	97.1%	[67]	2019
		-	-	10.6 s	96.8%	[71]	2019
		0.086 s	93.1%	4 s	-	[112]	2018
	flexible grasping	-	-	9.05 s	96.8%	[73]	2021
Strawberries		0.049 s	-	10.62 s	96.77%	[120]	2018
	-	-	86.58%	-	-	[83]	2022
		0.062 s	95.78%	-	-	[74]-	2019
	shearing-style	-	94.43%	-	84.35%	[79]	2020
		-	-	10.7 s	84%	[66]	2019
Sweet papers	shearing-style	-	91.84%	-	90.04%	[47]	2020
	-	-	96.91%	-	-	[84]	2021
		1.41 s	86.4%	-	-	[46]	2020
	shearing-style	0.154 s	93.5%	-	-	[121]	2021
Litchi fruits		0.464 s	83.33%	-	-	[89]	2020
	-	-	96.78%	-	-	[77]	2020
Cherry tomatoes	flexible grasping	-	-	6.4 s	84%	[41]	2022
	shearing-style	-	-	8 s	83%	[64]	2018
		-	-	12.51 s	99.81%	[49]	2021

5. Challenges and Prospects

Agricultural fruit and vegetable-harvesting robots represent one of the rapidly evolving domains in recent years, offering vast potential and opportunities for agricultural production. However, with the continuous development of agricultural harvesting robots, a series of challenges and problems that need to be addressed have also emerged. Following are the six challenges and six prospects identified in this article concerning agricultural fruit and vegetable harvesting robots.

5.1. Challenges

5.1.1. Multi-Species, Multi-Form Fruit, and Vegetable Picking Is More Difficult

Due to the wide variety of crops, the size, shape, hardness, and other physical characteristics of different kinds of fruit are often very different, so it is difficult to design a robot that can adapt to the needs of multiple kinds of fruit picking at the same time. Generally, one kind of picking robot can only pick one specific kind of fruit, while other types of fruits and vegetables can only be significantly modified or redesigned, which will waste a lot of time and resources for researchers.

5.1.2. Difficulty in Picking in Complex Environment

Even when picking the same variety of fruit, the operating environment of the robot in different orchards is also complex and changeable. First of all, most picking robots are designed based on the structured or semi-structured picking environment, and in the actual operation process, different orchard structures are different, which may have a certain impact on the working effect of the picking robot. Secondly, weather conditions can also have an impact on the operation of robots, such as excessive wind power that may affect the stability, safety, and harvesting efficiency of robots. Therefore, fruit and vegetable-picking robots need to face the challenges of adapting and operating in different operating environments.

5.1.3. High Real-Time Requirements

The fruit and vegetable-picking robot operates in a real-time environment, requiring rapid recognition, positioning, and grasping of targets. At the same time, it needs to dynamically adjust control parameters during the movement process to maintain a stable motion trajectory, so its real-time requirements are very high. In different harvesting environments, robots also need to handle a variety of complex situations in real time, such as avoiding obstacles and adapting to various light conditions to ensure the efficient completion of harvesting tasks. Therefore, real-time performance represents a significant challenge for fruit and vegetable-harvesting robots.

5.1.4. Few Research on Walking Platforms and Navigation

At present, the research on fruit and vegetable-picking robots mainly focuses on visual systems, mechanical arms, terminal actuators, and picking-path planning, while the research on walking platforms and their navigation algorithms in the environment is less frequent. Of course, such studies have limited impact on picking robots operating in a structured environment, and some of the robots moving through slides rarely require navigation. However, under a semi-structural or unstructured working environment, the stability, driving speed, power, and other performances of the picking platform will have a great impact on the picking accuracy, harvesting efficiency, and even the fruit quality in the collecting device. The navigation and route planning of the walking platform in the orchard environment will have a great impact on the working efficiency of long-time picking. These are the questions we need to consider in the future.

5.1.5. The Working Height of Picking Robot Is Generally Limited

The existing harvesting robots are mainly designed based on ground mobile platforms, but the structural design of ground harvesting robots is generally fixed, and they usually have limitations in height and size. These limitations determine that the main target of ground-picking robots is some low-fruit trees, which are not suitable for picking higher-fruit trees. Even though some researchers have raised the picking height of the robot to some extent by using a liftable platform, the lifting is very limited, considering the power and balance problems. In this case, how to complete the picking of higher fruit trees has become an urgent problem.

5.1.6. High Costs

Currently, the research, development, and production costs of fruit and vegetable harvesting robots are high, which restricts their large-scale deployment in agricultural production. The costs of fruit and vegetable harvesting robots mainly comprise the costs of robot production research and development, maintenance, and labor, among others. At present, most fruit and vegetable harvesting robots necessitate substantial research and development and material costs, along with regular maintenance, which escalates the cost of robot usage. Hence, ensuring the efficiency and quality of robotic harvesting while reducing its costs is a significant issue that needs to be addressed in future research and development of fruit and vegetable harvesting robots.

5.2. Prospects

5.2.1. Modular Harvesting Robot

In the face of the difficulty of picking fruits and vegetables with multiple varieties and forms, in addition to using adjustable grippers or flexible end-effectors, modular design can also be used to solve the problem. By implementing modular design to achieve product versatility, this has been successfully applied and achieved good results in multiple types of military or civilian products. In the future field of harvesting robots, in addition to modular end-effectors, modular walking platforms, and even modular robotic arms and sensing systems should be widely developed. Through modular design, researchers can not only flexibly match various components based on the characteristics of the target fruit and the picking environment but also promote the standardization of picking robots while reducing costs. This will effectively promote the further promotion and application of intelligent picking robots.

5.2.2. Sensor Fusion and Algorithm Optimization

In the face of complex and ever-changing operating environments and high real-time requirements for harvesting, multiple strategies can be adopted to solve the problem. The sensing ability of picking robots can be improved by adopting the strategy of multi-mode sensor fusion and multi-algorithm fusion sensing so as to identify and locate the fruit and vegetable more accurately in a complex environment. In addition, the application of the neural network model in the picking field can effectively shorten the sensing time, and the real-time picking robot can be further improved by selecting a better hardware processor and continuously optimizing the control algorithm. In the future, with the continuous development of various new technologies, we have every reason to believe that fruit and vegetable harvesting robots will become more intelligent, efficient, and flexible, bringing further benefits to agricultural production.

5.2.3. Strengthening Research on Walking Platform and Navigation Algorithm

For the walking platform, first of all, it shall be ensured that the platform has good terrain adaptability so that it can maintain good stability and reliability in different terrain environments such as flat ground, slope, and grassland. Second, the walking platform shall be equipped with the necessary sensors, or the sensing system of the robot shall be invoked to serve the walking platform when the mechanical arm and the end actuator are idle so that the platform has basic environmental awareness and obstacle avoidance functions when moving between the picking areas. In addition, the research on the navigation algorithm of the walking platform can refer to the path-planning technology of the terminal actuator and the mechanical arm introduced in Section 3.3 and the obstacle avoidance strategy mentioned in Section 3.2. Using optimization algorithms to optimize traditional path-planning algorithms or using deep learning-based path-planning algorithms can shorten the movement time of robots between different picking areas, thereby effectively improving the efficiency of picking robots, especially for long-term operations.

5.2.4. The Development of Picking Drones

Due to the limited picking height of common ground picking robots, it is necessary to explore new picking modes in order to meet the picking needs of higher fruit trees. Thanks to the great flexibility of multi-rotor drones in three-dimensional space, harvesting operations based on multi-rotor drone platforms are an ideal way to solve such problems. At the same time, the tethered power supply mode can greatly extend the operation time of the picking drones, and theoretically, it can achieve 24 h of uninterrupted operation. In addition, if multiple harvesting drones work together, it can greatly improve harvesting efficiency. In terms of drone harvesting, Israel's Tevel company has achieved good results, but overall, there is still relatively little research and application of drone harvesting. In the future, it should improve its sensing ability and balance control ability in a complex environment and promote its further development and application in light-weight, miniaturization, clustering, and non-destructive picking.

5.2.5. Multi-Robot Collaborative Operation

With the expansion of agricultural scale and the increasing complexity of harvesting tasks, a single harvesting robot may not be able to efficiently complete all tasks. Therefore, multi-robot collaboration has become a trend that can improve the overall production efficiency and quality of harvesting. Multi-robot collaboration can be based on a distributed concept, where different tasks are assigned to multiple robots and executed. These robots can work collaboratively, share information, and allocate tasks via wireless communication or LAN. This cooperation is not only limited to the cooperation between different picking areas but also, each robot can perform different picking operations according to the task requirements and its own capabilities. For example, one robot is responsible for identifying and locating the target, and the other robot is responsible for precise grabbing and shearing, which can avoid the limitation of the picking robot's own vision. In addition, multi-robot cooperation can also better realize obstacle avoidance and provide a better picking environment for the picking robot so as to optimize the picking path and improve the picking efficiency. Of course, the advantages of the multi-robot cooperative picking mode are not limited to this; it has a huge development space and prospects, and its development and application in the future are worthy of expectation.

5.2.6. Reducing Costs

Researchers have adopted various methods to address the high costs of fruit and vegetable harvesting robots. One approach involves utilizing modular design and manufacturing to reduce production costs and increase efficiency. Another method employs new materials and manufacturing technologies to lower material costs and ease manufacturing difficulties. Moreover, with continuous technological advancements and market expansion, the production scale of fruit and vegetable harvesting robots will continue to grow, further reducing costs. Thus, although high costs constitute a significant issue, it will gradually be addressed with ongoing technological developments and market expansion.

6. Conclusions

This paper systematically reviewed the research progress of the "eye, brain, and hand" picking system in the past six years and discussed its potential impact and innovation value in the field of modern agriculture. Through the gradual analysis of each part of the article, we can understand the technical realization and application prospects of this intelligent agricultural picking robot, which brings unprecedented opportunities for the future of agricultural production.

In the detailed discussion of each section, this review provides insight into the core elements of the "eye–brain–hand" picking system. From the "eye" part using advanced sensors and image processing technology to achieve the accurate judgment of crop maturity, to the "brain" part through advanced algorithms to achieve real-time decision-making and guidance, and finally to the "hand" part to achieve accurate picking. This intelligent picking system not only improves agricultural production efficiency but also reduces resource waste and human investment, playing a positive role in the green and sustainable development of global agriculture.

The main contribution of this review is a comprehensive analysis of the "eye–brain–hand" picking system. From hardware modules to technical approaches, from potential challenges to future trends, this review provides valuable guidance for researchers. Section 2 provides a detailed introduction to the perception hardware system of the intelligent picking "eye" system. In terms of target sensing methods, we compare a variety of methods, which provide important guidance for achieving high-precision target detection. In Section 3, the intelligent picking "brain" system deeply studies the key issues such as regional division, task allocation, obstacle avoidance strategy, and path planning. In this section, the importance of task allocation and obstacle avoidance strategies for robot agricultural operations is emphasized, providing key support for ensuring the efficient and safe operation of robots. Section 4 systematically reviews the performance indicators of four end-effectors, namely, negative pressure adsorption, shear, cavity trapping, and flexible grasping, for the intelligent harvesting "hand" system. Through analyzing the evaluation index of picking effect, we provide a valuable reference for the type selection and design of end-effectors for different kinds of fruit. In Section 5, "Challenges and Prospects", we identify the challenges faced by intelligent agricultural picking robots and also provide some prospects for their future development.

Through the comprehensive explanation of the above conclusions, this paper provides a deep insight into the agricultural picking robot field and provides important guidance and enlightenment for future research and application. In a word, the development of agricultural picking robots is not only the embodiment of technological progress but also the key step in the agricultural field toward sustainable development. We believe that the content of this review will make a beneficial contribution to the goals of agricultural modernization and sustainable development and promote the wide application and development of intelligent agriculture.

Author Contributions: Conceptualization, X.H. (Xiongkui He) and S.W.; methodology, W.J. and X.H. (Xianhao Huang); analysis, W.J. and X.H. (Xianhao Huang); investigation, S.W., W.J. and X.H. (Xianhao Huang); resources, X.H. (Xiongkui He), S.W., W.J. and X.H. (Xianhao Huang); data curation, W.J. and X.H. (Xianhao Huang); writing—original draft preparation, S.W., W.J. and X.H. (Xianhao Huang); writing—review and editing, X.H. (Xiongkui He), S.W., W.J. and X.H. (Xianhao Huang); visualization, W.J. and X.H. (Xianhao Huang); supervision, X.H. (Xiongkui He) and S.W.; project administration, X.H. (Xiongkui He) and S.W.; funding acquisition, X.H. (Xiongkui He) and S.W. All authors have read and agreed to the published version of the manuscript.

Funding: This study was supported by the earmarked fund for China Agriculture Research System (CARS-28), Chinese Universities Scientific Fund (Grant No. 2022TC128), Sanya Institute of China Agricultural University Guiding Fund Project, Grant No. SYND-2021-06, and the 2115 Talent Development Program of China Agricultural University and the CCF-Baidu Apollo Joint Development Project Fund.

Institutional Review Board Statement: Not applicable.

Informed Consent Statement: Not applicable.

Data Availability Statement: Not applicable.

Acknowledgments: The authors would like to thank Shubo Wang and all other staff of CCAT and CAUS, China Agricultural University, for their great contributions to this work.

Conflicts of Interest: The authors declare no conflict of interest.

References

1. Gil, G.; Casagrande, D.E.; Cortés, L.P.; Verschae, R. Why the Low Adoption of Robotics in the Farms? Challenges for the Establishment of Commercial Agricultural Robots. *Smart Agric. Technol.* **2023**, *3*, 100069. [CrossRef]
2. Suresh Kumar, M.; Mohan, S. Selective Fruit Harvesting: Research, Trends and Developments towards Fruit Detection and Localization—A Review. *Proc. Inst. Mech. Eng. C J. Mech. Eng. Sci.* **2023**, *237*, 1405–1444. [CrossRef]
3. Rose, D.C.; Bhattacharya, M. Adoption of Autonomous Robots in the Soft Fruit Sector: Grower Perspectives in the UK. *Smart Agric. Technol.* **2023**, *3*, 100118. [CrossRef]
4. Tang, Q.; Luo, Y.W.; Wu, X. Di Research on the Evaluation Method of Agricultural Intelligent Robot Design Solutions. *PLoS ONE* **2023**, *18*, e0281554. [CrossRef] [PubMed]
5. Kuta, Ł.; Li, Z.; Stopa, R.; Komarnicki, P.; Słupska, M. The Influence of Manual Harvesting on the Quality of Picked Apples and the Picker's Muscle Load. *Comput. Electron. Agric.* **2020**, *175*, 105511. [CrossRef]
6. Liu, J.; Peng, Y.; Faheem, M. Experimental and Theoretical Analysis of Fruit Plucking Patterns for Robotic Tomato Harvesting. *Comput. Electron. Agric.* **2020**, *173*, 105330. [CrossRef]
7. Xiong, Y.; Ge, Y.; From, P.J. Push and Drag: An Active Obstacle Separation Method for Fruit Harvesting Robots. In Proceedings of the IEEE International Conference on Robotics and Automation, Paris, France, 31 May–31 August 2020.

8. Li, K.; Qi, Y. Motion Planning of Robot Manipulator for Cucumber Picking. In Proceedings of the 2018 3rd International Conference on Robotics and Automation Engineering, ICRAE 2018, Guangzhou, China, 17–19 November 2018.
9. Wu, Z.; Du, H. Artificial Intelligence in Agricultural Picking Robot Displacement Trajectory Tracking Control Algorithm. *Wirel Commun. Mob. Comput.* **2022**, *2022*, 3105909. [CrossRef]
10. Wang, Z.; Xun, Y.; Wang, Y.; Yang, Q. Review of Smart Robots for Fruit and Vegetable Picking in Agriculture. *Int. J. Agric. Biol. Eng.* **2022**, *15*, 33–54. [CrossRef]
11. Li, Y.; Feng, Q.; Li, T.; Xie, F.; Liu, C.; Xiong, Z. Advance of Target Visual Information Acquisition Technology for Fresh Fruit Robotic Harvesting: A Review. *Agronomy* **2022**, *12*, 1336. [CrossRef]
12. Lin, G.; Tang, Y.; Zou, X.; Xiong, J.; Li, J. Guava Detection and Pose Estimation Using a Low-Cost RGB-D Sensor in the Field. *Sensors* **2019**, *19*, 428. [CrossRef]
13. Zheng, C.; Chen, P.; Pang, J.; Yang, X.; Chen, C.; Tu, S.; Xue, Y. A Mango Picking Vision Algorithm on Instance Segmentation and Key Point Detection from RGB Images in an Open Orchard. *Biosyst. Eng.* **2021**, *206*, 32–54. [CrossRef]
14. Garillos-Manliguez, C.A.; Chiang, J.Y. Multimodal Deep Learning and Visible-Light and Hyperspectral Imaging for Fruit Maturity Estimation. *Sensors* **2021**, *21*, 1288. [CrossRef]
15. Xu, N.; Song, Y.; Meng, Q. Application RFID and Wi-FI Technology in Design of IOT Sensor Terminal. In Proceedings of the Journal of Physics: Conference Series, Chongqing, China, 28–30 May 2021; Volume 1982.
16. Chen, M.; Tang, Y.; Zou, X.; Huang, Z.; Zhou, H.; Chen, S. 3D Global Mapping of Large-Scale Unstructured Orchard Integrating Eye-in-Hand Stereo Vision and SLAM. *Comput. Electron. Agric.* **2021**, *187*, 106237. [CrossRef]
17. Chen, M.; Tang, Y.; Zou, X.; Huang, K.; Huang, Z.; Zhou, H.; Wang, C.; Lian, G. Three-Dimensional Perception of Orchard Banana Central Stock Enhanced by Adaptive Multi-Vision Technology. *Comput. Electron. Agric.* **2020**, *174*, 105508. [CrossRef]
18. Mahanti, N.K.; Pandiselvam, R.; Kothakota, A.; Ishwarya, S.P.; Chakraborty, S.K.; Kumar, M.; Cozzolino, D. Emerging Non-Destructive Imaging Techniques for Fruit Damage Detection: Image Processing and Analysis. *Trends Food Sci. Technol.* **2022**, *120*, 418–438. [CrossRef]
19. Koirala, A.; Walsh, K.B.; Wang, Z.; McCarthy, C. Deep Learning—Method Overview and Review of Use for Fruit Detection and Yield Estimation. *Comput. Electron. Agric.* **2019**, *162*, 219–234. [CrossRef]
20. Tang, Y.; Chen, M.; Wang, C.; Luo, L.; Li, J.; Lian, G.; Zou, X. Recognition and Localization Methods for Vision-Based Fruit Picking Robots: A Review. *Front. Plant Sci.* **2020**, *11*, 510. [CrossRef]
21. Mohd Ali, M.; Hashim, N.; Abd Aziz, S.; Lasekan, O. Utilisation of Deep Learning with Multimodal Data Fusion for Determination of Pineapple Quality Using Thermal Imaging. *Agronomy* **2023**, *13*, 401. [CrossRef]
22. Yang, F.; Ma, Z.; Xie, M. Image Classification with Superpixels and Feature Fusion Method. *J. Electron. Sci. Technol.* **2021**, *19*, 100096. [CrossRef]
23. Shivendra; Chiranjeevi, K.; Tripathi, M.K. Detection of Fruits Image Applying Decision Tree Classifier Techniques. In *Lecture Notes on Data Engineering and Communications Technologies*; Springer Nature: Singapore, 2023; Volume 142.
24. Zhang, C.; Wang, H.; Fu, L.H.; Pei, Y.H.; Lan, C.Y.; Hou, H.Y.; Song, H. Three-Dimensional Continuous Picking Path Planning Based on Ant Colony Optimization Algorithm. *PLoS ONE* **2023**, *18*, e0282334. [CrossRef]
25. He, Z.; Ma, L.; Wang, Y.; Wei, Y.; Ding, X.; Li, K.; Cui, Y. Double-Arm Cooperation and Implementing for Harvesting Kiwifruit. *Agriculture* **2022**, *12*, 1763. [CrossRef]
26. Yang, C.; Liu, Y.; Wang, Y.; Xiong, L.; Xu, H.; Zhao, W. Research and Experiment on Recognition and Location System for Citrus Picking Robot in Natural Environment. *Nongye Jixie Xuebao/Trans. Chin. Soc. Agric. Mach.* **2019**, *50*, 14–22. [CrossRef]
27. Peng, H.; Shao, Y.; Chen, K.; Deng, Y.; Xue, C. Research on Multi-Class Fruits Recognition Based on Machine Vision and SVM. *IFAC-Pap.* **2018**, *51*, 817–821. [CrossRef]
28. Udhaya, K.; Miruthula, R.; Pavithra, G.; Revathi, R.; Suganya, M. FPGA-Based Hardware Acceleration for Fruit Recognition Using SVM. *Ir. Interdiscip. J. Sci. Res.* **2022**, *06*, 22–29. [CrossRef]
29. Xu, L.; Cao, M.; Song, B. A New Approach to Smooth Path Planning of Mobile Robot Based on Quartic Bezier Transition Curve and Improved PSO Algorithm. *Neurocomputing* **2022**, *473*, 98–106. [CrossRef]
30. Guo, Y.; Wang, W.; Wu, S. Modeling Method of Mobile Robot Workspace. *IEEE* **2017**, 2146–2150.
31. Chen, W.; Xu, T.; Liu, J.; Wang, M.; Zhao, D. Picking Robot Visual Servo Control Based on Modified Fuzzy Neural Network Sliding Mode Algorithms. *Electronics* **2019**, *8*, 605. [CrossRef]
32. Dai, Y.; Zhang, R.; Ma, L. Path Planning and Tracking Control of Picking Robot Based on Improved A* Algorithm. *J. Chin. Agric. Mech.* **2022**, *43*, 138. [CrossRef]
33. Ma, Y.; Zhang, W.; Qureshi, W.S.; Gao, C.; Zhang, C.; Li, W. Autonomous Navigation for a Wolfberry Picking Robot Using Visual Cues and Fuzzy Control. *Inf. Process. Agric.* **2021**, *8*, 15–26. [CrossRef]
34. Zhang, F.; Chen, Z.; Wang, Y.; Bao, R.; Chen, X.; Fu, S.; Tian, M.; Zhang, Y. Research on Flexible End-Effectors with Humanoid Grasp Function for Small Spherical Fruit Picking. *Agriculture* **2023**, *13*, 123. [CrossRef]
35. Xu, L.; Liu, X.; Zhang, K.; Xing, J.; Yuan, Q.; Chen, J.; Duan, Z.; Ma, S.; Yu, C. Design and Test of End-Effector for Navel Orange Picking Robot. *Nongye Gongcheng Xuebao/Trans. Chin. Soc. Agric. Eng.* **2018**, *34*, 53–61. [CrossRef]
36. Guo, T.; Zheng, Y.; Bo, W.; Liu, J.; Pi, J.; Chen, W.; Deng, J. Research on the Bionic Flexible End-Effector Based on Tomato Harvesting. *J. Sens.* **2022**, *2022*, 1–14. [CrossRef]

37. Gharakhani, H.; Thomasson, J.A.; Lu, Y. An End-Effector for Robotic Cotton Harvesting. *Smart Agric. Technol.* **2022**, *2*, 100043. [CrossRef]
38. Xiao, X.; Wang, Y.; Jiang, Y. End-Effectors Developed for Citrus and Other Spherical Crops. *Appl. Sci.* **2022**, *12*, 7945. [CrossRef]
39. Hu, G.; Chen, C.; Chen, J.; Sun, L.; Sugirbay, A.; Chen, Y.; Jin, H.; Zhang, S.; Bu, L. Simplified 4-DOF Manipulator for Rapid Robotic Apple Harvesting. *Comput. Electron. Agric.* **2022**, *199*, 107177. [CrossRef]
40. Chen, M.; Chen, F.; Zhou, W.; Zuo, R. Design of Flexible Spherical Fruit and Vegetable Picking End-Effector Based on Vision Recognition. In *Proceedings of the Journal of Physics: Conference Series*; IOP Publishing: Bristol, UK, 2022; Volume 2246.
41. Gao, J.; Zhang, F.; Zhang, J.; Yuan, T.; Yin, J.; Guo, H.; Yang, C. Development and Evaluation of a Pneumatic Finger-like End-Effector for Cherry Tomato Harvesting Robot in Greenhouse. *Comput. Electron. Agric.* **2022**, *197*, 106879. [CrossRef]
42. Lu, W.; Wang, P.; Du, X.; Ma, Z. *Design and Experiment of a Multi-Knuckle End-Effector for Tomato Picking Robot*; American Society of Agricultural and Biological Engineers: St. Joseph Charter Township, MI, USA, 2022.
43. Oliveira, F.; Tinoco, V.; Magalhaes, S.; Santos, F.N.; Silva, M.F. End-Effectors for Harvesting Manipulators-State Of The Art Review. In Proceedings of the 2022 IEEE International Conference on Autonomous Robot Systems and Competitions, ICARSC 2022, Santa Maria da Feira, Portugal, 29–30 April 2022.
44. Yang, Q.; Luo, S.; Chang, C.; Xun, Y.; Bao, G. Segmentation Algorithm for Hangzhou White Chrysanthemums Based on Least Squares Support Vector Machine. *Int. J. Agric. Biol. Eng.* **2019**, *12*, 127–134. [CrossRef]
45. Du, X.; Yang, X.; Ji, J.; Jin, X.; Chen, L. Design and Test of a Pineapple Picking End-Effector. *Appl. Eng. Agric.* **2019**, *35*, 1045–1055. [CrossRef]
46. Lin, G.; Tang, Y.; Zou, X.; Xiong, J.; Fang, Y. Color-, Depth-, and Shape-Based 3D Fruit Detection. *Precis. Agric.* **2020**, *21*, 1–17. [CrossRef]
47. Ning, Z.; Luo, L.; Ding, X.M.; Dong, Z.; Yang, B.; Cai, J.; Chen, W.; Lu, Q. Recognition of Sweet Peppers and Planning the Robotic Picking Sequence in High-Density Orchards. *Comput. Electron. Agric.* **2022**, *196*, 106878. [CrossRef]
48. Mu, L.; Cui, G.; Liu, Y.; Cui, Y.; Fu, L.; Gejima, Y. Design and Simulation of an Integrated End-Effector for Picking Kiwifruit by Robot. *Inf. Process. Agric.* **2020**, *7*, 58–71. [CrossRef]
49. Zhang, Q.; Liu, F.; Jiang, X.; Xiong, Z.; Xu, C. Motion Planning Method and Experiments of Tomato Bunch Harvesting Manipulator. *Nongye Gongcheng Xuebao/Trans. Chin. Soc. Agric. Eng.* **2021**, *37*, 149–156.
50. Yu, F.; Zhou, C.; Yang, X.; Guo, Z.; Chen, C. Design and Experiment of Tomato Picking Robot in Solar Greenhouse. *Nongye Jixie Xuebao/Trans. Chin. Soc. Agric. Mach.* **2022**, *53*, 41–49. [CrossRef]
51. Li, T.; Qiu, Q.; Zhao, C.; Xie, F. Task Planning of Multi-Arm Harvesting Robots for High-Density Dwarf Orchards. *Nongye Gongcheng Xuebao/Trans. Chin. Soc. Agric. Eng.* **2021**, *37*, 1–10. [CrossRef]
52. Kang, H.; Zhou, H.; Chen, C. Visual Perception and Modeling for Autonomous Apple Harvesting. *IEEE Access* **2020**, *8*, 62151–62163. [CrossRef]
53. Sarabu, H.; Ahlin, K.; Hu, A.P. Graph-Based Cooperative Robot Path Planning in Agricultural Environments. In Proceedings of the IEEE/ASME International Conference on Advanced Intelligent Mechatronics, AIM, Hong Kong, China, 8–12 July 2019; Volume 2019.
54. Gong, L.; Wang, W.; Wang, T.; Liu, C. Robotic Harvesting of the Occluded Fruits with a Precise Shape and Position Reconstruction Approach. *J. Field Robot.* **2022**, *39*, 69–84. [CrossRef]
55. Fu, L.; Majeed, Y.; Zhang, X.; Karkee, M.; Zhang, Q. Faster R–CNN–Based Apple Detection in Dense-Foliage Fruiting-Wall Trees Using RGB and Depth Features for Robotic Harvesting. *Biosyst. Eng.* **2020**, *197*, 245–256. [CrossRef]
56. Xiong, Y.; Ge, Y.; From, P.J. An Improved Obstacle Separation Method Using Deep Learning for Object Detection and Tracking in a Hybrid Visual Control Loop for Fruit Picking in Clusters. *Comput. Electron. Agric.* **2021**, *191*, 106508. [CrossRef]
57. Lv, J.; Wang, Y.; Xu, L.; Gu, Y.; Zou, L.; Yang, B.; Ma, Z. A Method to Obtain the Near-Large Fruit from Apple Image in Orchard for Single-Arm Apple Harvesting Robot. *Sci. Hortic.* **2019**, *257*, 108758. [CrossRef]
58. Wang, L.; Li, H.R.; Zhou, K.; Mu, B. Design of Binocular Vision System for Fruit and Vegetable Picking Based on Embedded Arm. *Guangdianzi Jiguang/J. Optoelectron. Laser* **2020**, *31*, 71–80. [CrossRef]
59. Yu, X.; Fan, Z.; Wang, X.; Wan, H.; Wang, P.; Zeng, X.; Jia, F. A Lab-Customized Autonomous Humanoid Apple Harvesting Robot. *Comput. Electr. Eng.* **2021**, *96*, 107459. [CrossRef]
60. Zhou, T.; Zhang, D.; Zhou, M.; Xi, H.; Chen, X. System Design of Tomatoes Harvesting Robot Based on Binocular Vision. In Proceedings of the 2018 Chinese Automation Congress, CAC 2018, Xi'an, China, 30 November 2019.
61. Jin, Z.; Sun, W.; Zhang, J.; Shen, C.; Zhang, H.; Han, S. Intelligent Tomato Picking Robot System Based on Multimodal Depth Feature Analysis Method. In *Proceedings of the IOP Conference Series: Earth and Environmental Science*; IOP Publishing: Bristol, UK, 2020; Volume 440.
62. Ye, L.; Duan, J.; Yang, Z.; Zou, X.; Chen, M.; Zhang, S. Collision-Free Motion Planning for the Litchi-Picking Robot. *Comput. Electron. Agric.* **2021**, *185*, 106151. [CrossRef]
63. Oktarina, Y.; Dewi, T.; Risma, P.; Nawawi, M. Tomato Harvesting Arm Robot Manipulator; A Pilot Project. In Proceedings of the Journal of Physics: Conference Series, South Sumatera, Indonesia, 9–10 October 2020; Volume 1500.
64. Feng, Q.; Zou, W.; Fan, P.; Zhang, C.; Wang, X. Design and Test of Robotic Harvesting System for Cherry Tomato. *Int. J. Agric. Biol. Eng.* **2018**, *11*, 96–100. [CrossRef]

65. Sepulveda, D.; Fernandez, R.; Navas, E.; Armada, M.; Gonzalez-De-Santos, P. Robotic Aubergine Harvesting Using Dual-Arm Manipulation. *IEEE Access* **2020**, *8*, 121889–121904. [CrossRef]
66. Feng, Q.; Chen, J.; Zhang, M.; Wang, X. Design and Test of Harvesting Robot for Table-Top Cultivated Strawberry. In Proceedings of the WRC SARA 2019—World Robot Conference Symposium on Advanced Robotics and Automation 2019, Beijing, China, 21–22 August 2019.
67. Xiong, Y.; Ge, Y.; Grimstad, L.; From, P.J. An Autonomous Strawberry-harvesting Robot: Design, Development, Integration, and Field Evaluation. *J. Field Robot.* **2020**, *37*, 202–224. [CrossRef]
68. Zhuang, J.; Hou, C.; Tang, Y.; He, Y.; Guo, Q.; Zhong, Z.; Luo, S. Computer Vision-Based Localisation of Picking Points for Automatic Litchi Harvesting Applications towards Natural Scenarios. *Biosyst. Eng.* **2019**, *187*, 1–20. [CrossRef]
69. Septiarini, A.; Hamdani, H.; Hatta, H.R.; Anwar, K. Automatic Image Segmentation of Oil Palm Fruits by Applying the Contour-Based Approach. *Sci. Hortic.* **2020**, *261*, 108939. [CrossRef]
70. Mao, S.; Li, Y.; Ma, Y.; Zhang, B.; Zhou, J.; Kai, W. Automatic Cucumber Recognition Algorithm for Harvesting Robots in the Natural Environment Using Deep Learning and Multi-Feature Fusion. *Comput. Electron. Agric.* **2020**, *170*, 105254. [CrossRef]
71. Xiong, Y.; Peng, C.; Grimstad, L.; From, P.J.; Isler, V. Development and Field Evaluation of a Strawberry Harvesting Robot with a Cable-Driven Gripper. *Comput. Electron. Agric.* **2019**, *157*, 392–402. [CrossRef]
72. Liu, D.; Shen, J.; Yang, H.; Niu, Q.; Guo, Q. Recognition and Localization of Actinidia Arguta Based on Image Recognition. *EURASIP J. Image Video Process.* **2019**, *2019*, 21. [CrossRef]
73. Kurpaska, S.; Bielecki, A.; Sobol, Z.; Bielecka, M.; Habrat, M.; Śmigielski, P. The Concept of the Constructional Solution of the Working Section of a Robot for Harvesting Strawberries. *Sensors* **2021**, *21*, 3933. [CrossRef]
74. Yu, Y.; Zhang, K.; Yang, L.; Zhang, D. Fruit Detection for Strawberry Harvesting Robot in Non-Structural Environment Based on Mask-RCNN. *Comput. Electron. Agric.* **2019**, *163*, 104846. [CrossRef]
75. Yang, C.H.; Xiong, L.Y.; Wang, Z.; Wang, Y.; Shi, G.; Kuremot, T.; Zhao, W.H.; Yang, Y. Integrated Detection of Citrus Fruits and Branches Using a Convolutional Neural Network. *Comput. Electron. Agric.* **2020**, *174*, 105469. [CrossRef]
76. Fu, L.; Duan, J.; Zou, X.; Lin, J.; Zhao, L.; Li, J.; Yang, Z. Fast and Accurate Detection of Banana Fruits in Complex Background Orchards. *IEEE Access* **2020**, *8*, 196835–196846. [CrossRef]
77. Liang, C.; Xiong, J.; Zheng, Z.; Zhong, Z.; Li, Z.; Chen, S.; Yang, Z. A Visual Detection Method for Nighttime Litchi Fruits and Fruiting Stems. *Comput. Electron. Agric.* **2020**, *169*, 105192. [CrossRef]
78. Suo, R.; Gao, F.; Zhou, Z.; Fu, L.; Song, Z.; Dhupia, J.; Li, R.; Cui, Y. Improved Multi-Classes Kiwifruit Detection in Orchard to Avoid Collisions during Robotic Picking. *Comput. Electron. Agric.* **2021**, *182*, 106052. [CrossRef]
79. Yu, Y.; Zhang, K.; Liu, H.; Yang, L.; Zhang, D. Real-Time Visual Localization of the Picking Points for a Ridge-Planting Strawberry Harvesting Robot. *IEEE Access* **2020**, *8*, 116556–116568. [CrossRef]
80. Xu, Z.F.; Jia, R.S.; Sun, H.M.; Liu, Q.M.; Cui, Z. Light-YOLOv3: Fast Method for Detecting Green Mangoes in Complex Scenes Using Picking Robots. *Appl. Intell.* **2020**, *50*, 4670–4687. [CrossRef]
81. Xu, Z.F.; Jia, R.S.; Liu, Y.B.; Zhao, C.Y.; Sun, H.M. Fast Method of Detecting Tomatoes in a Complex Scene for Picking Robots. *IEEE Access* **2020**, *8*, 55289–55299. [CrossRef]
82. Chen, J.; Wang, Z.; Wu, J.; Hu, Q.; Zhao, C.; Tan, C.; Teng, L.; Luo, T. An Improved Yolov3 Based on Dual Path Network for Cherry Tomatoes Detection. *J. Food Process. Eng.* **2021**, *44*, e13803. [CrossRef]
83. Wang, Y.; Yan, G.; Meng, Q.; Yao, T.; Han, J.; Zhang, B. DSE-YOLO: Detail Semantics Enhancement YOLO for Multi-Stage Strawberry Detection. *Comput. Electron. Agric.* **2022**, *198*, 107057. [CrossRef]
84. Li, X.; Pan, J.; Xie, F.; Zeng, J.; Li, Q.; Huang, X.; Liu, D.; Wang, X. Fast and Accurate Green Pepper Detection in Complex Backgrounds via an Improved Yolov4-Tiny Model. *Comput. Electron. Agric.* **2021**, *191*, 106503. [CrossRef]
85. Wang, L.; Zhao, Y.; Liu, S.; Li, Y.; Chen, S.; Lan, Y. Precision Detection of Dense Plums in Orchards Using the Improved YOLOv4 Model. *Front. Plant Sci.* **2022**, *13*, 839269. [CrossRef] [PubMed]
86. Yan, B.; Fan, P.; Lei, X.; Liu, Z.; Yang, F. A Real-Time Apple Targets Detection Method for Picking Robot Based on Improved YOLOv5. *Remote Sens.* **2021**, *13*, 1619. [CrossRef]
87. Qian, Y.; Jiacheng, R.; Pengbo, W.; Zhan, Y.; Changxing, G. Real-Time Detection and Localization Using SSD Method for Oyster Mushroom Picking Robot. In Proceedings of the 2020 IEEE International Conference on Real-Time Computing and Robotics, RCAR 2020, Asahikawa, Japan, 28–29 September 2020.
88. Liu, J.; Zhao, M.; Guo, X. A Fruit Detection Algorithm Based on R-FCN in Natural Scene. In Proceedings of the 32nd Chinese Control and Decision Conference, CCDC 2020, Hefei, China, 22–24 August 2020.
89. Li, J.; Tang, Y.; Zou, X.; Lin, G.; Wang, H. Detection of Fruit-Bearing Branches and Localization of Litchi Clusters for Vision-Based Harvesting Robots. *IEEE Access* **2020**, *8*, 117746–117758. [CrossRef]
90. Peng, H.; Xue, C.; Shao, Y.; Chen, K.; Xiong, J.; Xie, Z.; Zhang, L. Semantic Segmentation of Litchi Branches Using Deeplabv3+ Model. *IEEE Access* **2020**, *8*, 164546–164555. [CrossRef]
91. Xiong, Y.; Ge, Y.; From, P.J. An Obstacle Separation Method for Robotic Picking of Fruits in Clusters. *Comput. Electron. Agric.* **2020**, *175*, 105397. [CrossRef]
92. Mghames, S.; Hanheide, M.; Ghalamzan, E.A. Interactive Movement Primitives: Planning to Push Occluding Pieces for Fruit Picking. In Proceedings of the IEEE International Conference on Intelligent Robots and Systems, Las Vegas, NV, USA, 24 October 2020–24 January 2021.

93. Cao, X.; Yan, H.; Huang, Z.; Ai, S.; Xu, Y.; Fu, R.; Zou, X. A Multi-Objective Particle Swarm Optimization for Trajectory Planning of Fruit Picking Manipulator. *Agronomy* **2021**, *11*, 2286. [CrossRef]
94. Chen, J.; Qiang, H.; Wu, J.; Xu, G.; Wang, Z. Navigation Path Extraction for Greenhouse Cucumber-Picking Robots Using the Prediction-Point Hough Transform. *Comput. Electron. Agric.* **2021**, *180*, 105911. [CrossRef]
95. Colucci, G.; Botta, A.; Tagliavini, L.; Cavallone, P.; Baglieri, L.; Quaglia, G. Kinematic Modeling and Motion Planning of the Mobile Manipulator Agri.Q for Precision Agriculture. *Machines* **2022**, *10*, 321. [CrossRef]
96. Le Flécher, E.; Durand-Petiteville, A.; Cadenat, V.; Sentenac, T. Visual Predictive Control of Robotic Arms with Overlapping Workspace. In Proceedings of the ICINCO 2019—Proceedings of the 16th International Conference on Informatics in Control, Automation and Robotics, Prague, Czech Republic, 29–31 July 2019; 2019; Volume 1, pp. 130–137.
97. Ji, W.; Zhang, J.; Xu, B.; Tang, C.; Zhao, D. Grasping Mode Analysis and Adaptive Impedance Control for Apple Harvesting Robotic Grippers. *Comput. Electron. Agric.* **2021**, *186*, 106210. [CrossRef]
98. Wang, Y.; Zhang, H.; Wang, L.; Li, G.; Zhang, Y.; Liu, X. Development of Control System for Cotton Picking Test Bench Based on Fuzzy PID Control. *Nongye Gongcheng Xuebao/Trans. Chin. Soc. Agric. Eng.* **2018**, *34*, 23–32. [CrossRef]
99. Ramin Shamshiri, R.; Weltzien, C.; Hameed, I.A.; Yule, I.J.; Grift, T.; Balasundram, S.; Pitonakova, L.; Ahmad, D.; Chowdhary, G. Research and Development in Agricultural Robotics: A Perspective of Digital Farming. *Int. J. Agric. Biol. Eng.* **2018**, *11*, 1–11. [CrossRef]
100. Navas, E.; Fernández, R.; Sepúlveda, D.; Armada, M.; Gonzalez-de-Santos, P. Soft Grippers for Automatic Crop Harvesting: A Review. *Sensors* **2021**, *21*, 2689. [CrossRef]
101. Liu, J.; Yuan, Y.; Gao, Y.; Tang, S.; Li, Z. Virtual Model of Grip-and-Cut Picking for Simulation of Vibration and Falling of Grape Clusters. *Trans ASABE* **2019**, *62*, 603–614. [CrossRef]
102. Wei, J.; Yi, D.; Bo, X.; Guangyu, C.; Dean, Z. Adaptive Variable Parameter Impedance Control for Apple Harvesting Robot Compliant Picking. *Complexity* **2020**, *2020*, 1–15. [CrossRef]
103. Miao, Y.; Zheng, J. Optimization Design of Compliant Constant-Force Mechanism for Apple Picking Actuator. *Comput. Electron. Agric.* **2020**, *170*, 105232. [CrossRef]
104. Liu, C.H.; Chiu, C.H.; Chen, T.L.; Pai, T.Y.; Chen, Y.; Hsu, M.C. A Soft Robotic Gripper Module with 3d Printed Compliant Fingers for Grasping Fruits. In Proceedings of the IEEE/ASME International Conference on Advanced Intelligent Mechatronics, AIM, Auckland, New Zealand, 9–12 July 2018; Volume 2018.
105. Pi, J.; Liu, J.; Zhou, K.; Qian, M. An Octopus-Inspired Bionic Flexible Gripper for Apple Grasping. *Agriculture* **2021**, *11*, 1014. [CrossRef]
106. Hohimer, C.J.; Wang, H.; Bhusal, S.; Miller, J.; Mo, C.; Karkee, M. Design and Field Evaluation of a Robotic Apple Harvesting System with a 3D-Printed Soft-Robotic End-Effector. *Trans ASABE* **2019**, *62*, 405–414. [CrossRef]
107. Vu, Q.; Ronzhin, A. Models and algorithms for design robotic gripper for agricultural products. *Comptes Rendus De L'Academie Bulg. Des Sci.* **2020**, *73*, 103–110.
108. Chen, Z.; Yang, M.; Li, Y.; Yang, L. Design and Experiment of Tomato Picking End-Effector Based on Non-Destructive Pneumatic Clamping Control. *Nongye Gongcheng Xuebao/Trans. Chin. Soc. Agric. Eng.* **2021**, *37*, 27–35. [CrossRef]
109. Yung, I.; Maccarana, Y.; Maroni, G.; Previdi, F. Partially Structured Robotic Picking for Automation of Tomato Transplantation. In Proceedings of the 2019 IEEE International Conference on Mechatronics, ICM 2019, Ilmenau, Germany, 18–20 March 2019.
110. Zhang, J.; Lai, S.; Yu, H.; Wang, E.; Wang, X.; Zhu, Z. Fruit Classification Utilizing a Robotic Gripper with Integrated Sensors and Adaptive Grasping. *Math. Probl. Eng.* **2021**, *2021*, 1–15. [CrossRef]
111. Habegger, R.; Bergamo, E.; Schwab, W.; Berninger, T.; Rixen, D. Impact of Intensive Modification of Sweet Pepper Plants on Performance of End Effectors for Autonomous Harvesting. *Eur. J. Hortic. Sci.* **2021**, *86*, 354–359. [CrossRef]
112. De Preter, A.; Anthonis, J.; De Baerdemaeker, J. Development of a Robot for Harvesting Strawberries. *FAC-PapersOnLine* **2018**, *51*, 14–19. [CrossRef]
113. Li, Z.; Miao, F.; Yang, Z.; Wang, H. An Anthropometric Study for the Anthropomorphic Design of Tomato-Harvesting Robots. *Comput. Electron. Agric.* **2019**, *163*, 104881. [CrossRef]
114. Li, Z.; Miao, F.; Yang, Z.; Chai, P.; Yang, S. Factors Affecting Human Hand Grasp Type in Tomato Fruit-Picking: A Statistical Investigation for Ergonomic Development of Harvesting Robot. *Comput. Electron. Agric.* **2019**, *157*, 90–97. [CrossRef]
115. Hou, Z.; Li, Z.; Fadiji, T.; Fu, J. Soft Grasping Mechanism of Human Fingers for Tomato-Picking Bionic Robots. *Comput. Electron. Agric.* **2021**, *182*, 106010. [CrossRef]
116. Öz, E.; Jakob, M. Ergonomic Evaluation of Simulated Apple Hand Harvesting by Using 3D Motion Analysis. *Ege Üniversitesi Ziraat Fakültesi Derg.* **2020**, *57*, 249–256. [CrossRef]
117. Liu, X.; Xu, H.; Chen, F. Research on Vision and Trajectory Planning System for Tomato Picking Robots. In Proceedings of the 2020 5th International Conference on Mechanical, Control and Computer Engineering, ICMCCE 2020, Harbin, China, 25–27 December 2020.
118. Zhang, L.; Jia, J.; Gui, G.; Hao, X.; Gao, W.; Wang, M. Deep Learning Based Improved Classification System for Designing Tomato Harvesting Robot. *IEEE Access* **2018**, *6*, 67940–67950. [CrossRef]
119. Horng, G.J.; Liu, M.X.; Chen, C.C. The Smart Image Recognition Mechanism for Crop Harvesting System in Intelligent Agriculture. *IEEE Sens. J.* **2020**, *20*, 2766–2781. [CrossRef]

120. Xiong, Y.; From, P.J.; Isler, V. Design and Evaluation of a Novel Cable-Driven Gripper with Perception Capabilities for Strawberry Picking Robots. In Proceedings of the IEEE International Conference on Robotics and Automation, Brisbane, QLD, Australia, 21–25 May 2018.
121. Zhong, Z.; Xiong, J.; Zheng, Z.; Liu, B.; Liao, S.; Huo, Z.; Yang, Z. A Method for Litchi Picking Points Calculation in Natural Environment Based on Main Fruit Bearing Branch Detection. *Comput. Electron. Agric.* **2021**, *189*, 106398. [CrossRef]

agronomy

Review

Object Detection and Recognition Techniques Based on Digital Image Processing and Traditional Machine Learning for Fruit and Vegetable Harvesting Robots: An Overview and Review

Feng Xiao, Haibin Wang *, Yaoxiang Li, Ying Cao, Xiaomeng Lv and Guangfei Xu

College of Engineering and Technology, Northeast Forestry University, Harbin 150040, China
* Correspondence: whb_nefu@nefu.edu.cn

Abstract: The accuracy, speed, and robustness of object detection and recognition are directly related to the harvesting efficiency, quality, and speed of fruit and vegetable harvesting robots. In order to explore the development status of object detection and recognition techniques for fruit and vegetable harvesting robots based on digital image processing and traditional machine learning, this article summarizes and analyzes some representative methods. This article also demonstrates the current challenges and future potential developments. This work aims to provide a reference for future research on object detection and recognition techniques for fruit and vegetable harvesting robots based on digital image processing and traditional machine learning.

Keywords: digital image processing; traditional machine learning; harvesting robot; computer vision; object detection; object recognition; research overview; research review

Citation: Xiao, F.; Wang, H.; Li, Y.; Cao, Y.; Lv, X.; Xu, G. Object Detection and Recognition Techniques Based on Digital Image Processing and Traditional Machine Learning for Fruit and Vegetable Harvesting Robots: An Overview and Review. *Agronomy* **2023**, *13*, 639. https://doi.org/10.3390/agronomy13030639

Academic Editors: Baohua Zhang and Simon Pearson

Received: 4 January 2023
Revised: 18 February 2023
Accepted: 19 February 2023
Published: 23 February 2023

1. Introduction

Fruit harvesting is an important aspect of farming. It directly affects the yield and profitability of cultivation. With the increasing scale of global cultivation (e.g., global annual production of fruits and vegetables such as tomato, citrus, apple, and strawberry, has reached 182 million tons [1], 89 million tons [2], 86 million tons [3], and 9 million tons [4], respectively), the contradiction between the large amount of labor used in traditional production methods and labor shortages has become increasingly prominent. The labor cost of fruit and vegetable harvesting has reached 30–50% of the total production cost [5–9]. Fruit and vegetable harvesting robots have attracted broad attention in the agricultural field (as shown in Figure 1) because of their high productivity and low production cost [10,11]. As shown in Figure 2, taking typical fruits and vegetables such as plums [12], apples [13–16], sweet peppers [17–19], strawberries [6,7,20], litchis [21], tomatoes [22,23], and kiwifruits [24] as objects, a series of harvesting robots have been developed and applied in greenhouses and orchards. Fruit and vegetable harvesting robots have entered a critical period in the progression from laboratory research to industrial applications.

As an important part of vision systems of fruit and vegetable harvesting robots, the accuracy, speed, and robustness of object detection and recognition are directly related to the harvesting efficiency, quality, and speed. Vision systems of harvesting robots vary for different picking targets. Their characteristics mainly include the imaging sensor and the specific content of crop visual information. Black/white, RGB, spectral, and thermal cameras (as shown in Table 1) are widely used in harvesting robots to obtain color, shape, texture, and size information of fruits in a specific operational area. Different processes of object detection and recognition of fruits and vegetables are shown in Figure 3. Many researchers have conducted extensive and in-depth research on object detection and recognition techniques for fruit and vegetable harvesting robots based on digital image processing and traditional machine learning. The research can be subdivided into the following aspects:

(1) Techniques based on digital image processing, such as color features (RGB (Red, Green, Blue) [25–28], HSV (Hue, Saturation, Value) [29–31], HSI (Hue, Saturation Intensity) [32–34], Lab (Lightness, Green to Red and Blue to Yellow) [33,35,36], HSB (Hue, Saturation, Brightness), YCbCr)-based methods, shape feature-based methods [37–46], texture feature-based methods [44,47–52], and multi-feature fusion-based methods [17,28,39,52–67].

(2) Image segmentation and classifiers based on traditional machine learning, such as K-means clustering algorithm-based methods [68–75], SVM (Support Vector Machine) algorithm-based methods [54,57,69,73,76–84], KNN (K Nearest Neighbor) clustering algorithm-based methods [36,85–91], AdaBoost (Adaptive Boosting) algorithm-based methods [62,92–99], decision tree algorithm-based methods [100–107], and Bayesian algorithm-based methods [108–113].

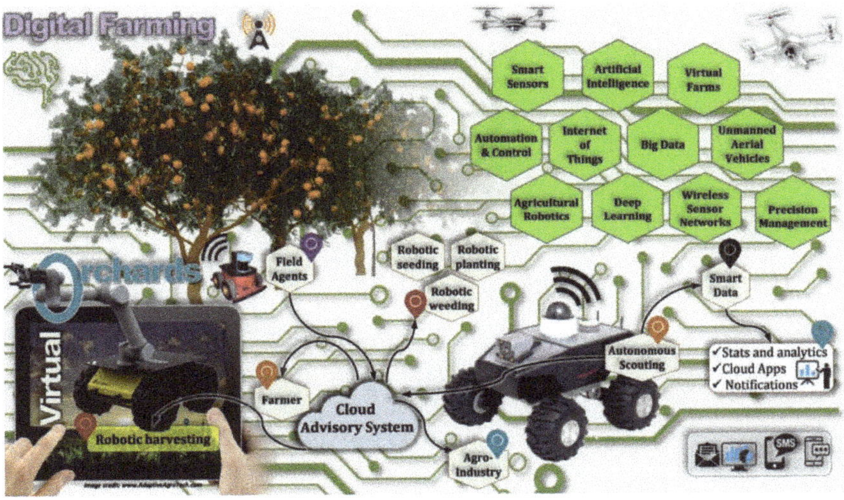

Figure 1. Digital farming with agricultural robotics (source: www.AdaptiveAgroTech.com (accessed on 1 October 2022)).

Table 1. Comparison of frequently used sensors for fruit and vegetable recognition.

Sensors	Features Exploited	Advantages	Disadvantages
Black/white camera	Shape and texture features	A negligible effect on changing lighting conditions	Lack of color information of target objects
RGB camera	Color, shape, and texture features	Exploits all the basic features of target objects	Highly sensitive to changing lighting conditions
Spectral camera	Color features and spectral information	Provides more information about reflectance	Computationally expensive for complete spectrum analysis
Thermal camera	Thermal signatures	Color Invariant	Dependency on minute thermal difference

This article provides an overview and review of the progress in object detection and recognition techniques for fruit and vegetable harvesting robots based on digital image processing and traditional machine learning. Although there have been some reviews of techniques for object detection and recognition of fruits and vegetables [114–135], the contributions of this work are to: (1) systematically summarize object detection and recognition techniques of fruit and vegetable harvesting robots based on digital image processing and traditional machine learning in recent years; (2) systematically analyze the advantages, disadvantages, and applicability of various techniques; and (3) demonstrate the current challenges and future potential developments. Through this clearer and more comprehensive overview and review, we aim to provide a reference for future research on object detection and recognition techniques of fruit and vegetable harvesting robots based on digital image processing and traditional machine learning.

Figure 2. Typical harvesting robots: (**a**) a plum harvesting robot (Photo: Reprinted with permission from Ref. [12]. 2021, Brown, J.); (**b**,**d–f**) apple harvesting robots (Photo: Reprinted with permission from Ref. [13]. 2021, Yan, B.; Ref. [14]. 2017, He, L.; Ref. [15]. 2012, Ji, W.; Ref. [16]. 2011, Zhao, D.); (**c**,**n–p**) sweet pepper harvesting robots (Photo: Reprinted with permission from Ref. [17]. 2020, Arad, B.; Ref. [18]. 2017, Lehnert, C.; Ref. [19]. 2014, Bac, C.W.); (**g–i**) strawberry harvesting robots (Photo: Reprinted with permission from Ref. [6]. 2020, Xiong, Y.; Ref. [7]. 2019, Xiong, Y.; Ref. [20]. 2010, Hayashi, S.); (**j**) a litchi harvesting robot (Photo: Reprinted with permission from Ref. [21]. 2018, Xiong, J.); (**k**,**m**) tomato harvesting robots (Photo: Reprinted with permission from Ref. [22]. 2018, Feng, Q.; Ref. [23]. 2010, Kondo, N.); (**l**) a kiwifruit harvesting robot (Photo: Reprinted with permission from Ref. [24]. 2019, Williams, H.A.M.).

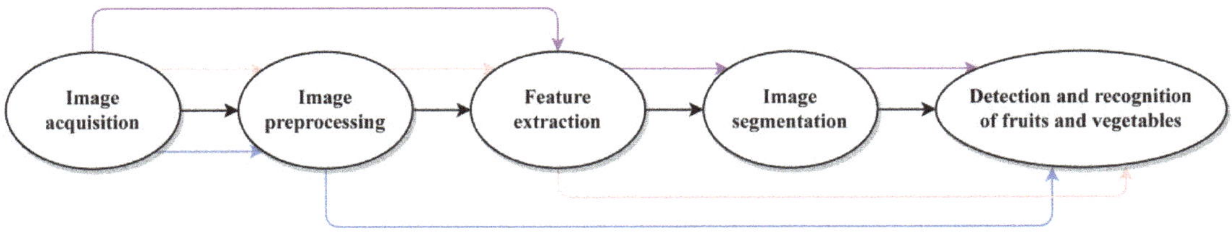

Figure 3. Different processes of object detection and recognition of fruits and vegetables.

The outline of this overview and review is shown in Figure 4. The organization of this paper is as follow: in Section 2, we provide an overview and review of the research and development in object detection and recognition techniques of fruits and vegetables based on digital image processing. We present separate discussions focused on color, shape, texture features, and multi-feature fusion-based methods.

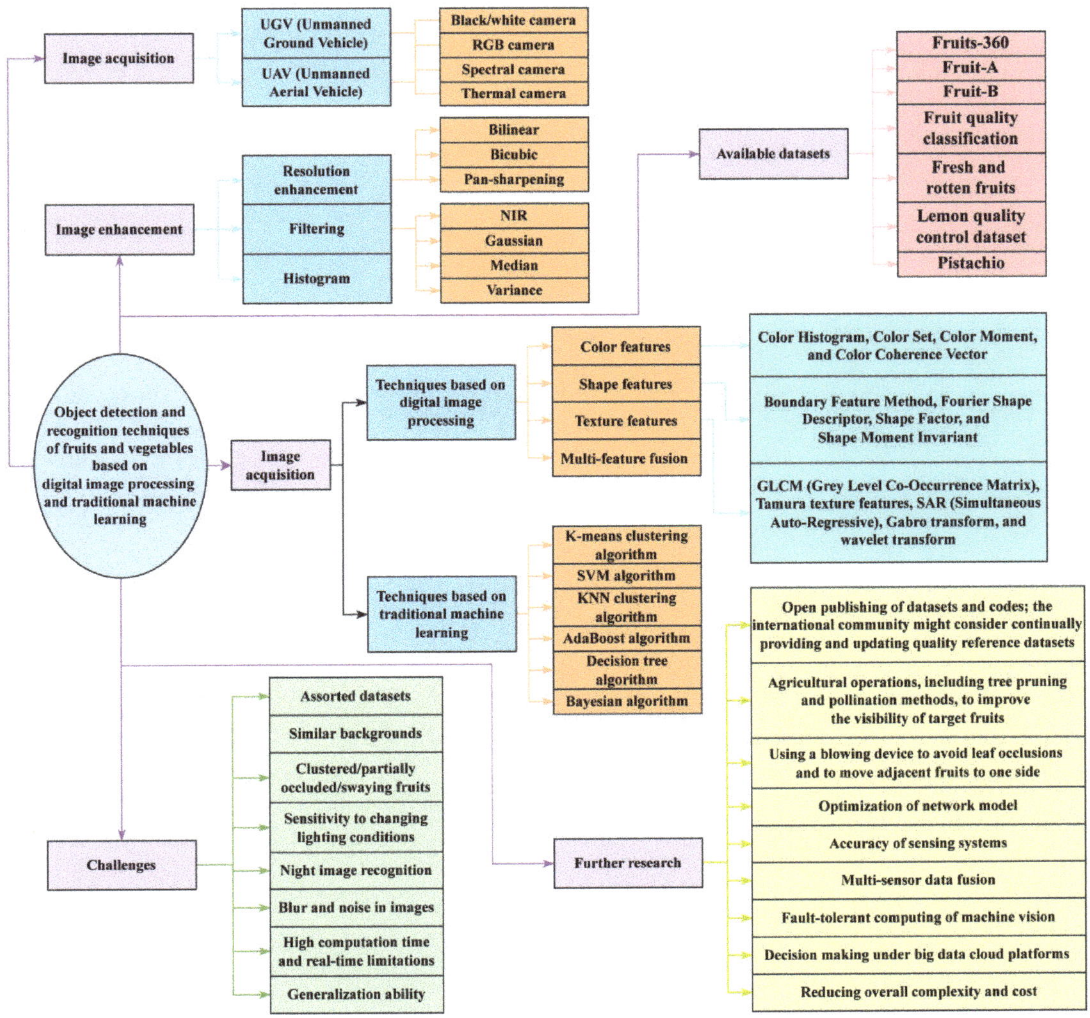

Figure 4. The outline of this overview and review.

In Section 3, we provide an overview and review of the research and development in object detection and recognition techniques of fruits and vegetables based on traditional machine learning. We present separate discussions focused on K-means clustering, SVM, KNN clustering, AdaBoost, decision tree, and Bayesian algorithm-based methods.

Section 4 extends our discussions to the challenges and further research of object detection and recognition techniques of fruits and vegetables. A summary of findings and conclusions are presented in Section 5.

2. Techniques Based on Digital Image Processing

Colors, shapes, and textures are important features used by fruit and vegetable harvesting robots for detecting and recognizing target objects. Many researchers have conducted extensive and in-depth research on object detection and recognition techniques of fruits and vegetables based on color features (RGB [25–28], HSV [29–31], HIS [32–34], Lab [33,35,36], HSB, YCbCr), shape features [37–46], texture features [44,47–52], and multi-feature fusion [17,28,39,52–67] (as shown in Figure 5). Table 2 compares the results of different techniques by different researchers, and presents analysis of the advantages, disadvantages, and applicability of various techniques.

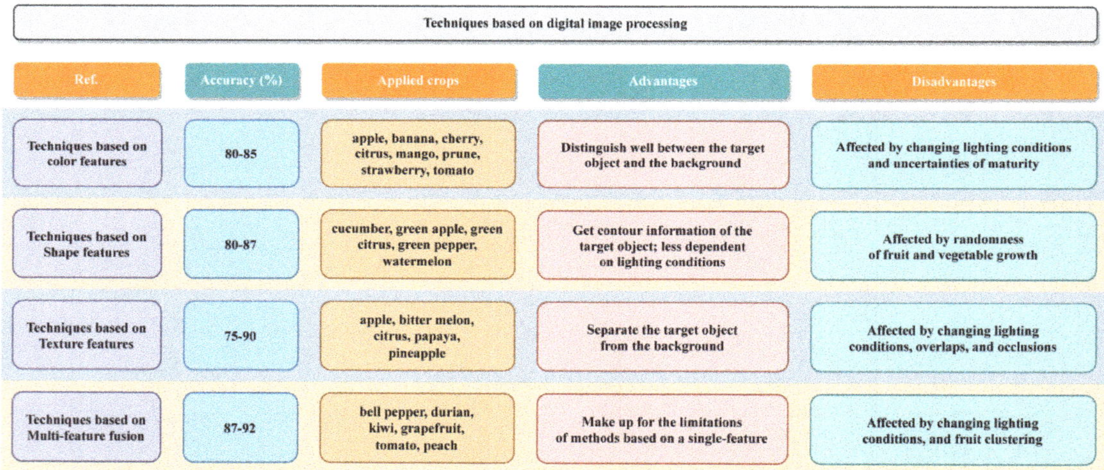

Figure 5. Techniques based on digital image processing.

Table 2. Comparison of techniques based on digital image processing.

Applied Crops	Description	Sensors	Advantages	Improvements	Value of Metrics Used %	Ref.
Apple	The near-large fruit from the apple image in orchards should be obtained	RGB camera	The R-channel and G-channel images of orchard apple RGB images are operated by the Adaptive Gamma Correction method	Future work may include improving the detection rate	70	[27]
Tomato	A new mature tomato detection algorithm based on the improved HSV color space and the improved watershed segmentation	RGB camera	Mature red tomatoes are detected successfully even with light effect	The accuracy of recognition needs to be improved	81.6	[31]
Apple	The potential use of close-range and low-cost terrestrial RGB imaging sensors for fruit detection in a high-density apple orchard	RGB camera	Band combinations are generated as additional parameters for fruit detection	Unripe fruits with poor lighting are not detected in the methodology	75	[35]
Blueberry	Recognizing blueberry fruit of different maturity using histogram-oriented gradients and color features in outdoor scenes	RGB camera	Using a* and b* features in the L*a*b* color space to discard non-fruit regions	The speed of detection needs to be improved	mature fruit: 96.1 intermediate fruit: 94.2 young fruit: 86	[36]
Apple	The Hough Circle Transformation algorithm is proposed to fit and extract apple shapes	RGB camera	In order to overcome the problem of Global Hough Transform, a local parameter Adaptive Hough Transform is used	When the recognition algorithm is faced with multiple overlapping apples, if the apples are not arranged in a straight line, it is easy to obtain recognition errors	91.3 (72 ms)	[25]

Table 2. *Cont.*

Applied Crops	Description	Sensors	Advantages	Improvements	Value of Metrics Used %	Ref.
Citrus, tomato, pumpkin, bitter gourd, towel gourd, and mango	Fruit detection in natural environments using Partial Shape Matching and Probabilistic Hough Transform	RGB camera	PSM and PHT are used for sub-fragment detection and aggregation without necessitating the painstaking design of specific features for each type of fruit. This makes the proposed algorithm a generalized method	PHT utilizes a scale-variant dissimilarity metric to determine the probability value of a vote. So, it may fail to detect fruits with large scale changes	78.3; 84.8; 74.5; 76.2; 80.7; 91.9	[37]
Orange	A machine vision algorithm combining adaptive segmentation and shape analysis for orange fruit detection	RGB camera	In the segmentation of the fruit, the orange is enhanced by using the red chromaticity coefficient, which enables adaptive segmentation under variable outdoor illumination	The speed of detection needs to be improved	93	[45]
Green fruits	A technique based on texture analysis is proposed for detecting green fruits	RGB camera	The method is sufficiently accurate for precise location and monitoring of textured fruit in the field	The method needs to be improved to better handle some disadvantageous conditions such as strong sunlight and occlusions	pineapple: 85 bitter melon: 100	[51]
Green apple	Detection of green apples in hyperspectral images of apple-tree foliage using machine vision	Spectral camera	The method uses several techniques, such as extraction and classification of homogenous objects for analyzing hyperspectral data	Independent studies need to be conducted in a variety of conditions and with a number of crop varieties to verify the robustness of the method	88.1	[46]
Green citrus	Green citrus detection using 'eigenfruit', color and circular Gabor texture features under natural outdoor conditions	RGB camera	The method proposes the use of color, shape, and texture features together to detect immature green citrus fruits, including scanning an image using a sub-window, and merging results of different classifiers with majority voting	Future work may include improving the detection rate, reducing the processing time, and accommodating more varied outdoor conditions	75.3	[44]

Table 2. *Cont.*

Applied Crops	Description	Sensors	Advantages	Improvements	Value of Metrics Used %	Ref.
Immature citrus	Immature citrus fruit detection based on local binary pattern features and hierarchical contour analysis	RGB camera	The good performance of occlusion tolerance of the proposed method is mainly due to the robust LBP texture descriptor and hierarchical contour analysis which uses the pattern of light intensity distribution on the fruit surface	The fruit occluded very seriously or even completely by leaves and other fruits couldn't be detected by the proposed method	82.3	[39]
Litchi	A method of ripe litchi recognition for two varieties of litchis using RGB-D images is proposed	RGB-D camera	The random forest binary classification model is trained employing color and texture features to recognize litchi fruits	Depth segmentation can effectively reduce the false positive rate of litchi recognition	green litchi: 89.92 red litchi: 94.5	[55]
Oil palm fresh fruit bunch	The maturity classification of oil palm fresh fruit bunches based on color and texture features	RGB camera	Forty features are extracted from several color spaces, which were reduced to five features using the PCA method to optimize the computation time	The speed of detection needs to be improved	98.3	[53]
Strawberry	A simple color thresholding algorithm based on the RGB channels for detecting strawberries	RGB-D camera	The vision system uses color thresholding combined with screening of the object area and the depth range to select ripe and reachable strawberries, which is fast for processing	Future work could merge the detection from multiple frames so that occluded strawberries can be visible from a different view	isolated strawberry:96.8 occluded strawberry:53.6	[18]

2.1. Techniques Based on Color Features

Mature fruits and vegetables usually have significant and stable color features. Color features provide a set of indicators for the detection and recognition of fruits and vegetables. Object detection and recognition techniques of fruits and vegetables based on color features, extract color features through Color Histogram, Color Set, Color Moment, and Color Coherence Vector. The techniques based on color features are mainly applicable to cases where the colors of fruits and vegetables are significantly different from the backgrounds (branches, leaves, trunks), such as tomatoes [28,31], apples [29,35], mangoes [34], bananas, cherries, citrus, prunes, and strawberries.

Goel and Sehgal [28] detected and recognized several ripening stages of tomatoes using RGB image information. This research has a positive implication for selecting the best ripening stage of fruits and vegetables. For example, fruits and vegetables that need to be transported over long distances can be harvested at an early stage of ripeness.

Zemmour et al. [26] analyzed different color spaces. The research results showed that evaluating different color spaces is very important, because for different kinds of fruits and vegetables, a different color space might be superior to the others. In order to improve the accuracy of the detection and recognition of tomatoes, marigold flowers, and apples, Malik et al. [31], Sethy et al. [30], Yu et al. [29], respectively, converted RGB images into HSV color space, and then separated the image luminance channels. Ratprakhon et al. [34] converted RGB images into HIS color space to detect and recognize the ripeness of mangoes. Tan et al. [36] and Biffi et al. [35], respectively, converted RGB images into Lab color space to detect and recognize blueberries and apples. Zemmour et al. [26] suggested that Lab color space could be used more for low quality images because it is more robust to noise in images. In challenging color conditions (for example, where fruit and vegetable colors are similar to the backgrounds), other features could be considered to improve the effectiveness of object detection and recognition for fruit and vegetable harvesting robots.

The detection and recognition time of fruits and vegetables based on color features is relatively long. In order to shorten the detection and recognition time, Yang et al. [25] proposed an Otsu's thresholding method based on the two times Red minus Green minus Blue (2R-G-B) color feature to segment images. Lv et al. [27] operated the R-channel and G-channel images of orchard apple RGB images using the Adaptive Gamma Correction method. This method not only shortened the detection and recognition time, but also overcame the influence of changing lighting conditions. Zemmour et al. [26] proposed an automatic parameter tuning procedure specially developed for the dynamic adaptive thresholding algorithm for object detection and recognition of fruits and vegetables. The thresholds were selected by quantifying the required relationship between the true and false positive rates.

In general, techniques for object detection and recognition of fruits and vegetables based on color features are less dependent on image size. However, the variability and uncertainty of fruit and vegetable maturity can affect the accuracy, speed, and robustness of detection and recognition. These techniques are mainly applicable to structured environments such as greenhouses.

2.2. Techniques Based on Shape Features

Mature fruits and vegetables usually have significant and stable shape features. Geometric shape features provide another set of indicators for the detection and recognition of fruits and vegetables. Techniques for object detection and recognition of fruits and vegetables based on shape features, extract shape features using the Boundary Feature Method, Fourier Shape Descriptor, Shape Factor, and Shape Moment Invariant. These techniques are mainly applied to cases where the shapes of fruits and vegetables are significantly different from the backgrounds. For example, the shapes of apples and citrus are usually rounded compared to the branches and leaves, and a cucumber shows an elongated fruit shape (as shown in Figure 6).

Figure 6. Samples of cucumbers in a natural complex environment (Photo: Reprinted with permission from Ref. [136]. 2020, Mao S.).

For round fruits, Hannan et al. [45] detected and recognized fruits in clusters by shape analysis. This method can better detect and recognize target objects in changing lighting conditions. Jana and Parekh [42] proposed a shape-based fruit detection and recognition method. It involves a pre-processing step to normalize a fruit image with respect to variations in translation, rotation, and scaling, and utilizes features that do not change due to varying distances, growth stages, or surface appearances of fruits. The method was applied to 210 images of 7 fruit classes. The overall recognition accuracy ranged from 88 to 95%. Lu et al. [39] proposed a new shape analysis method called Hierarchical Contour Analysis (HCA). The hierarchical contour maps around each local maximum were extracted and fitted with Circular Hough Transform, and the fitted circles were predicted as fruit targets if their radii were in a predetermined range. The HCA can effectively utilize shape information, and does not need to extract and analyze the edge in an image. Therefore, it is efficient and robust under various lighting conditions and occlusions in natural environments. Lin et al. [37] also proposed a method for the detection and recognition of fruits and vegetables based on shape features. The research results showed that the method is competitive for detecting most kinds (such as green, orange, circular, and non-circular) of fruits and vegetables in natural environments.

Since the shapes of fruits and vegetables are usually not affected by the colors, object detection and recognition techniques of fruits and vegetables based on shape features are more effective for cases where the colors of fruits and vegetables are similar to the backgrounds, while the shapes of fruits and vegetables are significantly different from the backgrounds, such as green citrus [37,40,44], green apples [38,43,46], cucumbers, green peppers, and watermelons.

In general, techniques for object detection and recognition of fruits and vegetables based on shape features are less dependent on lighting conditions. However, in unstructured environments, the randomness of fruit and vegetable growth can affect the accuracy, speed, and robustness of detection and recognition of fruits and vegetables. These techniques are mainly applicable to natural orchards with certain agricultural operations.

2.3. *Techniques Based on Texture Features*

Mature fruits and vegetables usually have significant and stable texture features, and the surface textures of fruits and vegetables are usually smoother than the backgrounds. Texture features provide another set of indicators for the detection and recognition of fruits and vegetables. Techniques for object detection and recognition of fruits and vegetables based on texture features, extract texture features through the GLCM (Grey Level Co-Occurrence Matrix), Tamura texture features, SAR (Simultaneous Auto-Regression), Gabro transform, and Wavelet transform. These techniques are mainly applicable to cases where the textures of fruits and vegetables are significantly different from the backgrounds, such as apples [52], bitter melons [51], citrus [44], papayas [110], and pineapples [51].

Trey et al. [49] used leaf texture features as parameters for plant family detection and recognition. The research results showed that the method gives a perfect classification of three plant families of the Ivorian flora. Rahman et al. [47] detected and recognized tomato leaf diseases through 13 different statistical features calculated from tomato leaves using the GLCM algorithm. The method was implemented in the form of a cell phone application. The research results showed that the method provides excellent annotation with an accuracy of 100% for healthy leaf, 95% for early blight, 90% for Septoria leaf spot, and 85% for late blight.

Since the surface textures of fruits and vegetables are usually not affected by the colors and shapes, techniques for object detection and recognition of fruits and vegetables based on texture features are more effective for cases where the colors and shapes of fruits and vegetables are similar to the backgrounds, while the textures of fruits and vegetables are significantly different from the backgrounds. Kurtulmus et al. [44] used circular Gabor texture analysis for the detection and recognition of green citrus. The method detected and recognized target fruits by scanning the whole image, but the correct rate was only 75.3%. To improve the accuracy of detection and recognition of fruits and vegetables, Chaivivatrakul and Dailey [51] proposed a texture-based feature detection and recognition method for green fruits. The method involves interest point feature extraction and descriptor computation, interest point classification using support vector machines, candidate fruit point mapping, and morphological closing and fruit region extraction. This approach can effectively improve the correct rate of detection and recognition of green fruits (more than 85%). In addition, Hameed et al. [48] proposed a texture-based latent space disentanglement method to enhance the learning of representations for novel data samples.

In general, the main problem of techniques for object detection and recognition of fruits and vegetables based on texture features is that changing lighting conditions and complex backgrounds can affect the accuracy, speed, and robustness of detection and recognition. These techniques are mainly applicable to greenhouse environments.

2.4. Techniques Based on Multi-Feature Fusion

Techniques for object detection and recognition of fruits and vegetables based on a kind of feature can recognize fruits from natural environments, but they usually have certain limitations. Techniques for object detection and recognition of fruits and vegetables that integrate two or more features to form multi-feature fusion can effectively improve the accuracy, speed, and robustness of detection and recognition [59,92,95,136–139].

In terms of color and shape features, Liu et al. [60] proposed a method for the detection and recognition of incomplete red apples (as shown in Figure 7). The research results are shown in Figure 8. The method can be used to detect not only apples, but can also be used to detect other fruits that have different colors from the backgrounds, such as oranges, kiwifruits, and tomatoes. However, the method only detects fruits using rectangular boxes. Pixel-wise segmentation is more accurate than detection boxes. Recognizing fruits at the pixel level could be the focus of further work. Arad et al. [17], and Liu et al. [58] extracted color features from RGB color channels of fruit and vegetable images, and morphological features were extracted from the images with detected fruit and vegetable borders using morphological operations. Then, they detected and recognized bell peppers, grapefruits, and peaches.

In terms of color and texture features, to solve segmentation problems, Lin and Zou [62] proposed a new segmentation method using color and texture features. This method incorporates HSV color features and Leung–Malik texture features to detect citrus using fixed-size sub-windows. Madgi and Danti [63] classified fruits and vegetables based on color features and GLCM texture features. The research results showed that the combination of color with GLCM texture features is more effective than combined color and LBP texture features.

Figure 7. Two kinds of apple fruits: (**a**) completely red fruits; (**b**) incompletely red fruits (Photo: Reprinted with permission from Ref. [60]. 2019, Liu X.).

Figure 8. Detection results of different images: (**a1–a4**) images taken under front light; (**b1–b4**) images taken under backlight; (**c1–c4**) images taken under side light; (**d1–d4**) images taken under artificial light (Photo: Reprinted with permission from Ref. [60]. 2019, Liu X.).

In terms of shape and texture features, Lu et al. [39], Mustaffa et al. [61], and Bhargava and Bansal [54] recognized fruits and vegetables by shape features including area, perimeter, and roundness, and constructed fruit and vegetable textures based on local binary patterns. Finally, they classified green citrus, multi-species durians, and multi-species apples.

In terms of color, shape, and texture features, Rakun et al. [52] achieved apple detection and recognition under uneven lighting conditions, partial fruit shading, and a similar background by combining color, shape, and texture features. Basavaiah and Anthony [56] proposed a detection and recognition method based on color, shape, and texture features for a variety of tomato diseases. Azarmdel et al. [57] and Septiarini et al. [53], respectively, achieved the detection and recognition of mulberries and oil palms based on multiple features such as color, shape, and texture features.

Currently, digital image processing techniques used by researchers for the detection and recognition of fruits and vegetables always require setting thresholds such as color, shape, and texture features, but the optimal thresholds often vary with images. In order to address this problem, Payne et al. [66] proposed using RGB and YCbCr color segmentation and texture segmentation based on the variability of neighboring pixels to divide pixels into target fruit and background pixels for high-accuracy detection and recognition. However, this method relies too much on the color features of images, and the recognition accuracy is low when the color features are not obvious. For this reason, Payne et al. [65], based on the previously proposed algorithm, reduced the reliance on color features by setting the boundary-constrained mean and edge detection filters, and increased the use of texture filtering. The research results showed that the recognition accuracy is significantly improved compared with before the improvement. Yamamoto et al. [64] used a multi-feature fusion method to simplify the tedious steps of setting thresholds for each image and improve the accuracy of detection and recognition.

3. Image Segmentation and Classifiers Based on Machine Learning

Since machine learning can derive laws from sample data that can hardly be summarized by theoretical analysis, many researchers have conducted extensive and in-depth research on techniques for object detection and recognition of fruits and vegetables based on the K-means clustering algorithm [68–75], SVM algorithm [54,57,69,73,76–84], KNN clustering algorithm [36,85–91], AdaBoost algorithm [62,92–99], decision tree algorithm [100–107], and Bayesian algorithm [108–113] (as shown in Figures 9 and 10). Table 3 compares the results of different techniques of different researchers, and presents analysis of the advantages, disadvantages, and applicability of various techniques.

Table 3. Comparison of techniques based on traditional machine learning.

Applied Crops	Description	Sensors	Advantages	Improvements	Value of Metric Used %	Ref.
Litchi	A litchi recognition algorithm based on K-means clustering is presented to separate litchi from leaves, branches and background	Two CCD color cameras	The method can be robust against the influences of varying illumination and precisely recognize litchi	Future research could improve the localization accuracy of litchi via hardware and software improvements	unoccluded: 98.8; partially occluded: 97.5	[75]
Apple	The development of a real-time machine vision recognition system to guide a harvesting robotic for picking apples in different conditions	CCD camera	The segmentation method based on seeded region growing methods and color features is applied, and color and shape features of color images are extracted	Reducing the recognition execution time is still a challenge	89 (352 ms)	[14]
Aubergine	To detect and locate the aubergines automatically, an algorithm based on SVM classifier is implemented	TOF camera	The occlusion algorithm is applied to aubergines that have low visibility due to leaf occlusions by planning a collaborative behavior between the arms to solve the problem of occlusion and proceed with dual-arm harvesting	Most of the failures are related to changing lighting conditions. So, future work to enhance the harvester robot should prioritize improvements to image acquisition	91.67 (26 ms)	[77]
Citrus	Identification of fruits and branches in natural scenes for a citrus harvesting robot using machine vision and support vector machine	Color CCD camera	A multi-class support vector machine, which succeeds by morphological operation, was used to simultaneously segment the fruits and branches	The effect on feature extraction, and real-time response of the identification method, have to be further optimized	92.4	[73]
Tomato	An algorithm is proposed for tomato detection in regular color images to reduce the influence of illumination and occlusion	RGB camera	The proposed method used a combination of shape, texture, and color information. HOG descriptors are adopted in this work. An SVM classifier is used to implement the classification task	Future research could focus on further improving the detection accuracy and extension to other stages of tomatoes	94.41 (950 ms)	[83]
Green pepper	A green pepper recognition method based on least-squares support vector machine optimized by improved particle swarm optimization	RGB camera	In order to reduce the complexity of data calculations and improve the efficiency, the extracted feature vectors are normalized. The feature vector is used as the input eigenvector of the least-squares support vector machine (LSSVM).	Due to the high rate of leak recognition, the correct recognition rate of green pepper needs to be improved	89.04 (320 ms)	[81]

Table 3. *Cont.*

Applied Crops	Description	Sensors	Advantages	Improvements	Value of Metric Used %	Ref.
Tomato	A dual-arm cooperative approach for a tomato harvesting robot using a binocular vision sensor	Stereo camera	A tomato detection algorithm combining an AdaBoost classifier and color analysis is proposed and employed by the harvesting robot	Future work could focus on the improvement in the successful harvesting rate under uncertain conditions	96	[93]
Tomato	Detecting tomatoes in greenhouse scenes by combining an AdaBoost classifier and color analysis	RGB camera	To use shape, texture, and color information, Haar-like features, an AdaBoost algorithm, and APV-based color analysis are implemented	Future work could include enhanced detection rates, reducing the processing time, and various cultivars of tomatoes, and accommodate more varied unstructured environments	96	[99]
Immature green citrus	Used only regular RGB images of the citrus canopy to detect immature green citrus fruit in natural environments	RGB camera	A local binary patterns feature-based Adaptive Boosting (AdaBoost) classifier is built to remove false positives. A sub-window is used to scan the difference image between the illumination-normalized image and the resulting image from CHT detection in order to detect small areas and partially occluded fruit	It can improve image processing speed by decreasing false positive removal time	85.6	[96]
Grain impurity of rice	Real-time grain impurity sensing for rice combines harvesters using image processing and decision tree algorithm	CMOS camera	The illumination method is optimized by histogram equalization. Decision tree classification is used	Future work may include improving the detection rate, reducing the processing time, and accommodating more varied outdoor conditions	76	[102]

Image segmentation and classifiers based on traditional machine learning			
Type	**Learning style**	**Input**	**The idea of segmentation and classification**
Techniques based on K-means clustering algorithm	Unsupervised learning	Images to be segmented; number of clusters; initial clustering center	Classify input data into identical and different classes based on their fixed distances from each other
Techniques based on SVM algorithm	Supervised learning	Training set and pixel features; selection of the kernel function; test images	Classification by solving the separated hyperplane that correctly partitions the training set and has the largest geometric separation
Techniques based on KNN clustering algorithm	Supervised learning	Training set and pixel features; number of clusters; test images	Classification by classifying unknown feature vectors into classes of the most common attributes in the K nearest neighbors using the training set
Techniques based on AdaBoost algorithm	Supervised learning	Training set and pixel features; test images	Different classifiers (weak classifiers) are trained using the same training set, and then these weak classifiers are pooled to form a stronger final classifier (strong classifier)
Techniques based on decision tree algorithm	Supervised learning	Training set and pixel features; test images	Start from the root node; the corresponding features in the item to be classified are tested and the output branches are selected according to their values until the leaf node is reached; the category stored in the leaf node is used as the result
Techniques based on Bayesian algorithm	Supervised learning	Training set and pixel features; test images	Classify based on minimizing Bayesian risk, minimizing probability error, or maximizing posterior probability

Figure 9. The idea of image segmentation and classifiers based on traditional machine learning.

Image segmentation and classifiers based on traditional machine learning				
Type & Ref.	**Accuracy (%)**	**Applied crops**	**Advantage**	**Disadvantage**
Techniques based on K-means clustering algorithm	80-90	apple, grape, litchi	Automatically classify the target object and the background; short computation time, fast response time, and good clustering effect	The randomly selected K-values affect the classification results; sensitive to abnormal data
Techniques based on SVM algorithm	82-96	apple, banana, green pepper	Good classification for data outside the training set; don't increase the computational complexity when mapping to high-dimensional space	Sensitive to the adjustment of the algorithm parameters and the selection of the kernel function
Techniques based on KNN clustering algorithm	85-90	blueberry, betel nut, oil palm, papaya, pomegranate	High classification accuracy; relative insensitivity to the abnormal data; no assumptions on the input data	Tedious in setting a reasonable scaling factor K; high time and space complexity; large computational effort
Techniques based on AdaBoost-based algorithm	78-96	grape, tomato	Make good use of weak classifiers for cascading; high detection and recognition accuracy	Vulnerable to noise interference; relying on weak classifiers that often have long training time
Techniques based on decision tree algorithm	80-85	grain; impurity of rice, kiwifruit, origin	Visually show the decision process of the whole decision problem at different stages in time and decision sequence	Easy to overfit and do not perform well when dealing with data that has relatively strong feature correlations
Techniques based on Bayesian algorithm	75-86	cherry, papaya, tomato	Simplicity of the recognition and classification process; ability to handle multiple classification tasks; better performance for small-scale data	Prior probabilities need to be calculated

Figure 10. Techniques based on traditional machine learning.

In general, compared to techniques based on digital image processing, techniques based on traditional machine learning have improved the speed, accuracy, and robustness of the detection and recognition of fruits and vegetables to different degrees. However, techniques based on traditional machine learning are sensitive to the inputs of abnormal data. Various parameters need to be set in advance before training, and the final classification effect is related to the setting of various parameters. Some parameters are also affected by changing lighting conditions, which make the tuning processes more complicated. At the same time, the current mainstream image segmentation and classifiers based on traditional machine learning are often solutions for specific scenes, so they usually lack generality. They are less effective for multiple classification problems, and are mainly applicable to the detection and recognition of a single species in greenhouse environments.

3.1. Techniques Based on K-Means Clustering Algorithm

The K-means clustering algorithm is a widely used unsupervised learning method. It can automatically classify input data into identical and different classes based on their fixed distances from each other. Techniques for object detection and recognition of fruits and vegetables based on the K-means clustering algorithm are widely used. Wang et al. [75] proposed a litchi detection and recognition algorithm based on K-means clustering. The research results showed that the method can be robust against the influence of changing lighting conditions. The highest average recognition rates of un-occluded and partially occluded litchi were 98.8% and 97.5%, respectively. Luo et al. [72] proposed a K-means clustering algorithm-based detection and recognition method for cutting points of double-overlapping grape clusters for harvesting robots in a complex vineyard environment. The recognition accuracy of the overlapping grape clusters was 88.33%. The success detection rate of the cutting points on the peduncles of double-overlapping grape clusters was 81.66%. Jiao et al. [70] also proposed a fast detection and localization method for overlapping apples based on K-means clustering and a local maximum algorithm.

In order to further resist the effect of changing lighting conditions, Wang et al. [74] improved the wavelet transform and used the K-means clustering algorithm to segment target images. The method not only accurately segments fruits with different colors, but also maintains high accuracy for the detection and recognition of fruits under changing lighting conditions.

In order to exclude the interference information in images as much as possible, Luo et al. [72] used the K-means clustering algorithm to obtain a complete closed target image region after segmentation, denoising, and filling operations on the captured image. To obtain more feature information of target fruits, Moallem et al. [73] applied the K-means clustering algorithm to the Cb component in YCbCr color space, and the defect segmentation was achieved using a Multi-Layer Perceptron (MLP) neural network. Then, statistical, textural, and geometric features from refined defected regions were extracted. Although the classification accuracy of this method is high, the weaknesses are obvious. First, the K-value must be given in advance, but it is difficult to do so. Second, the randomly selected K-centroids will have a large impact on the classification results.

In general, these techniques do not need to give labels, and can automatically classify target objects and backgrounds according to the fixed values between input data. Therefore, the advantages of techniques for object detection and recognition of fruits and vegetables based on the K-means clustering algorithm are short computation time, fast response time, and good clustering effect (especially when the clusters are dense and the differences are obvious). The disadvantages are that they are sensitive to abnormal data, and the randomly selected K-values have a large impact on the classification results.

3.2. Techniques Based on SVM Algorithm

The SVM algorithm is a widely used supervised learning method. It is commonly used in linear/nonlinear regression analysis and pattern classification. It achieves classification by solving the separated hyperplane that correctly partitions the training set and has the largest geometric separation. Techniques for object detection and recognition of fruits and vegetables based on the SVM algorithm are widely used. Bhargaval and Bansal [54], Patel and Chaudhari [78], Singh and Singh [82], and Moallem et al. [73] compared the performance of different classifiers (SVM, KNN, etc.) for the detection and recognition of different fruits and vegetables. The research results showed that, in their studies, the SVM classifier performs better than the other classifiers.

To improve the cooperative capability of fruit and vegetable harvesting robots, Sepúlveda et al. [77] implemented a cooperative operation between the arms of a two-armed eggplant harvesting robot based on the SVM algorithm. To address the problems of local occlusions, irregular shapes, and high similarity to backgrounds, Ji et al. [81] proposed a green pepper recognition method based on a least-squares support vector machine optimized by the improved particle swarm optimization (IPSO-LSSVM). The research results showed that the recognition rate of green peppers was 89.04%, and the average recognition time was 320 ms. This approach meets the requirements of accuracy and time of greenhouse green pepper harvesting robots.

To further improve the accuracy, speed, and robustness of detection and recognition of fruits and vegetables, Yang et al. [80] also proposed an image segmentation method for Hangzhou white chrysanthemum based on the least-square support vector machine (LS-SVM). The research results showed that the trained LS-SVM model and SVM model could effectively segment the images of Hangzhou white chrysanthemum from complicated backgrounds under three lighting conditions, namely, front lighting, back lighting, and overshadowing, with an accuracy of above 90%. When segmenting an image, the SVM algorithm required 1.3 s, while the proposed LS-SVM algorithm needed just 0.7 s. In addition, the implementation of the proposed segmentation algorithm on the harvesting robot achieved an 81% harvesting success rate.

In general, the advantages of techniques for object detection and recognition of fruits and vegetables based on SVM algorithm are that they simplify classification and regression problems, and can achieve good classification for the data outside the training set. At the same time, they can solve the problem of small samples of target fruits in natural environments, and do not increase the computational complexity when mapping to high-dimensional space. Therefore, the segmentation of fruit and vegetable images containing many high light points can be effectively realized by these techniques. The disadvantages are that they are too sensitive to the adjustment of the algorithm parameters and the selection of the kernel function. The kernel function and its parameters must be reselected for a new dataset. In addition, the accuracy is only high for binary classification tasks, but less effective for multi-classification problems.

3.3. *Technique Based on KNN Clustering Algorithm*

The KNN clustering algorithm is a widely used supervised learning method. It is commonly used in classification and regression models. It achieves classification by classifying unknown feature vectors into classes of the most common attributes of the K nearest neighbors using the training set. Techniques for object detection and recognition of fruits and vegetables based on the KNN clustering algorithm are more widely used. Based on the KNN clustering algorithm, Tan et al. [36], Astuti et al. [90], Suban et al. [89], Sarimole and Rosiana [85], and Sarimole and Fadillah [86] detected and recognized the ripeness of blueberries, oil palms, papayas, betel nuts, and pomegranates, respectively.

Tanco et al. [91] studied the detection and recognition of fruits and vegetables using three types of classifiers (SVM, KNN, and decision tree). The research results showed that the KNN clustering algorithm produced the best detection and recognition results. Ghazal et al. [88] trained and tested six supervised machine learning methods (SVM, KNN, decision tree, Bayesian, Linear Discriminant Analysis, and feed-forward back propagation neural network) on a publicly available Fruits 360 dataset. The research results showed that the methods based on the KNN clustering algorithm achieve relatively high classification accuracy.

In general, techniques based on the KNN clustering algorithm are able to classify the K nearest neighbors using functions to measure the distance between different eigenvalues. The advantages of techniques based on the KNN clustering algorithm are high classification accuracy, relative insensitivity to abnormal data, and no assumptions about input data. However, it is tedious to set a reasonable scaling factor of K in these methods. With a small value of K, the model complexity is high, overfitting is likely to occur, the estimation error of learning increases, and the prediction results are very sensitive to the instance points of the nearest neighbors. With a larger value of K, the complexity of the model and the estimation error of learning decreases, which is suitable for classification of a small dataset, but the approximation error of learning increases. The disadvantages are large computational effort, and high time and space complexity. Moreover, the detection and recognition accuracy of fruits and vegetables are easily affected by the growth environments and lighting conditions.

3.4. Techniques Based on AdaBoost Algorithm

The AdaBoost algorithm is a widely used supervised learning method. It is commonly used in two-class problems, multi-class single-label problems, multi-class multi-label problems, large-class single-label problems, and regression problems. Different classifiers (weak classifiers) are trained using the same training set, and then these weak classifiers are pooled to form a stronger final classifier (strong classifier). Techniques for object detection and recognition of fruits and vegetables based on the AdaBoost algorithm are widely used. Kumar et al. [93] introduced a novel plant species classifier based on the extraction of morphological features using a Multilayer Perceptron with the AdaBoost algorithm. In addition, they tested the classification accuracy of different classifiers, such as KNN, decision tree, and the Multilayer Perceptron. The research results showed that a precision rate of 95.42% was achieved using the proposed machine learning classifier, which is one of the state-of-the-art algorithms.

Ling et al. [94] proposed a tomato detection method combining an AdaBoost classifier and color analysis, and applied them to the harvesting robot. The research results showed that the ripe tomato detection success rate was about 95%., and 5% of the ripe tomatoes missed detection because of the occluding leaves. When the leaf occlusion area is more than 50% of the tomato area, the target tomato might not be detected. The method also has good robustness, and can meet the challenges of environmental factors such as changing lighting conditions and partial occlusions and overlaps. The speed of the method is about 10 fps, which is enough for the harvesting robot to operate in real time.

To further cope with challenges such as changing lighting conditions, cluttered backgrounds, and cluster occlusions, Lin and Zou [62] also proposed a novel segmentation method using the AdaBoost classifier and texture–color features. The research results showed that the method achieved a precision of 0.867 and recall of 0.768. However, the method may over-segment images because the LM filter bank tends to be influenced by illumination changes. A possible solution is to investigate an illumination invariant version of an LM filter bank.

In general, the advantages of techniques for object detection and recognition of fruits and vegetables based on the AdaBoost algorithm are that they can use different classification algorithms as weak classifiers and make good use of weak classifiers for cascading, with high detection and recognition accuracy. The disadvantages are that during the training process, the AdaBoost algorithm will cause the weight of difficult samples to exponentially increase, and the training will be biased towards such difficult samples, which makes the AdaBoost algorithm vulnerable to noise interference. In addition, the AdaBoost algorithm relies on weak classifiers, which often have a long training time.

3.5. Techniques Based on Decision Tree Algorithm

The decision tree algorithm is a widely used supervised learning method. It is commonly used in decision-making problems. It starts from the root node. Then, the corresponding features in the item to be classified are tested and the output branches are selected according to their values until the leaf node is reached. Finally, the category stored in the leaf node is used as the decision result. Wajid et al. [105] investigated the applicability and performance of various classification algorithms including Naïve Bayes, Artificial Neural Networks, and decision trees. The research results showed that the decision tree classification method performs better than the other methods for orange detection. The results recorded for the accuracy, precision, and sensitivity using this method were 93.13%, 93.45%, and 93.24%, respectively. In addition, in order to investigate the cost of implementation relative to the classification performance, Kuang, et al. [103] compared two types of machine learning algorithms (the multivariate alternating decision tree and the deep-learning-based kiwifruit classifiers). The research results showed that traditional decision tree classifiers can achieve comparable classification performance at a fraction of the cost.

Ma et al. [104] proposed a segmentation method based on a decision tree which is constructed by a two-step coarse-to-fine procedure. Firstly, a coarse decision tree is built by the CART (Classification and Regression Tree) algorithm with a feature subset. The feature subset consists of color features that are selected by Pearson's Rank correlations. Then, the coarse decision tree is optimized by pruning. Using the optimized decision tree, segmentation of images is achieved by conducting pixel-wise classification. Abd al karim and Karim [100] also proposed a decision tree classifier to classify fruit types. The Fruits 360 dataset was used, where 70% of the dataset was used in the training phase and 30% was used in the testing phase. Chen et al. [102] proposed a classification method for kernel and impurity particles using the decision tree algorithm.

In general, the advantages of techniques for object detection and recognition of fruits and vegetables based on the decision tree algorithm are that they enumerate the full range of feasible solutions to the decision problem, and the expected values of each feasible solution in various states. They can visually show the decision process of the whole decision problem at different stages in time and in the decision sequence. When applied to a complex multi-stage decision-making problem, the stages are obvious and the hierarchy is clear, so that various factors can be thoughtfully considered, which is conducive to making the right decision. The disadvantages are that they are easy to overfit and do not perform well when dealing with data that has relatively strong feature correlations. In addition, for data with an inconsistent number of samples in each category, the result gained in the decision tree is biased towards those features with more values.

3.6. Techniques Based on Bayesian Algorithm

The Bayesian algorithm is a widely used supervised learning method. It classifies based on minimizing Bayesian risk, minimizing probability of error, or maximizing posterior probability. It is commonly used in large-scale databases. The Bayesian algorithm was proposed because it has high accuracy and computational speed when applied to a large number of databases, is robust to isolated noise points, and only requires a small training set to estimate the parameters needed for classification.

Kusuma and Setiadi [113] proposed a classification method using feature histogram extraction and a Naïve Bayes Classifier for tomato recognition. In addition, Sari, et al. [110] proposed a classification method for papaya types based on leaf images using a Naive Bayes classifier and LBP feature extraction. In the research of Reyes et al. [108], the method based on the Bayesian algorithm, along with the off-the-shelf hardware, made it possible to perform an optimal classification of cherries in real time to meet international fruit quality standards.

In general, the advantages of techniques for object detection and recognition of fruits and vegetables based on Bayesian algorithms are the simplicity of recognition and classification processes, the fast response time, the better performance for small-scale data, the ability to handle multiple classification tasks, and the suitability for incremental training. The disadvantage is that the prior probabilities need to be calculated. Furthermore, the recognition performance is affected by the fact that the prior probabilities depend on the target image features. In addition, the recognition function may fail for data (variable features) that do not appear in the training set.

4. Challenges and Further Research

As summarized and reviewed in this article, various techniques for object detection and recognition of fruits and vegetables, each with their own pros and cons, have been investigated in the past. However, it is difficult to find studies reporting the absolute accuracy of each technique and comparisons of performance between those techniques in the same environment.

Therefore, open publishing of all reference datasets and all code is necessary. Some frequently used image databases of fruits and vegetables are shown in Table 4. As much as possible, further research should be carried out based on these open datasets to help compare different techniques. Moreover, the international community might consider continually providing and updating quality reference datasets.

Table 4. Some frequently used image databases of crops: fruits and vegetables.

Datasets	Samples			Species	Web-Link	Year
	Total	Training Sets	Testing Sets			
Fruits-360	90,380	67,692	22,688	131 (100 × 100 pixels)	https://www.kaggle.com/datasets/moltean/fruits (accessed on 16 February 2023)	2020
Fruit-A	22,495	16,854	5641	33 (100 × 100 pixels)	https://www.kaggle.com/datasets/sshikamaru/fruit-recognition (accessed on 16 February 2023)	2022
Fruit-B	21,000	15,000	vail: 3000 text: 3000	15 (224 × 224 pixels)	https://www.kaggle.com/datasets/misrakahmed/vegetable-image-dataset (accessed on 16 February 2023)	2021
Fruit quality classification	19,526	-	-	18 (256 × 256/192 pixels)	https://www.kaggle.com/datasets/ryandpark/fruit-quality-classification (accessed on 16 February 2023)	2022
Fresh and rotten fruits	13,599	10,901	2698	6	https://www.kaggle.com/datasets/sriramr/fruits-fresh-and-rotten-for-classification (accessed on 16 February 2023)	2019
Lemon quality control dataset	2533	-	-	3 (256 × 256 pixels)	https://github.com/robotduinom/lemon_dataset (accessed on 16 February 2023)	2022
Pistachio	2148	-	-	2	https://www.muratkoklu.com/datasets/ (accessed on 16 February 2023)	2022
Grapevine leaves dataset	500	-	-	5		2022
Apple	1300	1000	300	2	https://data.nal.usda.gov/search/type/dataset (accessed on 16 February 2023)	2020
Cauliflower	656	-	-	4	https://www.kaggle.com/datasets/noamaanabdulazeem/cauliflower-dataset (accessed on 16 February 2023)	2022
Sweet pepper and peduncle segmentation	620	-	-	8	https://www.kaggle.com/datasets/lemontyc/sweet-pepper (accessed on 16 February 2023)	2021

In addition, there are many factors leading to the low accuracy, slow speed, and poor robustness of object detection and recognition of fruit and vegetable harvesting robots. They can be summarized into the following aspects: (1) similar backgrounds; (2) clustered/partially occluded/swaying fruits; (3) sensitivity to changing lighting conditions; (4) night image recognition; (5) blur and noise in images; (6) high computation time and real-time limitations; and (7) generalization ability. To be more specific:

(1) Object detection and recognition of fruits and vegetables require fast response capability to improve the harvesting efficiency. The current mainstream object detection

and recognition techniques based on digital image processing and traditional machine learning have certain limitations, although they may have good accuracy performance. In complex environments, influenced by many factors such as changing lighting conditions and growth states of fruits, the more factors the method considers, the more complex the method, and the longer the running computation time. This will lead to low real-time performance for vision systems.

(2) When fruit and vegetable harvesting robots work, they can only detect and recognize the target objects according to the pre-trained model. In the actual harvesting process, there is often more than one kind of target object that needs to be harvested. In addition, the harvesting robots are only used during the harvesting season of the year, and are idle for the rest of the year, due to the obvious seasonality and timeliness of fruit harvesting, thus leading to the relatively poorer economics of harvesting robots. Therefore, the generalization ability of the algorithms still needs to be enhanced to achieve the detection and recognition of multiple kinds of fruits and vegetables. Future research could make the algorithms generalizable (i.e., derive the ability to recognize fruits with similar characteristics based on a kind of target object). In addition, the night image recognition algorithm could be required for vision systems, where the harvesting robots can work during the day, and then continue at night.

(3) Object detection and recognition of fruits and vegetables require the detection and recognition of clustered/partially occluded/swaying fruits. However, the presence of clustered/partially occluded/swaying parts may cause confusion in images, which is currently a greater challenge for detection and recognition in unstructured environments. A popular method is the Circular Hough Transform, which is more effective for round objects such as apples, oranges, and tomatoes. However, research results showed that this method is not only prone to false positives generated by the contours of other objects, such as leaves, but also has a long computation time. Another popular method is to use a blowing device to avoid leaf occlusions and to move adjacent fruits to one side. However, this method will increase the weight of end-effectors of harvesting robots, and may not be applicable to all kinds of crops. Future research could focus on agricultural operations, including tree pruning and pollination methods, to improve the visibility of target fruits, which may help to improve detection and recognition accuracy.

As summarized and reviewed in this article, methods based on multi-feature fusion and the SVM algorithm achieve a better accuracy rate in addressing these challenges. Furthermore, methods based on multi-algorithm fusion should be paid more attention. In addition, further research should focus on solving these challenges and improving the accuracy, speed, robustness, and generalization of vision systems, while reducing the overall complexity and cost. The optimization of network models, the accuracy of sensing systems, multi-sensor data fusion, fault-tolerant computing of machine vision, and decision making using a big data cloud platform may be key breakthroughs for further techniques for object detection and recognition of fruits and vegetables.

5. Conclusions

The intelligent harvesting robot is one of the most important artificial intelligence (AI) robots used for fruit and vegetable harvesting in modern agriculture. The excellent vision system can greatly promote the environmental perception ability of the harvesting robot. However, current visual systems of harvesting robots still cannot fully meet the requirements of commercialization. This article summarizes and reviews the progress in developing techniques for object detection and recognition of fruit and vegetable harvesting robots based on digital image processing and traditional machine learning. Although there previous reviews of techniques for object detection and recognition of fruits and vegetables have been published, the contributions of this work are: (1) systematic summary of the techniques developed in recent years for object detection and recognition of fruit and vegetable harvesting robots based on digital image processing and traditional machine learning; (2) systematic analysis of the advantages, disadvantages, and applicability of various techniques; and (3) demonstration of the current challenges and future potential developments. Through this clearer and more comprehensive overview and review, we aim to provide a reference for future research on techniques for object detection and recognition of fruit and vegetable harvesting robots based on digital image processing and traditional machine learning.

The current challenges of techniques for object detection and recognition of fruits and vegetables are mainly the similar backgrounds, clustered/partially occluded/swaying fruits, sensitivity to changing lighting conditions, night image recognition, blur and noise in images, high computation time and real-time limitations, and generalization ability.

Techniques for object detection and recognition of fruit and vegetable harvesting robots based on digital image processing can be subdivided into color feature (RGB, HSV, HSI, Lab, HSB, YCbCr)-based methods, shape feature-based methods, texture feature-based methods, and multi-feature fusion-based methods.

As summarized and reviewed in this article, techniques based on digital image processing require precise information about the target fruit features, which are usually used for object detection and recognition of fruits and vegetables based on features such as colors, shapes, and textures. However, in complex environments, these features of the target objects are affected by non-controllable factors, resulting in low accuracy, slow speed, and poor robustness of object detection and recognition of fruits and vegetables. Methods based on multi-feature fusion can improve the accuracy and robustness of object detection and recognition of fruits and vegetables. However, it is important to determine which features to integrate; for example, Lab color space could be used more for low-quality images because it is more robust to noise in images. In addition, the combination of color with GLCM texture features has proven to be more effective than combined color and LBP texture features.

Object detection and recognition techniques of fruit and vegetable harvesting robots based on traditional machine learning can be subdivided into K-means clustering algorithm-based methods, SVM algorithm-based methods, KNN clustering algorithm-based methods, AdaBoost algorithm-based methods, decision tree algorithm-based methods, and Bayesian algorithm-based methods.

In general, techniques based on traditional machine learning have good performance, but they require various parameters to be set in advance, where the parameters set in advance have a large impact on recognition accuracy. For classifiers, prior probabilities from the training set need to be obtained in advance, and the classification accuracy is affected by the weights of difficult to classify samples. As summarized and reviewed in this article, methods based on the SVM algorithm achieve a better accuracy rate. However, the current mainstream image segmentation approaches and classifiers based on traditional machine learning are often solutions for specific scenes. They usually lack generality and are less effective for multiple classification problems. They are mainly applicable to the detection and recognition of a single species in greenhouse environments. Methods based on multi-algorithm fusion should be paid more attention. This may be a breakthrough for future techniques for object detection and recognition of fruits and vegetables.

Further research into and development of techniques for object detection and recognition for fruit and vegetable harvesting robots are necessary. Commercial applications of harvesting robots need to be further addressed through integrated horticultural and engineering approaches for improved image segmentation, and for increased overall performance of crop detection and recognition.

Author Contributions: Conceptualization, F.X. and Y.C.; methodology, F.X. and Y.C.; analysis, F.X.; investigation, F.X., Y.C., X.L., G.X. and H.W.; resources, F.X., H.W. and Y.L.; data curation, F.X.; writing—original draft preparation, F.X.; writing—review and editing, F.X., H.W. and Y.L.; visualization, F.X.; supervision, H.W. and Y.L.; project administration, F.X., H.W. and Y.L.; funding acquisition, H.W. and Y.L. All authors have read and agreed to the published version of the manuscript.

Funding: This research was funded by Natural Science Foundation of Heilongjiang Province of China (LH2020C047), Northeast Forestry University Foundation (2572022DP01) and China Postdoctoral Science Foundation (2019T120248, 2017M611338).

Data Availability Statement: Not applicable.

Conflicts of Interest: The authors declare no conflict of interest.

References

1. Tomato Production. 2020. Available online: https://ourworldindata.org/grapher/tomato-production (accessed on 1 October 2022).
2. FAOSTAT. Available online: https://www.fao.org/faostat/en/#data/QCL (accessed on 1 October 2022).
3. Apple Production. 2020. Available online: https://ourworldindata.org/grapher/apple-production (accessed on 1 October 2022).
4. Strawberry—Wikipedia. Available online: https://en.wikipedia.org/wiki/Strawberry (accessed on 1 October 2022).
5. Zhang, K.; Lammers, K.; Chu, P.; Li, Z.; Lu, R. System Design and Control of an Apple Harvesting Robot. *Mechatronics* **2021**, *79*, 102644. [CrossRef]
6. Xiong, Y.; Ge, Y.; Grimstad, L.; From, P.J. An Autonomous Strawberry-Harvesting Robot: Design, Development, Integration, and Field Evaluation. *J. Field Robot.* **2020**, *37*, 202–224. [CrossRef]
7. Xiong, Y.; Peng, C.; Grimstad, L.; From, P.J.; Isler, V. Development and Field Evaluation of a Strawberry Harvesting Robot with a Cable-Driven Gripper. *Comput. Electron. Agric.* **2019**, *157*, 392–402. [CrossRef]
8. Anjom, F.K.; Vougioukas, S.G.; Slaughter, D.C. Development of a Linear Mixed Model to Predict the Picking Time in Strawberry Harvesting Processes. *Biosyst. Eng.* **2018**, *166*, 76–89. [CrossRef]
9. Silwal, A.; Davidson, J.R.; Karkee, M.; Mo, C.; Zhang, Q.; Lewis, K. Design, Integration, and Field Evaluation of a Robotic Apple Harvester. *J. Field Robot.* **2017**, *34*, 1140–1159. [CrossRef]
10. Wang, Z.; Xun, Y.; Wang, Y.; Yang, Q. Review of Smart Robots for Fruit and Vegetable Picking in Agriculture. *Int. J. Agric. Biol. Eng.* **2022**, *15*, 33–54. [CrossRef]
11. Zhou, H.; Wang, X.; Au, W.; Kang, H.; Chen, C. Intelligent Robots for Fruit Harvesting: Recent Developments and Future Challenges. *Precis. Agric.* **2022**, *23*, 1856–1907. [CrossRef]
12. Brown, J.; Sukkarieh, S. Design and Evaluation of a Modular Robotic Plum Harvesting System Utilizing Soft Components. *J. Field Robot.* **2021**, *38*, 289–306. [CrossRef]
13. Yan, B.; Fan, P.; Lei, X.; Liu, Z.; Yang, F. A Real-Time Apple Targets Detection Method for Picking Robot Based on Improved YOLOv5. *Remote Sens.* **2021**, *13*, 1619. [CrossRef]
14. He, L.; Fu, H.; Karkee, M.; Zhang, Q. Effect of Fruit Location on Apple Detachment with Mechanical Shaking. *Biosyst. Eng.* **2017**, *157*, 63–71. [CrossRef]
15. Ji, W.; Zhao, D.; Cheng, F.; Xu, B.; Zhang, Y.; Wang, J. Automatic Recognition Vision System Guided for Apple Harvesting Robot. *Comput. Electr. Eng.* **2012**, *38*, 1186–1195. [CrossRef]
16. Zhao, D.; Lv, J.; Ji, W.; Zhang, Y.; Chen, Y. Design and Control of an Apple Harvesting Robot. *Biosyst. Eng.* **2011**, *110*, 112–122. [CrossRef]
17. Arad, B.; Balendonck, J.; Barth, R.; Ben-Shahar, O.; Edan, Y.; Hellström, T.; Hemming, J.; Kurtser, P.; Ringdahl, O.; Tielen, T.; et al. Development of a Sweet Pepper Harvesting Robot. *J. Field Robot.* **2020**, *37*, 1027–1039. [CrossRef]
18. Lehnert, C.; English, A.; McCool, C.; Tow, A.W.; Perez, T. Autonomous Sweet Pepper Harvesting for Protected Cropping Systems. *IEEE Robot. Autom. Lett.* **2017**, *2*, 872–879. [CrossRef]
19. Bac, C.W.; Hemming, J.; Van Henten, E.J. Stem Localization of Sweet-Pepper Plants Using the Support Wire as a Visual Cue. *Comput. Electron. Agric.* **2014**, *105*, 111–120. [CrossRef]
20. Hayashi, S.; Shigematsu, K.; Yamamoto, S.; Kobayashi, K.; Kohno, Y.; Kamata, J.; Kurita, M. Evaluation of a Strawberry-Harvesting Robot in a Field Test. *Biosyst. Eng.* **2010**, *105*, 160–171. [CrossRef]
21. Xiong, J.; He, Z.; Lin, R.; Liu, Z.; Bu, R.; Yang, Z.; Peng, H.; Zou, X. Visual Positioning Technology of Picking Robots for Dynamic Litchi Clusters with Disturbance. *Comput. Electron. Agric.* **2018**, *151*, 226–237. [CrossRef]
22. Feng, Q.; Zou, W.; Fan, P.; Zhang, C.; Wang, X. Design and Test of Robotic Harvesting System for Cherry Tomato. *Int. J. Agric. Biol. Eng.* **2018**, *11*, 96–100. [CrossRef]
23. Kondo, N.; Yata, K.; Iida, M.; Shiigi, T.; Monta, M.; Kurita, M.; Omori, H. Development of an End-Effector for a Tomato Cluster Harvesting Robot. *Eng. Agric. Environ. Food* **2010**, *3*, 20–24. [CrossRef]
24. Williams, H.A.M.; Jones, M.H.; Nejati, M.; Seabright, M.J.; Bell, J.; Penhall, N.D.; Barnett, J.J.; Duke, M.D.; Scarfe, A.J.; Ahn, H.S.; et al. Robotic Kiwifruit Harvesting Using Machine Vision, Convolutional Neural Networks, and Robotic Arms. *Biosyst. Eng.* **2019**, *181*, 140–156. [CrossRef]
25. Yang, Q.; Chen, C.; Dai, J.; Xun, Y.; Bao, G. Tracking and Recognition Algorithm for a Robot Harvesting Oscillating Apples. *Int. J. Agric. Biol. Eng.* **2020**, *13*, 163–170. [CrossRef]
26. Zemmour, E.; Kurtser, P.; Edan, Y. Automatic Parameter Tuning for Adaptive Thresholding in Fruit Detection. *Sensors* **2019**, *19*, 2130. [CrossRef]
27. Lv, J.; Wang, Y.; Xu, L.; Gu, Y.; Zou, L.; Yang, B.; Ma, Z. A Method to Obtain the Near-Large Fruit from Apple Image in Orchard for Single-Arm Apple Harvesting Robot. *Sci. Hortic.* **2019**, *257*, 108758. [CrossRef]

28. Goel, N.; Sehgal, P. Fuzzy Classification of Pre-Harvest Tomatoes for Ripeness Estimation—An Approach Based on Automatic Rule Learning Using Decision Tree. *Appl. Soft Comput.* **2015**, *36*, 45–56. [CrossRef]
29. Yu, X.; Fan, Z.; Wang, X.; Wan, H.; Wang, P.; Zeng, X.; Jia, F. A Lab-Customized Autonomous Humanoid Apple Harvesting Robot. *Comput. Electr. Eng.* **2021**, *96*, 107459. [CrossRef]
30. Sethy, P.K.; Routray, B.; Behera, S.K. Detection and Counting of Marigold Flower Using Image Processing Technique. In *Advances in Computer, Communication and Control*, 2nd ed.; Biswas, U., Banerjee, A., Pal, S., Biswas, A., Sarkar, D., Haldar, S., Eds.; Springer: Singapore, 2019; Volume 41, pp. 87–93. [CrossRef]
31. Malik, M.H.; Zhang, T.; Li, H.; Zhang, M.; Shabbir, S.; Saeed, A. Mature Tomato Fruit Detection Algorithm Based on Improved HSV and Watershed Algorithm. *IFAC-Paper* **2018**, *51*, 431–436. [CrossRef]
32. Muthukrishnan, V.; Ramasamy, S.; Damodaran, N. Disease Recognition in Philodendron Leaf Using Image Processing Technique. *Environ. Sci. Pollut. Res.* **2021**, *28*, 67321–67330. [CrossRef]
33. Nanehkaran, Y.A.; Zhang, D.; Chen, J.; Tian, Y.; Al-Nabhan, N. Recognition of Plant Leaf Diseases Based on Computer Vision. *J. Ambient. Intell. Humaniz. Comput.* **2020**, 1–18. [CrossRef]
34. Ratprakhon, K.; Neubauer, W.; Riehn, K.; Fritsche, J.; Rohn, S. Developing an Automatic Color Determination Procedure for the Quality Assessment of Mangos (Mangifera Indica) Using a CCD Camera and Color Standards. *Foods* **2020**, *9*, 1709. [CrossRef] [PubMed]
35. Biffi, L.J.; Mitishita, E.A.; Liesenberg, V.; Centeno, J.A.S.; Schimalski, M.B.; Rufato, L. Evaluating the Performance of a Semi-Automatic Apple Fruit Detection in a High-Density Orchard System Using Low-Cost Digital RGB Imaging Sensor. *Bull. Geod. Sci.* **2021**, *27*, 1–20. [CrossRef]
36. Tan, K.; Lee, W.S.; Gan, H.; Wang, S. Recognising Blueberry Fruit of Different Maturity Using Histogram Oriented Gradients and Colour Features in Outdoor Scenes. *Biosyst. Eng.* **2018**, *176*, 59–72. [CrossRef]
37. Lin, G.; Tang, Y.; Zou, X.; Cheng, J.; Xiong, J. Fruit Detection in Natural Environment Using Partial Shape Matching and Probabilistic Hough Transform. *Precis. Agric.* **2020**, *21*, 160–177. [CrossRef]
38. Sun, S.; Jiang, M.; He, D.; Long, Y.; Song, H. Recognition of Green Apples in an Orchard Environment by Combining the GrabCut Model and Ncut Algorithm. *Biosyst. Eng.* **2019**, *187*, 201–213. [CrossRef]
39. Lu, J.; Lee, W.S.; Gan, H.; Hu, X. Immature Citrus Fruit Detection Based on Local Binary Pattern Feature and Hierarchical Contour Analysis. *Biosyst. Eng.* **2018**, *171*, 78–90. [CrossRef]
40. Zhuang, J.J.; Luo, S.M.; Hou, C.J.; Tang, Y.; He, Y.; Xue, X.Y. Detection of Orchard Citrus Fruits Using a Monocular Machine Vision-Based Method for Automatic Fruit Picking Applications. *Comput. Electron. Agric.* **2018**, *152*, 64–73. [CrossRef]
41. Oo, L.M.; Aung, N.Z. A Simple and Efficient Method for Automatic Strawberry Shape and Size Estimation and Classification. *Biosyst. Eng.* **2018**, *170*, 96–107. [CrossRef]
42. Jana, S.; Parekh, R. Shape-Based Fruit Recognition and Classification. In Proceedings of the International Conference on Computational Intelligence, Communications, and Business Analytics, Kolkata, India, 24–25 March 2017. [CrossRef]
43. Linker, R.; Cohen, O.; Naor, A. Determination of the Number of Green Apples in RGB Images Recorded in Orchards. *Comput. Electron. Agric.* **2012**, *81*, 45–47. [CrossRef]
44. Kurtulmus, F.; Lee, W.S.; Vardar, A. Green Citrus Detection Using 'Eigenfruit', Color and Circular Gabor Texture Features under Natural Outdoor Conditions. *Comput. Electron. Agric.* **2011**, *78*, 140–149. [CrossRef]
45. Hannan, M.W.; Burks, T.F.; Bulanon, D.M. A Machine Vision Algorithm Combining Adaptive Segmentation and Shape Analysis for Orange Fruit Detection. *Agric. Eng. Int. CIGR J.* **2009**, *XI*, 1281.
46. Safren, O.; Alchanatis, V.; Ostrovsky, V.; Levi, O. Detection of Green Apples in Hyperspectral Images of Apple-Tree Foliage Using Machine Vision. *Trans. Am. Soc. Agric. Biol. Eng.* **2007**, *50*, 2303–2313. [CrossRef]
47. Rahman, S.U.; Alam, F.; Ahmad, N.; Arshad, S. Image Processing Based System for the Detection, Identification and Treatment of Tomato Leaf Diseases. *Multimed. Tools Appl.* **2022**, *82*, 9431–9445. [CrossRef]
48. Hameed, K.; Chai, D.; Rassau, A. Texture-Based Latent Space Disentanglement for Enhancement of a Training Dataset for ANN-Based Classification of Fruit and Vegetables. *Inf. Process. Agric.* **2021**, *10*, 85–105. [CrossRef]
49. Trey, Z.F.; Goore, B.T.; Bagui, K.O.; Tiebre, M.S. Classification of Plants into Families Based on Leaf Texture. *Int. J. Comput. Sci. Netw. Secur.* **2021**, *21*, 205–211. [CrossRef]
50. Pulido, C.; Solaque, L.; Velasco, N. Weed Recognition by SVM Texture Feature Classification in Outdoor Vegetable Crops Images. *Ing. E Investig.* **2017**, *37*, 68–74. [CrossRef]
51. Chaivivatrakul, S.; Dailey, M.N. Texture-Based Fruit Detection. *Precis. Agric.* **2014**, *15*, 662–683. [CrossRef]
52. Rakun, J.; Stajnko, D.; Zazula, D. Detecting Fruits in Natural Scenes by Using Spatial-Frequency Based Texture Analysis and Multiview Geometry. *Comput. Electron. Agric.* **2011**, *76*, 80–88. [CrossRef]
53. Septiarini, A.; Sunyoto, A.; Hamdani, H.; Kasim, A.A.; Utaminingrum, F.; Hatta, H.R. Machine Vision for the Maturity Classification of Oil Palm Fresh Fruit Bunches Based on Color and Texture Features. *Sci. Hortic.* **2021**, *286*, 110245. [CrossRef]
54. Bhargava, A.; Bansal, A. Classification and Grading of Multiple Varieties of Apple Fruit. *Food Anal. Methods* **2021**, *14*, 1359–1368. [CrossRef]
55. Yu, L.; Xiong, J.; Fang, X.; Yang, Z.; Chen, Y.; Lin, X.; Chen, S. A Litchi Fruit Recognition Method in a Natural Environment Using RGB-D Images. *Biosyst. Eng.* **2021**, *204*, 50–63. [CrossRef]
56. Basavaiah, J.; Anthony, A.A. Tomato Leaf Disease Classification Using Multiple Feature Extraction Techniques. *Wirel. Pers. Commun.* **2020**, *115*, 633–651. [CrossRef]

57. Azarmdel, H.; Jahanbakhshi, A.; Mohtasebi, S.S.; Muñoz, A.R. Evaluation of Image Processing Technique as an Expert System in Mulberry Fruit Grading Based on Ripeness Level Using Artificial Neural Networks (ANNs) and Support Vector Machine (SVM). *Postharvest Biol. Technol.* **2020**, *166*, 111201. [CrossRef]
58. Liu, T.; Ehsani, R.; Toudeshki, A.; Zou, X.; Wang, H. Identifying Immature and Mature Pomelo Fruits in Trees by Elliptical Model Fitting in the Cr–Cb Color Space. *Precis. Agric.* **2019**, *20*, 138–156. [CrossRef]
59. Wu, J.; Zhang, B.; Zhou, J.; Xiong, Y.; Gu, B.; Yang, X. Automatic Recognition of Ripening Tomatoes by Combining Multi-Feature Fusion with a Bi-Layer Classification Strategy for Harvesting Robots. *Sensors* **2019**, *19*, 612. [CrossRef] [PubMed]
60. Liu, X.; Zhao, D.; Jia, W.; Ji, W.; Sun, Y. A Detection Method for Apple Fruits Based on Color and Shape Features. *IEEE Access* **2019**, *7*, 67923–67933. [CrossRef]
61. Mustaffa, M.R.; Yi, N.X.; Abdullah, L.N.; Nasharuddin, N.A. Durian Recognition Based on Multiple Features and Linear Discriminant Analysis. *Malays. J. Comput. Sci.* **2018**, 57–72. [CrossRef]
62. Lin, G.; Zou, X. Citrus Segmentation for Automatic Harvester Combined with AdaBoost Classifier and Leung-Malik Filter Bank. *IFAC-Paper* **2018**, *51*, 379–383. [CrossRef]
63. Madgi, M.; Danti, A. An Enhanced Classification of Indian Vegetables Using Combined Color and Texture Features. *Int. J. Comput. Eng. Appl.* **2018**, *XII(III)*, 1–8.
64. Yamamoto, K.; Guo, W.; Yoshioka, Y.; Ninomiya, S. On Plant Detection of Intact Tomato Fruits Using Image Analysis and Machine Learning Methods. *Sensors* **2014**, *14*, 12191–12206. [CrossRef]
65. Payne, A.; Walsh, K.; Subedi, P.; Jarvis, D. Estimating Mango Crop Yield Using Image Analysis Using Fruit at 'Stone Hardening' Stage and Night Time Imaging. *Comput. Electron. Agric.* **2014**, *100*, 160–167. [CrossRef]
66. Payne, A.B.; Walsh, K.B.; Subedi, P.P.; Jarvis, D. Estimation of Mango Crop Yield Using Image Analysis—Segmentation Method. *Comput. Electron. Agric.* **2013**, *91*, 57–64. [CrossRef]
67. Stajnko, D.; Rakun, J.; Blanke, M. Modelling Apple Fruit Yield Using Image Analysis for Fruit Colour, Shape and Texture. *Eur. J. Hortic. Sci.* **2009**, *74*, 260–267.
68. Fan, P.; Lang, G.; Guo, P.; Liu, Z.; Yang, F.; Yan, B.; Lei, X. Multi-Feature Patch-Based Segmentation Technique in the Gray-Centered RGB Color Space for Improved Apple Target Recognition. *Agriculture* **2021**, *11*, 273. [CrossRef]
69. Habib, M.T.; Majumder, A.; Jakaria, A.Z.M.; Akter, M.; Uddin, M.S.; Ahmed, F. Machine Vision Based Papaya Disease Recognition. *J. King Saud Univ. Comput. Inf. Sci.* **2020**, *32*, 300–309. [CrossRef]
70. Jiao, Y.; Luo, R.; Li, Q.; Deng, X.; Yin, X.; Ruan, C.; Jia, W. Detection and Localization of Overlapped Fruits Application in an Apple Harvesting Robot. *Electronics* **2020**, *9*, 1023. [CrossRef]
71. Sun, S.; Song, H.; He, D.; Long, Y. An Adaptive Segmentation Method Combining MSRCR and Mean Shift Algorithm with K-Means Correction of Green Apples in Natural Environment. *Inf. Process. Agric.* **2019**, *6*, 200–215. [CrossRef]
72. Luo, L.; Tang, Y.; Lu, Q.; Chen, X.; Zhang, P.; Zou, X. A Vision Methodology for Harvesting Robot to Detect Cutting Points on Peduncles of Double Overlapping Grape Clusters in a Vineyard. *Comput. Ind.* **2018**, *99*, 130–139. [CrossRef]
73. Moallem, P.; Serajoddin, A.; Pourghassem, H. Computer Vision-Based Apple Grading for Golden Delicious Apples Based on Surface Features. *Inf. Process. Agric.* **2017**, *4*, 33–40. [CrossRef]
74. Wang, C.; Tang, Y.; Zou, X.; SiTu, W.; Feng, W. A Robust Fruit Image Segmentation Algorithm against Varying Illumination for Vision System of Fruit Harvesting Robot. *Optik* **2017**, *131*, 626–631. [CrossRef]
75. Wang, C.; Zou, X.; Tang, Y.; Luo, L.; Feng, W. Localisation of Litchi in an Unstructured Environment Using Binocular Stereo Vision. *Biosyst. Eng.* **2016**, *145*, 39–51. [CrossRef]
76. Zhang, Z.; Zhou, J.; Yan, Z.; Wang, K.; Mao, J.; Jiang, Z. Hardness Recognition of Fruits and Vegetables Based on Tactile Array Information of Manipulator. *Comput. Electron. Agric.* **2021**, *181*, 105959. [CrossRef]
77. Sepúlveda, D.; Fernández, R.; Navas, E.; Armada, M.; González-De-Santos, P. Robotic Aubergine Harvesting Using Dual-Arm Manipulation. *IEEE Access* **2020**, *8*, 121889–121904. [CrossRef]
78. Patel, C.C.; Chaudhari, V.K. Comparative Analysis of Fruit Categorization Using Different Classifiers. *Adv. Eng. Optim. Through Intell. Tech.* **2020**, *949*, 153–164. [CrossRef]
79. Dhakshina Kumar, S.; Esakkirajan, S.; Bama, S.; Keerthiveena, B. A Microcontroller Based Machine Vision Approach for Tomato Grading and Sorting Using SVM Classifier. *Microprocess. Microsyst.* **2020**, *76*, 103090. [CrossRef]
80. Yang, Q.; Luo, S.; Chang, C.; Xun, Y.; Bao, G. Segmentation Algorithm for Hangzhou White Chrysanthemums Based on Least Squares Support Vector Machine. *Int. J. Agric. Biol. Eng.* **2019**, *12*, 127–134. [CrossRef]
81. Ji, W.; Chen, G.; Xu, B.; Meng, X.; Zhao, D. Recognition Method of Green Pepper in Greenhouse Based on Least-Squares Support Vector Machine Optimized by the Improved Particle Swarm Optimization. *IEEE Access* **2019**, *7*, 119742–119754. [CrossRef]
82. Singh, S.; Singh, N.P. Machine Learning-Based Classification of Good and Rotten Apple. In *Recent Trends in Communication, Computing, and Electronics*, 2nd ed.; Khare, A., Tiwary, U.S., Sethi, I.K., Singh, N., Eds.; Springer: Singapore, 2019; Volume 524, pp. 377–386. [CrossRef]
83. Liu, G.; Mao, S.; Kim, J.H. A Mature-Tomato Detection Algorithm Using Machine Learning and Color Analysis. *Sensors* **2019**, *19*, 2023. [CrossRef]
84. Lv, Q.; Cai, J.; Liu, B.; Deng, L.; Zhang, Y. Identification of Fruit and Branch in Natural Scenes for Citrus Harvesting Robot Using Machine Vision and Support Vector Machine. *Int. J. Agric. Biol. Eng.* **2014**, *7*, 115–121. [CrossRef]

85. Sarimole, F.M.; Rosiana, A. Classification of Maturity Levels in Areca Fruit Based on HSV Image Using the KNN Method. *J. Appl. Eng. Technol. Sci.* **2022**, *4*, 64–73. [CrossRef]
86. Sarimole, F.M.; Fadillah, M.I. Classification of Guarantee Fruit Murability Based on HSV Image With K-Nearest Neighbor. *J. Appl. Eng. Technol. Sci.* **2022**, *4*, 48–57. [CrossRef]
87. Behera, S.K.; Rath, A.K.; Sethy, P.K. Maturity Status Classification of Papaya Fruits Based on Machine Learning and Transfer Learning Approach. *Inf. Process. Agric.* **2021**, *8*, 244–250. [CrossRef]
88. Ghazal, S.; Qureshi, W.S.; Khan, U.S.; Iqbal, J.; Rashid, N.; Tiwana, M.I. Analysis of Visual Features and Classifiers for Fruit Classification Problem. *Comput. Electron. Agric.* **2021**, *187*, 106267. [CrossRef]
89. Suban, I.B.; Paramartha, A.; Fortwonatus, M.; Santoso, A.J. Identification the Maturity Level of Carica Papaya Using the K-Nearest Neighbor. In Proceedings of the International Conference on Electronics Representation and Algorithm "Innovation and Transformation for Best Practices in Global Community", Yogyakarta, Indonesia, 12–13 December 2019. [CrossRef]
90. Astuti, I.F.; Nuryanto, F.D.; Widagdo, P.P.; Cahyadi, D. Oil Palm Fruit Ripeness Detection Using K-Nearest Neighbour. In Proceedings of the International Conference on Mathematics, Science and Computer Science, Balikpapan, Indonesia, 24 October 2018. [CrossRef]
91. Tanco, M.M.; Tejera, G.; Martino, J.M.D. Computer Vision Based System for Apple Detection in Crops. In Proceedings of the International Joint Conference on Computer Vision, Imaging and Computer Graphics Theory and Applications, Funchal, Madeira, Portugal, 27–29 January 2018. [CrossRef]
92. Tu, S.; Pang, J.; Liu, H.; Zhuang, N.; Chen, Y.; Zheng, C.; Wan, H.; Xue, Y. Passion Fruit Detection and Counting Based on Multiple Scale Faster R-CNN Using RGB-D Images. *Precis. Agric.* **2020**, *21*, 1072–1091. [CrossRef]
93. Kumar, M.; Gupta, S.; Gao, X.Z.; Singh, A. Plant Species Recognition Using Morphological Features and Adaptive Boosting Methodology. *IEEE Access* **2019**, *7*, 163912–163918. [CrossRef]
94. Ling, X.; Zhao, Y.; Gong, L.; Liu, C.; Wang, T. Dual-Arm Cooperation and Implementing for Robotic Harvesting Tomato Using Binocular Vision. *Robot. Auton. Syst.* **2019**, *114*, 134–143. [CrossRef]
95. Fu, L.; Duan, J.; Zou, X.; Lin, G.; Song, S.; Ji, B.; Yang, Z. Banana Detection Based on Color and Texture Features in the Natural Environment. *Comput. Electron. Agric.* **2019**, *167*, 105057. [CrossRef]
96. Wang, C.; Lee, W.S.; Zou, X.; Choi, D.; Gan, H.; Diamond, J. Detection and Counting of Immature Green Citrus Fruit Based on the Local Binary Patterns (LBP) Feature Using Illumination-Normalized Images. *Precis. Agric.* **2018**, *19*, 1062–1083. [CrossRef]
97. Fernandes, A.; Utkin, A.; Eiras-Dias, J.; Silvestre, J.; Cunha, J.; Melo-Pinto, P. Assessment of Grapevine Variety Discrimination Using Stem Hyperspectral Data and AdaBoost of Random Weight Neural Networks. *Appl. Soft Comput.* **2018**, *72*, 140–155. [CrossRef]
98. Luo, L.; Tang, Y.; Zou, X.; Wang, C.; Zhang, P.; Feng, W. Robust Grape Cluster Detection in a Vineyard by Combining the AdaBoost Framework and Multiple Color Components. *Sensors* **2016**, *16*, 2098. [CrossRef]
99. Zhao, Y.; Gong, L.; Zhou, B.; Huang, Y.; Liu, C. Detecting Tomatoes in Greenhouse Scenes by Combining AdaBoost Classifier and Colour Analysis. *Biosyst. Eng.* **2016**, *148*, 127–137. [CrossRef]
100. Abd al karim, M.H.; Karim, A.A. Using Texture Feature in Fruit Classification. *Eng. Technol. J.* **2021**, *39*, 67–79. [CrossRef]
101. Abasi, S.; Minaei, S.; Jamshidi, B.; Fathi, D. Development of an Optical Smart Portable Instrument for Fruit Quality Detection. *IEEE Trans. Instrum. Meas.* **2021**, *70*, 1–9. [CrossRef]
102. Chen, J.; Lian, Y.; Li, Y. Real-Time Grain Impurity Sensing for Rice Combine Harvesters Using Image Processing and Decision-Tree Algorithm. *Comput. Electron. Agric.* **2020**, *175*, 105591. [CrossRef]
103. Kuang, Y.C.; Streeter, L.; Cree, M.J.; Ooi, M.P.L. Evaluation of Deep Neural Network and Alternating Decision Tree for Kiwifruit Detection. In Proceedings of the IEEE International Instrumentation and Measurement Technology Conference, Auckland, New Zealand, 20–23 May 2019. [CrossRef]
104. Ma, J.; Du, K.; Zheng, F.; Zhang, L.; Sun, Z. A Segmentation Method for Processing Greenhouse Vegetable Foliar Disease Symptom Images. *Inf. Process. Agric.* **2019**, *6*, 216–223. [CrossRef]
105. Wajid, A.; Singh, N.K.; Junjun, P.; Mughal, M.A. Recognition of Ripe, Unripe and Scaled Condition of Orange Citrus Based on Decision Tree Classification. In Proceedings of the International Conference on Computing, Mathematics and Engineering Technologies, Sukkur, Pakistan, 3–4 March 2018. [CrossRef]
106. Ilic, M.; Ilic, S.; Jovic, S.; Panic, S. Early Cherry Fruit Pathogen Disease Detection Based on Data Mining Prediction. *Comput. Electron. Agric.* **2018**, *150*, 418–425. [CrossRef]
107. Ishikawa, T.; Hayashi, A.; Nagamatsu, S.; Kyutoku, Y.; Dan, I.; Wada, T.; Oku, K.; Saeki, Y.; Uto, T.; Tanabata, T.; et al. Classification of Strawberry Fruit Shape by Machine Learning. In Proceedings of the International Archives of the Photogrammetry, Remote Sensing and Spatial Information Sciences Technical Commission II Mid-term Symposium "Towards Photogrammetry 2020", Riva del Garda, Italy, 4–7 June 2018. [CrossRef]
108. Reyes, J.F.; Contreras, E.; Correa, C.; Melin, P. Image Analysis of Real-Time Classification of Cherry Fruit from Colour Features. *J. Agric. Eng.* **2021**, *52*, 1–6. [CrossRef]
109. Chithra, P.L.; Henila, M. Apple Fruit Sorting Using Novel Thresholding and Area Calculation Algorithms. *Soft Comput.* **2021**, *25*, 431–445. [CrossRef]

110. Sari, C.A.; Puspa Sari, I.; Rachmawanto, E.H.; Rosal Ignatius Moses Setiadi, D.; Proborini, E.; Bijanto; Ali, R.R.; Rizqa, I. Papaya Fruit Type Classification Using LBP Features Extraction and Naive Bayes Classifier. In Proceedings of the International Seminar on Application for Technology of Information and Communication (iSemantic), Semarang, Indonesia, 19–20 September 2020. [CrossRef]
111. Muhathir; Santoso, M.H.; Muliono, R. Analysis Naïve Bayes in Classifying Fruit by Utilizing Hog Feature Extraction. *J. Inform. Telecommun. Eng.* **2020**, *4*, 151–160. [CrossRef]
112. Abdelghafour, F.; Rosu, R.; Keresztes, B.; Germain, C.; Da Costa, J.P. A Bayesian Framework for Joint Structure and Colour Based Pixel-Wise Classification of Grapevine Proximal Images. *Comput. Electron. Agric.* **2019**, *158*, 345–357. [CrossRef]
113. Kusuma, A.; Setiadi, D.R.I.M.; Putra, M.D.M. Tomato Maturity Classification Using Naive Bayes Algorithm and Histogram Feature Extraction. *J. Appl. Intell. Syst.* **2018**, *3*, 39–48. [CrossRef]
114. Lv, J.; Xu, H.; Xu, L.; Zou, L.; Rong, H.; Yang, B.; Niu, L.; Ma, Z. Recognition of Fruits and Vegetables with Similar-Color Background in Natural Environment: A Survey. *J. Field Robot.* **2022**, *39*, 888–904. [CrossRef]
115. Ukwuoma, C.C.; Qin, Z.; Heyat, M.B.B.; Ali, L.; Almaspoor, Z.; Monday, H.N. Recent Advancements in Fruit Detection and Classification Using Deep Learning Techniques. *Math. Probl. Eng.* **2022**, *2022*, 9210947. [CrossRef]
116. Li, Y.; Feng, Q.; Li, T.; Xie, F.; Liu, C.; Xiong, Z. Advance of Target Visual Information Acquisition Technology for Fresh Fruit Robotic Harvesting: A Review. *Agronomy* **2022**, *12*, 1366. [CrossRef]
117. Aslam, F.; Khan, Z.; Tahir, A.; Parveen, K.; Albasheer, F.O.; Abrar, S.U.; Khan, D.M. A Survey of Deep Learning Methods for Fruit and Vegetable Detection and Yield Estimation. *Big Data Anal. Comput. Intell. Cybersecur.* **2022**, *111*, 299–323. [CrossRef]
118. Li, Z.; Yuan, X.; Wang, C. A Review on Structural Development and Recognition–Localization Methods for End-Effector of Fruit–Vegetable Picking Robots. *Int. J. Adv. Robot. Syst.* **2022**, *19*, 172988062211049. [CrossRef]
119. Darwin, B.; Dharmaraj, P.; Prince, S.; Popescu, D.E.; Hemanth, D.J. Recognition of Bloom/Yield in Crop Images Using Deep Learning Models for Smart Agriculture: A Review. *Agronomy* **2021**, *11*, 646. [CrossRef]
120. Maheswari, P.; Raja, P.; Apolo-Apolo, O.E.; Pérez-Ruiz, M. Intelligent Fruit Yield Estimation for Orchards Using Deep Learning Based Semantic Segmentation Techniques—A Review. *Front. Plant Sci.* **2021**, *12*, 684328. [CrossRef]
121. Bhargava, A.; Bansal, A. Fruits and Vegetables Quality Evaluation Using Computer Vision: A Review. *J. King Saud Univ. Comput. Inf. Sci.* **2021**, *33*, 243–257. [CrossRef]
122. Saleem, M.H.; Potgieter, J.; Arif, K.M. Automation in Agriculture by Machine and Deep Learning Techniques: A Review of Recent Developments. *Precis. Agric.* **2021**, *22*, 2053–2091. [CrossRef]
123. Tang, Y.; Chen, M.; Wang, C.; Luo, L.; Li, J.; Lian, G.; Zou, X. Recognition and Localization Methods for Vision-Based Fruit Picking Robots: A Review. *Front. Plant Sci.* **2020**, *11*, 510. [CrossRef]
124. Jia, W.; Zhang, Y.; Lian, J.; Zheng, Y.; Zhao, D.; Li, C. Apple Harvesting Robot under Information Technology: A Review. *Int. J. Adv. Robot. Syst.* **2020**, *17*, 1729881420925310. [CrossRef]
125. Tripathi, M.K.; Maktedar, D.D. A Role of Computer Vision in Fruits and Vegetables among Various Horticulture Products of Agriculture Fields: A Survey. *Inf. Process. Agric.* **2020**, *7*, 183–203. [CrossRef]
126. Naranjo-Torres, J.; Mora, M.; Hernández-García, R.; Barrientos, R.J.; Fredes, C.; Valenzuela, A. A Review of Convolutional Neural Network Applied to Fruit Image Processing. *Appl. Sci.* **2020**, *10*, 3443. [CrossRef]
127. Koirala, A.; Walsh, K.B.; Wang, Z.; McCarthy, C. Deep Learning—Method Overview and Review of Use for Fruit Detection and Yield Estimation. *Comput. Electron. Agric.* **2019**, *162*, 219–234. [CrossRef]
128. Rehman, T.U.; Mahmud, M.S.; Chang, Y.K.; Jin, J.; Shin, J. Current and Future Applications of Statistical Machine Learning Algorithms for Agricultural Machine Vision Systems. *Comput. Electron. Agric.* **2019**, *156*, 585–605. [CrossRef]
129. Shamshiri, R.R.; Weltzien, C.; Hameed, I.A.; Yule, I.J.; Grift, T.E.; Balasundram, S.K.; Pitonakova, L.; Ahmad, D.; Chowdhary, G. Research and Development in Agricultural Robotics: A Perspective of Digital Farming. *Int. J. Agric. Biol. Eng.* **2018**, *11*, 1–14. [CrossRef]
130. Kamilaris, A.; Prenafeta-Boldú, F.X. Deep Learning in Agriculture: A Survey. *Comput. Electron. Agric.* **2018**, *147*, 70–90. [CrossRef]
131. Zhu, N.; Liu, X.; Liu, Z.; Hu, K.; Wang, Y.; Tan, J.; Huang, M.; Zhu, Q.; Ji, X.; Jiang, Y.; et al. Deep Learning for Smart Agriculture: Concepts, Tools, Applications, and Opportunities. *Int. J. Agric. Biol. Eng.* **2018**, *11*, 32–44. [CrossRef]
132. Iqbal, Z.; Khan, M.A.; Sharif, M.; Shah, J.H.; Rehman, M.H.U.; Javed, K. An Automated Detection and Classification of Citrus Plant Diseases Using Image Processing Techniques: A Review. *Comput. Electron. Agric.* **2018**, *153*, 12–32. [CrossRef]
133. Hameed, K.; Chai, D.; Rassau, A. A Comprehensive Review of Fruit and Vegetable Classification Techniques. *Image Vis. Comput.* **2018**, *80*, 24–44. [CrossRef]
134. Zhao, Y.; Gong, L.; Huang, Y.; Liu, C. A Review of Key Techniques of Vision-Based Control for Harvesting Robot. *Comput. Electron. Agric.* **2016**, *127*, 311–323. [CrossRef]
135. Gongal, A.; Amatya, S.; Karkee, M.; Zhang, Q.; Lewis, K. Sensors and Systems for Fruit Detection and Localization: A Review. *Comput. Electron. Agric.* **2015**, *116*, 8–19. [CrossRef]
136. Mao, S.; Li, Y.; Ma, Y.; Zhang, B.; Zhou, J.; Wang, K. Automatic Cucumber Recognition Algorithm for Harvesting Robots in the Natural Environment Using Deep Learning and Multi-Feature Fusion. *Comput. Electron. Agric.* **2020**, *170*, 105254. [CrossRef]
137. Zhao, S.; Liu, J.; Wu, S. Multiple Disease Detection Method for Greenhouse-Cultivated Strawberry Based on Multiscale Feature Fusion Faster R_CNN. *Comput. Electron. Agric.* **2022**, *199*, 107176. [CrossRef]

138. Wu, G.; Li, B.; Zhu, Q.; Huang, M.; Guo, Y. Using Color and 3D Geometry Features to Segment Fruit Point Cloud and Improve Fruit Recognition Accuracy. *Comput. Electron. Agric.* **2020**, *174*, 105475. [CrossRef]
139. Lin, G.; Tang, Y.; Zou, X.; Xiong, J.; Fang, Y. Color-, Depth-, and Shape-Based 3D Fruit Detection. *Precis. Agric.* **2020**, *21*, 1–17. [CrossRef]

agronomy

MDPI

Review

Fruit Detection and Recognition Based on Deep Learning for Automatic Harvesting: An Overview and Review

Feng Xiao, Haibin Wang *, Yueqin Xu and Ruiqing Zhang

College of Mechanical and Electrical Engineering, Northeast Forestry University, Harbin 150040, China; xiaofeng@nefu.edu.cn (F.X.)
* Correspondence: whb_nefu@nefu.edu.cn

Abstract: Continuing progress in machine learning (ML) has led to significant advancements in agricultural tasks. Due to its strong ability to extract high-dimensional features from fruit images, deep learning (DL) is widely used in fruit detection and automatic harvesting. Convolutional neural networks (CNN) in particular have demonstrated the ability to attain accuracy and speed levels comparable to those of humans in some fruit detection and automatic harvesting fields. This paper presents a comprehensive overview and review of fruit detection and recognition based on DL for automatic harvesting from 2018 up to now. We focus on the current challenges affecting fruit detection performance for automatic harvesting: the scarcity of high-quality fruit datasets, fruit detection of small targets, fruit detection in occluded and dense scenarios, fruit detection of multiple scales and multiple species, and lightweight fruit detection models. In response to these challenges, we propose feasible solutions and prospective future development trends. Future research should prioritize addressing these current challenges and improving the accuracy, speed, robustness, and generalization of fruit vision detection systems, while reducing the overall complexity and cost. This paper hopes to provide a reference for follow-up research in the field of fruit detection and recognition based on DL for automatic harvesting.

Keywords: computer vision; deep learning; fruit detection; fruit recognition; automatic harvesting; current challenge; development trend; research review

Citation: Xiao, F.; Wang, H.; Xu, Y.; Zhang, R. Fruit Detection and Recognition Based on Deep Learning for Automatic Harvesting: An Overview and Review. *Agronomy* **2023**, *13*, 1625. https://doi.org/10.3390/agronomy13061625

Academic Editor: Shubo Wang

Received: 11 May 2023
Revised: 13 June 2023
Accepted: 14 June 2023
Published: 16 June 2023

1. Introduction

In recent years, the application of artificial intelligence (AI) techniques and robotic systems to automate agricultural processes has garnered significant interest (as shown in Figure 1). Fruits usually grow in complex environments with many uncertainties. Powerful fruit vision detection systems are necessary for intelligent agriculture and automatic harvesting. Fruit vision detection systems' characteristics mainly include imaging sensors and visual information about fruits. Fruit vision detection systems generally operate through the five stages (as shown in Figure 2): fruit image acquisition, fruit image preprocessing, fruit feature extraction, fruit image segmentation, and fruit image recognition. Black and white cameras, red–green–blue (RGB) cameras, spectral cameras, thermal cameras, and RGB-depth map (RGB-D) cameras (as shown in Figure 3) are commonly used for fruit vision detection systems to obtain color, shape, texture, and size information of fruits in specific operational areas. A comparison of different types of imaging sensors is shown in Table 1. Fruit images acquired through different imaging methods are shown in Figure 4. The main research processes of fruit detection and recognition methods are shown in Figure 5. Since DL has a strong ability to extract high-dimensional features from fruit images, researchers have proposed many fruit detection and recognition methods based on DL (you only look once (YOLO), single shot multibox detector (SSD), Alex Krizhevsky networks (AlexNet), visual geometry group networks (VGGNet), residual networks (ResNet), faster region-convolutional neural networks (Faster R-CNN), fully convolutional networks (FCN), SegNet, and mask region-convolutional neural networks (Mask R-CNN)) for automatic harvesting (as shown in Table 2). Despite much research, many challenges need to be overcome to build an effective fruit vision detection and harvesting system.

Figure 1. Typical harvesting robots. (**a**) A plum-harvesting robot (photo reprinted with permission from ref. [1]. 2021, Brown, J.); (**b**,**d**–**f**) Apple-harvesting robots (photo reprinted with permission from ref. [2]. 2021, Yan, B.; ref. [3]. 2017, He, L.; ref. [4]. 2012, Ji, W.; ref. [5]. 2011, Zhao, D.); (**c**,**n**–**p**) Sweet pepper-harvesting robots (photo reprinted with permission from ref. [6]. 2020, Arad, B.; ref. [7]. 2017, Lehnert, C.; ref. [8]. 2014, Bac, C.W.); (**g**–**i**) Strawberry-harvesting robots (photo reprinted with permission from ref. [9]. 2020, Xiong, Y.; ref. [10]. 2019, Xiong, Y.; ref. [11]. 2010, Hayashi, S.); (**j**) A lychee-harvesting robot (photo reprinted with permission from ref. [12]. 2018, Xiong, J.); (**k**,**m**) Tomato-harvesting robots (photo reprinted with permission from ref. [13]. 2018, Feng, Q.; ref. [14]. 2010, Kondo, N.); (**l**) A kiwifruit-harvesting robot (photo reprinted with permission from ref. [15]. 2019, Williams, H.A.M.).

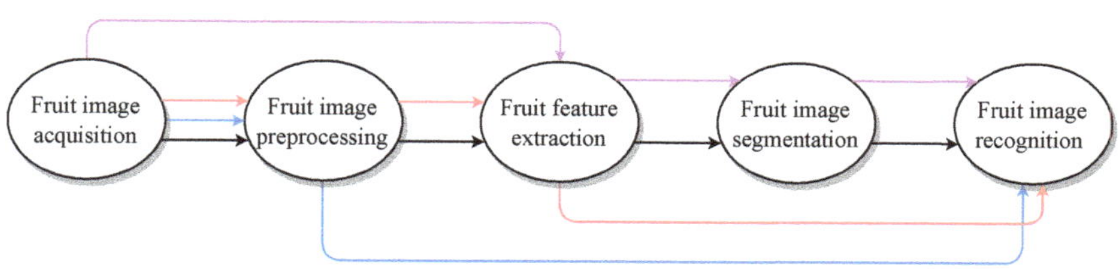

Figure 2. Different processes of fruit detection and recognition based on DL (image reprinted with permission from ref. [16]. 2023, Xiao F.).

(**a**) Black and white camera
(Leica Q2 Monochrom)

(**b**) RGB camera
(Leica SL2)

source: www.leica-camera.cn

(**c**) Thermal camera
(Fluke TiX580)
source: www.fluke.com

(**d**) Multispectral camera
(SmartMore SMI-SC-1080P-02)

(**e**) Hyperspectral camera
(SmartMore SMI-SC-600C-01)

source: inside.smartmore.com

(**f**) RGB-D camera
(Microsoft Azure Kinect DK)
source: www.microsoftstore.com.cn

Figure 3. Different types of imaging sensors commonly used for fruit vision detection systems (accessed on 5 January 2023).

(**a**)

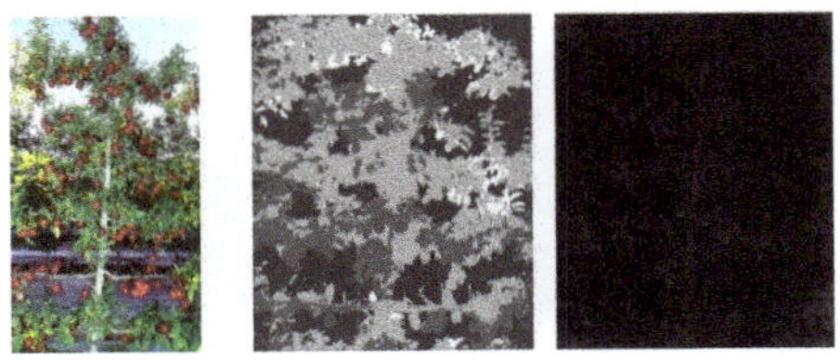

(**b**)

(**c**)

(**d**)

Figure 4. Fruit images acquired using different imaging methods. (**a**) Black and white image; (**b**) RGB, depth, and infrared images (photos reprinted with permission from ref. [17]. 2020, Fu L.); (**c**) spectral image (photo reprinted with permission from ref. [18]. 2009, Okamoto H.); (**d**) color and thermal-registered image (photo reprinted with permission from ref. [19]. 2010, Wachs J.P.).

Some review articles have been published encompassing diverse agricultural applications, such as crop recognition, fruit counting, weed discrimination, and plant disease detection, with or without a robotic system, by considering AI/computer vision (CV)/other advanced vision control techniques. For example, Rehman, T.U. et al., (including researchers based in America and Canada) (2019) [20] provided a comprehensive summary of ML algorithms that have been utilized in diverse agricultural operations. Brazilian researchers Patrício, D.I. and Rieder, R. (2018) [21] investigated potential applications of machine vision (MV) for diverse agricultural tasks, such as crop disease/pest detection, grain quality evaluation, and automatic plant phenotyping. Narvaez, F.Y. et al., (including researchers based in Chile, Italy, and America) (2017) [22] summarized various sensing techniques, along with their limitations, to categorize fruits/plants. Indian researchers Jha, K. et al., (2019) [23] outlined the latest smart methodologies, such as the Internet of Things (IoT), for agricultural purposes. Dutch researchers Wolfert, S. et al., (2017) [24] reviewed the application of big data in agriculture. There are also some review articles that have been published incorporating only a particular type of agricultural application or scenario. For example, we reviewed fruit detection and recognition techniques based on digital image processing and traditional ML for fruit harvesters in [16]. New Zealand researchers Saleem, M.H. et al., (2019) [25] summarized and explained DL models for the identification and classification of plant diseases, along with the application of DL with advanced imaging techniques, including hyperspectral/multispectral imaging. Wang, D. et al., (including American researchers and a researcher based in Israel) (2019) [26] and Chinese researchers Wang, A. et al., (2019) [27] reviewed procedures for weed detection using various classification methods, including ML and DL. The review literature on AI/ML/DL/MV/CV/other advanced vision control techniques for intelligent agriculture and automatic harvesting also includes [28–41]. However, unlike the articles mentioned above, our work focuses on providing an overview and review of the use of DL applied to fruit image recognition (mainly in the areas of detection and classification) for automatic harvesting. In order to further define the study areas of our paper, we identify fruit detection and classification tasks such as the determination of classes based on their specific types.

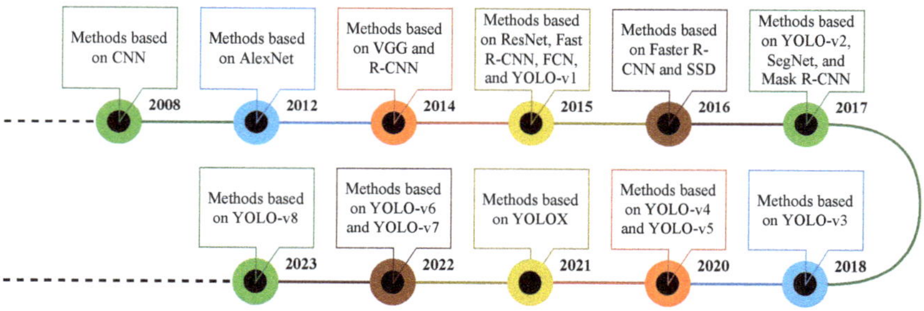

Figure 5. Main research processes of fruit detection and recognition methods based on DL.

Table 1. Comparison of different types of imaging sensors commonly used in fruit vision detection systems.

Fruit Imaging Sensors	Types	Information	Advantages	Limitations
RGB-D camera and LSS (Lift, Splat, Shoot)	Active	RGB and depth images	Complete fruit scene characteristics	Lack of feature descriptors
Black and white camera	Passive	Shape and texture features	Little effect of changes in lighting conditions	Lack of color information
RGB camera		Color, shape, and texture features	Exploiting all the basic features of target fruits	Highly sensitive to changing lighting conditions
Spectral camera		Color features and spectral information	Providing more information about reflectance	Computationally expensive for complete spectrum analysis
Thermal camera		Thermal signatures	Color-invariant	Dependency on minute thermal difference

Table 2. Fruit detection and recognition methods based on DL.

Types	Accuracy	Applied Crops	Advantages	Disadvantages
YOLO	84–98%	cabbage, citrus, lychee, mango, tomato	High fruit detection speed; it can meet real-time requirements well for automatic harvesting	Fruit detection accuracy under severe occlusion, low resolution, and changing lighting conditions is low
SSD	75–92%	apple, mango, pear, sour lemon	High detection accuracy and speed; good robustness and generalization	Fruit images need to be preprocessed; detection accuracy for small targets is low
AlexNet	86–96%	apple, strawberry, sugar beet, tomato	Using dropout to avoid overfitting; good generalization ability	Network convergence takes a little longer
VGGNet	92–99%	jujube, potato, sugar beet, tomato	Simple structure of fruit vision detection models	Network convergence takes a little longer; using more network parameters
ResNet	90–95%	apple, banana	Using residual blocks to deepen network layers and reduce network parameters	Too deep network layers may result in vanishing gradients, poor training effectiveness, and low detection accuracy
Faster R-CNN	90–99%	apple, mango, orange	High detection accuracy	Fruit detection speed is slow, and it cannot meet real-time requirements well
FCN	89–98%	cotton, grape, guava, kiwifruit	Accepting fruit image inputs with arbitrary sizes; high efficiency and low computational effort	Insensitive to the details of fruits in fruit images; fruit classification does not consider inter-pixel relationships
SegNet	83–95%	apple, tomato	Obtaining edge contours and maintaining the integrity of high-frequency details in segmentation	Neighboring information may be ignored when fruit feature maps with low resolution are unpooled
Mask R-CNN	80–94%	apple, strawberry, tomato	Combining semantic segmentation with fruit detection by outputting mask images	Fruit detection speed is slow, and it cannot meet real-time requirements well

The contributions of this work are as follows: (1) systematically summarizes and explains all kinds of fruit detection and recognition methods based on DL for automatic harvesting from 2018 up to now; (2) systematically compares and analyzes the advantages, disadvantages, and applicability of various fruit detection and recognition methods based on DL for automatic harvesting; (3) systematically demonstrates the current challenges affecting fruit detection performance for automatic harvesting and proposes feasible solutions and prospective future potential developments. Through this clearer and more comprehensive overview and review, we aim to provide a reference for follow-up research in the field of fruit detection and recognition based on DL for automatic harvesting.

According to Martín-Martín, A. et al., (including Spanish researchers and a researcher based in the UK) (2018) [42], Google Scholar citation data encompass a larger set of publications than Web of Science and Scopus. In order to comprehensively survey the literature relevant to the scope of this article, the Google Scholar database has been selected as the source. In the first step, combinations of keywords such as "fruit detection", "fruit recognition", "deep learning", "computer vision", and "fruit harvesting" were utilized in the initial search process. All retrieved papers were subsequently evaluated for their relevance to the subject matter. The second step included the examination of the references from step one for a more thorough review. In the final step, to ensure that our study focuses on the most current research, all papers published before 2018 were excluded. Only the recent literature from 2018 to the present was considered. The final set of papers regarding fruit detection and recognition based on DL for automatic harvesting included 53 research articles. Figure 6 displays the distribution of articles per year, network models used, and crops detected.

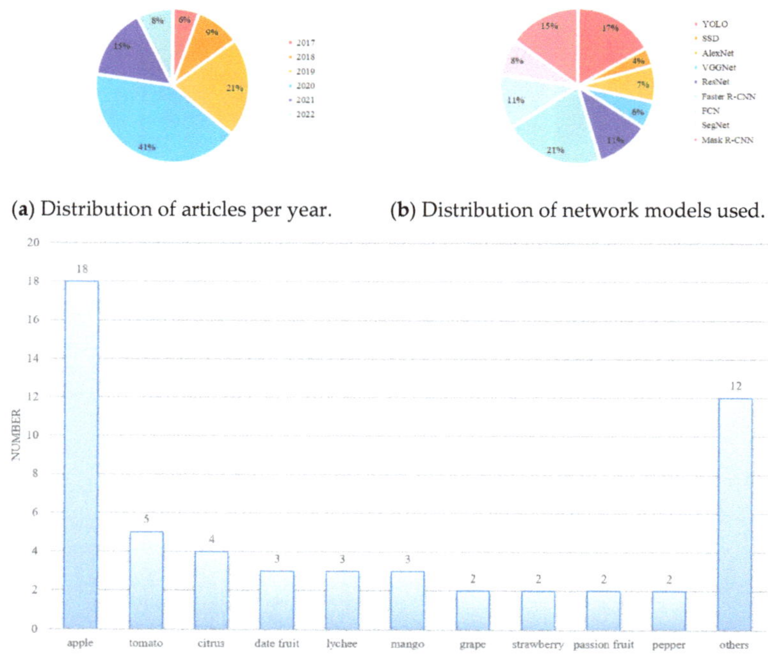

(**a**) Distribution of articles per year. (**b**) Distribution of network models used.

(**c**) Distribution of crops detected and recognized.

Figure 6. Summary of literature search.

As shown in Figure 6, in recent years, the application of DL techniques and robotic systems to automate agricultural processes has garnered significant interest. Improvement and application research based on Faster R-CNN (21%) is currently a hotspot. The recognition accuracy of fruit detection methods based on Faster R-CNN is high, but recognition speed is limited by complex anchor frame mechanisms. When there are mobile deployment and high recognition speed requirements, fruit detection methods based on YOLO (17%) are used most frequently. Their recognition speed is fast, but the recognition effect for small target fruits is not very good. In addition, ResNet (11%) is the most popular backbone network, followed by AlexNet (7%).

Most of the research focuses on apples (32.14%), followed by tomatoes (8.93%), and citrus (7.14%). These three kinds of fruits are in high demand and yield globally. There are some reasons that make them ideal candidates for automatic harvesting. Firstly, they individually hang from plants, making them easily detectable based on their distinctive features. Secondly, they have no extreme variations in size or weight. Lastly, they are relatively hard and not easily damaged in mechanical operations. However, in terms of fruit dimensions and peduncle length, different cultivars may exhibit different characteristics, which can affect fruit detection and recognition performance. This poses challenges for adapting fruit detection and recognition methods for different cultivars. Future work could aim to identify cultivars that are more suitable for automatic harvesting.

The outline of this article is shown in Figure 7. The organization of the rest of the paper is as follows: Section 2 summarizes and explains previous research articles about DL applied to fruit detection and recognition for automatic harvesting. We compare and analyze the advantages, disadvantages, and applicability of various fruit detection and recognition methods based on DL (YOLO, SSD, AlexNet, VGGNet, ResNet, Faster R-CNN, FCN, SegNet, and Mask R-CNN) for automatic harvesting; Section 3 discusses the current challenges affecting fruit detection and recognition performance for automatic harvesting (scarcity of high-quality fruit datasets, fruit detection of small targets, fruit detection in occluded and dense scenarios, fruit detection of multiple scales and multiple species, and lightweight fruit detection models) and proposes feasible solutions and prospective future development trends; Section 4 concludes this article.

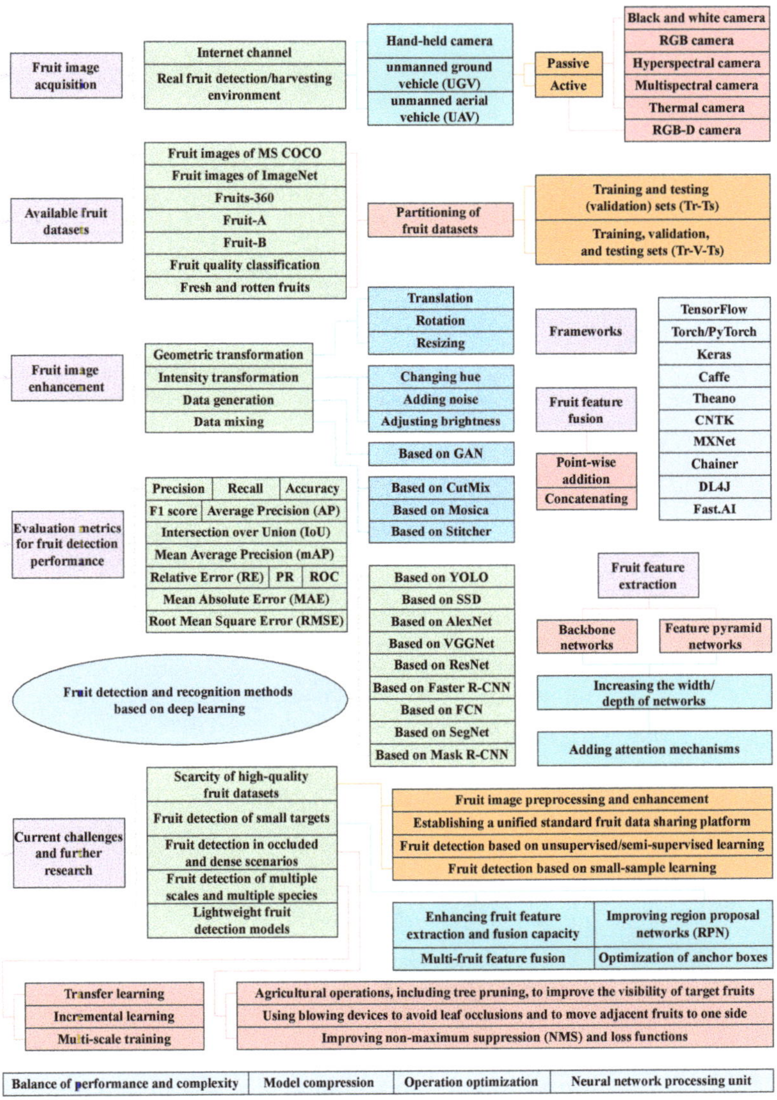

Figure 7. Outline of the article.

2. Fruit Detection and Recognition Based on DL

The concept of DL originated from research on artificial neural networks (ANN), proposed by Canadian researchers Hinton, G.E. and Salakhutdinov, R.R. in 2006 [43]. Since DL has a strong ability to extract high-dimensional features from fruit images, many researchers have conducted extensive and in-depth research on fruit detection and recognition based on DL for automatic harvesting. The basic architecture of DL-based ANN for fruit detection and recognition is shown in Figure 8.

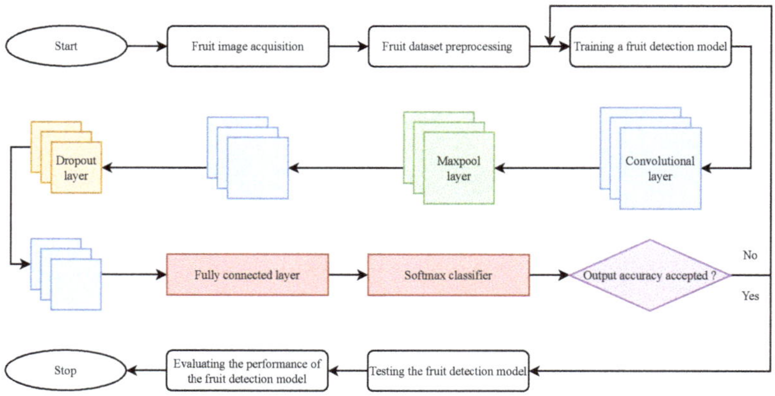

Figure 8. Basic architecture of DL-based ANN for fruit detection and recognition.

CNNs were proposed by American researchers LeCun, Y. et al. in the 1980s [44,45]. They can efficiently capture patterns in multidimensional space. A typical CNN framework for fruit detection and recognition is shown in Figure 9. It includes the convolutional layer (Conv), pooling layer (Pool), nonlinear activation function, and fully connected layer (FC). The convolutional layer is the core of the CNN for fruit feature extraction. Depending on the designed convolution kernel, convolution operations capture fruit image contours and generate corresponding fruit feature maps. In order to reduce the spatial size of the fruit feature maps, the pooling layer performs down-sampling operations by sampling the maximum or average value in a neighborhood range. The nonlinear activation function uses activation functions to process the input data. Neurons in the fully connected layer are connected to all activated neurons in the layer above it. When training the CNN, the model scores categories of predicted images, calculates training loss using selected loss functions, and updates weights through backpropagation functions and gradient descent. The cross-entropy loss function is one of the most widely used loss functions, and the stochastic gradient descent method is the most popular method to address gradient descent.

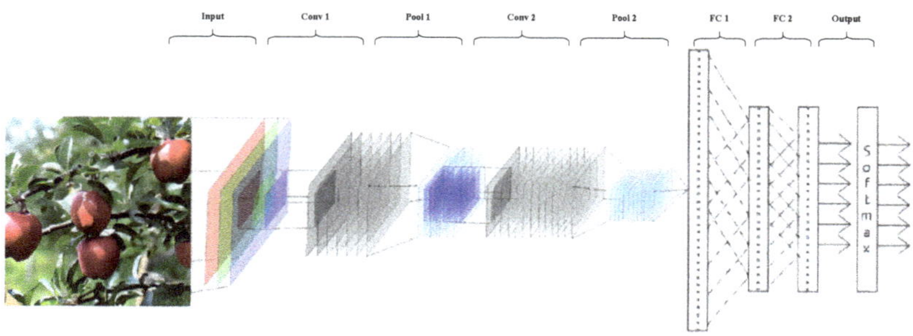

Figure 9. Typical CNN framework for fruit detection and recognition.

Compared with digital image processing and traditional ML techniques, fruit detection and recognition methods based on CNN have great advantages in terms of accuracy. Jahanbakhshi, A. et al., (including Iranian researchers and a researcher based in the UK) (2020) [46] proposed an improved CNN (15, 16, and 18 layers) to detect apparent defects in sour lemons. In comparison to traditional fruit feature extraction methods, such as histogram of oriented gradient (HOG), local binary pattern (LBP), support vector machine (SVM), k-nearest neighbor (KNN), decision tree, and fuzzy classification, the improved CNN was found to outperform these methods, achieving an accuracy of 100%. Bangladeshi researchers Sakib, S. et al., (2019) [47] proposed a fruit detection system using CNN. The Fruits-360 dataset was utilized to evaluate the proposed system. The training accuracy and testing accuracy are 99.79% and 100%, respectively. In general, fruit detection and recognition methods based on CNN can achieve state-of-the-art (SOTA) accuracy for detecting and recognizing any type of fruit on any background.

Current fruit detection and recognition methods based on DL for automatic harvesting can be classified into two categories: single-stage fruit detection and recognition methods (such as YOLO and SSD) based on regression, and two-stage fruit detection and recognition methods (AlexNet, VGGNet, ResNet, Faster R-CNN, FCN, SegNet, and Mask R-CNN) based on candidate regions. Single-stage methods define fruit detection tasks as regression problems of class confidence and bounding box locations (as shown in Figure 10). They divide input fruit images into a grid of cells, extract fruit feature information through the convolutional layer, and predict object class probabilities and bounding box coordinates for each cell. In contrast, as shown in Figure 11, for two-stage methods, in the first stage, a set of target fruit proposals is generated by the RPN on fruit feature maps produced by the convolutional layer. The RPN generates region of interest (RoI) proposals for each location on the fruit feature maps. Each proposal consists of a fixed-size bounding box and a probability score of containing a target fruit. Based on the scores assigned to these proposals, the top N highest-scoring regions are selected as final RoI proposals. To generate RoI proposals, the RPN applies sliding windows of different scales and aspect ratios to fruit feature maps. In the second stage, each final RoI proposal is cropped into a fixed-size feature map using RoI pooling. The maps are then fed into a separate CNN for fruit classification and bounding box regression.

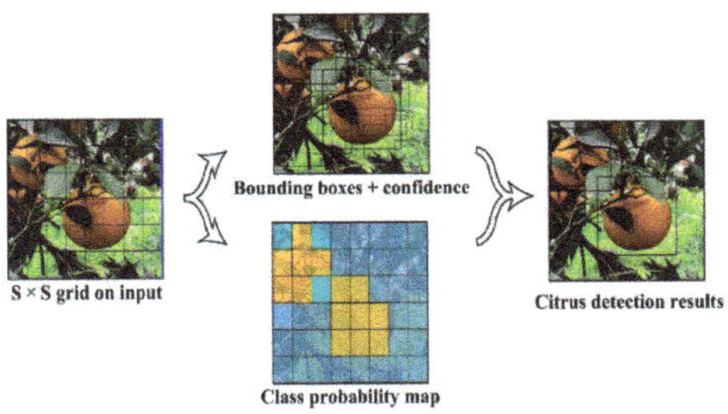

Figure 10. Modeling fruit detection as a regression problem (photos reprinted with permission from ref. [48]. 2022, Chen J.).

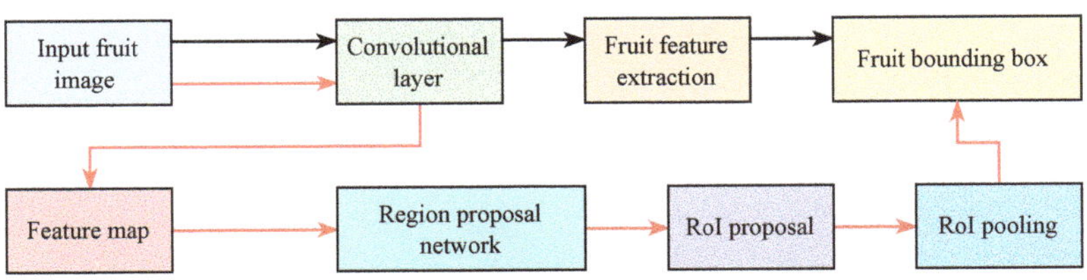

Figure 11. Comparison of one-stage and two-stage fruit detection and recognition methods.

Table 3 compares and analyzes different fruit detection and recognition methods used by various researchers. In the section on "crop, description, and merit", we explain the innovation. In the section on "improvement", we identify the weaknesses and potential improvements. In general, two-stage fruit detection and recognition methods have been shown to achieve higher accuracy than single-stage fruit detection and recognition methods due to their ability to propose more accurate fruit locations. However, they are slower and computationally more intensive than single-stage fruit detection and recognition methods. On the other hand, while single-stage fruit detection and recognition methods are faster and simpler than two-stage fruit detection and recognition methods, they may be less accurate, especially for small target fruits.

Table 3. Comparison of different fruit detection and recognition methods based on DL.

Crops, Description, and Merit	En*	Datasets	Pixels	Sensors	Condition	Improvements	Value (%)
Olive (CNN)/Indian researchers Khosravi, H. et al., (2021) [49] propose a real-time detection method for two olive cultivars in four ripening stages. Adagrad, SGD, SGDM, RMSProp, Adam, and Nadam are evaluated. Nadam shows the best efficiency	√	Train: 14,017; test: 878	256 × 256	Galaxy J6 smartphone camera	Natural lighting	Lighting conditions and fruit image-capturing settings are not considered	Overall accuracy: 91.91; CPU: 12.64 ms; GPU: 4.10 ms
Blueberry (CNN)/Chilean researcher Quiroz, I.A. and Mexican researcher Alférez, G.H. (2020) [50] present a DL solution for image recognition of legacy blueberries in rooting stages	×	Total: 258; train: 168; val: 54; pre: 36	1920 × 1080	Microsoft Lifecam Studio digital camera	Good lighting conditions, not blurred, and without distracting objects in the background	It could use GANs to generate synthetic images that closely resemble real ones, minimizing the need for accessing real data	Accuracy: 86; precision: 86; recall: 88; F1 score: 86
Sour lemon (CNN)/Jahanbakhshi, A. et al., (including Iranian researchers and a researcher based in the UK) (2020) [46] detect apparent defects in sour lemons. Data augmentation and stochastic pooling mechanisms are used to improve detection performance	√	Total: 5456; healthy: 2960; damaged: 2496; train: 70%; val: 30%	16 × 16; 32 × 32; 64 × 64	Camera (Canon, Japan)	A lighting box including two LED lamps	Future work may include accommodating more varied fruit detection conditions	Accuracy: 100
56 diseases infecting 12 plant species (CNN)/Brazilian researcher Barbedo, J.G.A. (2018) [51] studies the effectiveness of DL and TL for plant disease classification	×	Total: 1383; train: 80%; val: 20%	224 × 224 × 3	A variety of digital cameras and mobile devices	Under controlled conditions: 15%; under real conditions: 85%	The number of samples is too small for the CNN to thoroughly capture the characteristics and variations associated with each class	It is a challenge to build fruit databases comprehensive enough for the creation of robust fruit detection models
Strawberry (AlexNet)/Chinese researchers Ni, J. et al., (2021) [52] propose an enhanced AlexNet for strawberry quality evaluation. The size of the convolution kernel is modified. The single convolutional layer is divided into three convolutional layers with different convolution kernels. The BN layer and L2 regularization are used	×	Total: 3006; unripe: 778; medium: 382; fully: 787; bad: 847; malformed: 212; train: 80%; val: 10%; test: 10%	227 × 227	HUAWEI mobile phone	Two different scenes of a field and a laboratory	It is not certain which augmentation method will help improve fruit detection performance	Average accuracy: 90.70; after augmentation: 95.75

Table 3. *Cont.*

Crops, Description, and Merit	En*	Datasets	Pixels	Sensors	Condition	Improvements	Value (%)
Grape bunch (AlexNet)/Italian researchers Marani, R. et al., (2021) [53] investigate the use of DL for grape bunch segmentation in natural fruit images captured using a consumer-grade camera. It is based on the optimal threshold selection of bunch probability maps as an alternative to the conventional minimization of cross-entropy loss for mutually exclusive classes	×	Total: 84; train: 60; val: 24	640 × 480	Intel RealSense R200 RGB-D camera	Fruit images under direct (opposite) sunlight are not considered since they become overexposed, and their colors saturate to white	Depth data could be used to guide the selection of the size N of the moving window for the proposed processing	Mean segmentation accuracy on the bunch class: 80.58; IoU: 45.64
Date fruit (VGGNet)/Saudi Arabian researchers Altaheri, H. et al., (2019) [54] propose an efficient MV framework for date fruit-harvesting robots	×	Total: 8072; 5 date types in different pre-maturity and maturity stages; more than 350 date bunches; belong to 29 date palms	--	RGB video camera	The dataset reflects the challenges, including variations in angles, scales, and illumination conditions	It may lead to confusion in the detection of date fruit maturity, including labeling rules and interference between maturity stages	Type, maturity, and harvesting decision classification accuracies: 99.01, 97.25, 98.59; classification times: 20.6, 20.7, 35.9 ms
Apple (ResNet)/Chinese researchers Wang, D. et al., (2020) [55] develop a remote apple horizontal diameter detection system to achieve automatic measurement of apple growth throughout the entire growth period. The fused convolutional feature network developed can effectively remove complex backgrounds and accurately detect apple edges with near real-time performance	√	Total: 903; train: 743; val: 160; test: 170; 5944 images are eventually obtained through data augmentation; mature red, immature green, semimature	403 × 303	iPhone 7 plus	To prevent distinct edges from forming on the surfaces of apples due to intense natural light, the images are captured on cloudy days or at dusk when the light is not as intense	Future improvements are needed to track the monitored apple in order to achieve the goal of adjusting the camera's shooting angle and selecting seed points automatically	F1 score: 53.1; average run time: 75 ms; mean average absolute error of the apples' horizontal diameters detected: 0.90 mm

Table 3. *Cont.*

Crops, Description, and Merit	En*	Datasets	Pixels	Sensors	Condition	Improvements	Value (%)
Passion fruit (Faster R-CNN)/Chinese researchers Tu, S. et al., (2020) [56] propose a multiple-scale Faster R-CNN approach based on RGB-D images for small passion fruit detection and counting. It detects lower-level features by incorporating feature maps from shallower convolutional feature maps for RoI pooling	√	Total RGB images: 8651; train: 6055; test: 2596; total depth images: 3352; train: 2346; test 1006	1920 × 1080; 512 × 424	Kinect V2	The Kinect V2 sensor is used to avoid strong sunlight and work in shady areas because the ToF technique is unsuitable in strong sunlight conditions	The detection performance of passion fruit in different growth stages could be evaluated and analyzed	Recall: 96.2; precision: 93.1; F1-score: 94.6
Young tomato fruit (Faster R-CNN)/Chinese researchers Wang, P. et al., (2021) [57] propose a method for detecting young tomatoes on near-color backgrounds based on an improved Faster R-CNN with attention mechanisms. Soft non-maximum suppression is used to reduce the missed detection rate of overlapping fruits	×	Total: 2235; train: 80%; val: 10%; test: 10%	3000 × 3000	MI 9 smartphone	Different weather conditions (sunny and cloudy) and different time periods (morning, noon, and evening)	Future work could include accommodating various cultivars of tomatoes and more unstructured environments	mAP: 98.46; average detection time: 84 ms
Lychee (YOLO)/To improve the efficiency of lychee harvesting, Chinese researchers Li, C. et al., (2022) [58] propose a column-comb litchi harvesting method based on K-means 3D clustering partitioning	×	Total: 1049; train: 840; test: 209	1280 × 800; 1280 × 720	Intel RealSense depth camera	Orchard environments (strong light and backlight, sunny and cloudy days, and far and near distances)	Current detection performance are obtained by testing on well-defined fruit images with a limited sample size	Recall: 78.99; precision: 87.43; F1 score: 0.83
Tomato (YOLO)/Chinese researchers Miao, Z. et al. (2022) [59] integrate classic image processing methods with YOLOv5 to increase fruit detection accuracy and robustness	×	Total: 1000; train: 800; val: 200	1920 × 1080; 1280 × 720	Intel RealSense depth camera	Artificial experimental environments	Extended tests and improvements in a real orchard and greenhouse will be the main focus	Average deviation: 2 mm; average operating time: 9 s/cluster

Table 3. *Cont.*

Crops, Description, and Merit	En*	Datasets	Pixels	Sensors	Condition	Improvements	Value (%)
Hass avocado, lemon, apples (SSD)/Vasconez, J.P. et al., (including Chilean researchers and a researcher based in America) (2020) [60] test two of the most common architectures: Faster R-CNN with Inception V2 and SSD with MobileNet. To address the problem of video-based fruit counting, it uses multi-object tracking based on Gaussian estimation	√	Avocado train: 1021; val: 211; test: 211; apple train: 694; val: 191; test: 191; lemon train: 539; val: 202; test: 202	360 × 640	Commercial RGB camera; acquiring at 30 FPS	Hass avocado, lemon, and apple datasets acquired under illumination levels ranging from 1890 to 43,600, 4800 to 52,000, and 3500 to 38,000 lux, respectively	The CNN architectures are highly dependent on the quality of the training set. The results might not be conclusive for other groves with different fruits	SSD with MobileNet, the minimum relative error: 7 (avocados); 13 (apples); 20 (lemons); computing time: 220 ms
Guava (FCN)/Chinese researchers Lin, G. et al., (2019) [61] use a low-cost RGB-D sensor to achieve guava detection and pose estimation. It uses Euclidean clustering to detect all the 3D fruits from the fruit binary maps output by FCN. It also establishes a 3D line segment detection method to reconstruct the branches from the branch binary maps	×	Total: 437; train: 80%; val: 20%	424 × 512	Kinect V2	All kinds of illuminations	Branch is a little difficult to segment	Precision: 98.3; recall: 94.8; 3D pose error: 23.43° ± 14.18°; execution time: 56.5 ms
Lychee clusters (SegNet)/Chinese researchers Li, J. et al., (2020) [62] develop a reliable algorithm based on RGB-D cameras to accurately detect and locate the fruit-bearing branches of multiple lychee clusters. It revises density clustering-based branch extraction and optimal clustering-based parameter analysis	√	Total: 452; train: 80%; val: 20%	1920 × 1080; 512 × 424	Kinect V2	All kinds of illuminations; no artificial shade or lighting interference	Future studies could focus on improving the success rate of picking tasks	Detection accuracy: 83.33; positioning accuracy: 17.29° ± 24.57°; execution time: 464 ms
Apple (SegNet)/Majeed, Y. et al., (including American researchers and researchers based in China) (2020) [63] develop a DL-based semantic segmentation method. Both simple and foreground RGB images are used for training SegNet to segment trunks and branches	√	Total: 509; train: 70%; test: 30%	960 × 540	Kinect V2	Different lighting conditions (sunny, cloudy, and night)	Optimal branches will be selected for training by estimating the essential parameters desired for canopy architecture	Mean accuracy: 89; IoU: 52; boundary-F1-score: 81

Table 3. *Cont.*

Crops, Description, and Merit	En*	Datasets	Pixels	Sensors	Condition	Improvements	Value (%)
Cherry tomato (Mask R-CNN)/Chinese researchers Xu, P. et al., (2022) [64] propose an improved Mask R-CNN for the visual recognition of cherry tomatoes by using depth information and considering the prior adjacent constraint between fruits and stems	√	Total: 3444; train: 80%; val: 20%	640 × 480	Intel RealSense depth camera	Natural conditions	Future work may include reducing the processing time and accommodating more varied conditions	Detection accuracy of fruits: 93.76; accuracy and recall of stems: 89.34 and 94.47; computing time: 40 ms
Strawberry (Mask R-CNN)/Chinese researchers Yu, Y. et al., (2019) [65] perform a visual localization method for strawberry picking points after generating mask images of ripe fruits using Mask R-CNN. ResNet-50 is adopted as the backbone network, combined with the FPN for fruit feature extraction. The RPN is trained end-to-end to create region proposals for each feature map	×	Total: 1900; train: 1520; val: 380; test: 100	640 × 480	Hand-held digital camera	Different periods (morning and afternoon); under varying light intensity (sunny and cloudy conditions); different levels of interference (overlap, occlusion, and oscillation)	Although the average processing frames per second is 8, the speed of the embedded mobile harvesting robot is lower than this result. Therefore, the real-time performance of the model needs to be further improved	Average detection precision: 95.78; recall: 95.41; IoU of instance segmentation: 89.85; average error of picking points: ±1.2 mm

En* represents data enhancement.

2.1. Single-Stage Fruit Detection and Recognition Methods Based on Regression

2.1.1. Fruit Detection and Recognition Methods Based on YOLO

YOLO is one of the most classic and advanced fruit detection algorithms. It can detect and classify target fruits simultaneously in a single image. As shown in Figure 12, YOLO-v1 was the beginning. YOLO-v1 was proposed by American researchers Redmon, J. et al. in 2015 [66]. YOLO-v2 was proposed by American researchers Redmon, J. and Farhadi, A. in 2017 [67]. It included improvements to the structure of YOLO-v1. The K-means clustering algorithm was used to determine the optimal number of anchor boxes and to analyze the relationship between recognition accuracy and speed. Then, they also proposed YOLO-v3 [68], which featured improvements such as the Darknet-53 backbone network and multi-scale prediction. Bochkovskiy, A. et al., (2020) [69] systematically analyzed the processes of data preprocessing and the design of detection and prediction networks. Based on the analysis, they designed an efficient target detector (YOLO-v4) suitable for a single graphics card. YOLO-v5 [70] provided four different sizes of target detectors to meet the needs of different applications. YOLOR [71], YOLOX [72], YOLO-v6 [73], YOLO-v7 [74], and YOLO-v8 [75] also appeared one after another. YOLO-v8 is a SOTA model. It was open-sourced on January 10, 2023. The framework is shown in Figure 13. Specific innovations include a new backbone network, a new anchor-free detection head, and a new loss function that can run on various hardware platforms from CPU to GPU.

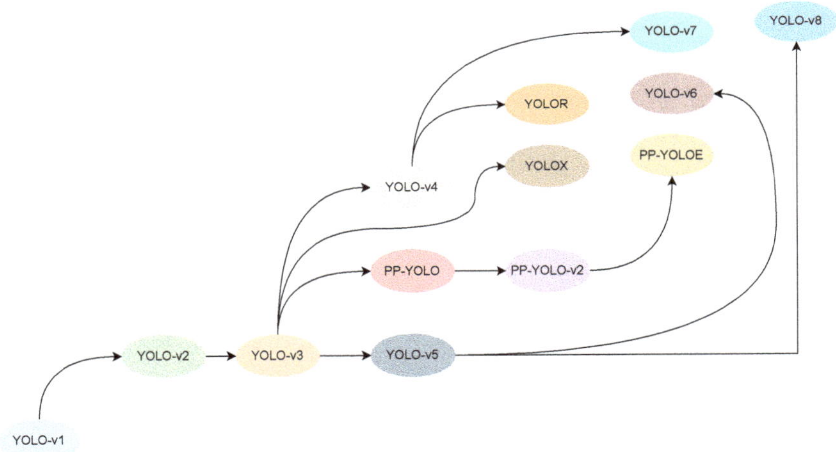

Figure 12. Main research processes of YOLO.

Fruit detection and recognition methods based on YOLO are widely used, by virtue of their advantages. Chinese researchers Xiong, J. et al., (2020) [76] proposed a method based on YOLO-v2 to detect and count mangoes in fruit images taken by an UAV. The processing time is 80ms, and the average detection accuracy is 96.1%. British researchers Birrell, S. et al., (2020) [77] proposed a method based on YOLO-v3 to detect and classify cabbages in four growth stages, achieving a total detection accuracy of 91% and a classification accuracy of 82%. In order to create an even more lightweight fruit detection model, Chinese researchers Li, C. et al., (2022) [58] proposed an improved YOLO-v3-tiny fruit detection model based on K-means 3D clustering partitioning for small and densely packed lychee fruits, and compared it with other fruit detection networks (YOLO-v3-tiny, YOLO-v4, YOLO-v5, and Faster R-CNN). The improved YOLOv3-tiny can recognize lychee fruits more accurately. The check-all rate, check-accuracy rate, and F1 score are 78.99%, 87.43%, and 0.83, respectively. However, fruit detection and recognition methods based on YOLO do not use prior information when predicting fruit positions. This results in a loss of fruit location accuracy. In addition, when YOLO predicts detection results corresponding to each bounding box, it requires that the target fruit's center point must be located inside the bounding box. This imposes a strong spatial constraint on the prediction process of YOLO and makes fruit detection and recognition methods based on YOLO less effective at detecting small target fruits that appear in groups. In the future, we can input and fuse semantic information (such as fruit scene and context-related information) into fruit detection algorithms to greatly improve fruit detection accuracy. For example, Chinese researchers Miao, Z. et al., (2022) [59] integrated classic image processing methods with YOLO-v5 to increase fruit detection accuracy and robustness. A tomato-harvesting robot can be guided to efficiently harvest truss tomatoes, with an average operating time of 9 s per cluster.

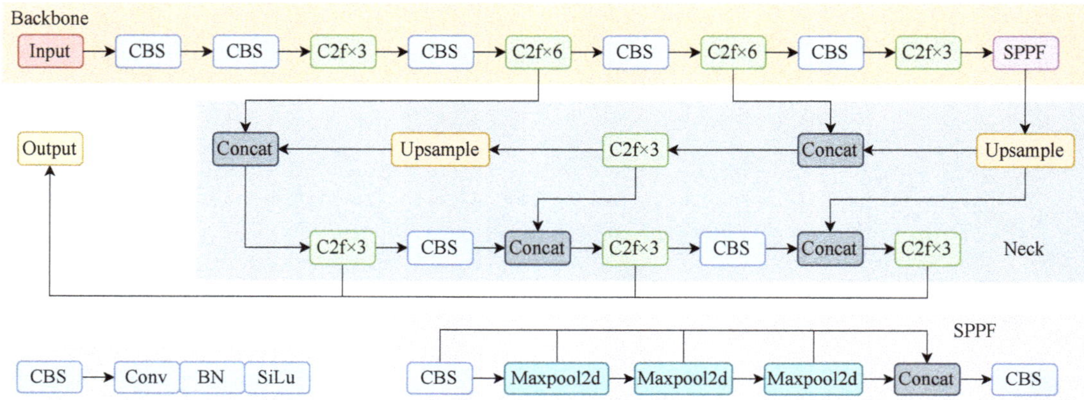

Figure 13. YOLO-v8 framework (image reprinted with permission from ref. [78]. 2023, Lou, H.).

2.1.2. Fruit Detection and Recognition Methods Based on SSD

SSD was proposed by American researchers Liu, W. et al. in 2016 [79]. A typical SSD framework for fruit detection and recognition is shown in Figure 14. It consists of a base network (such as VGG-16) and an additional set of convolutional and pooling layers for fruit feature extraction and detection. It also includes an NMS layer for filtering and selecting the detection results. It borrows the idea of multi-scale fruit detection. Fruit detection tasks are accomplished by generating multiple fruit feature maps of different scales during the fruit detection process. The network model calculates confidence scores for each category in predicted boxes and ground truth boxes, respectively. Then, an NMS operation is performed on the calculated scores of each prediction boxes. Finally, top-ranked prediction boxes are outputted as the final result of fruit detection.

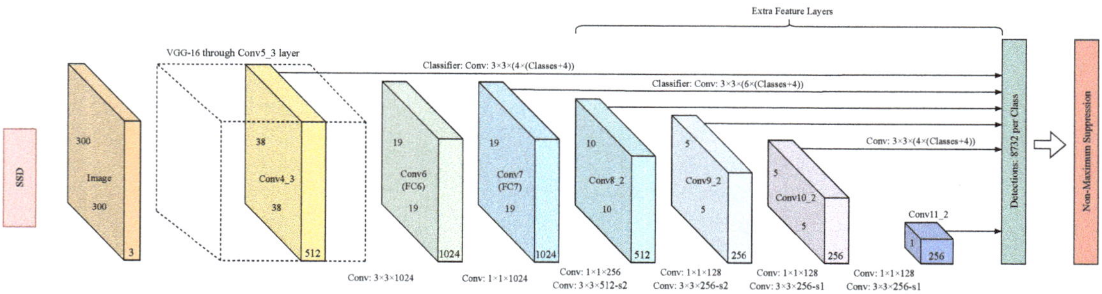

Figure 14. Typical SSD framework for fruit detection and recognition.

Validated on multiple fruit datasets, fruit detection and recognition methods based on SSD have high accuracy and speed. Vasconez, J.P. et al. (including Chilean researchers and a researcher based in America) (2020) [60] evaluated two of the most widely used architectures (Faster R-CNN with Inception V2 and SSD with MobileNet) for fruit detection. The former achieves 4.55 FPS, whereas the latter achieves a significantly higher performance of approximately 16.67 FPS. However, it is worth noting that fruit detection and recognition methods based on SSD preprocess input fruit images, which may lead to lower fruit detection accuracy for relatively small target fruits when passing through deeper convolutional layers. Chinese researchers Liang, Q. et al., (2018) [80] proposed a real-time detection method for on-tree mangoes based on SSD. New sampling strategies were designed to optimize data augmentation techniques. With optimized data augmentation techniques and default box proposals, SSD outperforms Faster R-CNN in mango detection. Detection results for an almond dataset further confirm the effectiveness of the proposed method. However, it is important to note that the proposed method has deeper layers and a larger number of parameters. This results in slower operation speed and longer computation time.

In general, fruit detection and recognition methods based on SSD also have certain disadvantages. They independently input fruit image features, extracted by different convolutional layers, into corresponding network detection branches. This means that the same fruits in detected images may be identified by bounding boxes of different sizes simultaneously, which can easily lead to the problem of repeated detection. Additionally, each detection branch only operates on target fruits in its respective field, making it difficult to consider the relationship between target fruits of different layers and scales. Therefore, the detection effect of fruit detection and recognition methods based on SSD on small target fruits is not good. Further research could improve SSD in detector frameworks, prediction mechanisms, matching mechanisms, and loss functions.

2.2. *Two-Stage Fruit Detection and Recognition Methods Based on Candidate Regions*

2.2.1. Fruit Detection and Recognition Methods Based on AlexNet, VGGNet, and ResNet

Typical AlexNet, VGGNet, and ResNet frameworks for fruit detection and recognition are shown in Figure 15. AlexNet was proposed by American researchers Krizhevsky, A. et al. in 2012 [81]. It is the first DL framework that extends CNN to the field of CV. Compared with techniques based on digital image processing and traditional ML, fruit detection and recognition methods based on AlexNet have great advantages in terms of accuracy. Chinese researchers Zhu, L. et al., (2018) [82] proposed a highly effective method for vegetable classification based on AlexNet. The accuracy achieved in the testing set was significantly improved compared to the BP neural network (78%) and SVM classifier method (80.5%), with a remarkable accuracy of 92.1%. Indian researchers Rangarajan, A.K. et al., (2018) [83] demonstrated that the classification accuracy of 13,262 fruit images was 97.49% for AlexNet. Fruit detection and recognition methods based on AlexNet have gained widespread acceptance due to their advantages. By modifying the size of the convolutional kernel and convolutional layer, fruit detection accuracy can be effectively improved. For example, Chinese researchers Ni, J. et al., (2021) [52] improved AlexNet by proposing a new architecture—E-AlexNet. The new architecture enhanced the convolutional layer, reduced kernel size, and used L2 regularization and a BN layer instead of LRN layer. E-AlexNet was compared with the original AlexNet by classifying five strawberry varieties with different qualities. The average recognition accuracy of E-AlexNet was 90.70%, while that of the original AlexNet was 84.50%.

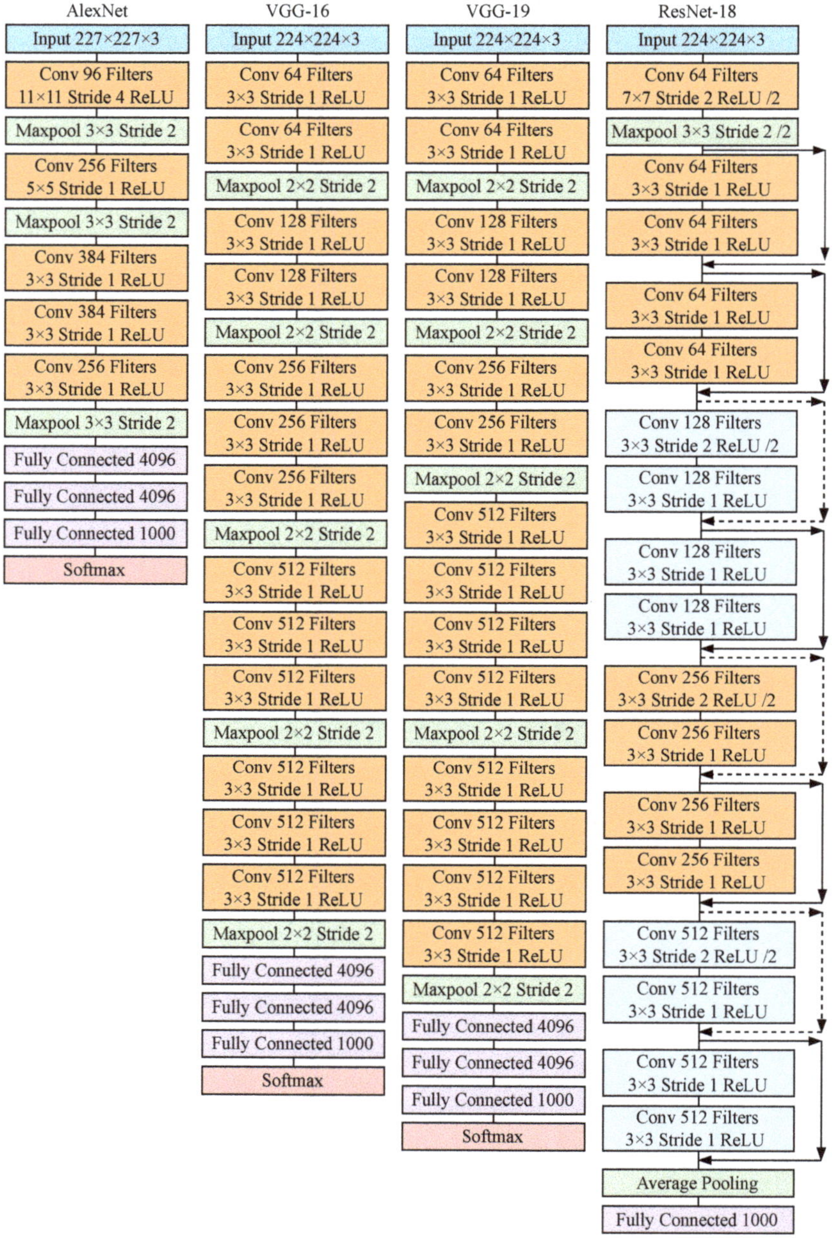

Figure 15. Typical AlexNet, VGGNet, and ResNet frameworks for fruit detection and recognition.

VGGNet was proposed by American researchers Simonyan, K. and Zisserman, A. in 2014 [84]. It has high accuracy in fruit detection and recognition. The biggest improvement of VGGNet is the depth of the network, which has been increased from 8 layers to 16 or 19 layers. Additionally, VGGNet uses a 3 × 3 convolution kernel to replace the large convolution kernels (11 × 11, 7 × 7, 5 × 5) in AlexNet. In the case of the same receptive field, the accumulation effect of the small convolution kernel is better than that of the large convolution kernel. For example, Indian researchers Mahmood, A. et al., (2022) [85] assessed the effectiveness of two CNN paradigms (AlexNet and VGG-16) in classifying jujube fruits based on their maturity level (unripe, ripe, and overripe). The best accuracy achieved by VGG-16 was 97.65%. Indian researchers Begum, N. and Hazarika, M.K. (2022) [86] used several fruit detection models (VGG-16, VGG-19, Inception V3, ResNet-101, and ResNet-152) to classify three tomato classes (immature, partially mature, and mature). VGG-19 had the best classification accuracy of 97.37% at epoch 50 and batch size 32. Chinese researchers Pérez-Pérez, B.D. et al., (2021) [87] pre-trained seven CNN architectures (AlexNet, VGG-16, VGG-19, ResNet-50, ResNet-101, ResNet-152, and Inception V3) using the ImageNet dataset. VGG-19, with the Adam optimizer, is the one that reported the best accuracy (99.32%).

In order to further improve the accuracy and speed of fruit detection and recognition, Chinese researchers Li, Z. et al., (2020) [88] proposed a fruit recognition and classification method based on VGG-M and VGG-M-BN. On the basis of the original VGG, VGG-M combined the output features of the first two fully connected layers. VGG-M-BN had the BN layer added. The convergence rate of VGG-M-BN is nearly three times faster. The quality of datasets, batch size, and different activation functions also influence fruit recognition and classification accuracy. Firstly, they used VGG-M-BN to train different numbers of vegetable datasets. Recognition accuracy decreases as the quality of datasets decreases. Secondly, by contrasting activation functions, they verified that the rectified linear unit (ReLU) activation function is better than the traditional Sigmoid and Tanh functions in VGG-M-BN. Finally, they verified that the fruit recognition and classification accuracy of VGG-M-BN increases as the batch size increases.

ResNet was proposed by American researchers He, K. et al. in 2015 [89]. It has a high pattern recognition capability. According to the number of backbone layers, ResNet can be further subdivided into ResNet-18, ResNet-50, ResNet-101, and ResNet-152. Fruit detection and recognition methods based on ResNet are widely used, by virtue of their advantages. Helwan, A. et al., (including Lebanese researchers and researchers based in Turkey) (2019) [90] performed automatic segmentation of bananas based on ResNet. Wang, D. et al., (including Chinese researchers and a researcher based in America) (2020) [55] developed a remote apple horizontal diameter detection system based on ResNet to achieve automatic measurement of apples throughout the entire growth period.

Capturing fruit feature information on multiple scales is one way to address the problem that target fruits are overlapped and occluded by branches and leaves. American researchers Rahnemoonfar, M. and Sheppard, C. (2017) [91] optimized the structure of Inception-ResNet. The Improved-Inception-ResNet can count efficiently, even if fruits are under shadow, overlapped, and occluded by leaves. However, although the above fruit detection and recognition methods have high accuracy, they are slow. To address this problem, Australian researchers Kang, H. and Chen, C. (2020) [92] introduced an enhanced deep neural network DaSNet-v2 with ResNet. It has the ability to carry out both detection and instance segmentation of fruits, alongside semantic segmentation of branches. To further improve the speed of fruit detection and meet the real-time requirements of harvesters, Australian researchers Kang, H. and Chen, C. (2019) [93] constructed a multifunctional network for the real-time detection and semantic segmentation of apples and branches. They combined it with the lightweight backbone of ResNet-101 to improve the real-time computational performance of the fruit detection model.

2.2.2. Fruit Detection and Recognition Methods Based on R-CNN, Fast R-CNN, and Faster R-CNN

Typical R-CNN, Fast R-CNN, and Faster R-CNN frameworks for fruit detection and recognition are shown in Figure 16. R-CNN was proposed by American researchers Girshick, R. et al. in 2014 [94]. It is the first algorithm to successfully apply DL to object detection and recognition. Fast R-CNN was proposed by American researcher Girshick, R, one of the creators of R-CNN, in 2015 [95]. It solves some problems of its predecessor, such as slow speed and a large overlap of proposal boxes. One of the key innovations of Fast R-CNN is the "RoI pooling layer", which operates by taking CNN feature maps and regions of interest as inputs and providing the corresponding features for each region. This allows Fast R-CNN to extract fruit features from all regions of interest in fruit images in a single pass, instead of R-CNN processing each region separately. It significantly improves the speed of fruit detection and recognition. However, Fast R-CNN still requires regions of fruit images to be extracted and provided as inputs to fruit detection models.

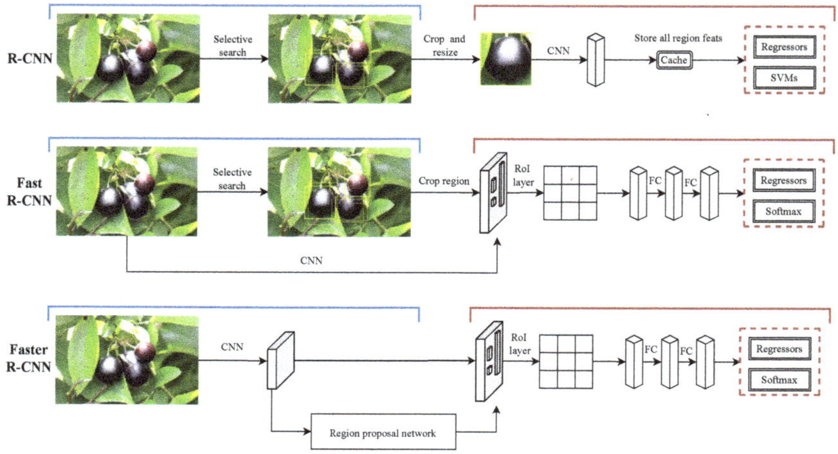

Figure 16. Typical R-CNN, Fast R-CNN, and Faster R-CNN frameworks for fruit detection and recognition.

Faster R-CNN was proposed by American researchers Ren, S. et al. in 2016 [96]. It takes images of fruits as inputs and returns a list of fruit classes, along with their corresponding bounding boxes. Its main innovation is the "RPN". By integrating region detection into the main neural network structure, Faster R-CNN achieves near real-time detection speed with high accuracy and generalization capability.

The fruit detection performance obtained by Faster R-CNN may outperform other networks (YOLOv3, SSD, and ReFCN) [97]. Therefore, fruit detection and recognition methods based on Faster R-CNN are widely used. Chinese researcher Wan, S. and Greek researcher Goudos, S. (2020) [98] proposed a multi-class fruit (apple, mango, and orange) detection method based on Faster R-CNN. The average detection accuracy was 90.72%, and the image processing time was 58ms. Fu, L. et al., (including Chinese researchers and researchers based in America) (2018) [99] proposed a kiwifruit detection method based on Faster R-CNN and evaluated it on kiwifruit images collected in field environments. Zhang, J. et al., (including Chinese researchers and researchers based in America) (2020) [100] used Faster R-CNN to improve a multi-class fruit detection method. They aimed to automatically detect apples, branches, and tree trunks in natural environments and estimate the bobbing locations of collected and captured apples.

Under changing lighting conditions, with low resolution, and with severe occlusion by adjacent fruits and leaves, fruit detection and recognition are very challenging tasks. To solve the problem, Chinese researchers Wang, P. et al., (2021) [57] proposed an improved Faster R-CNN with an attention mechanism based on a near-color background for young tomato detection and recognition. Small target fruit detection and recognition are also very challenging tasks. To solve this problem, in the localization phase, Chinese researchers Cao, C. et al., (2019) [101] proposed an improved loss function based on intersection and ratio for bounding box regression. Additionally, in the recognition phase, the bilinear interpolation method is used to improve the pooling operation of interest regions.

2.2.3. Fruit Detection and Recognition Methods Based on FCN, SegNet, and Mask R-CNN

FCN was proposed by American researchers Long, J. et al. in 2015 [102]. A typical FCN framework for fruit detection and recognition is shown in Figure 17. FCN classifies fruit images at the pixel level and solves the problem of semantic image segmentation. FCN replaces the fully connected layer of the original CNN with the convolutional layer so that the output will be a heatmap instead of a category. Meanwhile, to solve the problem of smaller image size due to convolution and pooling, up-sampling is used to recover image size. Chinese researchers Lin, G. et al., (2019) [61], German researchers Zabawa, L. et al., (2019) [103], and Li, Y. et al., (including Chinese researchers and a researcher based in Germany) (2017) [104] used FCN for the semantic segmentation of guava, grape, and cotton, respectively. Although guava can be segmented easily, the branch is a little difficult to segment. They also compared FCN with SegNet and classification and regression tree classifier (CART). FCN outperforms the other two methods. However, FCN makes some false predictions due to the effects of overlaps and changing lighting conditions. American researchers Chen, S.W. et al., (2017) [105] proposed a method based on FCN for accurate fruit counting in complex natural environments. The method works well even under highly shaded conditions. Furthermore, American researchers Liu, X. et al., (2018) [106] combined deep convolutional segmentation to accurately count sequential images of visible fruits.

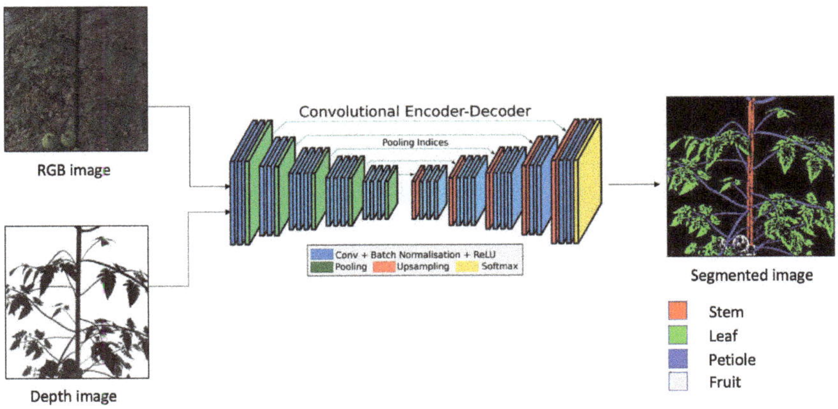

Figure 17. Typical FCN framework for fruit detection and recognition (Source: https://github.com/Alpharouk (accessed on 5 January 2023).

In general, fruit detection and recognition methods based on FCN can accept fruit image inputs of arbitrary size, and the recognition efficiency is higher. They avoid the problem of repeated storage and computational convolution caused by the use of pixel blocks. They reduce the computational effort of the whole fruit detection operation. However, the recognition accuracy is not high because they are insensitive to the details in fruit images, and the classification does not consider inter-pixel relationships.

SegNet was proposed by British researchers Badrinarayanan, V. et al. in 2017 [107]. A typical SegNet framework for fruit detection and recognition is shown in Figure 18. It follows the segmentation idea of FCN and is a symmetric network model with a supervised coding and decoding structure. SegNet can handle fruit image inputs of arbitrary sizes. The coding part reduces the size of input fruit images and the number of parameters stage by stage through maximum pooling, and records the pooling index positions in the fruit images. In order to ensure consistency in resolution between input and output fruit images, decoding processes recover fruit image information through up-sampling. Finally, it outputs semantic segmentation results through the SoftMax classifier. The major difference between SegNet and FCN is the method used for up-sampling low-resolution feature maps to high-resolution feature maps.

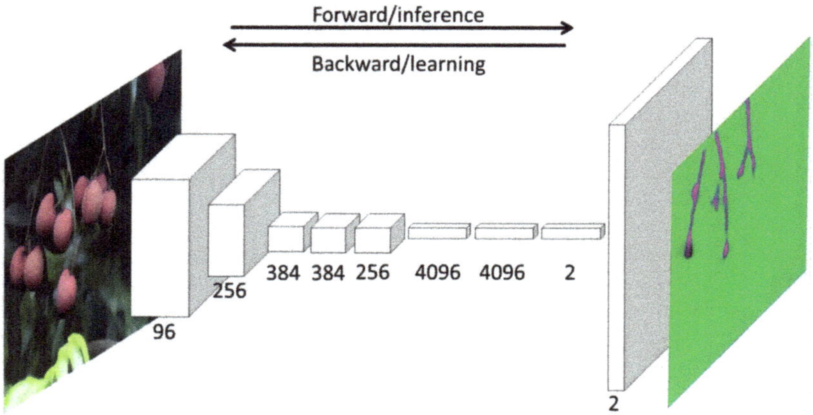

Figure 18. Typical SegNet framework for fruit detection and recognition (image reprinted with permission from ref. [108]. 2020, Peng, H.).

Harvesting robots usually operate in complex natural environments, and the random growth of trunks and branches poses a challenge for fruit detection and recognition. Majeed, Y. et al., (including American researchers and a researcher based in China) (2018) [109] developed a trunk and branch segmentation method using a Kinect V2 sensor. Harvesting robots need to optimize the position of the end effector based on the position and angle between fruits and robot components before approaching, grasping, and cutting target fruits. For this purpose, Dutch researchers Barth, R. et al., (2019) [110] proposed inferring the position of fruits and stems through sparse semantic segmentation in the image plane. In addition, to improve the efficiency of fruit detection and enhance real-time performance, Australian researchers Kang, H. and Chen, C. (2019) [93] used a semantic segmentation network to detect and segment apples and branches in an orchard in real-time. Meanwhile, in order to enable harvesting robots to simultaneously recognize and locate multiple target fruit clusters, Chinese researchers Li, J. et al., (2020) [62] proposed a semantic segmentation method to segment fruit RGB images into three categories: background, fruit, and branch. The method achieved accurate and automatic detection of fruits and branches of multiple lychee clusters in complex natural environments and guided robots to complete continuous harvesting tasks.

Mask R-CNN was proposed by American researchers He, K. et al. in 2017 [111]. A typical Mask R-CNN framework for fruit detection and recognition is shown in Figure 19. It consists of three parts. Firstly, the backbone network extracts fruit feature maps from input fruit images. Secondly, the fruit feature maps outputted by the backbone network are sent to the RPN to generate proposals. Finally, the proposals outputted by the RPN are mapped, and the corresponding target fruit features are extracted from the shared feature maps. These features are outputted to the FC and FCN for fruit classification and instance segmentation, respectively. The process generates classification confidence, bounding boxes, and mask images.

Mask R-CNN combines semantic segmentation with object detection by outputting mask images. This improves the localization accuracy of small target fruits, as well as the prediction accuracy of mask images. Fruit detection and recognition methods based on Mask R-CNN have better robustness and generality for fruit detection and recognition, especially in situations of clustered fruit growth. Chinese researchers Yu, Y. et al., (2019) [65] and Jia, W. et al., (2020) [112] used a Mask R-CNN instance segmentation network model to recognize overlapping strawberries and apples, respectively. They can determine not only categories but also individuals. Since some ripe green tomatoes are similar in color to branches and leaves, shaded by branches and leaves, or overlapped by other tomatoes, accurate detection and localization of these tomatoes is difficult. Chinese researchers

Zu, L. et al., (2021) [113] proposed using Mask R-CNN for the detection and segmentation of ripe green tomatoes. The research results showed the effectiveness of the method. The best model performance was achieved when the IoU was 0.5, and the F1-score of both the testing set bounding box and the masked region reached 92%. Chinese researchers Xu, P. et al., (2022) [64] proposed an improved Mask R-CNN network model for the recognition of cherry tomatoes, considering the prior neighborhood constraint between fruits and stalks.

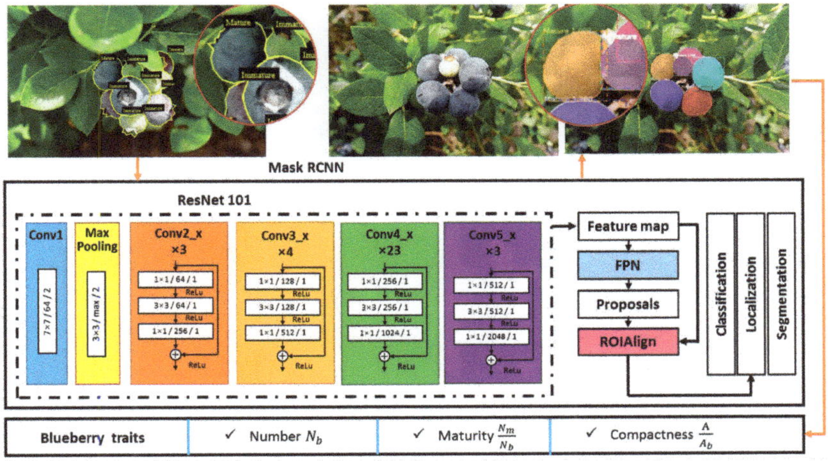

Figure 19. Typical Mask R-CNN framework for fruit detection and recognition (image reprinted with permission from ref. [114]. 2020, Ni X.).

In the future, improvements in fruit detection and recognition methods based on Mask R-CNN should focus on integrating more convolutions to improve performance, and reducing the computational complexity of multi-head attention in the transformer. In addition, multimodal fruit detection methods could be adopted to design Mask R-CNN fruit detection models based on vision, LiDAR, millimeter-wave radar, and other multisensor fusion technologies.

3. Discussion

Currently, there are many factors leading to low accuracy, slow speed, and poor robustness of fruit detection and recognition. They can be summarized in the following aspects: scarcity of high-quality fruit datasets, detection of small target fruits, fruit detection in occluded and dense scenarios, detection of multi-scale and multi-species fruits, and lightweight fruit detection models.

(1) Scarcity of high-quality fruit datasets. Fruit datasets, as signal sources to guide fruit detection algorithms based on DL for information understanding [41], largely determine the final performance of trained fruit detection models. Fruit detection and recognition methods based on DL have two requirements for datasets. One is the sufficiency of data, and the other is the richness of data categories. Fruit datasets are mainly collected in real field environments and through internet channels. A comparison of the advantages and shortcomings of the two collection methods is shown in Table 4. In order to objectively compare the performance of fruit detection and recognition methods, as shown in Table 5, international communities provide some public benchmark datasets. Different fruit datasets have significant differences in the number, quality, and category of images. Researchers can choose compatible fruit datasets for experiments according to their needs. The Fruits-360 dataset is the most commonly used public benchmark dataset. The total number of categories in this dataset is as high as 131, and the total amount of images is considerable.

However, it suffers from the problems of single-image backgrounds, insufficient data diversity, and category imbalance.

Table 4. Comparison of fruit image collection methods.

Types	Methods	Advantages	Shortcomings
Real fruit detection environment	Hand-held camera	High image quality of fruits; close to real scenes	The process of shooting is time-consuming and laborious; fruit image quality is unstable; fruit image quantization and contrast are difficult
	UGV		
	UAV		
Internet channel	--	No need for a camera; easy and fast collection	There are situations such as blurred images and incorrect labels; data cleaning and inspection are required

Table 5. Some frequently used fruit image databases.

Datasets	Samples			Species	Web-Link	Year
	Total	Training Set	Testing Set			
Fruit images of MS COCO	-	-	-	-	https://cocodataset.org/#download (accessed on 6 March 2023)	2017
Fruit images of ImageNet	-	-	-	-	https://image-net.org/challenges/LSVRC/index.php (accessed on 6 March 2023)	2012
Fruits-360	90,380	67,692	22,688	131 (100 × 100 pixels)	www.kaggle.com/datasets/moltean/fruits (accessed on 16 February 2023)	2020
Fruit-A	22,495	16,854	5641	33 (100 × 100 pixels)	www.kaggle.com/datasets/sshikamaru/fruit-recognition (accessed on 16 February 2023)	2022
Fruit-B	21,000	15,000	vail: 3000 text: 3000	15 (224 × 224 pixels)	www.kaggle.com/datasets/misrakahmed/vegetable-image-dataset (accessed on 16 February 2023)	2021
Fruit quality classification	19,526	-	-	18 (256 × 256/192 pixels)	www.kaggle.com/datasets/ryandpark/fruit-quality-classification (accessed on 16 February 2023)	2022
Fresh and rotten fruits	13,599	10,901	2698	6	www.kaggle.com/datasets/sriramr/fruits-fresh-and-rotten-for-classification (accessed on 16 February 2023)	2019

When public benchmark fruit detection datasets cannot meet practical needs, some scholars have created individual fruit datasets to train a fruit detection model for fruit detection and recognition in specific environments. In particular, most of the existing public benchmark fruit detection datasets, such as fruit images of MS COCO and ImageNet, are collected through internet channels. Many of these images differ greatly from actual fruit recognition and harvesting situations. They consist of data from simple scenes, mainly for large and medium-sized fruits. Additionally, datasets for small target fruit detection in complex scenes are especially scarce. International communities might consider continually providing and updating quality public benchmark fruit detection datasets, for example, establishing a unified standard fruit data-sharing platform. The public can upload their fruit images to the platform, and the platform organizes personnel to identify and annotate them.

Due to the scarcity of high-quality fruit datasets, there are potential directions for development in the future: (1) Fruit detection and recognition methods based on small-sample learning may be a key breakthrough. For certain fruit categories for which it is difficult to obtain a large number of samples, this method allows a small number of fruit samples to be selected as representative of new fruit categories. Then, the inherent internal connection between the base fruit class and the new fruit class is used to realize effective knowledge transfer. (2) Fruit detection and recognition methods based on unsupervised learning/semi-supervised learning may be another key breakthrough. Current methods are mainly based on supervised learning, in which performance relies on a large amount of labeled fruit data. In unsupervised/semi-supervised learning, the model is pre-trained on the data with no or little labeled information.

(2) Detection of small target fruits. Fruits always grow in complex environments with many uncertainties (as shown in Figure 20). We usually define a small target fruit as being smaller than 32 × 32 pixels relative to the absolute size of the image it is in. The difficulties of small target fruit detection are as follows: (1) Limited features that can be extracted. Small target fruits have a small area share and low resolution in images, and they contain limited features themselves. (2) Convolution operations can cause loss of small target features. Fruit detection and recognition methods based on DL extract information of interest about fruits by performing convolution operations on fruit images containing a large amount of redundant information [50]. Fruit feature maps keep shrinking as the number of convolutions increases. If the down-sampling rate is too high, a lot of detailed information for small target fruit detection will be lost. (3) Requirements for the positioning accuracy of the small target fruit bounding boxes are higher. Compared with large target fruits, small target fruits are more sensitive to the offset of prediction boxes and less tolerant of errors. (4) The scale of anchor boxes has not been designed properly. When the scale of anchor boxes is too large, the area of small target fruits is reduced. Therefore, even if small target fruits are within anchor frames, the IoU may not reach the threshold value, resulting in missed detection. In addition, when the receptive field is too large, the fruit detection results are easily disturbed by a large number of other features. When the preset scale of anchor boxes is too close, the spatial difference after down-sampling cannot be guaranteed, resulting in small target fruits being ignored. (5) Sample imbalance. The IoU-based positive and negative samples are considered negative if the IoU is smaller than the threshold. This may lead to small target fruits being ignored in the process of model learning due to the small number of positive samples. Small target fruits usually grow in clusters, which may further cause occlusion and dense detection problems. When small target fruits appear together with other scaled fruits, this gives rise to multi-scale detection problems.

Figure 20. Examples of fruit images acquired under different conditions (photos reprinted with permission from ref. [48]. 2022, Chen J.).

Current solutions for small target fruit detection mainly include: (1) Increasing the number of small target fruit samples through data preprocessing and enhancement, such as in the research presented in [46,49,55,56,60,62–64]; (2) Generating higher-quality small target fruit candidate regions by improving the RPN, such as in the research presented in [56,57,65]; (3) Ensuring the sensory field and small target fruit matching by optimizing anchor boxes, such as in the research presented in [58,61,62]. Combining traditional methods for small target fruit image detection may be a trend for future development. Some small target fruit images contain little information, and they lack the necessary semantic information. Fruit detection and recognition methods based on DL have limited feature extraction ability for small target fruits at the pixel level. Therefore, traditional feature extractors can be introduced to make them more capable of representing features of fruit images. In addition, depth features extracted using CNN can also be combined with traditional methods, such as saliency detection and superpixel segmentation, to obtain a more effective fruit feature representation.

(3) Fruit detection in occluded and dense scenarios. Fruit growth environments are usually complex. There are cases of inter-fruit occlusion and occlusion by shadows or other distractions, such as branches and leaves. The difficulty of detecting fruits in occluded and dense scenarios lies in improving the recall rate of the occluded target fruits [29]. In overlapping cases, the main reasons for missing the detection of obscured target fruits are: (1) Fruits are incomplete, and extractable fruit features are sharply reduced. (2) Overlapping target fruits usually have highly similar features, and it is difficult for fruit detection models to determine whether they belong to different individuals. (3) The NMS post-processing method directly discards objects with lower scores in overlapping regions. For fruit detection in occluded and dense scenarios, the main method of improvement is to enhance fruit feature extraction.

Commonly used methods to enhance the feature extraction capability of fruits are: (1) increasing the width or depth of networks, such as in the research presented in [52,53,57,59]. However, this method will increase the computational load of models. This requires us to strike a balance between performance improvement and computational cost increase. (2) Adding attention mechanisms, such as in the research presented in [55–57]. The introduction of attention mechanisms can help fruit detection models fully consider the connection between each position of target fruits, effectively enhancing the ability of fruit detection models to learn fruit features. Current scholars divide them into the channel attention mechanism and spatial attention mechanism, according to the way the attention acts on feature maps. In fruit detection models, common implementations of attention mechanisms include squeeze-and-excitation networks (SENet) and the convolutional block attention module (CBAM). However, adding an attention mechanism will make fruit detection models more complex and increase convergence time. At the same time, adding an attention mechanism requires careful consideration of whether the design principle of attention, as well as the position and method of action, are suitable for current tasks. Otherwise, it may have a negative impact on fruit detection models. How to reasonably design and implement attention mechanisms, and efficiently use a wide range of environmental features, are important research directions for the future.

(4) Detection of multi-scale and multi-species fruit. Most current fruit detection models are solutions for specific crops. When the detected fruits appear on multiple scales or in multiple types, it is difficult to guarantee the model's generalization ability. For the multi-scale fruit detection problem, the multi-scale training method may be a key breakthrough. It can enable fruit detection models to process fruit information at different scales and improve their ability to capture cross-scale fruit information. Overall, the use of multi-scale fruit prediction networks can make full use of receptive fields, which can effectively alleviate the lack of scale invariance in convolutional neural networks. However, this also increases the number of calculations, resulting in higher demand for hardware facilities. For multi-category fruit detection problems, a common solution is to use transfer learning technology to fine-tune existing models. However, this may result in a loss of detection accuracy for the original fruit categories. Adding a large amount of new category data to the original dataset and retraining a new model can ensure the detection effectiveness of

the original fruit categories. However, every time a new category appears, it needs to be trained from scratch. This not only consumes time and resources, but also cannot satisfy complex, dynamic environments such as farmland orchards. For this problem, we believe that introducing the idea of incremental learning can improve the generalization ability and adaptive learning ability of fruit detection models.

(5) Lightweight fruit detection models. With the continuous development of the field of fruit detection and recognition, researchers are also committed to improving the accuracy of fruit detection models, and the fruit detection models are gradually becoming more complex. For example, some researchers add a super-resolution module to the localization part of fruit detection networks. This may increase the computational load, which, in turn, makes the fruit detection model more dependent on high-performance computing resources. How to further optimize network structures, reduce the number of model parameters, decrease computational complexity, improve running speed, and deploy them on mobile devices are currently hot research topics. Model pruning, quantization, knowledge distillation, and matrix decomposition are effective ways to achieve lightweight and high efficiency. For example, the lightweight MobileNet [11] or ResNet-101 [23] are used to replace the original backbone feature extraction network for fruit feature extraction. At the same time, optimizing operators within frameworks and using AI chips in hardware can greatly accelerate the running speed and parallelism of fruit detection models.

4. Conclusions

Fruit detection and recognition methods based on DL are the mainstream methods for accurate, fast, and robust fruit detection and recognition. These methods are also an important development trend. They are relatively less affected by environments. Our work focuses on providing an overview and review of DL applied to fruit image recognition, mainly in the areas of detection and classification. In order to further define the study areas of this paper, we identify fruit detection and classification tasks as the determination of the class based on their specific types. In general, current fruit detection and recognition methods based on DL can be divided into the following areas: methods based on YOLO, SSD, AlexNet, VGGNet, ResNet, Faster R-CNN, FCN, SegNet, and Mask R-CNN. These methods can also be classified into two categories: single-stage fruit detection and recognition methods (YOLO, SSD) based on regression, and two-stage fruit detection and recognition methods (AlexNet, VGGNet, ResNet, Faster R-CNN, FCN, SegNet, and Mask R-CNN) based on candidate regions.

Most of the current research work is based on two-stage fruit detection and recognition methods. Improvement and application research based on Faster R-CNN (21%) is currently a hotspot. The recognition accuracy of fruit detection and recognition methods based on Faster R-CNN is high, but the recognition speed is limited by complex anchor frame mechanisms. When there are mobile deployment and high recognition speed requirements, fruit detection and recognition methods based on YOLO (17%) are used most frequently. Their recognition speed is fast, but the recognition effect on small target fruits is not very good. In addition, ResNet (11%) is the most popular backbone network, followed by AlexNet (7%). Most of the research focuses on apples (32.14%), followed by tomatoes (8.93%), and citrus (7.14%). These three kinds of fruits are in high demand and yield globally. There are some reasons that make them ideal candidates for automatic harvesting. Firstly, they hang from plants individually, making them easily detectable based on their distinctive features. Secondly, they have no extreme variations in size or weight. Lastly, they are relatively hard and not easily damaged in mechanical operations. However, in terms of fruit dimensions and peduncle length, different cultivars may exhibit different characteristics that can affect fruit detection and recognition performance. This poses challenges for adapting fruit detection and recognition methods for different cultivars. Future work could aim to identify cultivars that are more suitable for automatic harvesting.

The scarcity of high-quality fruit datasets, detection of small target fruits, fruit detection in occluded and dense scenarios, detection of multiple scales and multiple species of fruits, and lightweight fruit detection models are the current challenges of fruit detection and recognition based on DL for automatic harvesting. The quality and scale of fruit datasets, appropriate improvement strategies, and underlying model architectures all have a significant impact on the detection and recognition performance. For example, fruit data preprocessing can standardize data by cleaning and adjusting them. Fruit data augmentation can effectively expand data and increase data diversity, thereby reducing the dependence on specific factors and improving model robustness. Fruit feature fusion is conducive to alleviating the problem of fruit feature disappearance and improving the detection effect of small target fruits and multi-scale fruits. Building a multi-task learning model, the original fruit detection framework is beneficial for obtaining more fruit information by combining other learning tasks. Moreover, establishing a parameter-sharing mechanism through multi-task learning can significantly improve the performance of fruit detection and recognition. Two-stage fruit detection and recognition methods pursue faster speeds and lighter weights while ensuring fruit detection accuracy. Single-stage fruit detection and recognition methods improve fruit detection accuracy while maintaining the advantages of detection speed and model size. Achieving higher fruit detection performance and a balance between fruit detection precision and speed are current development trends.

Future research should prioritize addressing these current challenges and improving the accuracy, speed, robustness, and generalization of fruit vision detection systems, while reducing the overall complexity and cost. This paper hopes to provide a reference for follow-up research in the field of fruit detection and recognition based on DL for automatic harvesting.

Author Contributions: Conceptualization, F.X.; methodology, F.X.; analysis, F.X. and Y.X.; investigation, F.X., Y.X., H.W. and R.Z.; resources, F.X. and H.W.; data curation, F.X.; writing—original draft preparation, F.X.; writing—review and editing, F.X. and H.W.; visualization, F.X.; supervision, H.W.; project administration, F.X. and H.W.; funding acquisition, H.W. All authors have read and agreed to the published version of the manuscript.

Funding: This research was funded by the Natural Science Foundation of Heilongjiang Province of China (LH2020C047) and China Postdoctoral Science Foundation (2019T120248).

Data Availability Statement: Not applicable.

Conflicts of Interest: The authors declare no conflict of interest.

References

1. Brown, J.; Sukkarieh, S. Design and Evaluation of a Modular Robotic Plum Harvesting System Utilizing Soft Components. *J. Field Robot.* **2021**, *38*, 289–306. [CrossRef]
2. Yan, B.; Fan, P.; Lei, X.; Liu, Z.; Yang, F. A Real-Time Apple Targets Detection Method for Picking Robot Based on Improved YOLOv5. *Remote Sens.* **2021**, *13*, 1619. [CrossRef]
3. He, L.; Fu, H.; Karkee, M.; Zhang, Q. Effect of Fruit Location on Apple Detachment with Mechanical Shaking. *Biosyst. Eng.* **2017**, *157*, 63–71. [CrossRef]
4. Ji, W.; Zhao, D.; Cheng, F.; Xu, B.; Zhang, Y.; Wang, J. Automatic Recognition Vision System Guided for Apple Harvesting Robot. *Comput. Electr. Eng.* **2012**, *38*, 1186–1195. [CrossRef]
5. Zhao, D.; Lv, J.; Ji, W.; Zhang, Y.; Chen, Y. Design and Control of an Apple Harvesting Robot. *Biosyst. Eng.* **2011**, *110*, 112–122. [CrossRef]
6. Arad, B.; Balendonck, J.; Barth, R.; Ben-Shahar, O.; Edan, Y.; Hellström, T.; Hemming, J.; Kurtser, P.; Ringdahl, O.; Tielen, T.; et al. Development of a Sweet Pepper Harvesting Robot. *J. Field Robot.* **2020**, *37*, 1027–1039. [CrossRef]
7. Lehnert, C.; English, A.; McCool, C.; Tow, A.W.; Perez, T. Autonomous Sweet Pepper Harvesting for Protected Cropping Systems. *IEEE Robot. Autom. Lett.* **2017**, *2*, 872–879. [CrossRef]
8. Bac, C.W.; Hemming, J.; Van Henten, E.J. Stem Localization of Sweet-Pepper Plants Using the Support Wire as a Visual Cue. *Comput. Electron. Agric.* **2014**, *105*, 111–120. [CrossRef]
9. Xiong, Y.; Ge, Y.; Grimstad, L.; From, P.J. An Autonomous Strawberry-Harvesting Robot: Design, Development, Integration, and Field Evaluation. *J. Field Robot.* **2020**, *37*, 202–224. [CrossRef]
10. Xiong, Y.; Peng, C.; Grimstad, L.; From, P.J.; Isler, V. Development and Field Evaluation of a Strawberry Harvesting Robot with a Cable-Driven Gripper. *Comput. Electron. Agric.* **2019**, *157*, 392–402. [CrossRef]

11. Hayashi, S.; Shigematsu, K.; Yamamoto, S.; Kobayashi, K.; Kohno, Y.; Kamata, J.; Kurita, M. Evaluation of a Strawber-ry-Harvesting Robot in a Field Test. *Biosyst. Eng.* **2010**, *105*, 160–171. [CrossRef]
12. Xiong, J.; He, Z.; Lin, R.; Liu, Z.; Bu, R.; Yang, Z.; Peng, H.; Zou, X. Visual Positioning Technology of Picking Robots for Dynamic Litchi Clusters with Disturbance. *Comput. Electron. Agric.* **2018**, *151*, 226–237. [CrossRef]
13. Feng, Q.; Zou, W.; Fan, P.; Zhang, C.; Wang, X. Design and Test of Robotic Harvesting System for Cherry Tomato. *Int. J. Agric. Biol. Eng.* **2018**, *11*, 96–100. [CrossRef]
14. Kondo, N.; Yata, K.; Iida, M.; Shiigi, T.; Monta, M.; Kurita, M.; Omori, H. Development of an End-Effector for a Tomato Cluster Harvesting Robot. *Eng. Agric. Environ. Food* **2010**, *3*, 20–24. [CrossRef]
15. Williams, H.A.M.; Jones, M.H.; Nejati, M.; Seabright, M.J.; Bell, J.; Penhall, N.D.; Barnett, J.J.; Duke, M.D.; Scarfe, A.J.; Ahn, H.S.; et al. Robotic Kiwifruit Harvesting Using Machine Vision, Convolutional Neural Networks, and Robotic Arms. *Biosyst. Eng.* **2019**, *181*, 140–156. [CrossRef]
16. Xiao, F.; Wang, H.; Li, Y.; Cao, Y.; Lv, X.; Xu, G. Object Detection and Recognition Techniques Based on Digital Image Processing and Traditional Machine Learning for Fruit and Vegetable Harvesting Robots: An Overview and Review. *Agronomy* **2023**, *13*, 639. [CrossRef]
17. Fu, L.; Gao, F.; Wu, J.; Li, R.; Karkee, M.; Zhang, Q. Application of Consumer RGB-D Cameras for Fruit Detection and Localization in Field: A Critical Review. *Comput. Electron. Agric.* **2020**, *177*, 105687. [CrossRef]
18. Okamoto, H.; Lee, W.S. Green Citrus Detection Using Hyperspectral Imaging. *Comput. Electron. Agric.* **2009**, *66*, 201–208. [CrossRef]
19. Wachs, J.P.; Stern, H.I.; Burks, T.; Alchanatis, V. Low and High-Level Visual Feature-Based Apple Detection from Multi-Modal Images. *Precis. Agric.* **2010**, *11*, 717–735. [CrossRef]
20. Rehman, T.U.; Mahmud, M.S.; Chang, Y.K.; Jin, J.; Shin, J. Current and Future Applications of Statistical Machine Learning Algorithms for Agricultural Machine Vision Systems. *Comput. Electron. Agric.* **2019**, *156*, 585–605. [CrossRef]
21. Patrício, D.I.; Rieder, R. Computer Vision and Artificial Intelligence in Precision Agriculture for Grain Crops: A Systematic Review. *Comput. Electron. Agric.* **2018**, *153*, 69–81. [CrossRef]
22. Yandun Narvaez, F.; Reina, G.; Torres-Torriti, M.; Kantor, G.; Cheein, F.A. A Survey of Ranging and Imaging Techniques for Precision Agriculture Phenotyping. *IEEE/ASME Trans. Mechatron.* **2017**, *22*, 2428–2439. [CrossRef]
23. Jha, K.; Doshi, A.; Patel, P.; Shah, M. A Comprehensive Review on Automation in Agriculture Using Artificial Intelligence. *Artif. Intell. Agric.* **2019**, *2*, 1–12. [CrossRef]
24. Wolfert, S.; Ge, L.; Verdouw, C.; Bogaardt, M.J. Big Data in Smart Farming—A Review. *Agric. Syst.* **2017**, *153*, 69–80. [CrossRef]
25. Saleem, M.H.; Potgieter, J.; Arif, K.M. Plant Disease Detection and Classification by Deep Learning. *Plants* **2019**, *8*, 468. [CrossRef]
26. Wang, D.; Vinson, R.; Holmes, M.; Seibel, G.; Bechar, A.; Nof, S.; Tao, Y. Early Detection of Tomato Spotted Wilt Virus by Hyperspectral Imaging and Outlier Removal Auxiliary Classifier Generative Adversarial Nets (OR-AC-GAN). *Sci. Rep.* **2019**, *9*, 4377. [CrossRef]
27. Wang, A.; Zhang, W.; Wei, X. A Review on Weed Detection Using Ground-Based Machine Vision and Image Processing Techniques. *Comput. Electron. Agric.* **2019**, *158*, 226–240. [CrossRef]
28. Lv, J.; Xu, H.; Xu, L.; Zou, L.; Rong, H.; Yang, B.; Niu, L.; Ma, Z. Recognition of Fruits and Vegetables with Similar-Color Background in Natural Environment: A Survey. *J. Field Robot.* **2022**, *39*, 888–904. [CrossRef]
29. Li, Y.; Feng, Q.; Li, T.; Xie, F.; Liu, C.; Xiong, Z. Advance of Target Visual Information Acquisition Technology for Fresh Fruit Robotic Harvesting: A Review. *Agronomy* **2022**, *12*, 1336. [CrossRef]
30. Aslam, F.; Khan, Z.; Tahir, A.; Parveen, K.; Albasheer, F.O.; Ul Abrar, S.; Khan, D.M. A Survey of Deep Learning Methods for Fruit and Vegetable Detection and Yield Estimation. In *Big Data Analytics and Computational Intelligence for Cybersecurity*, 2nd ed.; Ouaissa, M., Boulouard, Z., Ouaissa, M., Khan, I.U., Kaosar, M., Eds.; Springer: Cham, Switzerland, 2022; Volume 111, pp. 299–323. [CrossRef]
31. Li, Z.; Yuan, X.; Wang, C. A Review on Structural Development and Recognition-Localization Methods for End-Effector of Fruit-Vegetable Picking Robots. *Int. J. Adv. Robot. Syst.* **2022**, *19*, 17298806221104906. [CrossRef]
32. Darwin, B.; Dharmaraj, P.; Prince, S.; Popescu, D.E.; Hemanth, D.J. Recognition of Bloom/Yield in Crop Images Using Deep Learning Models for Smart Agriculture: A Review. *Agronomy* **2021**, *11*, 646. [CrossRef]
33. Maheswari, P.; Raja, P.; Apolo-Apolo, O.E.; Pérez-Ruiz, M. Intelligent Fruit Yield Estimation for Orchards Using Deep Learning Based Semantic Segmentation Techniques-A Review. *Front. Plant Sci.* **2021**, *12*, 684328. [CrossRef] [PubMed]
34. Bhargava, A.; Bansal, A. Fruits and Vegetables Quality Evaluation Using Computer Vision: A Review. *J. King Saud Univ. Comput. Inf. Sci.* **2021**, *33*, 243–257. [CrossRef]
35. Saleem, M.H.; Potgieter, J.; Arif, K.M. Automation in Agriculture by Machine and Deep Learning Techniques: A Review of Recent Developments. *Precis. Agric.* **2021**, *22*, 2053–2091. [CrossRef]
36. Tang, Y.; Chen, M.; Wang, C.; Luo, L.; Li, J.; Lian, G.; Zou, X. Recognition and Localization Methods for Vision-Based Fruit Picking Robots: A Review. *Front. Plant Sci.* **2020**, *11*, 510. [CrossRef]
37. Jia, W.; Zhang, Y.; Lian, J.; Zheng, Y.; Zhao, D.; Li, C. Apple Harvesting Robot under Information Technology: A Review. *Int. J. Adv. Robot. Syst.* **2020**, *17*, 1729881420925310. [CrossRef]
38. Tripathi, M.K.; Maktedar, D.D. A Role of Computer Vision in Fruits and Vegetables among Various Horticulture Products of Agriculture Fields: A Survey. *Inf. Process. Agric.* **2020**, *7*, 183–203. [CrossRef]

39. Naranjo-Torres, J.; Mora, M.; Hernández-García, R.; Barrientos, R.J.; Fredes, C.; Valenzuela, A. A Review of Convolutional Neural Network Applied to Fruit Image Processing. *Appl. Sci.* **2020**, *10*, 3443. [CrossRef]
40. Koirala, A.; Walsh, K.B.; Wang, Z.; McCarthy, C. Deep Learning-Method Overview and Review of Use for Fruit Detection and Yield Estimation. *Comput. Electron. Agric.* **2019**, *162*, 219–234. [CrossRef]
41. Zhu, N.; Liu, X.; Liu, Z.; Hu, K.; Wang, Y.; Tan, J.; Huang, M.; Zhu, Q.; Ji, X.; Jiang, Y.; et al. Deep Learning for Smart Agriculture: Concepts, Tools, Applications, and Opportunities. *Int. J. Agric. Biol. Eng.* **2018**, *11*, 32–44. [CrossRef]
42. Martín-Martín, A.; Orduna-Malea, E.; Thelwall, M.; Delgado López-Cózar, E. Google Scholar, Web of Science, and Scopus: A Systematic Comparison of Citations in 252 Subject Categories. *J. Informetr.* **2018**, *12*, 1160–1177. [CrossRef]
43. Hinton, G.E.; Salakhutdinov, R.R. Reducing the Dimensionality of Data with Neural Networks. *Science* **2006**, *313*, 504–507. [CrossRef] [PubMed]
44. LeCun, Y.; Boser, B.; Denker, J.S.; Henderson, D.; Howard, R.E.; Hubbard, W.; Jackel, L.D. Backpropagation Applied to Handwritten Zip Code Recognition. *Neural Comput.* **1989**, *1*, 541–551. [CrossRef]
45. Lecun, Y.; Bottou, L.; Bengio, Y.; Haffner, P. Gradient-Based Learning Applied to Document Recognition. *Proc. IEEE* **1998**, *86*, 2278–2324. [CrossRef]
46. Jahanbakhshi, A.; Momeny, M.; Mahmoudi, M.; Zhang, Y.D. Classification of Sour Lemons Based on Apparent Defects Using Stochastic Pooling Mechanism in Deep Convolutional Neural Networks. *Sci. Hortic.* **2020**, *263*, 109133. [CrossRef]
47. Sakib, S.; Ashrafi, Z.; Sidique, A.B. Implementation of Fruits Recognition Classifier Using Convolutional Neural Network Algorithm for Observation of Accuracies for Various Hidden Layers. *arXiv* **2019**, arXiv:1904.00783. [CrossRef]
48. Chen, J.; Liu, H.; Zhang, Y.; Zhang, D.; Ouyang, H.; Chen, X. A Multiscale Lightweight and Efficient Model Based on YOLOv7: Applied to Citrus Orchard. *Plants* **2022**, *11*, 3260. [CrossRef]
49. Khosravi, H.; Saedi, S.I.; Rezaei, M. Real-Time Recognition of on-Branch Olive Ripening Stages by a Deep Convolutional Neural Network. *Sci. Hortic.* **2021**, *287*, 110252. [CrossRef]
50. Quiroz, I.A.; Alférez, G.H. Image Recognition of Legacy Blueberries in a Chilean Smart Farm through Deep Learning. *Comput. Electron. Agric.* **2020**, *168*, 105044. [CrossRef]
51. Barbedo, J.G.A. Impact of Dataset Size and Variety on the Effectiveness of Deep Learning and Transfer Learning for Plant Disease Classification. *Comput. Electron. Agric.* **2018**, *153*, 46–53. [CrossRef]
52. Ni, J.; Gao, J.; Li, J.; Yang, H.; Hao, Z.; Han, Z. E-AlexNet: Quality Evaluation of Strawberry Based on Machine Learning. *Food Meas.* **2021**, *15*, 4530–4541. [CrossRef]
53. Marani, R.; Milella, A.; Petitti, A.; Reina, G. Deep Neural Networks for Grape Bunch Segmentation in Natural Images from a Consumer-Grade Camera. *Precis. Agric.* **2021**, *22*, 387–413. [CrossRef]
54. Altaheri, H.; Alsulaiman, M.; Muhammad, G. Date Fruit Classification for Robotic Harvesting in a Natural Environment Using Deep Learning. *IEEE Access* **2019**, *7*, 117115–117133. [CrossRef]
55. Wang, D.; Li, C.; Song, H.; Xiong, H.; Liu, C.; He, D. Deep Learning Approach for Apple Edge Detection to Remotely Monitor Apple Growth in Orchards. *IEEE Access* **2020**, *8*, 26911–26925. [CrossRef]
56. Tu, S.; Pang, J.; Liu, H.; Zhuang, N.; Chen, Y.; Zheng, C.; Wan, H.; Xue, Y. Passion Fruit Detection and Counting Based on Multiple Scale Faster R-CNN Using RGB-D Images. *Precis. Agric.* **2020**, *21*, 1072–1091. [CrossRef]
57. Wang, P.; Niu, T.; He, D. Tomato Young Fruits Detection Method under Near Color Background Based on Improved Faster R-CNN with Attention Mechanism. *Agriculture* **2021**, *11*, 1059. [CrossRef]
58. Li, C.; Lin, J.; Li, B.; Zhang, S.; Li, J. Partition Harvesting of a Column-Comb Litchi Harvester Based on 3D Clustering. *Comput. Electron. Agric.* **2022**, *197*, 106975. [CrossRef]
59. Miao, Z.; Yu, X.; Li, N.; Zhang, Z.; He, C.; Li, Z.; Deng, C.; Sun, T. Efficient Tomato Harvesting Robot Based on Image Processing and Deep Learning. *Precis. Agric.* **2022**, *24*, 254–287. [CrossRef]
60. Vasconez, J.P.; Delpiano, J.; Vougioukas, S.; Auat Cheein, F. Comparison of Convolutional Neural Networks in Fruit Detection and Counting: A Comprehensive Evaluation. *Comput. Electron. Agric.* **2020**, *173*, 105348. [CrossRef]
61. Lin, G.; Tang, Y.; Zou, X.; Xiong, J.; Li, J. Guava Detection and Pose Estimation Using a Low-Cost RGB-D Sensor in the Field. *Sensors* **2019**, *19*, 428. [CrossRef]
62. Li, J.; Tang, Y.; Zou, X.; Lin, G.; Wang, H. Detection of Fruit-Bearing Branches and Localization of Litchi Clusters for Vision-Based Harvesting Robots. *IEEE Access* **2020**, *8*, 117746–117758. [CrossRef]
63. Majeed, Y.; Zhang, J.; Zhang, X.; Fu, L.; Karkee, M.; Zhang, Q.; Whiting, M.D. Deep Learning Based Segmentation for Automated Training of Apple Trees on Trellis Wires. *Comput. Electron. Agric.* **2020**, *170*, 105277. [CrossRef]
64. Xu, P.; Fang, N.; Liu, N.; Lin, F.; Yang, S.; Ning, J. Visual Recognition of Cherry Tomatoes in Plant Factory Based on Improved Deep Instance Segmentation. *Comput. Electron. Agric.* **2022**, *197*, 106991. [CrossRef]
65. Yu, Y.; Zhang, K.; Yang, L.; Zhang, D. Fruit Detection for Strawberry Harvesting Robot in Non-Structural Environment Based on Mask-RCNN. *Comput. Electron. Agric.* **2019**, *163*, 104846. [CrossRef]
66. Redmon, J.; Divvala, S.; Girshick, R.; Farhadi, A. You Only Look Once: Unified, Real-Time Object Detection. In Proceedings of the 2016 IEEE Conference on Computer Vision and Pattern Recognition (CVPR), Las Vegas, NV, USA, 26 June–1 July 2016. [CrossRef]
67. Redmon, J.; Farhadi, A. YOLO9000: Better, Faster, Stronger. In Proceedings of the 2017 IEEE Conference on Computer Vision and Pattern Recognition (CVPR), Honolulu, HI, USA, 21–26 July 2017. [CrossRef]
68. Redmon, J.; Farhadi, A. YOLOv3: An Incremental Improvement. *arXiv* **2018**, arXiv:1804.02767. [CrossRef]

69. Bochkovskiy, A.; Wang, C.Y.; Liao, H.Y.M. YOLOv4: Optimal Speed and Accuracy of Object Detection. *arXiv* **2020**, arXiv:2004.10934. [CrossRef]
70. YOLOv5. Available online: https://github.com/ultralytics/yolov5 (accessed on 7 February 2023).
71. Wang, C.Y.; Yeh, I.H.; Liao, H.Y.M. You Only Learn One Representation: Unified Network for Multiple Tasks. *arXiv* **2021**, arXiv:2105.04206. [CrossRef]
72. Ge, Z.; Liu, S.; Wang, F.; Li, Z.; Sun, J. YOLOX: Exceeding YOLO Series in 2021. *arXiv* **2021**, arXiv:2107.08430. [CrossRef]
73. Li, C.; Li, L.; Jiang, H.; Weng, K.; Geng, Y.; Li, L.; Ke, Z.; Li, Q.; Cheng, M.; Nie, W.; et al. YOLOv6: A Single-Stage Object Detection Framework for Industrial Applications. *arXiv* **2022**, arXiv:2209.02976. [CrossRef]
74. Wang, C.Y.; Bochkovskiy, A.; Liao, H.Y.M. YOLOv7: Trainable Bag-of-Freebies Sets New State-of-the-Art for Real-Time Object Detectors. *arXiv* **2022**, arXiv:2207.02696. [CrossRef]
75. YOLOv8. Available online: https://github.com/ultralytics/ultralytics (accessed on 7 February 2023).
76. Xiong, J.; Liu, Z.; Chen, S.; Liu, B.; Zheng, Z.; Zhong, Z.; Yang, Z.; Peng, H. Visual Detection of Green Mangoes by an Unmanned Aerial Vehicle in Orchards Based on a Deep Learning Method. *Biosyst. Eng.* **2020**, *194*, 261–272. [CrossRef]
77. Birrell, S.; Hughes, J.; Cai, J.Y.; Iida, F. A Field-Tested Robotic Harvesting System for Iceberg Lettuce. *J. Field Robot.* **2020**, *37*, 225–245. [CrossRef] [PubMed]
78. Lou, H.; Duan, X.; Guo, J.; Liu, H.; Gu, J.; Bi, L.; Chen, H. DC-YOLOv8: Small-Size Object Detection Algorithm Based on Camera Sensor. *Electronics* **2023**, *12*, 2323. [CrossRef]
79. Liu, W.; Anguelov, D.; Erhan, D.; Szegedy, C.; Reed, S.; Fu, C.Y.; Berg, A.C. SSD: Single Shot MultiBox Detector. In Proceedings of the Computer Vision–ECCV 2016: 14th European Conference, Amsterdam, The Netherlands, 11–14 October 2016. [CrossRef]
80. Liang, Q.; Zhu, W.; Long, J.; Wang, Y.; Sun, W.; Wu, W. A Real-Time Detection Framework for On-Tree Mango Based on SSD Network. In Proceedings of the International Conference on Intelligent Robotics and Applications, Newcastle, NSW, Australia, 9–11 August 2018. [CrossRef]
81. Krizhevsky, A.; Sutskever, I.; Hinton, G.E. ImageNet Classification with Deep Convolutional Neural Networks. *Commun. ACM* **2017**, *60*, 84–90. [CrossRef]
82. Zhu, L.; Li, Z.; Li, C.; Wu, J.; Yue, J. High Performance Vegetable Classification from Images Based on AlexNet Deep Learning Model. *Int. J. Agric. Biol. Eng.* **2018**, *11*, 217–223. [CrossRef]
83. Rangarajan, A.K.; Purushothaman, R.; Ramesh, A. Tomato Crop Disease Classification Using Pre-Trained Deep Learning Algorithm. *Procedia Comput. Sci.* **2018**, *133*, 1040–1047. [CrossRef]
84. Simonyan, K.; Zisserman, A. Very Deep Convolutional Networks for Large-Scale Image Recognition. *arXiv* **2015**, arXiv:1409.1556. [CrossRef]
85. Mahmood, A.; Singh, S.K.; Tiwari, A.K. Pre-Trained Deep Learning-Based Classification of Jujube Fruits According to Their Maturity Level. *Neural Comput. Appl.* **2022**, *34*, 13925–13935. [CrossRef]
86. Begum, N.; Hazarika, M.K. Maturity Detection of Tomatoes Using Transfer Learning. *Meas. Food* **2022**, *7*, 100038. [CrossRef]
87. Pérez-Pérez, B.D.; García Vázquez, J.P.; Salomón-Torres, R. Evaluation of Convolutional Neural Networks' Hyperparameters with Transfer Learning to Determine Sorting of Ripe Medjool Dates. *Agriculture* **2021**, *11*, 115. [CrossRef]
88. Li, Z.; Li, F.; Zhu, L.; Yue, J. Vegetable Recognition and Classification Based on Improved VGG Deep Learning Network Model. *Int. J. Comput. Intell. Syst.* **2020**, *13*, 559–564. [CrossRef]
89. He, K.; Zhang, X.; Ren, S.; Sun, J. Deep Residual Learning for Image Recognition. In Proceedings of the 2016 IEEE Conference on Computer Vision and Pattern Recognition (CVPR), Las Vegas, NV, USA, 26 June–1 July 2016. [CrossRef]
90. Helwan, A.; Sallam Ma'aitah, M.K.; Abiyev, R.H.; Uzelaltinbulat, S.; Sonyel, B. Deep Learning Based on Residual Networks for Automatic Sorting of Bananas. *J. Food Qual.* **2021**, *2021*, 5516368. [CrossRef]
91. Rahnemoonfar, M.; Sheppard, C. Deep Count: Fruit Counting Based on Deep Simulated Learning. *Sensors* **2017**, *17*, 905. [CrossRef]
92. Kang, H.; Chen, C. Fruit Detection, Segmentation and 3D Visualisation of Environments in Apple Orchards. *Comput. Electron. Agric.* **2020**, *171*, 105302. [CrossRef]
93. Kang, H.; Chen, C. Fruit Detection and Segmentation for Apple Harvesting Using Visual Sensor in Orchards. *Sensors* **2019**, *19*, 4599. [CrossRef] [PubMed]
94. Girshick, R.; Donahue, J.; Darrell, T.; Malik, J. Rich Feature Hierarchies for Accurate Object Detection and Semantic Segmentation. In Proceedings of the 2014 IEEE Conference on Computer Vision and Pattern Recognition (CVPR), Columbus, OH, USA, 24–27 June 2014. [CrossRef]
95. Girshick, R. Fast R-CNN. In Proceedings of the 2015 IEEE International Conference on Computer Vision (ICCV), Santiago, Chile, 7–13 December 2015. [CrossRef]
96. Ren, S.; He, K.; Girshick, R.; Sun, J. Faster R-CNN: Towards Real-Time Object Detection with Region Proposal Networks. *IEEE Trans. Pattern Anal. Mach. Intell.* **2017**, *39*, 1137–1149. [CrossRef]
97. Parvathi, S.; Tamil Selvi, S. Detection of Maturity Stages of Coconuts in Complex Background Using Faster R-CNN Model. *Biosyst. Eng.* **2021**, *202*, 119–132. [CrossRef]
98. Wan, S.; Goudos, S. Faster R-CNN for Multi-Class Fruit Detection Using a Robotic Vision System. *Comput. Netw.* **2020**, *168*, 107036. [CrossRef]
99. Fu, L.; Feng, Y.; Majeed, Y.; Zhang, X.; Zhang, J.; Karkee, M.; Zhang, Q. Kiwifruit Detection in Field Images Using Faster R-CNN with ZFNet. *IFAC Pap.* **2018**, *51*, 45–50. [CrossRef]

100. Zhang, J.; Karkee, M.; Zhang, Q.; Zhang, X.; Yaqoob, M.; Fu, L.; Wang, S. Multi-Class Object Detection Using Faster R-CNN and Estimation of Shaking Locations for Automated Shake-and-Catch Apple Harvesting. *Comput. Electron. Agric.* **2020**, *173*, 105384. [CrossRef]
101. Cao, C.; Wang, B.; Zhang, W.; Zeng, X.; Yan, X.; Feng, Z.; Liu, Y.; Wu, Z. An Improved Faster R-CNN for Small Object Detection. *IEEE Access* **2019**, *7*, 106838–106846. [CrossRef]
102. Long, J.; Shelhamer, E.; Darrell, T. Fully Convolutional Networks for Semantic Segmentation. In Proceedings of the 2015 IEEE Conference on Computer Vision and Pattern Recognition (CVPR), Boston, MA, USA, 7–12 June 2015. [CrossRef]
103. Zabawa, L.; Kicherer, A.; Klingbeil, L.; Milioto, A.; Topfer, R.; Kuhlmann, H.; Roscher, R. Detection of Single Grapevine Berries in Images Using Fully Convolutional Neural Networks. In Proceedings of the 2019 IEEE/CVF Conference on Computer Vision and Pattern Recognition Workshops (CVPRW), Long Beach, CA, USA, 16–17 June 2019. [CrossRef]
104. Li, Y.; Cao, Z.; Xiao, Y.; Cremers, A.B. DeepCotton: In-Field Cotton Segmentation Using Deep Fully Convolutional Network. *J. Electron. Imaging* **2017**, *26*, 16. [CrossRef]
105. Chen, S.W.; Shivakumar, S.S.; Dcunha, S.; Das, J.; Okon, E.; Qu, C.; Taylor, C.J.; Kumar, V. Counting Apples and Oranges with Deep Learning: A Data-Driven Approach. *IEEE Robot. Autom. Lett.* **2017**, *2*, 781–788. [CrossRef]
106. Liu, X.; Chen, S.W.; Aditya, S.; Sivakumar, N.; Dcunha, S.; Qu, C.; Taylor, C.J.; Das, J.; Kumar, V. Robust Fruit Counting Combining Deep Learning, Tracking, and Structure from Motion. In Proceedings of the 2018 IEEE/RSJ International Conference on Intelligent Robots and Systems (IROS), Madrid, Spain, 1–5 October 2018. [CrossRef]
107. Badrinarayanan, V.; Kendall, A.; Cipolla, R. SegNet: A Deep Convolutional Encoder-Decoder Architecture for Image Segmentation. *IEEE Trans. Pattern Anal. Mach. Intell.* **2017**, *39*, 2481–2495. [CrossRef]
108. Peng, H.; Xue, C.; Shao, Y.; Chen, K.; Xiong, J.; Xie, Z.; Zhang, L. Semantic Segmentation of Litchi Branches Using DeepLabV3+ Model. *IEEE Access* **2020**, *8*, 164546–164555. [CrossRef]
109. Majeed, Y.; Zhang, J.; Zhang, X.; Fu, L.; Karkee, M.; Zhang, Q.; Whiting, M.D. Apple Tree Trunk and Branch Segmentation for Automatic Trellis Training Using Convolutional Neural Network Based Semantic Segmentation. *IFAC Pap.* **2018**, *51*, 75–80. [CrossRef]
110. Barth, R.; Hemming, J.; Van Henten, E.J. Angle Estimation between Plant Parts for Grasp Optimisation in Harvest Robots. *Biosyst. Eng.* **2019**, *183*, 26–46. [CrossRef]
111. He, K.; Gkioxari, G.; Dollar, P.; Girshick, R. Mask R-CNN. In Proceedings of the 2017 IEEE International Conference on Computer Vision (ICCV), Venice, Italy, 22–29 October 2017. [CrossRef]
112. Jia, W.; Tian, Y.; Luo, R.; Zhang, Z.; Lian, J.; Zheng, Y. Detection and Segmentation of Overlapped Fruits Based on Optimized Mask R-CNN Application in Apple Harvesting Robot. *Comput. Electron. Agric.* **2020**, *172*, 105380. [CrossRef]
113. Zu, L.; Zhao, Y.; Liu, J.; Su, F.; Zhang, Y.; Liu, P. Detection and Segmentation of Mature Green Tomatoes Based on Mask R-CNN with Automatic Image Acquisition Approach. *Sensors* **2021**, *21*, 7842. [CrossRef]
114. Ni, X.; Li, C.; Jiang, H.; Takeda, F. Deep Learning Image Segmentation and Extraction of Blueberry Fruit Traits Associated with Harvestability and Yield. *Hortic. Res.* **2020**, *7*, 110. [CrossRef]

agronomy

Article

A Path Planning System for Orchard Mower Based on Improved A* Algorithm

Mengke Zhang [1], Xiaoguang Li [1], Ling Wang [2], Liujian Jin [2] and Shubo Wang [1,*]

1 School of Automation, Institute of Intelligent Unmanned System, Qingdao University, Qingdao 266071, China; zhangmengke@qdu.edu.cn (M.Z.); lixiaoguang@qdu.edu.cn (X.L.)
2 College of Engineering, China Agricultural University, Beijing 100083, China; wangling2017068@cau.edu.cn (L.W.); sy20213071443@cau.edu.cn (L.J.)
* Correspondence: shubowang@qdu.edu.cn

Abstract: The application of intelligent mobile robots in agriculture has emerged as a new research frontier, with the integration of autonomous navigation technology and intelligent agricultural robots being the key to the widespread adoption of smart agricultural machinery. This paper investigates comprehensive coverage path planning for tracked lawnmowers within orchard environments and addresses challenges related to task allocation and battery life. Firstly, in this study, the motion model of the tracked lawnmower was initially simplified based on assumptions about the orchard environment. Force analyses were conducted on each of its motion mechanisms. For the known orchard environment, a grid-based mapping technique was employed to model the orchard environment. Then, in order to improve the algorithm speed and reduce the number of turns during the lawnmower's traversal, the A* search algorithm was enhanced by combining the method of robot cluster traversal in the orchard environment. Finally, the improved method was simulated and verified in the MATLAB platform to investigate the influence of the number of lawnmower clusters on the path planning in the connected and non-connected orchards. Furthermore, two sets of on-site field trials were meticulously designed to validate the reliability, practicality, and efficacy of the simulation experiments.

Keywords: clustered lawnmowers; grid-based method; task allocation; comprehensive coverage path planning

Citation: Zhang, M.; Li, X.; Wang, L.; Jin, L.; Wang, S. A Path Planning System for Orchard Mower Based on Improved A* Algorithm. *Agronomy* **2024**, *14*, 391. https://doi.org/10.3390/agronomy14020391

Academic Editor: Paul Kwan

Received: 19 January 2024
Revised: 15 February 2024
Accepted: 16 February 2024
Published: 18 February 2024

1. Introduction

Traditional manual weeding is plagued by issues such as low labor efficiency, high production costs, and heavy workloads [1,2]. Therefore, mechanized weeding, particularly in conjunction with the emerging technology of robotic lawnmowers [3], is poised to become a primary solution. Robotic lawnmowers have the potential to alleviate labor burdens, reduce pesticide usage, protect orchard ecosystems, and enhance orchard productivity [4–6]. In the application of lawnmowers, the focal point of research lies in how to plan the travel paths of robotic lawnmowers. Commonly used path planning methods are categorized into traditional and intelligent approaches. Traditional path planning methods include free space methods [7], graph search methods, grid methods, and artificial potential field methods [8], whereas intelligent path planning methods are built upon theories such as genetic algorithms [9,10] and neural networks [11].

The problem of orchard traversal and coverage by robotic lawnmowers is an extension of path planning, falling under the umbrella of full-coverage path planning. Scholars have explored various methods for full-coverage path planning, with the A* algorithm having first been introduced by P. Hart, N. Nilsson, and B. Raphael in 1968 [12]. This algorithm, an enhancement of Dijkstra's algorithm, incorporates a heuristic function, estimating the cost function value, to effectively search for the optimal path. However, it suffers from an abundance of expanded nodes and large turning angles during path search. T. Tsuji et al. proposed the artificial potential field method (APF) [13], treating the movement of robots in the environment as particles in a potential field to achieve obstacle avoidance and navigation. Nevertheless, in complex environments, the robot may fail to find a globally optimal path, and obstacle avoidance is less effective in densely populated obstacle layouts.

Moreover, integrating full-coverage path planning technology with a robotic swarm system [6] not only addresses issues of insufficient work capacity and poor fault tolerance in single-robot coverage but also adds flexibility to the task execution process, greatly improving work efficiency. In this context, ref. [14] combined a single-robot path planning algorithm with unit decomposition and heuristic search algorithms, proposing a multi-robot coverage planning method for floor-cleaning robots. Although this algorithm segmented the target area, it did not further investigate multi-robot task allocation. Refs. [15,16] introduced a heuristic function into the spanning tree algorithm and introduced a backtracking mechanism. This method was applied to multi-robot coverage path planning, enhancing robot coverage efficiency and algorithm robustness. However, it exhibits lower coverage efficiency when the starting points of the robots are unevenly distributed.

In the working scenario of a lawnmower cluster system, the problem of multi-robot collaborative task allocation is also a research hotspot [17]. Currently, multi-robot task allocation methods are mainly centralized, distributed [18], and hybrid, with corresponding algorithms including market mechanism-based allocation, linear programming-based task allocation, and swarm intelligence-based task allocation [19]. The market mechanism-based task allocation method proposed in [20] effectively addresses task allocation issues in dynamic or uncertain environments where a robot swarm works. However, it is not suitable for complex task scenarios in the system. The combinatorial auction-based task allocation method suggested in [21] resolves insoluble problems in task allocation and approximates the optimal solution. Still, this method typically involves a large number of combinatorial and auction processes, leading to high computational complexity, especially when the number of tasks and robots is large, impacting the real-time efficiency of the algorithm. The hybrid integer linear programming allocation method proposed in [21] can search for the optimal solution but exhibits lower allocation efficiency. Task allocation methods based on [22,23] address the issues of low task allocation efficiency and poor system scalability. When tackling the problem of multi-robot task allocation, swarm intelligence algorithms such as ant colony algorithms [24,25] or neural networks [26,27] exhibit high allocation efficiency, strong applicability, and ease of implementation, garnering significant attention from researchers [28].

Currently, there is relatively limited research by scholars on the path planning of orchard lawnmower robots, mainly focusing on flat areas and trimming fields of lawns. Furthermore, due to the overall large structure and high cost of lawnmower robots, their widespread application is constrained. This paper, set against the backdrop of cluster lawnmower operation, focuses on the traversal and coverage technology of cluster lawnmower robots in known orchard environments. It considers area decomposition and path planning under different constraints, analyzes the impact of different cluster lawnmower quantities on path planning, and identifies the optimal number of cluster lawnmowers for full-coverage path planning in the target area. The research objectives of this project are as follows:

(1) To investigate the motion dynamics of a tracked lawnmower, this study simplifies its motion model based on specific assumptions regarding the orchard environment. Subsequently, a detailed force analysis is conducted on each individual motion mechanism. Employing a grid-based mapping technique, the known orchard environment is accurately modeled.
(2) An improved A* algorithm is proposed to address existing issues in current lawnmower robot full-coverage path planning, thereby ensuring better traversal and coverage of the target area, with the lawnmower robot completely avoiding fruit

trees and other obstacles during traversal. This paper validates the practicality of the improved algorithm through simulation experiments and field tests.

(3) A comprehensive investigation is undertaken concerning the clustering of lawnmowers, involving an in-depth analysis of the influence of cluster quantity on path planning within both connected and non-connected orchards. The validation of simulation experiments is accomplished through meticulous field trials, ensuring the rigor of the research findings.

The remaining sections of this paper are organized as follows. Section 2 simplifies the motion model of the tracked lawnmower and conducts force analysis. The grid diagram method is used to model the known orchard environment. Section 3 introduces the task allocation method, which evenly distributes the workload based on the lawnmower robot's energy capacity, and presents the improved A* path planning approach. In Section 4, simulations and on-site experiments are conducted to validate the improved A* path planning algorithm for varying numbers of lawnmower robots in the cluster. The performance of the proposed algorithm is assessed and analyzed. Finally, Section 5 summarizes and provides an outlook on the presented work.

2. Orchard Environmental Information Processing and Motion Analysis of Tracked Lawnmower

2.1. Orchard Environmental Information Rasterization

To ensure the accuracy of path planning, this paper employs a grid model to describe the two-dimensional workspace of the lawnmower robot, using binary information to represent orchard layout information. Initially, the grid size is determined, dividing the actual working environment into equally sized grid cells based on the cutting width of the lawnmower's blade. Grid cells without obstacles are designated as white grids, indicating areas where the lawnmower can navigate. Grid cells with obstacles are represented as black grids, indicating areas the lawnmower must avoid and cannot traverse. The following assumptions are made:

(1) As fruit trees within the orchard constitute the primary impediments in grid-based environmental modeling, requiring the lawnmower to navigate around them during operation, it is essential to represent obstacle dimensions in the grid map as larger than their real-world counterparts (employing fuzzy processing for modeling fruit trees). This entails incorporating a designated safety margin into the actual dimensions of obstacles within the environment (illustrated in Figure 1). Consequently, safety considerations are not imperative during operation of the lawnmower, and the lawnmower robot can be treated as a point mass while in motion.

(2) Additional impediments within the orchard, such as large rocks and wells, similarly undergo fuzzy processing. In essence, obstacle dimensions in the grid map are magnified compared to their actual dimensions (as depicted in Figure 1).

(3) In this study, the lawnmower robot operates solely within a two-dimensional workspace, and the impact of the lawnmower robot's height is not considered.

(4) The positions of fruit trees in the orchard are static, so the locations of obstacles on the grid map remain constant throughout the lawnmower robot's movement.

On a two-dimensional plane, the orchard environment map where the lawnmower operates is divided into numerous equally sized grid cells. Each grid is annotated as 0 or 1 based on the presence of obstacles within the grid. In the grid map, the coordinates in the grid coordinate system can be correlated with latitude and longitude coordinates, resulting in more precise localization and thus enabling more accurate path planning. To objectively represent the grid-based environmental modeling, the following definitions are established:

Definition 1. *Target Workspace:*

$$M = \{(x,y)|x,y \text{ are points within the mowing range of the lawnmower}\}$$

Definition 2. *Traversable Area:*

$$M_{\varnothing} = \{x, y | obs(x, y) = 0\} \subseteq M;$$

where $obs(x, y)$ is the occupancy function:

$$obs(x, y) = \left\{ \begin{array}{l} 0, \text{No obstacles} \\ 1, \text{With obstacles} \end{array} \right\}$$

The no-obstacles area can be further divided into the normal driving area, uphill area, downhill area, and overgrown grass area.

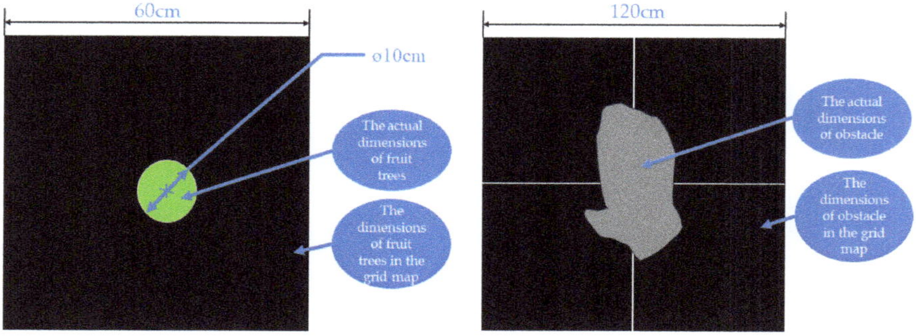

(a) (b)

Figure 1. Actual dimensions of the object and its dimensions in the grid map. (**a**) Actual dimensions of fruit trees and their dimensions in the grid map; (**b**) actual dimensions of obstacles and their dimensions in the grid map.

Definition 3. *Non-traversable Area:*

$$M_{\Delta} = \{x, y | obs(x, y) = 1\} \subseteq M;$$

Definition 4. *After the lawnmower starts working, grids that have been traversed and do not contain obstacles are labeled* $obs(x, y) = 2$.

Definition 5. *The maximum length of the environmental map is denoted as* L_{long}, *the maximum width is denoted as* W_{wide}, *and the dimensions (length and width) of each unit grid, denoted as* D_s, *are equal to the mowing width* W_{mower} *of the lawnmower.*

The grass-cutting robot requires effective paths to navigate in the simulated grid map. Therefore, it is necessary to define valid paths to enable the robot to move properly. During the journey from the starting point to the destination point, the point robot can move in eight directions: up, down, left, right, upper left, upper right, lower left, and lower right. The point robot is not allowed to cross over obstacles; any attempt to traverse through the center of an obstacle is considered an invalid path. Instead, the point robot must follow valid paths by maneuvering around obstacles and moving into adjacent grid cells.

2.2. Motion Analysis of Tracked Lawnower

When establishing the motion model of the tracked lawnmower, the ground environment of the orchard should be considered first, taking into account the area of contact between the tracks and the ground. Simultaneously, it is necessary to confirm the operational state of the tracked lawnmower. In this study, the focus is primarily on establishing a motion model for the steady-state movement of the tracked lawnmower in orchards located in hilly and mountainous areas. Therefore, the following conditional assumptions are made:

(1) Orchards in hilly regions mainly refer to those planted on undulating, continuous low hills. These orchards have an absolute height of 500 m, and the relative height does not exceed 200 m. Among them, orchard terrains with slopes of less than 15 degrees, such as flat or gently sloping areas, are suitable for mechanized cultivation and planting. However, when the slope of the orchard terrain exceeds 25 degrees, mechanized operations in the orchard become challenging. Therefore, it is assumed that the tracked lawnmower operates on slopes between 0 and 15 degrees, and the unevenness of the orchard ground is not considered.
(2) The assumption is made that the tracked lawnmower behaves as a rigid body during its motion.
(3) Deformation of the rubber tracks during motion and changes in track tension are not considered.

In this study, the walking model of the tracked lawnmower adopts a crawler walking mechanism. The main components constituting the crawler walking mechanism are the bogies on both sides of the tracked lawnmower. These bogies consist of a bogie frame and a "four-wheel and one-belt" arrangement (as shown in Figure 2). During operation of the lawnmower, the tracks come into direct contact with the soil. At this point, the weight of the tracked lawnmower is primarily supported by the supporting wheels. When the driving wheel operates, the tracks undergo a winding motion relative to the bogie frame.

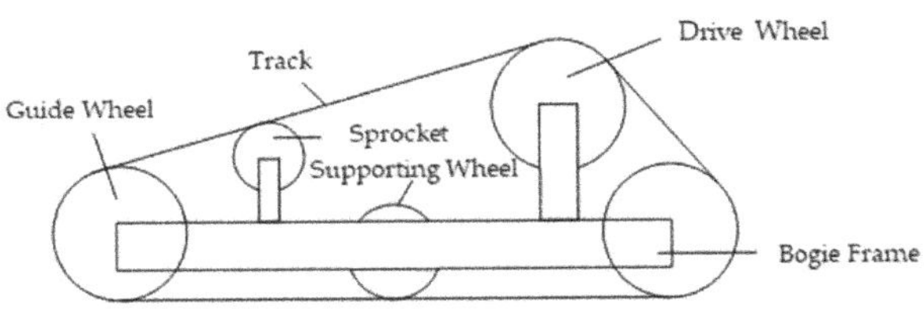

Figure 2. Tracked lawnmower bogie mechanism.

The overall structure of the tracked lawnmower is primarily composed of two systems: the chassis and the track, forming a complex multibody system. When the lawnmower travels on a certain slope without undergoing steering or pitching motion, both the frame and the tracks undergo translational motion, while the wheels undergo rotational motion. Therefore, when calculating the energy of the lawnmower, it needs to be simplified into a multibody model. When the lawnmower is working in an orchard and does not undergo steering motion, in order to ensure that the lawnmower continues to travel in the given direction along a specified path, coordinate $s_{(x)} = x$ is created for the given path. Thus, the motion of the lawnmower is subject to the following constraints:

$$(1 - s)r_1\theta_1 = x \tag{1}$$

where s represents the relative sliding rate between the lawnmower's track and the ground, r_1 denotes the radius of the drive wheel, and θ_1 represents the angular velocity of the drive wheel.

The resistance encountered by the lawnmower during motion includes acceleration resistance F_a, air resistance F_w, rolling resistance F_R, and slope resistance F_G. Due to the relatively low speed of the lawnmower during movement in the orchard, the impact of air resistance is not considered in this study. Assuming that the road conditions and forces acting on the lawnmower are symmetrical during travel, the forces on the tracks and the motion of the tracks are also symmetrical. For force analysis, one track is selected (as shown in Figure 3).

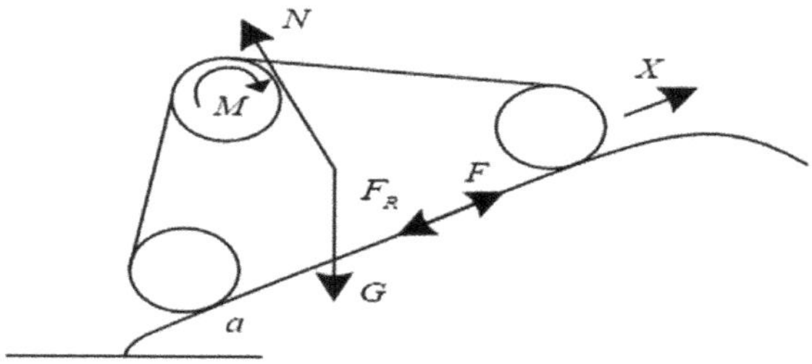

Figure 3. Force analysis of the lawnmower tracks.

In Figure 3, N is the vertical reaction force from the ground to the lawnmower, G is the weight of the lawnmower itself, and m is the mass of the lawnmower. Assuming that the motor of the lawnmower imparts a driving torque M to the drive wheel, the rated power of the motor is P, and the transmission efficiency of the motor is η, then the driving force of the lawnmower is given by:

$$F = \frac{\eta P}{\dot{x}} \tag{2}$$

where $\dot{x}$ represents the coefficient for the conversion of motor power output to mechanical force.

During the driving process, the lawnmower undergoes translational motion due to its own mass, resulting in a translational component $F_{a,x} = ma_x$. The rotational component generated by the acceleration and deceleration of rotating components in the lawnmower is given by:

$$F_{a,r} = -\frac{M_a}{r_1} = \frac{I\ddot{\theta}_1}{r_1} \tag{3}$$

where $\ddot{\theta}_1 = \frac{\ddot{x}}{r_1}$, $F_{a,x}$, and $F_{a,r}$ constitute the acceleration resistance F_a of the lawnmower during travel, where $a_x = \ddot{x}$ represents the acceleration of the lawnmower along the x-direction. M_a represents the inertia torque generated by the non-uniform motion of rotating components, and I is the total equivalent rotational inertia generated by all non-uniformly moving rotating components of the lawnmower.

To simplify the analysis process, considering equal masses for the tracks borne by each wheel, then:

$$I = \frac{1}{2}\sum_{i=1}^{n} \lambda_i^2 (m_i + \frac{m'}{n}) r_i^2 \tag{4}$$

where m_i is the radius of each wheel, r_i is the radius of each wheel, m' is the mass of the track, n is the number of wheels, and the transmission coefficient ratio is λ_i. The rotational component is then given by:

$$F_{a,r} = \frac{\ddot{x}}{2r_1^2}\sum_{i=1}^{n} \lambda_i^2 (m_i + \frac{m'}{n}) r_i^2 \tag{5}$$

The total resistance during mowing operation is:

$$F_D = G\sin\alpha + fG\cos\alpha + m\ddot{x} + \frac{\ddot{x}}{2r_1^2}\sum_{i=1}^{n} \lambda_i^2 (m_i + \frac{m'}{n}) r_i^2 \tag{6}$$

Under rated conditions, the motor will experience a certain torque loss $M_L = (1-\eta)\frac{P}{2\pi n_0}$, where n_0 is the rated speed of the motor. The magnitude of the torque loss is related to the speed. Assuming δ is a constant and s is the slip ratio, the transmission efficiency is given by $\eta = 1 - \delta r_1 \dot{\theta}_1 = 1 - \delta \frac{1}{1-s} \dot{x}$. When the slip ratio $s = 0$, the loss force during the lawnmower's travel is:

$$F_\Phi = \delta \frac{p}{2\pi n_0 r_1} \dot{x} \tag{7}$$

For a given path, the lawnmower is studied as a whole during the travel process since the system has only one degree of freedom. As the system only has one degree of freedom, the velocity $v = x$. While the lawnmower is traveling in the orchard, apart from the tracks and wheels, the other components undergo translational motion. Therefore, the velocities of these components are the same, i.e., $v = r_1 \theta_1 (1 - s)$. The total energy of the system includes the translational kinetic energy $\frac{1}{2} m v^2$, rotational kinetic energy $\frac{1}{2} \sum_{i=1}^{n} J_i \dot{\theta}_i^2$, gravitational potential energy $G x \sin\alpha$, and torsion spring elastic potential energy $\frac{1}{2} k \gamma^2$. The total energy of the lawnmower system is expressed as:

$$L = \frac{1}{2} m v^2 + \frac{1}{2} \sum_{i=1}^{n} J_i \dot{\theta}_i^2 - G x \sin\alpha - \frac{1}{2} k \gamma^2 \tag{8}$$

where J_i is the rotational inertia of the wheel and track system, and γ is the torsion angle of the spring.

The Lagrange equation for the system is given by:

$$\frac{d}{dt}\left(\frac{\partial L}{\partial q}\right) - \frac{\partial L}{\partial q} = Q + F_\Phi \tag{9}$$

where F_Φ is the loss force, $Q = F - F_D$ is the active generalized force, and L, Q, and F_Φ are substituted into the Lagrange equation to obtain the motion differential equation:

$$w\ddot{x} + a\dot{x} + c = 0 \tag{10}$$

where $w = 2m + \sum_{i=1}^{n} (m_i + \frac{m'}{n})$, $a = \frac{\delta M_0}{r_1}$, $c = 2G\sin\alpha + \mu G\cos\alpha - \frac{M_o}{r_1}$, and $M_O = \frac{P}{2\pi n_o}$.

Integrating the above equations, the kinematic equation for the tracked lawnmower in orchard travel is obtained as:

$$\begin{cases} x = -c_1 \frac{w}{a} e^{-\frac{a}{w}t} - \frac{c}{a} t + c_2 \\ v = c_1 \frac{w}{a} e^{-\frac{a}{w}t} - \frac{c}{a} \end{cases} \tag{11}$$

where c_1 and c_2 are constants. Assuming the initial travel conditions for the lawnmower are $t = 0$, $x = 0$, and $v = v_0$ and substituting them into the equation, the results are: $c_1 = v_0 + \frac{c}{a}$, $c_2 = \frac{w}{a}(v_0 + \frac{c}{a})$.

Therefore, the velocity of the tracked lawnmower in orchard travel is given by:

$$v = (v_0 + \frac{c}{a}) e^{-\frac{a}{w}t} - \frac{c}{a} \tag{12}$$

3. Clustered Lawnmower Task Allocation and Path Planning Methods

3.1. Problem Description and System Framework

For the path planning of multiple lawnmowers in the orchard, a global path planning approach is employed based on the known external working environment. This paper focuses on planning the movement paths of tracked lawnmowers in a two-dimensional plane and employs a grid-based mapping method to create a simulated working environment for the lawnmowers (as shown in Figure 4), comprising M*N grid cells.

In Figure 4, white grids represent grids where the lawnmower can travel, gray grids indicate obstacle grids (i.e., impassable grids), pink grids represent uphill areas, blue grids denote downhill areas, red grids represent overgrown grass areas, and yellow grids represent areas with sparse weeds. The positions of obstacles on the grid map are fixed.

In this section, the working area is divided into subareas, and each subarea is assigned to individual lawnmowers based on specific constraints. This allocation strategy ensures efficient obstacle avoidance as the lawnmowers execute their tasks.

Figure 4. Robotic simulated working environment. (White grids represent grids where the lawnmower can travel, gray grids indicate obstacle grids (i.e., impassable grids), pink grids represent uphill areas, blue grids denote down-hill areas, red grids represent overgrown grass areas, and yellow grids represent areas with sparse weeds.)

In this section, the research on lawnmowers and orchard environment is characterized by the following two points:

(1) The lawnmowers in the cluster are homogeneous robots, meaning each lawnmower carries the same amount of energy.

(2) The environmental information of the orchard is known. The positions of the trees and other obstacles in the orchard are fixed and unchanging, and there are no dynamic obstacles in the orchard.

The problem of orchard traversal coverage for the cluster of lawnmowers can be summarized as a multi-robot task allocation problem in obstacle environments. This includes constructing the environmental map, decomposing and assigning target areas, and planning subregion paths.

Subsequently, a mathematical description of the complete coverage path planning is presented. The lawnmower operates within a two-dimensional bounded environment. The lawnmower's specific location in the real environment can be represented using coordinates (x, y) in a two-dimensional coordinate system, denoted as follows:

$$M = \{(x,y)|\text{"}x,\ y\text{" represent the lateral and longitudinal coordinates in the actual space}\}$$

To indicate the specific status of each grid in the simulated environmental map, the function $obs(x,y)$ is defined. When $obs(x,y) = 0$, it indicates that the area is a grass-covered region in the orchard, which means it is the area that the lawnmower robot needs to traverse (the traversable area), and it can be represented as: $M_1 = \{(x,y)|obs(x,y) = 0\} \subseteq M$. When $obs(x,y) = 1$, it indicates that this grid is either an obstacle area or an area where fruit trees are planted, which means it is an area that the lawnmower robot does not need to traverse (non-traversable area), and it can be represented as: $M_2 = \{(x,y)|obs(x,y) = 1\} \subseteq M$. The geometric relationship among the environmental map region M, the area that the lawnmower robot needs to traverse M_1, and the area that the lawnmower robot does not need to traverse M_2 can be described as follows: $M_1 \cap M_2 = \varnothing$, $M_1 \cup M_2 = M$.

In this way, the travel path of the lawnmower robot in the orchard environment is composed of the trajectories of each point traversed by the lawnmower robot in the covered area l_i, represented by the set L:

$$L = \{l_i(x, y, x_s, y_s, x_g, y_g) | i = 1, 2, \ldots, n\}$$

where l_i represents the path traveled by each lawnmower robot to completely traverse a subregion of the covered area, the coordinates (x_s, y_s) are the starting positions of the lawnmower, and the coordinates (x_g, y_g) are the endpoints for each lawnmower's travel, so each lawnmower's travel path is a sub-path of L.

3.2. Target Area Decomposition

In this section, the optimal travel direction for the lawnmower is determined. One of the factors that must be considered in the full-coverage path planning problem is the turning situation of the lawnmower while traveling. To reduce the lawnmower's working time and its energy consumption, it is necessary to minimize the number of turns the lawnmower makes. Therefore, determining the direction in which the lawnmower traverses is aimed at reducing the total working time and conserving the overall energy of the lawnmower.

The optimization objective is to reduce the total working time and energy consumption of the lawnmower. Therefore, the direction with the minimum number of turns for the lawnmower is chosen as the optimal travel direction. To determine the lawnmower's travel direction, a common approach is to traverse and identify the minimum width direction of the area of the orchard to be covered. Subsequently, the vertical direction of the orchard's minimum width direction is chosen as the lawnmower's optimal travel direction. The specific steps are as follows:

(1) Place the lowest point of the convex polygon representing the area to be covered at the x-axis of the two-dimensional coordinate system, ensuring that the y-coordinate of this lowest point is 0.
(2) Rotate the polygon while maintaining its lowest point on the x-axis. Simultaneously, measure the height of the polygon in real time, i.e., the y-coordinate value of the highest point of the polygon.
(3) Select the direction with the smallest y-coordinate value for the highest point of the polygon, which corresponds to the minimum width direction of the convex polygon. This direction determines the optimal travel direction for the lawnmower, which is the angle in the vertical direction from the x-axis to the minimum width direction of the convex polygon, as illustrated in Figure 5, which represents the search process.

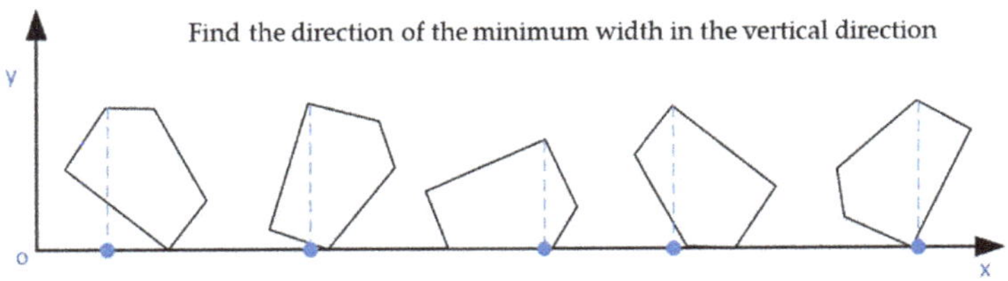

Figure 5. Determination of the optimal travel direction using a traversal search method. (The blue dash line is the height of each polygon, where the minimum value is the best driving direction.)

Partitioning the target area into multiple subareas is a prerequisite for task allocation. Rationally dividing the target area into subareas can reduce the problem of unsolvable task allocation and enhance the efficiency of robot task completion, improving system robustness. For the case of grass-cutting robots traversing the orchard for weed control, how to partition the area of the orchard waiting to be covered into corresponding subareas based on the robot's available energy is critical for solving the allocation problem. In this study, the proposed models for estimated energy consumption P_y and actual energy consumption P_z are introduced to constrain the subdivision of subareas. Estimating energy consumption calculates the total energy consumed by the lawnmower while traversing each row of the grid map. During the traversal process, the lawnmower may revisit certain grid cells, and actual energy consumption includes the cost of revisiting these cells in addition to the estimated energy consumption. The specific steps are as follows:

(1) Calculate the energy consumption to traverse each row in the grid map and iteratively determine the estimated energy consumption P_y to cover the entire grid map.
(2) Begin by partitioning the area based on the estimated energy consumption, making sure it is below the energy capacity of each lawnmower P, and perform initial path planning for the segmented areas.
(3) By traversing the initially partitioned regions, the actual walking path is determined, which is used to calculate the actual energy consumption, denoted as P_z, required to traverse the target area.
(4) Divide the areas based on actual energy consumption, ensuring it stays below the energy capacity of each lawnmower, to determine the final subdivision of the area.

The specific process of regional division is illustrated in Figure 6.

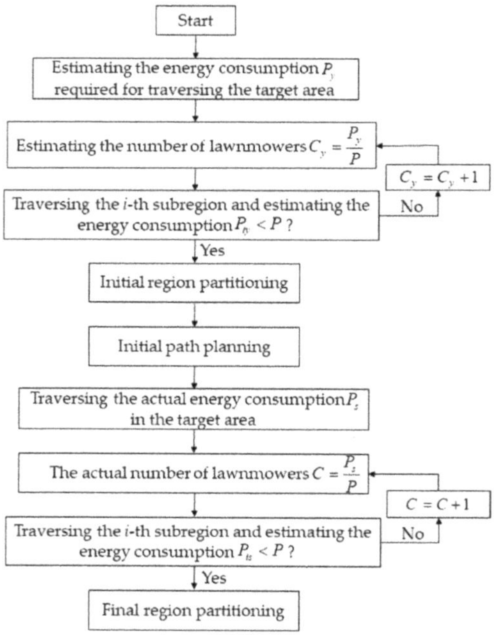

Figure 6. Flowchart of the region partitioning process.

3.3. Task Assignment under Multiple Constraints

Multi-Robot Task Allocation (MRTA) for Multiple Robots is a critical issue for the functionality of multi-robot systems. It involves assigning tasks to each robot in a way that improves system efficiency. In the face of complex orchard terrain, individual lawnmowers may experience insufficient coverage due to their own carried energy limitations and the influence of intricate topography. This issue is addressed by creating logically constrained work zones and allocating segmented sub-regions to each lawnmower to ensure complete coverage of the target area. In a multi-robot system, the shortest working time is determined by the longest working time of a single robot. This section focuses on optimizing objectives such as reducing the working time of the cluster of lawnmowers and lowering the total cost. It considers the energy carried by each lawnmower and the energy consumption required for the work. The goal is to balance the working time of each lawnmower, aiming to enhance the overall efficiency of the lawnmower cluster while minimizing costs. This section discusses the task allocation problem for the lawnmower robot cluster from two perspectives: task allocation objectives and mathematical models for task allocation. When assigning tasks to the lawnmower robot cluster, the following principles are followed:

(1) Rationality: This ensures that all the pending areas in the orchard are allocated to the corresponding lawnmower robots, guaranteeing the complete coverage of all pending areas in the orchard.
(2) Minimization of travel distance: The goal is to minimize the travel distance for each lawnmower robot to reduce their energy consumption.
(3) Fairness: Task allocation should be fair, ensuring that each lawnmower robot has an equal opportunity to perform tasks. (The principle of fairness employed in this study is instead integral as a strategic element for enhancing system efficiency. In this context, fairness refers to the equitable distribution of workloads among the robots, thereby optimizing the overall longevity and maintenance of the system. This approach ensures that no individual robot is subjected to continuous overload, which in turn mitigates the rate of wear and tear on each unit. Consequently, this strategy extends the operational lifespan of the entire fleet and maintains uniform performance levels across all units. Additionally, this principle enhances the system's adaptability in dynamic operational environments. By ensuring a balanced task allocation, the system is better equipped to manage unforeseen changes, such as varying terrain challenges or individual robot malfunctions, thereby maintaining operational resilience and efficiency.)

The mathematical description of the cluster lawnmower task allocation problem is as follows: The simulated grid map of the orchard is divided into n pending areas based on the energy of the lawnmower robot itself. In the cluster lawnmower system, which consists of m lawnmower robots and n pending areas, the m lawnmower robots are represented by the set $R = \{R_i \mid i = 1,2,3\ldots m\}$, and the n pending areas are represented by the set $T = \{T_i \mid i = 1,2,3\ldots n\}$. Each pending area in the system only requires the allocation of one lawnmower robot to complete the task. A single lawnmower robot can only complete one task within a certain time period. However, after completing a task and receiving energy replenishment, the lawnmower robot can traverse and cover another area. The system's objective is to allocate the n pending areas to the m lawnmower robots in a rational manner to improve work efficiency and minimize overall energy consumption.

Mathematical description of lawnmower task execution capability: The lawnmower's ability to execute tasks refers to whether the lawnmower has the energy to perform a specific task. Let $P_{ij} = 1$ represent that lawnmower i can execute a task in target area j; likewise, $P_{ij} = 0$ indicates that lawnmower i cannot execute a task in target area j.

Energy consumption for individual lawnmower tasks: During the execution of a specific task, a lawnmower consumes a certain amount of energy, where the cost of an individual task is represented by a two-dimensional matrix D with dimensions $(m+n)*(m+n)$. When a lawnmower moves from point a to point b, the distance between points a and b is L_{ab}. Therefore, the energy consumption for an individual task is $D_{ab} = L_{ab}$.

Total energy cost incurred by the system (cost): Define a binary variable $x_{ij}\begin{Bmatrix}1\\0\end{Bmatrix}$, where x_{ij} indicates whether the jth target coverage area T_j has been assigned to the ith lawnmower R_i. In this case, $x_{ij} = 0$ denotes that the jth target coverage area T_j has not been assigned to the ith lawnmower R_i. Given the task set Q_i for the ith lawnmower R_i, which is $Q_i = \{q_{i1}, q_{i2}, q_{ig}\}$, it represents that the ith lawnmower R_i has been assigned to g target coverage areas, and the tasks are executed sequentially from q_1 to q_2 all the way to q_g. Therefore, the energy cost incurred by the ith lawnmower R_i in executing tasks in the assigned subareas is:

$$C_i = D_{i,q_{i1}} + \sum_{k=1}^{g-1} D_{q_{ik},q_{ik+1}} \tag{13}$$

In a cluster lawnmower system comprising m lawnmowers and n target coverage areas, the total energy expenditure of the system is as follows:

$$\mathrm{Cost} = \sum_{i=1}^{m} C(i) \tag{14}$$

From the above analysis, it is evident that the cluster lawnmower task allocation problem can be transformed into finding a suitable task allocation strategy x_{ij} and providing a set of target coverage areas $Q_i = \{q_{i1}, q_{i2}, q_{ig}\}$ for each lawnmower to minimize the total energy expenditure cost within the system. Additionally, the system must adhere to several constraint conditions to refine the specific task allocation. These specific constraint conditions include:

Constraint 1. *The constraint condition in Formula (15) signifies that each lawnmower can only traverse and cover a single subregion at a time, meaning that an individual lawnmower can only operate within its assigned task area.*

$$\sum_{i=1}^{i=m} x_{ij} = 1, j = 1, 2 \ldots n \tag{15}$$

Constraint 2. *The constraint condition in Formula (16) ensures that each pending coverage area is allocated a lawnmower to guarantee the coverage of all pending subregions.*

$$\sum_{i=1}^{i=m} \sum_{j=1}^{j=n} x_{ij} = n \tag{16}$$

Constraint 3. *The constraint condition in Formula (17) implies that the ith lawnmower is assigned to g pending subregions (in this study, each lawnmower corresponds to one pending coverage area, i.e., g = 1).*

$$\sum_{j=1}^{j=n} x_{ij} = g, i = 1, 2 \ldots m \tag{17}$$

Constraint 4. *The constraint condition in Formula (18) implies that the lawnmower with the ith identifier has the capability to execute g pending subregions (g = 1) with the energy it carries.*

$$\sum_{k=1}^{g} P_{ij} = g, i = 1, 2 \ldots m \tag{18}$$

3.4. Improved A Algorithm and Experimental Verification*

3.4.1. Improved A* Algorithm

Common path planning algorithms, such as Dijkstra's algorithm, ensure the discovery of the shortest path from the starting point to the destination. However, it employs a greedy strategy, which makes it incapable of handling graphs with negative-weight edges, as these edges can result in incorrect shortest paths. Furthermore, the search speed is relatively slow due to the large number of explored nodes. Although the traditional A* algorithm is effective in path planning, it often expands numerous nodes during the search process, leading to a significant accumulation of turning angles. Therefore, this paper proposes an improved A* algorithm by weighting the heuristic cost function and introducing a turning cost into the total cost function. An eight-neighbor expansion method for pathfinding is employed based on the actual working conditions of the lawnmower. This improvement aims to enhance the search speed of the algorithm, increase the optimality of the global path, and reduce the number of turns made by the robot during traversal, thereby shortening the overall travel time.

Firstly, this study employs the Manhattan distance as the heuristic function, which calculates the sum of the distances along the x-axis and the y-axis between the current point and the target point as the heuristic cost function's value.

$$h(n) = |x_d - x_n| + |y_d - y_n| \tag{19}$$

where (x_d, y_d) refers to the coordinates of the target point, whereas (x_n, y_n) corresponds to the coordinates of the current point.

In the actual search process, as the estimated cost function $h(n)$ cannot be known in advance, it is assumed that the estimated distance from the current node n to the goal node is represented by $H(n)$. When replacing $h(n)$ with $H(n)$ in the A* algorithm, the actual shortest path in the A* algorithm must always be greater than or equal to the value of the heuristic function $h(n)$, i.e., $h(n) \geq H(n)$. If $h(n) > H(n)$, the efficiency of the A* algorithm is slightly reduced and it cannot guarantee the optimal path will be found, but it does maintain high accuracy. If $h(n) = H(n)$, the A* algorithm strictly follows the shortest path during the search, ensuring the optimal path while reducing the expansion of nodes, significantly improving efficiency. However, this scenario is not suitable when obstacles are present in the environment. If $h(n) < H(n)$, the A* algorithm expands more nodes in the search, but it ensures the optimal path will be found. Since the orchard environment has certain obstacles, when the Manhattan distance is applied in environments with obstacles, its estimated cost function $h(n)$ is often less than the estimated distance $H(n)$ from the current point to the target point. To address this issue, this paper introduces a weighting mechanism to the A* algorithm's estimated cost function $h(n)$. This involves multiplying $h(n)$ by a coefficient k ($k > 1$), resulting in the modified overall cost function expression:

$$f(n) = g(n) + k \times h(n) \tag{20}$$

where $g(n)$ represents the current cost from the starting point to the current node.

The choice of the constant k is based on balancing two distance estimates to ensure that the influence of the Manhattan distance is not overly significant when considering obstacles. Selecting an appropriate value for k requires experimentation and testing. In the MATLAB platform, a grid map size of 20 × 20 is set with an arbitrary distance l of 2 between any two adjacent grid cells. Through calculations, the weighted value k in the heuristic cost function is chosen as 1.3 to achieve a balanced point for obtaining the optimal path. At this point, the value of the weighted heuristic cost function $k \times h(n)$ from the current node to the target node tends to infinity, approaching the estimated distance $H(n)$ from the current node to the target node. This modification aims to reduce the expansion of nodes in the A* algorithm, thereby improving search speed and yielding an improved A* path.

Secondly, to reduce the issue of the lawnmower making frequent turns during actual movement, this paper introduces a turning cost to the current cost function $g(n)$. Additionally, on top of the weighted heuristic cost function $h(n)$, a turning cost is also added to the term $k \times h(n)$. Therefore, when calculating the current cost $g(n)$, the following steps should be followed (here, $g(n)$ refers to the cost from the current node n to the node to be expanded $r_i, i = 1, 2 \ldots 7$):

(1) Check if the current node n is the initial node, i.e., if there is a parent node m that precedes the current node.

(2) If the current node n is the initial node, meaning there is no parent node m for the current node n, then use the coordinates (x_n, y_n) of the current node n and the coordinates (x_r, y_r) of the node to be expanded, denoted as r, to calculate the turning cost. When $(x_n - x_r) \times (y_n - y_r) \neq 0$, indicating that the current node n and the node to be expanded r are not on a horizontal or vertical line, and the node r is in the left-up,

left-down, right-up, or right-down direction from the current node, the current cost function $g(r)$ from the current node n to the node to be expanded r is expressed as:

$$g(r) = g(n) + \sqrt{2}l \tag{21}$$

where l represents the distance between two adjacent nodes $g(n)$ in the horizontal or vertical direction. When $(x_n - x_r) \times (y_n - y_r) = 0$, indicating that the current node n and the node to be expanded r are aligned in the horizontal or vertical direction, the current cost function $g(r)$ from the current node n to the node to be expanded r is expressed as:

$$g(r) = g(n) + l \tag{22}$$

(3) If the current node n is not the initial node, meaning that the current node n has a parent node m, then the cost of turning is determined using the coordinates of the current node n (x_n, y_n) and its parent node m (x_m, y_m). When $(x_n - x_m) \times (y_n - y_m) \neq 0$, indicating that the current node n and the parent node m are not collinear in the horizontal or vertical direction, the current cost function $g(n)$ from the parent node m to the current node n is expressed as $g(r) = g(n) + \sqrt{2}l$; when $(x_n - x_m) \times (y_n - y_m) = 0$, meaning that the parent node m and the current node n are collinear in the horizontal or vertical direction, the current cost function $g(n)$ is expressed as $g(r) = g(n) + l$. After these improvements, the overall cost function is $f(n) = g(n) + k \times h(n)$, where $g(n)$ is calculated using the formula $g(r) = g(n) + l$ or $g(r) = g(n) + \sqrt{2}l$ depending on different situations.

3.4.2. Experiments and Conclusions

In this section, the improved A* algorithm was simulated and experimentally validated on a platform with 8 GB of computer memory, a processor frequency of 3.2 GHz, and based on MATLAB 2020b. Comparative analyses were conducted among the traditional A* algorithm, Dijkstra's algorithm, and the improved A* algorithm proposed in this section in different application scenarios. The grid map was set to a size of 20 × 20, with the distance between any adjacent two grid cells defined as l = 2. After calculation, the weighted factor k in the heuristic cost function was selected as 1.3. The algorithm's iteration count was set to 400 times. Twenty-five percent of the grid cells in the map were designated as obstacle cells, and the positions of these obstacles were randomly distributed on the grid map. The starting point of the lawnmower robot was set to [1,1], and the target endpoint was set to [20,20]. The simulation results are shown in Figure 7. The comparison of path length, turn counts, and search time of the three algorithms is shown in Table 1.

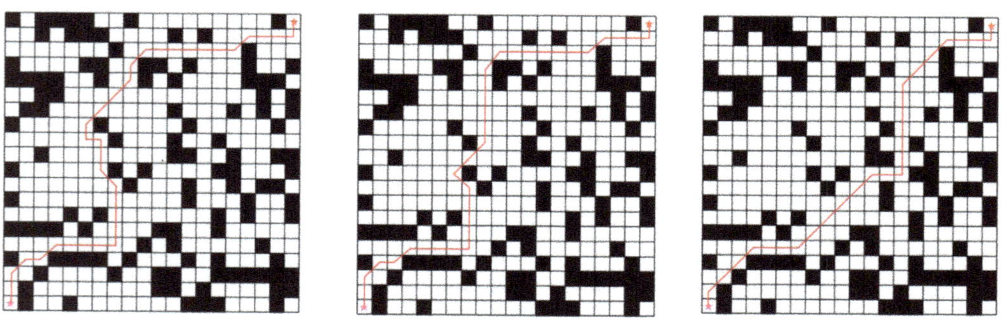

(**a**) Dijkstra's algorithm (**b**) A* Algorithm (**c**) Improved A* algorithm

Figure 7. Dijkstra's algorithm, A* algorithm, and improved A* algorithm search path simulation. (White grids represent grids where the lawnmower can travel, black grids indicate obstacle grids and the red lines are the optimal paths generated by the different algorithms).

Table 1. Comparison of path length, turn counts, and search time.

Algorithm	Path Length (m)	Turn Counts (Times)	Search Time (s)
A* algorithm	71.79	13	0.34
Dijkstra's algorithm	74.63	16	0.42
Improved A* algorithm	62.97	8	0.30

From Table 1, it can be concluded that the improved A* algorithm had fewer turns, a shorter path for robot travel, and a more superior planned path compared to the first two algorithms. This advantage was more pronounced in areas with many obstacles.

The three algorithms were subjected to 60 simulation experiments on different grid maps. The experimental results are shown in Figures 8 and 9. Through multiple experiments, the improved A* algorithm proposed in this study demonstrated superior performance in terms of the number of turns and the length of the path compared to the other two algorithms.

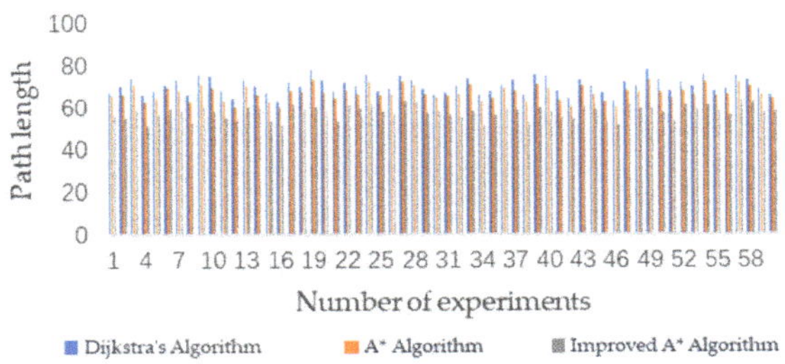

Figure 8. Comparison of path lengths among the A* algorithm, Dijkstra's algorithm, and the improved A* algorithm.

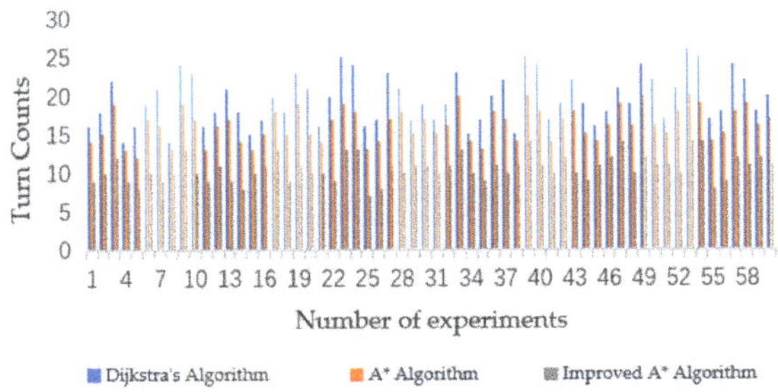

Figure 9. Comparison of turn counts between the A* algorithm, Dijkstra's algorithm, and the improved A* algorithm.

The average path length and turn counts for the three algorithms in these 60 simulation experiments are shown in Table 2.

Table 2. Comparison of average path length, turn counts, and search time.

Algorithm	Average Path Length (m)	Average Turn Counts (Times)	Average Search Time (s)
A* algorithm	66.53	14.22	0.38
Dijkstra's algorithm	70.34	16.57	0.35
Improved A* algorithm	57.61	12.36	0.32

From Table 2, it can be observed that in terms of the average path length, the A* algorithm was close to the Dijkstra algorithm. The average path length for the improved A* algorithm was 57.61, whereas for the traditional A* algorithm, it was 66.53, representing a reduction of approximately 13.4%. In terms of average turning times, the traditional A* algorithm had 14.22 turns, whereas the improved A* algorithm had 12.36 turns, a reduction of approximately 13.1%. Additionally, the average search time for the improved A* algorithm was reduced by 14.19%. The data indicate that the improved A* algorithm exhibited superior performance relative to its predecessor due to its enhanced heuristic approach and the incorporation of turning cost analysis into its cost function. The algorithm's heuristic functions are weighted to prioritize paths that seem more efficient, thereby reducing the search for suboptimal paths. This approach speeds up the pathfinding process and leads to shorter path lengths. Concurrently, incorporating turning cost considerations into the algorithm's cost function promotes the identification of more linear and direct routes, reduces the number of turn counts, and optimizes both path length and computational efficiency. The experimental results strongly suggest that the benefits of these algorithmic improvements in terms of path length and number of turn counts clearly outweigh the potential limitations of the algorithm, such as increased computational complexity and reduced applicability.

4. Path Planning Simulation and Field Experiment Verification with Different Numbers of Clustered Lawnmowers

4.1. Evaluation Metrics for Lawnmower Full-Coverage Path Planning

In the process of traversing the orchard, the lawnmower needs to cover the entire orchard at the smallest possible cost. Therefore, this section proposes quantitative evaluation metrics to assess the performance of the algorithm. In this study, the full-coverage path planning algorithm is based on a grid map, and it is assumed that the total number of grid cells in the simulated environment map is $m \times n$, where O represents the number of grid cells occupied by obstacles, S_l represents the number of grid cells missed by the lawnmower, S_r represents the number of grid cells revisited by the lawnmower, P_{sum} represents the total energy consumption of the lawnmower while traversing the orchard, C represents the total number of lawnmowers, and C_{sum} represents the total cost of the lawnmowers traversing the orchard.

(1) Repetition rate: Upon completing the orchard traversal, the percentage of all repeated grid cell occurrences to the total number of grid cells, excluding obstacles, is determined.

$$R_{re} = \frac{S_r}{m \times n - O} \tag{23}$$

(2) Omission rate: Upon completing the orchard traversal, the percentage of all missed grid cell occurrences to the total number of grid cells, excluding obstacles, is determined.

$$R_{lg} = \frac{S_l}{m \times n - O} \tag{24}$$

In this study, the omission rate of the lawnmower during its travel process mainly includes two types: The first type is the missed coverage during right-angle turns, and the second type is the missed coverage when the lawnmower circumvents fruit trees. These two types of missed mowing areas are shown in Figure 10.

$$S_1 = S_3 = \frac{(L^2 - \pi \times R^2)}{8} + \frac{\left(\frac{\sqrt{2}}{2} \times L - R\right)^2}{2} \tag{25}$$

$$S_2 = L \times \sqrt{2 \times (\frac{\sqrt{2}}{2}L - R)^2} + \frac{1}{4} \times L^2 \tag{26}$$

$$S_4 = \frac{L^2 - \pi R^2}{2} \tag{27}$$

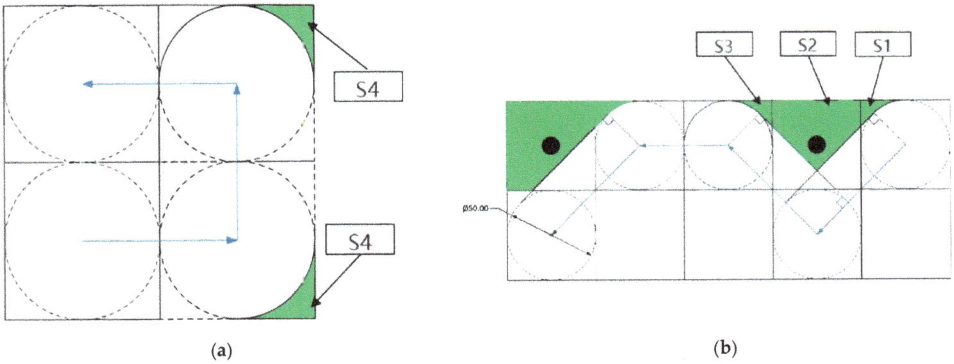

Figure 10. The area of missed mowing. (**a**) The missed mowing areas during right-angle turns. (**b**) The missed mowing areas when circumventing fruit trees. (S_1, S_2, S_3, S_4 are missing cut regions).

(3) Total cost: This refers to the sum of the fuel consumption of the lawnmower and the cost of the lawnmower itself after the lawnmower completes its traversal of the orchard.

$$C_{sum} = 6000 \times C + P_{sum} \times 8 \tag{28}$$

4.2. Connected Orchard Coverage Path Planning

The grid map was set to a size of 40 × 40, and the distance between any two adjacent grid cells was denoted as l = 0.6 m. The obstacle occupancy rate (the ratio of the number of grid cells occupied by obstacles to the total number of grid cells) was o = 6%, with fruit trees denoted as fixed obstacles on the grid map. The grid map was divided into the following regions: White cells represent areas accessible by the lawnmower, indicating regions where the lawnmower can achieve the highest travel speed during operation, with a total of g cells. Black cells represent obstacles, i.e., impassable cells. Pink cells represent uphill areas, with a total of g_1 cells. Blue cells represent downhill areas, with a total of g_2 cells. Red cells represent areas with dense weeds, with a total of g_3 cells. Yellow cells represent areas with sparse weeds, which have a negligible impact on the operation of the lawnmowers. The simulated grid map created based on Section 2 is shown in Figure 11.

The improved A* algorithm was defined with the following parameters: The weighted value in the heuristic function was chosen as k = 1.3, and the maximum number of iterations was set to 500. The coordinates of the initial position of the lawnmower robot were set at [1,1], and the endpoint was determined by the optimization goal. Each lawnmower carried an energy capacity of p = 28,800 kJ. The normal speed of the lawnmower when traveling in each white grid cell was v = 1.5 m/s, with an energy consumption rate of $p_{con} = 25,228.8\text{kJ/h}$. In the uphill region (pink grids), the speed constraint coefficient was set to $f_1 = 0.75$. For downhill regions (blue grids), the constraint coefficient was $f_2 = 1.05$. In areas with dense weeds (red grids), the constraint coefficient was $f_3 = 0.6$. As shown in Figure 12, different colors represent the paths of different lawnmowers. When the number of lawnmowers significantly exceeds the minimum required for complete coverage of the grid map, it is essential to determine the initial minimum number of lawnmowers. This is accomplished by using a formula to estimate the minimum number of lawnmowers required for the grid map.

$$\begin{cases} P_{sum} = \sum_{i=1}^{i=N} \left(\frac{g_1 \times l}{f_1 \times v} + \frac{g_2 \times l}{f_2 \times v} + \frac{g_3 \times l}{f_3 \times v} \right) \times p_{con} + \frac{g \times l}{v} \\ C = P_{sum} / p \end{cases} \tag{29}$$

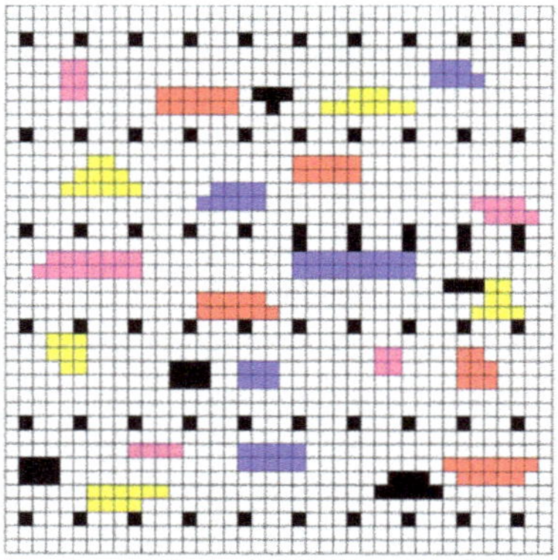

Figure 11. Grid map of the connected orchard. (White grids represent grids where the lawnmower can travel, gray grids indicate obstacle grids (i.e., impassable grids), pink grids represent uphill areas, blue grids denote down-hill areas, red grids represent overgrown grass areas, and yellow grids represent areas with sparse weeds).

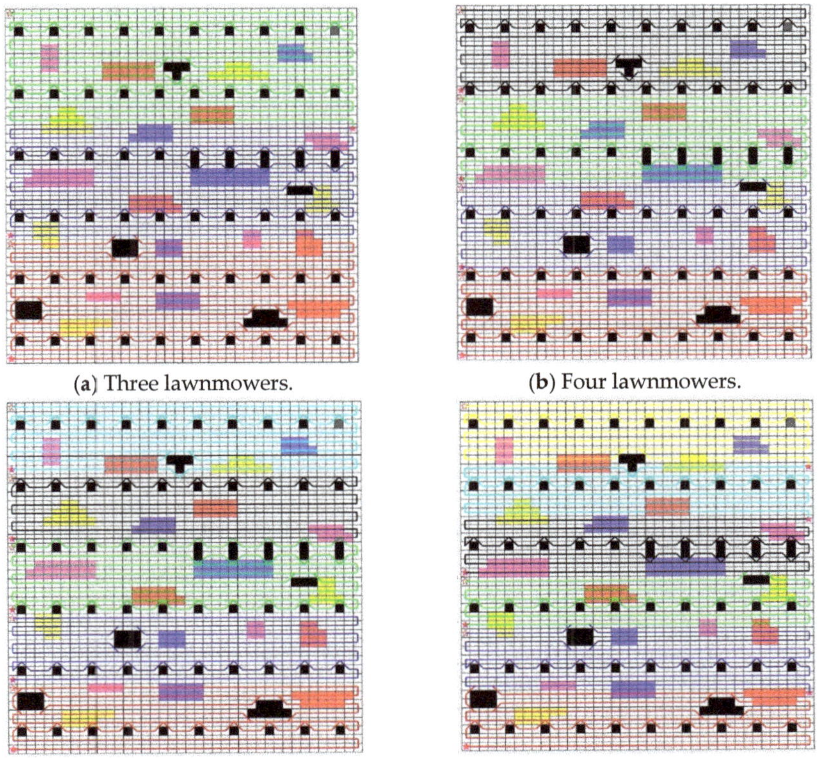

(**a**) Three lawnmowers. (**b**) Four lawnmowers. (**c**) Five lawnmowers. (**d**) Six lawnmowers.

Figure 12. The travel route maps for 3, 4, 5, and 6 lawnmowers in the connected orchard. (White grids represent grids where the lawnmower can travel, gray grids indicate obstacle grids (i.e., impassable grids), pink grids represent uphill areas, blue grids denote down-hill areas, red grids represent overgrown grass areas, and yellow grids represent areas with sparse weeds. The lines in the picture show the routes of the robots).

To evaluate the impact of different numbers of lawnmowers on path planning, four sets of experiments were conducted for path planning in the connected orchard. In the first set of experiments, the number of lawnmowers was set to three; in the second set, it was set to four; in the third set, it was set to five, and in the fourth set, it was set to six. This simulation experiment evaluated the influence of different numbers of lawnmowers on path planning under identical grid map conditions. The simulation was performed on a computer with 8 GB of memory and an AMD R7-5800 CPU, using MATLAB 2020b as the simulation software.

Therefore, this simulation experiment compared cases where the number of lawnmowers was set to three, four, five, and six, analyzing the repetition rate, omission rate, energy consumption, and cost of the lawnmowers. The improved A* path for each lawnmower in scenarios with different numbers of lawnmowers is shown in Figure 12 below.

In this simulation experiment, lawnmower travel directions were determined based on Figure 5. The total operation time, individual lawnmower working time, total energy consumption, and energy consumption per lawnmower for configurations with three, four, five, and six lawnmowers are summarized in Table 3.

Table 3. Comparison of lawnmower working time, energy consumption, and cost in the connected orchard.

Number of Lawnmowers	Time Consumption per Lawnmower (s)	Minimum Time Consumption (s)	Energy Consumption per Lawnmower (kJ)	Total Energy Consumption (kJ)	Total Cost (USD)
3	t1 = 461.15; t2 = 499.57; t3 = 465.49	499.57	p1 = 2459.52; p2 = 2664.48; p3 = 2482.56	7606.56	2880.41
4	t1 = 352.89; t2 = 357.68; t3 = 359.22; t4 = 356.33	359.22	p1 = 1882.056; p2 = 1908.72; p3 = 1915.824; p4 = 1899.84	7606.44	3840.55
5	t1 = 283.38; t2 = 286.04; t3 = 282.13; t4 = 288.24; t5 = 286.71	288.24	p1 = 1511.328; p2 = 1525.512; p3 = 1504.224; p4 = 1536.216; p5 = 1529.112	7606.392	4800.42
6	t1 = 246.95; t2 = 214.20; t3 = 249.62; t4 = 247.92; t5 = 215.86; t6 = 250.62	250.62	p1 = 1317.096; p2 = 1142.4; p3 = 1331.304; p4 = 1333.08; p5 = 1145.952; p6 = 1336.632	7406.464	5760.41

For this simulation experiment, the repetition rate and omission rate of the grid map traversal for three, four, five, and six lawnmowers are summarized in the Table 4.

Table 4. The experimental results for the lawnmower robot omission rate and repetition rate in the connected orchard.

Number of Lawnmowers	Repetition Rate	Omission Rate
3	6.17%	0.2131%
4	6.10%	0.2131%
5	6.00%	0.2129%
6	6.30%	0.2128%

From Tables 3 and 4, it can be observed that, with a constant grid map, as the number of lawnmower clusters increased, the minimum working time of lawnmowers in this grid map gradually decreased. However, the decreasing trend of the minimum working time diminished with the increasing number of lawnmowers. In terms of omission rate, as the number of lawnmower clusters increased from three to six, the omission rate gradually decreased. Similarly, the decreasing trend continued with the increasing number of lawnmower clusters. Regarding the repetition rate, when the number of lawnmowers increased from three to five, the repetition rate decreased. However, when the number of lawnmowers increased from five to six, the repetition rate increased by 0.30%. Simultaneously, with an increase in the number of lawnmowers, the total working cost of the lawnmowers increased. Therefore, in this simulation experiment, when the lowest repetition rate and omission rate of lawnmowers were considered as evaluation indicators, the optimal number of lawnmower clusters was five.

4.3. Covering Path Planning in Non-Connected Orchards

For the non-connected orchards, the following definitions were established. The orchards were denoted as two independent entities (labeled a and b). The size of the grid map for orchard a was set to 100 × 60, and for orchard b, it was set to 80 × 74. The obstacle occupancy rate in the grid map for orchard a (the ratio of the number of grid cells occupied by tree obstacles to the total number of grid cells) was o = 6.55%, and for orchard b, it was 6.93%. The distance between any two adjacent grid cells was denoted as l = 50. The grid map was categorized into the following regions: White cells represent areas where the lawnmower can move, achieving its maximum speed in operational conditions; black cells represent obstacles, indicating impassable areas; pink cells represent uphill regions, with the total number of uphill cells being g_{a1} for orchard a and g_{b1} for orchard b; blue cells represent downhill regions, with g_{a2} cells for orchard a and g_{b2} cells for orchard b; and red cells represent areas with dense grass, with g_{a3} cells for orchard a and g_{b3} cells for orchard b. The grid maps for the non-connected orchards are illustrated in Figure 13.

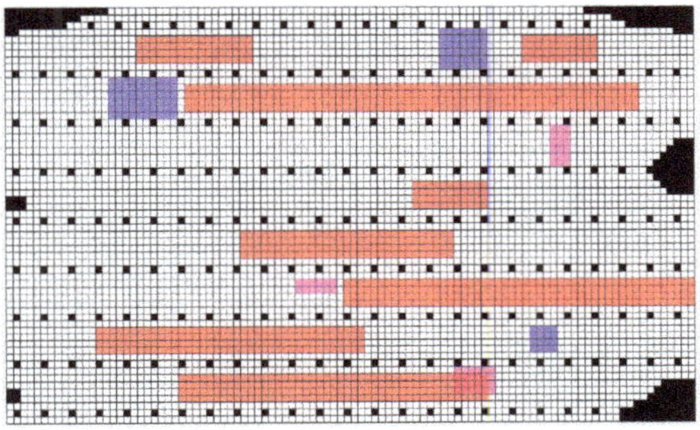

(**a**) The grid map of orchard a.

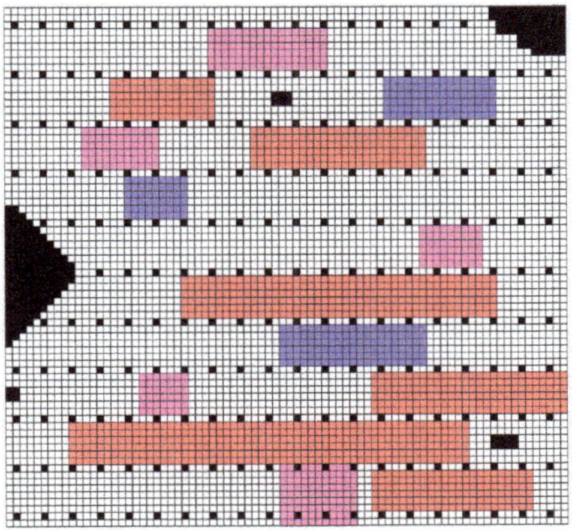

(**b**) The grid map of orchard b.

Figure 13. The grid map of the non-connected orchards. (White grids represent grids where the lawnmower can travel, gray grids indicate obstacle grids (i.e., impassable grids), pink grids represent uphill areas, blue grids denote down-hill areas, red grids represent overgrown grass areas).

To validate the impact of different numbers of lawnmowers on path planning, experiments were conducted on the non-connected orchards. Prior to these experiments, the minimum number of lawnmowers required to traverse and cover orchard a was determined to be two, and for orchard b, it was also determined to be two. Therefore, for this simulation experiment, four and five lawnmowers were set up, and four sets of experiments were conducted.

In the first set of experiments, the four lawnmowers were divided into two clusters, each containing two lawnmowers. Both clusters of lawnmowers simultaneously traversed orchards a and b. The experiment recorded the lawnmowers' shortest working time, repetition rate, omission rate, and cost during the traversal process. The improved A* paths for each lawnmower are shown in Figure 14.

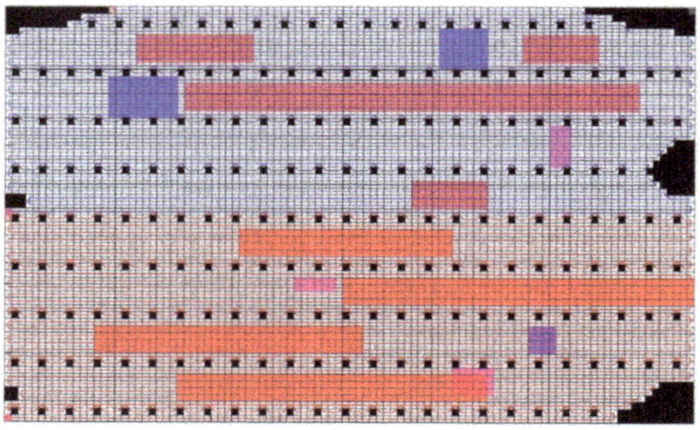

(a) Travel routes for 2 lawnmowers traversing orchard a.

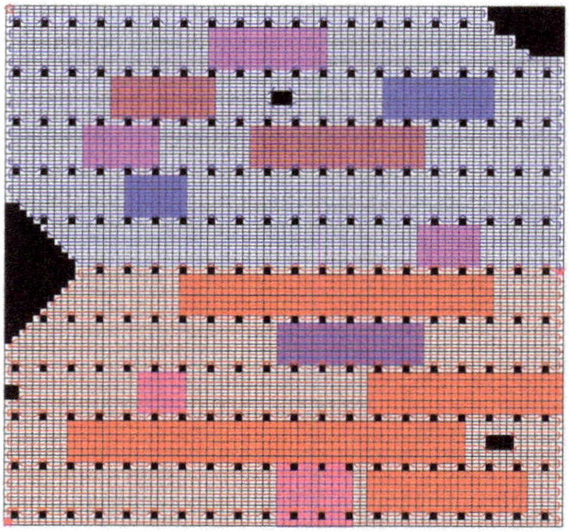

(b) Travel routes for 2 lawnmowers traversing orchard b.

Figure 14. Travel route maps for the lawnmowers in Experiment 1. (White grids represent grids where the lawnmower can travel, gray grids indicate obstacle grids (i.e., impassable grids), pink grids represent uphill areas, blue grids denote down-hill areas, red grids represent overgrown grass areas. The lines in the picture show the routes of the robots).

The data for the lawnmowers in Experiment 1 during the traversal of orchards a and b in terms of time, energy consumption, and cost are recorded in Table 5.

Table 5. The data for time, energy consumption, and cost in Experiment 1.

	Time Consumption per Lawnmower (s)	Minimum Time Consumption (s)	Energy Consumption per Lawnmower (kJ)	Total Energy Consumption (kJ)	Total Cost (USD)
Orchard a	t1 = 2098.21; t2 = 2068.07;	2098.21	p1 = 11,190.48; p2 = 11,029.68;	22,220.16	1877.55
Orchard b	t1 = 2060.47; t2 = 2042.82;	2060.47	p1 = 10,989.12; p2 = 10,895.04;	21,884.16	1877.38
Whole orchard	4158.68	4158.68	44,104.32	44,104.32	3754.93

The data for the repetition rate and the omission rate of the lawnmower during the traversal of grid maps for orchards a and b in the first set of simulation experiments are recorded in Table 6.

Table 6. Experimental results for repetition rate and omission rate in Experiment 1.

	Repetition Rate	Omission Rate
Orchard a	4.12%	0.24%
Orchard b	4.25%	0.19%

From Tables 5 and 6, it can be observed that in the first set of simulation experiments, the shortest working time for traversing and covering orchard a was 2098.21 s, and for orchard b, it was 2060.47 s. Since the two groups of lawnmowers simultaneously traversed and covered orchards a and b, the shortest time for traversing and covering the non-connected orchards was 2098.21 s. As shown in the table, the repetition rate of lawnmower traversal and coverage for orchard a was 4.12%, and for orchard b, it was 4.25%. The total repetition rate for non-connected orchards was calculated using the following formula:

$$\frac{\text{Repetition rate for orchard a} * \text{Area of orchard a} + \text{Repetition rate for orchard b} * \text{Area of orchard a}}{\text{Area of orchard a} + \text{Area of orchard b}} \quad (30)$$

The repetition rate for the non-connected orchard was 4.18%. The mowing machine's traversal coverage omission rate for orchard a was 0.24%, and for orchard b, it was 0.19%. The total omission rate for non-connected orchards was calculated using the following formula:

$$\frac{\text{Omission rate for Orchard a} * \text{Area of orchard a} + \text{Omission rate for Orchard b} * \text{Area of orchard a}}{\text{Area of orchard a} + \text{Area of orchard b}} \quad (31)$$

The total omission rate for the non-connected orchards was 0.21%.

Experiment 2 involved grouping the four lawnmowers together, resulting in a cluster count of four. The lawnmowers in this group sequentially traversed orchards a and b. After completing the first orchard, they required an energy refill, with the assumed refueling time being 300 s. The recorded data include the lawnmowers' minimum working time, repetition rate, omission rate, and cost during the traversal process. The improved A* paths for each lawnmower are depicted in Figure 15.

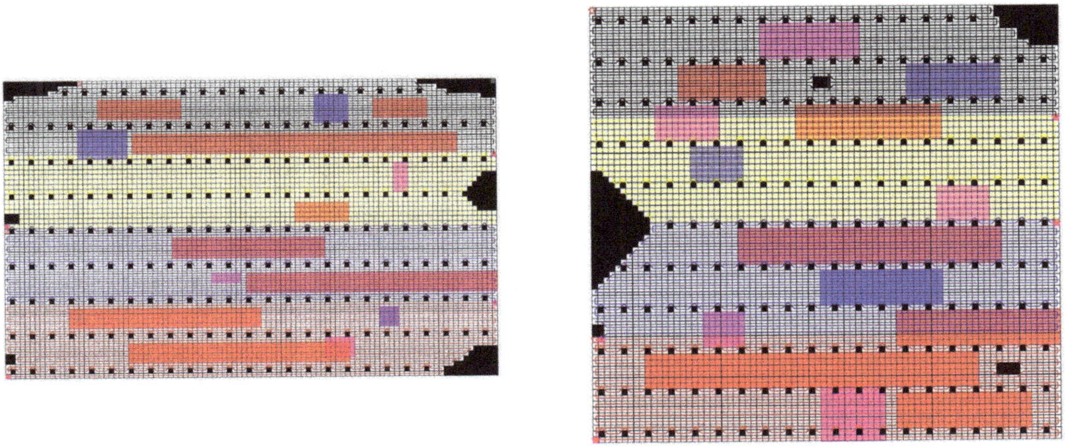

(**a**) Travel routes for 4 lawnmowers traversing orchard a. (**b**) Travel routes for 4 lawnmowers traversing orchard b.

Figure 15. Travel route maps for the lawnmowers in Experiment 2. (White grids represent grids where the lawnmower can travel, gray grids indicate obstacle grids (i.e., impassable grids), pink grids represent uphill areas, blue grids denote down-hill areas, red grids represent overgrown grass areas, and yellow grids represent areas with sparse weeds. The lines in the picture show the routes of the robots).

The data for the lawnmowers in Experiment 2 during the traversal of orchards a and b in terms of time, energy consumption, and cost are recorded in Table 7.

The data for the lawnmowers in Experiment 2 during the traversal of orchards a and b in terms of repetition rate and omission rate are recorded in Table 8.

From Tables 7 and 8, it can be observed that in the second set of simulation experiments, the shortest working time for traversing and covering orchard a was 1058.30 s, and for orchard b, it was 1039.78 s. Since the two groups of lawnmowers traversed and covered orchards a and b sequentially, and energy replenishment was required after completing the traversal and coverage of the first orchard, the shortest time for the lawnmowers to traverse and cover the non-connected orchards in Experiment 2 was 2098.07 s. According to Table 7, the repetition rate of the lawnmowers for orchard a was 4.16% and for orchard b it was 4.27%, and the total repetition rate for non-connected orchards was calculated using Formula (30) to be 4.21%. The omission rate of the lawnmowers for orchard a was 0.24% and for orchard b it was 0.19%, and the total omission rate for non-connected orchards was calculated using Formula (31) to be 0.21%.

Table 7. The data for time, energy consumption, and cost in Experiment 2.

	Time Consumption per Lawnmower (s)	Minimum Time Consumption (s)	Energy Consumption per Lawnmower (kJ)	Total Energy Consumption (kJ)	Total Cost (USD)
Orchard a	t1 = 1039.92; t2 = 1058.30; t3 = 1044.33; t4 = 1023.74	1058.30	p1 = 5565.12; p2 = 5644.32; p3 = 5571.6; p4 = 5460	22,241.04	3755.17
Orchard b	t1 = 1021.43; t2 = 1039.04; t3 = 1003.05; t4 = 1039.78	1039.78	p1 = 5447.52; p2 = 5561.04; p3 = 5349.6; p4 = 5545.44	21,903.6	3754.99
Whole orchard	2098.07	2098.07	44,144.64	44,144.64	7510.16

Table 8. Experimental results for repetition rate and omission rate in Experiment 2.

	Repetition Rate	Omission Rate
Orchard a	4.16%	0.24%
Orchard b	4.27%	0.19%

Comparing the data from the first and second sets of experiments, for non-connected orchards, when the number of lawnmowers was constant, dividing the lawnmowers into two groups resulted in a shorter shortest working time of 2060.47 s compared to 2098.07 s when the lawnmowers were in one group. Furthermore, dividing the lawnmowers into two groups led to a lower repetition rate than having them in one group, whereas the omission rate shows little variation.

Experiment 3 involved dividing the four lawnmowers into two groups, with one lawnmower in one group and three lawnmowers in the other. The lawnmower cluster with a size of three traversed the larger area of orchard a, whereas the cluster with a size of one covered the smaller area of orchard b. Both groups of lawnmowers simultaneously traversed orchards a and b, and the experiment recorded the lawnmowers' shortest working time, repetition rate, omission rate, and cost during the traversal process. The improved A* paths for each lawnmower are illustrated in Figure 16.

The data for the lawnmowers in experiment 3 during the traversal of orchards a and b in terms of time, energy consumption, and cost are recorded in Table 9.

Table 9. The data for time, energy consumption, and cost in experiment 3.

	Time Consumption per Lawnmower (s)	Minimum Time Consumption (s)	Energy Consumption per Lawnmower (kJ)	Total Energy Consumption (kJ)	Total Cost (USD)
Orchard a	t1 = 1392.68; t2 = 1399.76; t3 = 1374.30;	1399.76	p1 = 7427.52; p2 = 7463.04; p3 = 7329.6;	22,220.16	2813.49
Orchard b	t1 = 4103.29	4103.29	p1 = 21,884.16	21,883.2	937.11
Whole orchard	5503.05	5503.05	44,103.36	44,103.36	3750.6

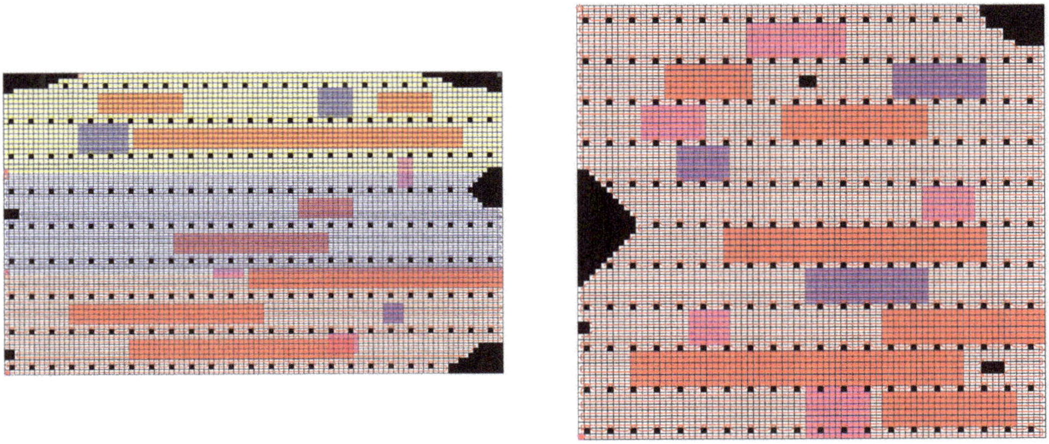

(**a**) Travel routes for 3 lawnmowers traversing orchard a. (**b**) Travel routes for 1 lawnmower traversing orchard b.

Figure 16. Travel route maps for the lawnmowers in experiment 3. (White grids represent grids where the lawnmower can travel, gray grids indicate obstacle grids (i.e., impassable grids), pink grids represent uphill areas, blue grids denote down-hill areas, red grids represent overgrown grass areas, and yellow grids represent areas with sparse weeds. The lines in the picture show the routes of the robots).

The data for the lawnmowers in experiment 3 during the traversal of orchards a and b in terms of repetition rate and omission rate are recorded in Table 10.

Table 10. Experimental results for repetition rate and omission rate in experiment 3.

	Repetition Rate	Omission Rate
Orchard a	4.12%	0.24%
Orchard b	4.23%	0.19%

From Tables 9 and 10, it can be observed that in the third set of simulation experiments, the shortest working time for traversing and covering orchard a was 1399.76 s and for orchard b it was 4103.29 s. Since the two groups of lawnmowers simultaneously traversed and covered orchards a and b, the shortest traversal time for the non-connected orchards was 4103.29 s. According to Table 9, the repetition rate of lawnmower traversal coverage for orchard a was 4.12% and for orchard b it was 4.23%. Using Formula (30), the total repetition rate for non-connected orchards was determined to be 4.17%. The omission rate of lawnmower traversal coverage for orchard a was 0.24% and for orchard b it was 0.19%. Using Formula (31), the total omission rate for non-connected orchards was determined to be 0.21%.

Comparing the results of the experiments in groups 1, 2, and 3, it was observed that dividing the lawnmower cluster into smaller clusters, especially when the cluster number could be evenly distributed, yielded better performance in terms of the shortest working time, repetition rate, and omission rate compared to treating the lawnmowers as a single collective cluster. This improvement was particularly evident in the case of the shortest working time. However, when the cluster number could not be evenly distributed, the performance in terms of the shortest working time, repetition rate, and omission rate was worse compared to treating the lawnmowers as a single collective cluster, with significant differences in the shortest working time. To validate this conclusion, 10 sets of simulation experiments were conducted on different non-connected orchards. For each non-connected orchard, four and five lawnmowers were selected for verification. The simulation experiments were conducted with four lawnmowers grouped according to the grouping in the first and second experiments and five lawnmowers grouped according to the grouping in the second and third experiments. The simulation results are shown in Figure 17.

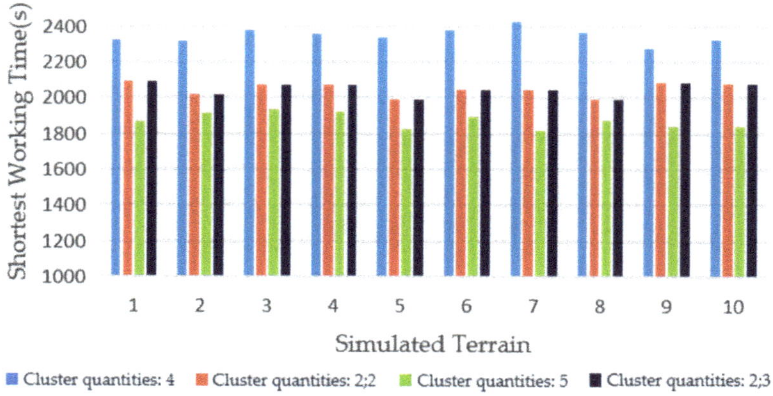

Figure 17. Shortest working time simulation experiment data.

Therefore, for non-connected orchards, when the number of lawnmowers was fixed, the most optimal working effect was achieved by evenly distributing the number of lawnmowers in a cluster according to the number of components in the non-connected orchard. When the number of lawnmowers in a cluster could not be evenly distributed according to the number of components in the non-connected orchard, the optimal working effect was achieved by treating the lawnmowers as a single group.

4.4. On-Site Experiment Validation

In the actual orchard environment, due to the differences between the real operating conditions of the lawnmower and the simulated operating conditions of the lawnmower, there were discrepancies between the results of the simulation experiments and the actual experiments. Therefore, to validate the rationality of the path planning algorithm, field experiments were designed for research and analysis.

4.4.1. Selection of Experimental Site and Equipment

To better align the simulation conditions of the lawnmower with its actual working environment, the selected orchard had a terrain slope ranging from 0 to 10 degrees, meeting the experimental requirements. The satellite image of the orchard is shown in Figure 18a, and the prescription map of the orchard based on Section 2 is shown in Figure 18b. The basic data of the experimental site are presented in Table 11.

The basic data of the experimental site are presented in Table 11.

Table 11. The basic data of the experimental site.

Coordinate Positions	(40.18° N, 116.97° E)
Orchard area (hectares)	80.937
Tree row spacing (m)	3
Tree vertical spacing (m)	1.5

The robotic lawnmower used in the experiment was designed and assembled by our team. Its main operating principle is as follows: The locomotion system is independently powered by a 48 V DC power supply. It employs a tracked dual-motor driving mechanism, providing stable movement and obstacle-crossing capabilities in hilly orchard terrains. The differential rotation of the dual motors enables the lawnmower to achieve minimum radius turns. The cutting system is powered by a gasoline engine with a rated speed of 3600 rpm. The power is transmitted from the engine's output shaft, passing through a first-stage gear reducer for appropriate speed reduction before being transmitted to the blade disc output shaft. This, in turn, drives the rapid rotational cutting motion of the blade disc and the blades. The lifting mechanism primarily consists of a parallelogram mechanism and an electric push rod. The platform, linkages, and linkages with the frame adopt articulated connections to ensure relative motion space. The telescopic movement of the electric push rod is converted into the platform's upward and downward motion, facilitating weed removal for different ground conditions and cutting heights. Additionally, the engine employed in the lawnmower robot for the experiments is a dedicated power source for the cutting system. It utilizes a vertical-axis gasoline engine, chosen for its cost-effectiveness and low energy consumption. If an alternative type of engine were used, issues such as the need for additional devices to alter the power direction during the transmission to the cutting system could arise. This process would entail a certain power loss, occupy internal space within the machine, and be detrimental to the design and operation of the lawnmower.

(**a**) Satellite image of the experimental site.

(**b**) Prescription map of the experimental site.

Figure 18. Satellite image and prescription map of the experimental site.

The Beidou Navigation Autonomous Driving System integrates the Beidou Navigation Satellite System, inertial navigation technology, and a variety of sensors, including laser radar and cameras, to enable global positioning and environmental perception for the lawnmower. During operation, the positioning of our autonomous lawnmowers combines GPS data with onboard sensors. Whereas the GPS system provides initial location and a navigational framework, onboard sensors offer real-time adjustments, ensuring accurate navigation through varied terrains and around obstacles. The challenging aspects of navigating uneven terrains and negotiating weed-infested orchards are effectively addressed through the deployment of advanced sensor systems. In terms of control mechanisms, each lawnmower is centrally operated through a wireless control system. This centralized system communicates seamlessly with individual units, issuing precise navigational commands based on our proprietary path planning algorithm. This integrated approach ensures synchronized and efficient operations, contributing to the overall effectiveness of our autonomous lawnmower fleet.

The parameters of the tracked small lawnmower used for the experiment are shown in Table 12 (a total of three units, as shown in Figure 19a), and the DJI Phantom 4 was chosen as the aerial photography equipment for this experiment.

In the experiment, the measuring equipment included two 50 m-long tape measures, two 5 m-long steel tape measures, a timer for recording the working time of the lawnmower, three sets of pens and notebooks for recording experimental data, a fuel tank (10 L) filled with fuel, a mobile power supply, a DJI waypoint marker, and a 500 mL measuring cup, along with several meters of nylon line.

Table 12. The basic parameters of the tracked small orchard lawnmower.

Name	Value
Dimensions (mm)	900 × 830 × 520
Total weight (kg)	110
Maximum travel speed (m/s)	1.5
Cutting width (m)	0.5
Climbing ability (°)	25
Fuel tank capacity (L)	1.2
Working efficiency	1.01175 hectares/h

(**a**) Lawnmower. (**b**) BeiDou Navigation Automatic Driving System.

Figure 19. Experimental equipment.

4.4.2. Experimental Plan

This experiment was designed with two groups to validate the accuracy of the optimal number of lawnmowers for covering orchards in both connected and non-connected orchards. The field test plan was as follows:

Step 1: Prepare the experimental equipment and select the experimental site.

Step 2: Collect experimental site data and set simulation parameters. Use a drone for aerial photography of the orchard terrain, obtaining information on the orchard and obstacle locations. Process the obtained data, including prescription maps, and create a simulated grid map. Test the fuel consumption of the lawnmowers in the orchard environment and their actual driving speeds. Set simulation experiment parameters based on this information.

Step 3: Conduct simulation experiments and obtain simulation data. Allocate the designated areas to lawnmowers based on the created simulated grid map. Use the improved A* algorithm to plan the lawnmowers' optimal driving paths. Simultaneously, collect experimental data on the lawnmowers' shortest working time, repetition rate, and omission rate during orchard traversal.

Step 4: Conduct on-site experiments and obtain actual experimental data. Input the paths planned during the simulation experiment into the control console, which sends signals to the lawnmowers to follow the planned paths. Record experimental data during the actual operation of the lawnmowers and any issues encountered.

Step 5: Analyze and compare the experimental results. Compare and analyze the repetition rate, omission rate, and shortest working time data of the lawnmowers between the simulation and the actual experiments to validate the feasibility of the plan.

4.4.3. Analysis of Simulation and Field Test Results

Three lawnmowers were used for field tests, and each lawnmower, when fully fueled, could work for approximately 1.5 h in the actual orchard environment, covering an area of around 0.6035 hectares. Through field testing, the average driving speed of the lawnmower in relatively flat orchard areas was determined to be v = 0.74 m/s. Since the slope of the orchard in the test area was mostly around 5 degrees, it had almost no impact on the lawnmower's speed, so the slope of the orchard was not considered in this experiment regarding its effect on the lawnmower's speed. However, in areas with dense weeds, the speed of the lawnmower would significantly decrease. After multiple measurements and averaging, the lawnmower's speed in such areas was determined to be v = 0.48 m/s. The fuel consumption of the lawnmower in the working condition was:

$$P_{con} = 25{,}228.8 \text{ kJ/h}.$$

Experiment 1. *A connected region of the orchard was selected as the experimental area (as shown in Figure 20a). The total area of the experimental region was 1645.34 square meters, with obstacles primarily concentrated in the boundary areas of the orchard. Dense weed areas in this region were mostly located in the middle of the tree rows. A grid map was established based on the prescription map (Figure 20c). The size of grid map of experimental area 1 was set to 100 × 66.*

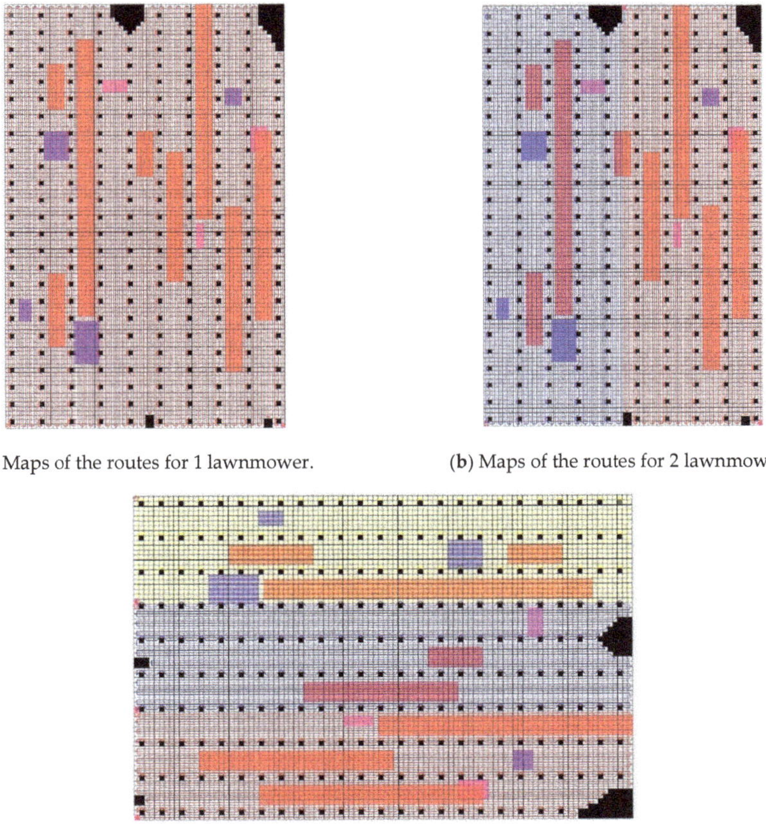

(**a**) Maps of the routes for 1 lawnmower.

(**b**) Maps of the routes for 2 lawnmowers.

(**c**) Maps of the routes for 3 lawnmowers.

Figure 20. Route maps for the lawnmowers in Experiment 1. (White grids represent grids where the lawnmower can travel, gray grids indicate obstacle grids (i.e., impassable grids), pink grids represent uphill areas, blue grids denote down-hill areas, red grids represent overgrown grass areas, and yellow grids represent areas with sparse weeds. The lines in the picture show the routes of the robots).

Experiment 2. *Two non-connected regions of the orchard were selected as experimental areas. The first experimental area was the same as that in Experiment 1. The second experimental area had an area of 1289.84 square units, with obstacles mainly concentrated in the boundary areas of the orchard. A grid map was established based on the prescription map (Figure 21b). The size of grid map of experimental area 2 was set to 120 × 45.*

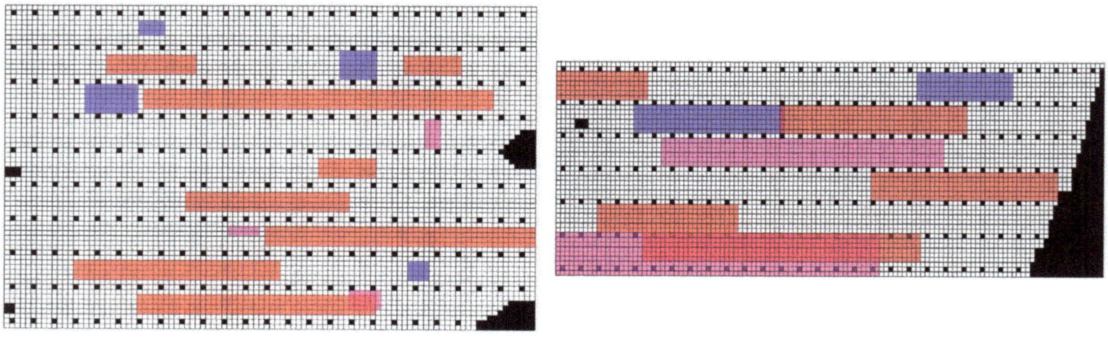

Figure 21. Grid maps of experimental areas 1 and 2 for experiments 1 and 2. (**a**) Grid map of experimental area 1 (rotated counterclockwise by 90 degrees); (**b**) grid map of experimental area 2 (rotated clockwise by 90 degrees). (White grids represent grids where the lawnmower can travel, gray grids indicate obstacle grids (i.e., impassable grids), pink grids represent uphill areas, blue grids denote down-hill areas, red grids represent overgrown grass areas).

(1) Simulation Experiment Results

Using the grid map created for the first experimental orchard environment as input, the improved A* path was planned through MATLAB simulation software, as shown in Figure 20.

The data on the omission rate, repetition rate, energy consumption, and minimum working time of one, two, and three lawnmowers during the traversal of the orchard obtained from the simulation software are presented in Table 13.

Table 13. Simulation test data for Experiment 1.

Number of Lawnmowers	1	2	3

Duration of operation for each lawnmower (s)	4214.86	t1 = 2060.81;
		t2 = 2154.05
			t1 = 1351.35;
			t2 = 1452.70;
			t3 = 1410.82

Minimum working time for each lawnmower (s)	4214.86	2154.05	1452.72

Energy consumption per lawnmower (kJ)	22,588.8	p1 = 11,239.2;
		p2 = 11,304
			p1 = 7884;
			p2 = 7634.4;
			p3 = 7502.4

Total energy consumption (kJ)	22,588.8	22,543.2	23,020.8
Total cost (USD)	937.45	1874.29	2811.68
Omission rate	0.155%	0.154%	0.153%
Repetition rate	4.013%	4.029%	4.012%

The conclusion that can be drawn from Table 13 is that in this experimental area, when using a different number of lawnmowers to traverse and cover the orchard, using three lawnmowers compared to one lawnmower resulted in a decrease in the omission rate of 0.002%, a decrease in the repetition rate of 0.001%, and a reduction in the shortest working time of 2762.14 s. Compared to using two lawnmowers, the omission rate decreased by 0.001%, the repetition rate decreased by 0.017%, and the shortest working time decreased by 701.33 s. However, the overall fuel consumption was slightly higher than when using one or two lawnmowers, resulting in the highest total cost. Therefore, in this simulation experiment, using three lawnmowers corresponds to the shortest working time, the lowest omission rate, and the lowest repetition rate in the planned path, whereas one lawnmower corresponded to the path with the lowest cost.

For Experiment 2, based on the grid map created according to the non-connected orchard environment as input, as mentioned in Section 4.3, three lawnmowers were used as a cluster to cover the orchard, with the omission rate, repetition rate, lawnmower energy consumption, and shortest working time recorded for each lawnmower. The improved A* path of the lawnmower was planned using MATLAB simulation software, as shown in Figure 22.

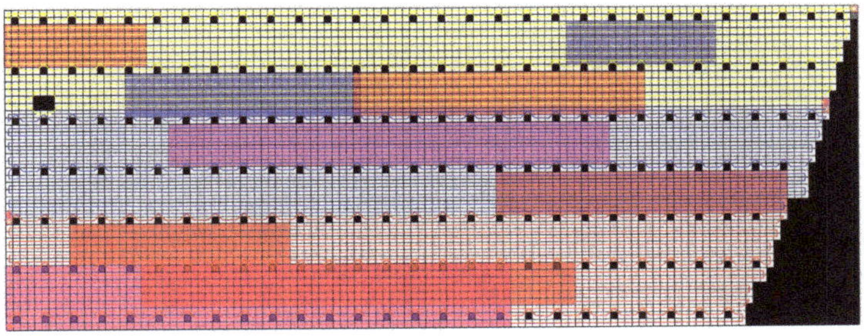

Figure 22. Lawnmower trajectory map for the second experimental area 2 in Experiment 2. (White grids represent grids where the lawnmower can travel, gray grids indicate obstacle grids (i.e., impassable grids), pink grids represent uphill areas, blue grids denote down-hill areas, red grids represent overgrown grass areas, and yellow grids represent areas with sparse weeds. The lines in the picture show the routes of the robots).

The data for the lawnmower's traversal in experimental area 2, including the omission rate, repetition rate, energy consumption for each lawnmower, and the minimum working time, are presented in Table 14.

Table 14. Simulation experimental data for experimental area 2 in Experiment 2.

Number of Lawnmowers	3
Duration of operation for each lawnmower (s)	t1 = 1209.73 t2 = 1221.53 t3 = 1175.25
Minimum working time for each lawnmower (s)	1221.53

Energy consumption per lawnmower (kJ)	p1 = 6451.92;
	p2 = 6514.8;
	p3 = 6268.08

Total energy consumption (kJ)	19,234.8
Total cost (USD)	2811.20
Omission rate	0.36%
Repetition rate	4.15%

From Tables 13 and 14, it can be observed that the lawnmower's minimum working time for traversing experimental area 1 of the orchard was 1452.72 s, and for experimental area 2, it was 1221.53 s. After completing the traversal of experimental area 1, the lawnmower required energy replenishment, with a duration of 300 s. Therefore, the total minimum working time for the lawnmower in this non-connected orchard was 2974.25 s. Calculations based on Formulas (18) and (19) yielded a total repetition rate of 4.07% and a total omission rate of 0.24%.

(2) Field Experiment

Experiment 1 was conducted in orchard test area 1. The process of Experiment 1 is illustrated in Figure 23.

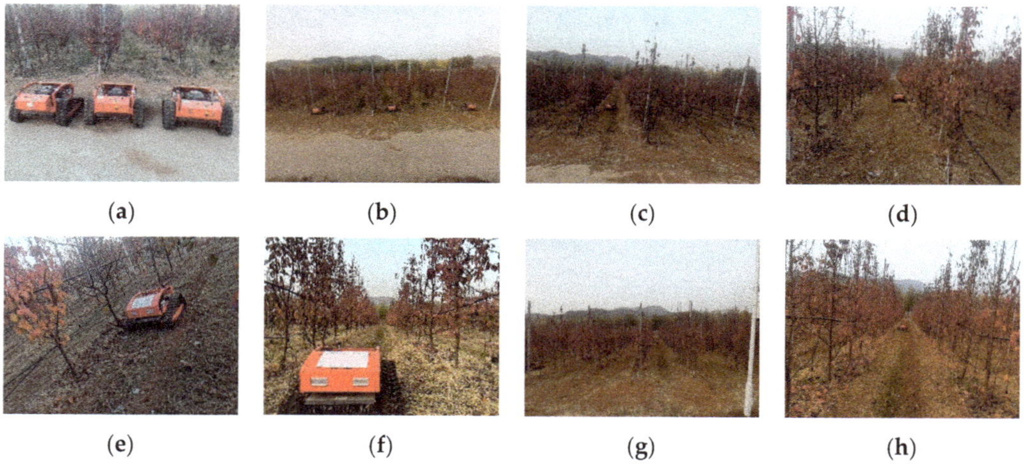

Figure 23. The field experiment for Experiment 1. (Figures (**a–h**) show part of the field test process in Experiment 1).

The resulting data of the coverage and traversal of experimental area 1 in the connected orchard by three lawnmowers are shown in Table 15.

Table 15. Field experiment data for Experiment 1.

Number of Lawnmowers	3
Duration of operation for each lawnmower (s)	t1 = 1642.76 t2 = 1721.43 t3 = 1667.28
Minimum working time for each lawnmower (s)	1721.43
Energy consumption per lawnmower (kJ)	p1 = 9866.4; p2 = 10,065.6 p3 = 9422.4
Total energy consumption (kJ)	29,354.4
Total cost (USD)	2812.22
Omission rate	0.30%
Repetition rate	5.63%

In experimental area 2 of the orchard, Experiment 2 was conducted, as illustrated in Figure 24. The energy replenishment time for the lawnmowers after completing the traversal of experimental area 1 was 258 s.

The resulting data of the coverage and traversal of experimental area 2 in the non-connected orchard by three lawnmowers in Experiment 2 are shown in Table 16.

The results from Tables 15 and 16 show that the lawnmower's shortest working time to traverse experimental area 1 in the orchard was 1721.43 s, and for experimental area 2, it was 1486.67 s. After traversing experimental area 1, the lawnmower required an energy replenishment time of 258 s. Therefore, the lawnmower's total shortest working time for this non-connected orchard was 3466.1 s. Calculations using Formulas (18) and (19) yielded a total repetition rate of 5.71% and a total omission rate of 0.40%.

Figure 24. Field test work of lawnmowers in area 2 in Experiment 2. (Figures (a–f) show part of the field test process in Experiment 2).

Table 16. Field test data for area 2 in Experiment 2.

Number of Lawnmowers	3
Duration of operation for each lawnmower (s)	t1 = 1463.58 t2 = 1486.67 t3 = 1422.95
Minimum working time for each lawnmower (s)	1486.67
Energy consumption per lawnmower (kJ)	p1 = 7286.4; p2 = 7567.2 p3 = 7161.6
Total energy consumption (kJ)	22,015.2
Total cost (USD)	2811.43
Omission rate	0.53%
Repetition rate	5.81%

During the field test process, the following issues were recorded:

(1) The unevenness of the orchard soil caused the lawnmower to experience fluctuations in its path during traversal.
(2) The lawnmower's traversal in the orchard compacted the weeds, leading to an increase in the lawnmower's omission rate.
(3) When turning, the lawnmower was affected by weeds, resulting in an increase in turning time.
(4) Measurement errors: The inherent error in the positioning system caused the lawnmower to deviate from the planned path. There were also errors in measuring the lawnmower's speed using a tape measure and a stopwatch.

Considering the actual conditions of the lawnmower's traversal in the orchard, and taking into account that the area of the connected orchard experimental zone was 1645.34 square meters, a deviation of up to 5 min in the shortest working time was considered reasonable, an omission rate deviation of up to 1% was considered reasonable, and a repetition rate deviation of up to 2% was considered reasonable. For the non-connected orchard area with a total area of 2935.18 square meters, a deviation of up to 10 min in the shortest working time was considered reasonable, an omission rate deviation of up to 1%

was considered reasonable, and a repetition rate deviation of up to 3% was considered reasonable.

(3) Analysis of Experimental Results

A comparison between the actual data and simulation data for the lawnmower's coverage path in the connected orchard experimental area is presented in Table 17 and was derived from Tables 13 and 15. It constitutes a comparative analysis of the actual and simulated data from Experiment 1.

Table 17. Comparative analysis of actual and simulated data from Experiment 1.

	Simulated Data	Actual Data
Minimum working time for each lawnmower (s)	1452.72	1865.43
Total energy consumption (kJ)	23,020.8	29,354.4
Omission rate	0.15%	0.30%
Repetition rate	4.01%	5.63%

As can be seen in Table 17, the lawnmower's actual data for traversing and covering the orchard show some discrepancies with the simulation data. There was a slight increase in the actual shortest working time compared to the simulated shortest working time of 412.71 s. The total energy consumption increased by 6333.6 kJ, the omission rate increased by 0.15%, and the repetition rate increased by 1.62%.

Tables 14 and 16 provide the actual and simulated data for the lawnmower's coverage path planning in the non-connected orchard experimental area, and a comparative analysis is presented in Table 18.

Table 18. Comparative analysis of actual and simulated data for Experiment 2.

	Simulated Data	Actual Data
Minimum working time for each lawnmower (s)	2974.25	3466.1
Total energy Consumption (kJ)	42,255.6	51,369.6
Omission rate	0.24%	0.40%
Repetition rate	4.07%	5.71%

As can be seen in Table 18, the lawnmower's actual data for traversing and covering the non-connected orchard exhibit some discrepancies with the simulation data. The actual shortest working time increased by 491.84 s. The total energy consumption increased by 9114 kJ, the omission rate increased by 0.15%, and the repetition rate increased by 1.62%.

Based on the above analysis, the main reasons for the discrepancies between the simulation data and actual experimental data are as follows:

(1) The orchard's uneven terrain caused fluctuations in the lawnmower's path during traversal.
(2) The lawnmower's movement in the orchard compressed the weeds, leading to an increase in the omission rate.
(3) While turning, the lawnmower was affected by weeds, resulting in increased turning time.
(4) Measurement errors in the positioning system contributed to deviations from the planned path. Measurement of the lawnmower's speed using a tape measure and stopwatch introduced errors.

In summary, the analysis indicates that the errors between the simulation and actual experimental data in both connected and non-connected orchards were relatively small, validating the rationality of orchard coverage path planning.

This section validates the practicality and reliability of orchard traversal path planning by conducting field experiments and comparing the data from field experiments with simulation test data. The section began by introducing the preparatory work for field experiments, describing the experimental site, equipment, methods, and objectives. Connected and non-connected areas within the orchard were selected to establish simulated grid maps. Subsequently, the selected orchard terrain underwent both simulation experiments and field experiments to collect data on the lawnmower's omission rate, repetition rate, shortest working time, and energy consumption during orchard traversal. A comparative analysis was then conducted to identify the reasons for the discrepancies between the actual and simulated data. Through the comparative analysis of the simulation and actual experimental data, the errors in the lawnmower's shortest working time, repetition rate, and omission rate during coverage in both connected and non-connected orchard areas were found to be within a reasonable range. Therefore, the path planning for this orchard was deemed rational.

5. Conclusions

This paper investigates the comprehensive coverage path planning method for tracked lawnmowers in orchards, exploring task allocation methods for lawnmower clusters. An improved A* algorithm is proposed, and the impact of lawnmower cluster quantity on path planning in different orchard scenarios is studied. Simulation verification of connected and non-connected areas, along with field experiments, demonstrates the practicality of the A* algorithm. However, future research requires further exploration and improvement. Firstly, extending path planning to three-dimensional space is necessary to consider the height of obstacles in the environment and the lawnmower's height on path planning. Secondly, in complex agricultural systems, path planning algorithms need further refinement and optimization to fully leverage the potential of comprehensive coverage path planning technology in agriculture. Therefore, future research is expected to expand in these areas to achieve a more comprehensive application of path planning technology in agriculture.

Author Contributions: Conceptualization, M.Z. and S.W.; methodology, M.Z. and S.W.; software, M.Z. and S.W.; validation, M.Z. and S.W.; formal analysis, M.Z. and S.W.; investigation, M.Z.; resources, S.W.; data curation, S.W.; writing—original draft preparation, M.Z. and S.W.; writing—review and editing, X.L. and S.W.; visualization, X.L. and S.W.; supervision, L.W. and L.J.; project administration, L.W.; funding acquisition, S.W. All authors have read and agreed to the published version of the manuscript.

Funding: This work was funded by the National Natural Science Foundation of China under grant 32372592 and the CCF-Baidu Apollo Joint Development Project Fund.

Data Availability Statement: Data are contained within the article.

Conflicts of Interest: The authors declare no conflicts of interest.

References

1. Li, Y.; Guo, Z.; Shuang, F.; Zhang, M.; Li, X. Key technologies of machine vision for weeding robots: A review and benchmark. *Comput. Electron. Agric.* **2022**, *196*, 106880. [CrossRef]
2. McCool, C.; Beattie, J.; Firn, J.; Lehnert, C.; Kulk, J.; Bawden, O.; Perez, T. Efficacy of mechanical weeding tools: A study into alternative weed management strategies enabled by robotics. *IEEE Robot. Autom. Lett.* **2018**, *3*, 1184–1190. [CrossRef]
3. Santos Valle, S.; Kienzle, J. *Agriculture 4.0—Agricultural Robotics and Automated Equipment for Sustainable Crop Production*; FAO: Rome, Italy, 2020.
4. Saidani, M.; Pan, Z.; Kim, H.; Wattonville, J.; Greenlee, A.; Shannon, T.; Yannou, B.; Leroy, Y.; Cluzel, F. Comparative life cycle assessment and costing of an autonomous lawn mowing system with human-operated alternatives: Implication for sustainable design improvements. *Int. J. Sustain. Eng.* **2021**, *14*, 704–724. [CrossRef]
5. Liao, W.; Zeng, F.; Chanieabate, M. Mechanization of small-scale agriculture in China: Lessons for enhancing smallholder access to agricultural machinery. *Sustainability* **2022**, *14*, 7964. [CrossRef]
6. Wang, S.; Song, J.; Qi, P.; Yuan, C.; Wu, H.; Zhang, L.; Liu, W.; Liu, Y.; He, X. Design and Development of Orchard Autonomous Navigation Spray System. *Front. Plant Sci.* **2022**, *13*, 960686. [CrossRef] [PubMed]

7. Patle, B.K.; Pandey, A.; Parhi, D.R.K.; Jagadeesh, A.J.D.T. A review: On path planning strategies for navigation of mobile robot. *Def. Technol.* **2019**, *15*, 582–606. [CrossRef]
8. Orozco-Rosas, U.; Montiel, O.; Sepúlveda, R. Mobile robot path planning using membrane evolutionary artificial potential field. *Appl. Soft Comput.* **2019**, *77*, 236–251. [CrossRef]
9. Lamini, C.; Benhlima, S.; Elbekri, A. Genetic algorithm based approach for autonomous mobile robot path planning. *Procedia Comput. Sci.* **2018**, *127*, 180–189. [CrossRef]
10. Nazarahari, M.; Khanmirza, E.; Doostie, S. Multi-objective multi-robot path planning in continuous environment using an enhanced genetic algorithm. *Expert Syst. Appl.* **2019**, *115*, 106–120. [CrossRef]
11. Wang, J.; Chi, W.; Li, C.; Wang, C.; Meng, M.Q.H. Neural RRT*: Learning-based optimal path planning. *IEEE Trans. Autom. Sci. Eng.* **2020**, *17*, 1748–1758. [CrossRef]
12. Hart, P.E.; Nilsson, N.J.; Raphael, B. A formal basis for the heuristic determination of minimum cost paths. *IEEE Trans. Syst. Sci. Cybern.* **1968**, *4*, 100–107. [CrossRef]
13. Tsuji, T.; Tanaka, Y.; Morasso, P.G.; Sanguineti, V.; Kaneko, M. Bio-mimetic trajectory generation of robots via artificial potential field with time base generator. *IEEE Trans. Syst. Man Cybern. Part C* **2002**, *32*, 426–439. [CrossRef]
14. Janchiv, A.; Batsaikhan, D.; hwan Kim, G.; Lee, S.G. Complete coverage path planning for multi-robots based on. In Proceedings of the International Conference on Control, Automation and Systems, Gyeonggi, Republic of Korea, 26–29 October 2011; pp. 824–827.
15. Hazon, N.; Kaminka, G.A. Redundancy, efficiency and robustness in multi-robot coverage. In Proceedings of the 2005 IEEE International Conference on Robotics and Automation, Barcelona, Spain, 18–22 April 2005; pp. 735–741.
16. Hazon, N.; Kaminka, G.A. On redundancy, efficiency, and robustness in coverage for multiple robots. *Robot. Auton. Syst.* **2008**, *56*, 1102–1114. [CrossRef]
17. Mathi, D.K.; Chinthamalla, R. A hybrid global maximum power point tracking method based on butterfly particle swarm optimization and perturb and observe algorithms for a photovoltaic system under partially shaded conditions. *Int. Trans. Electr. Energy Syst.* **2020**, *30*, e12543. [CrossRef]
18. Mazdin, P.; Barciś, M.; Hellwagner, H.; Rinner, B. Distributed task assignment in multi-robot systems based on information utility. In Proceedings of the 2020 IEEE 16th International Conference on Automation Science and Engineering (CASE), Hong Kong, China, 20–21 August 2020; pp. 734–740.
19. Bai, X.; Fielbaum, A.; Kronmüller, M.; Knoedler, L.; Alonso-Mora, J. Group-based distributed auction algorithms for multi-robot task assignment. *IEEE Trans. Autom. Sci. Eng.* **2022**, *20*, 1292–1303. [CrossRef]
20. Zlot, R.; Stentz, A.; Dias, M.B.; Thayer, S. Multi-robot exploration controlled by a market economy. In Proceedings of the 2002 IEEE International Conference on Robotics and Automation (Cat. No. 02CH37292), Washington, DC, USA, 11–15 May 2002; Volume 3, pp. 3016–3023.
21. Lin, B.; Zhao, Y.; Lin, R.; Liu, C. Integrating traffic routing optimization and train formation plan using simulated annealing algorithm. *Appl. Math. Model.* **2021**, *93*, 811–830. [CrossRef]
22. Hustiu, I.; Kloetzer, M.; Mahulea, C. Distributed path planning of mobile robots with ltl specifications. In Proceedings of the 24th International Conference on System Theory, Control and Computing (ICSTCC), Sinaia, Romania, 8–10 October 2020; pp. 60–65.
23. Atay, N.; Bayazit, B. *Mixed-Integer Linear Programming Solution to Multi-Robot Task Allocation Problem*; Washington University in St. Louis: St. Louis, MI, USA, 2006.
24. Liu, J.; Yang, J.; Liu, H.; Tian, X.; Gao, M. An improved ant colony algorithm for robot path planning. *Soft Comput.* **2017**, *21*, 5829–5839. [CrossRef]
25. Zhang, Y.; Pang, D. Research on path planning of mobile robot based on improved ant colony algorithm. In Proceedings of the 2022 IEEE 6th Information Technology and Mechatronics Engineering Conference (ITOEC), Chongqing, China, 4–6 March 2022; Volume 6, pp. 558–563.
26. Wang, S.; Han, Y.; Chen, J.; He, X.; Zhang, Z.; Liu, X.; Zhang, K. Weed Density Extraction Based on Few-shot Learning through UAV Remote Sensing RGB and Multi-spectral Images in Ecological Irrigation Area. *Front. Plant Sci.* **2022**, *12*, 735230. [CrossRef] [PubMed]
27. Medvedev, M.; Pshikhopov, V. Path planning of mobile robot group based on neural networks. In *Proceedings of the Trends in Artificial Intelligence Theory and Applications. Artificial Intelligence Practices: 33rd International Conference on Industrial, Engineering and Other Applications of Applied Intelligent Systems, IEA/AIE 2020, Kitakyushu, Japan, 22–25 September 2020*; Proceedings 33; Springer International Publishing: Berlin/Heidelberg, Germany, 2020; pp. 51–62.
28. Wang, S.; Qi, P.; Zhang, W.; He, X. Development and Application of an Intelligent Plant Protection Monitoring System. *Agronomy* **2022**, *12*, 1046. [CrossRef]

agronomy

MDPI

Article

Construction and Optimization of a Collaborative Harvesting System for Multiple Robotic Arms and an End-Picker in a Trellised Pear Orchard Environment

Hewen Zhang [1], Xiaoguang Li [1], Ling Wang [2], Dian Liu [2] and Shubo Wang [1,*]

[1] School of Automation, Institute of Intelligent Unmanned System, Qingdao University, Qingdao 266071, China; lixiaoguang@qdu.edu.cn (X.L.)

[2] College of Engineering, China Agricultural University, Beijing 100083, China; wangling2017068@cau.edu.cn (L.W.); sy20213071447@cau.edu.cn (D.L.)

* Correspondence: shubowang@qdu.edu.cn

Abstract: In order to meet the needs of intensive mechanized picking in trellised pear orchards, this paper designed a pick-place integrated end-picker based on the analysis of agronomic characteristics of trellised pear gardens and fruit. In order to realize the accurate positioning of pears in picking, based on the kinematic analysis of robot arms and the construction of a private dataset, the YOLOv5s object detection algorithm was used in conjunction with a depth camera to achieve fruit positioning. The hand–eye system calibration was carried out. Aiming at solving the problems of redundancy, inefficiency, and uneven distribution of task volume in the conventional multiple robot arms algorithm, a simulated annealing algorithm was introduced to optimize the picking sequence, and a task allocation method was proposed. On the basis of studying several key parameters affecting the performance of the algorithm, the picking efficiency was greatly optimized. And the effectiveness of the proposed multi-robot collaborative picking method in a trellised pear orchard environment was demonstrated through experiments and simulation verification. The experiments showed that the picking efficiency of the integrated end-picker was increased by about 30%, and the success rate was significantly higher than that of the flexible grippers. The results of this study can be utilized to advance robotic pear-picking research and development.

Keywords: fruit picking; multiple robotic arms; end-picker; target detection; task planning

Citation: Zhang, H.; Li, X.; Wang, L.; Liu, D.; Wang, S. Construction and Optimization of a Collaborative Harvesting System for Multiple Robotic Arms and an End-Picker in a Trellised Pear Orchard Environment. *Agronomy* **2024**, *14*, 80. https://doi.org/10.3390/agronomy14010080

Academic Editor: Baohua Zhang

Received: 1 December 2023
Revised: 25 December 2023
Accepted: 27 December 2023
Published: 28 December 2023

1. Introduction

The trellised pear orchard, which originated in Japan, was introduced to China in the 1990s and was popularized rapidly. The trellised pear orchard has the following five advantages: good fruit quality; convenient operation and management; easy to implement standardized cultivation; being convenient for mechanized operation; and preventing wind and bird damage. The harvest time of pear, the target crop in the trellised orchard in this study, is affected by a variety of characteristics and climatic conditions, so it is very important to harvest fruits at the right time according to specific requirements. Since the efficiency of a single robot arm is low, it is not enough to meet the demand of fruit harvest, and thus this study used multiple-robot-arm cooperative picking to improve efficiency and avoid the problems of quality degradation and low picking efficiency caused by not picking in time.

In general, picking work is divided into three aspects: perception, decision making, and control execution. The environment and fruit information are perceived by sensors. The decision system judges the picking target, and the execution system is driven to complete the picking action. The whole system is known as the "hand–eye–brain" picking system. This research direction has had wide concern worldwide in recent years.

Visual perception target detection in fruit and vegetable picking research is mainly divided into one-stage detection algorithms and two-stage detection algorithms. One stage detection algorithms, such as YOLO, SSD, SqueezeDet, and DetectNet, can directly extract features from the network to predict the classification and position of objects, thus eliminating the need for regional candidate networks (PRNs). The outstanding feature is that the detection speed is fast and only requires one step. On the contrary, two-stage detection algorithms, such as RCNN, FasterR-CNN, and MaskR-CNN, need to first create a proposal box that may contain the object to be detected and then perform further detection based on the object characteristics to complete the recognition and positioning of the target, which is more accurate. Mu et al. [1] used FasterR-CNN to identify kiwifruit, wherein they input the image and depth image obtained by a Kinectv2 camera (Microsoft, Redmond, WA, USA) into the convolution neural network to detect and locate the kiwifruit in the picture. Yang et al. [2] proposed a citrus fruit and branch recognition model based on MaskRCNN for fruit recognition and location under different occlusion conditions, and they constructed training datasets under a variety of complex conditions, including single fruit, multiple fruits, covered fruits, branches, and trunks. Qian et al. [3] proposed a method for mushroom detection and location based on SSD that optimizes the backbone network of the original SSD model to improve the real-time detection performance in embedded devices. The model has good detection performance for Pleurotus ostreatus. For apple fruit and branch segmentation, Kang et al. [4] adopted the Dasnet network model. Peng et al. [5] used the DeepLabV3+ semantic segmentation model based on the Xception_65 feature extraction network to detect litchi fruit. The experimental results showed that the model had 0.765 MIoU, which is 0.144 higher than the original DeepLabV3+ model. In order to adapt to the complex growth environment of litchi and simultaneously detect and locate the fruit branches of multiple litchi clusters, Li et al. [6] proposed a semantic segmentation method based on Deeplabv3 to segment the fruit, branches, and background in RGB images. However, due to the large differences in agronomic characteristics between fruits, we need to develop semantic segmentation and a target detection algorithm for the trellised pear orchard scene on the basis of the above fruit recognition algorithm.

In order to meet the needs of recognition and location of litchi fruit and stem at night, Liang et al. [7] proposed a litchi fruit detection method based on YOLOv3. In order to verify whether different classification modes will affect the detection effect of the kiwifruit detection model, Suo et al. [8] collected and classified 1160 kiwifruit images according to picking strategy and occlusion conditions, and they inputted them into YOLOv4 and YOLOv3 network models for training and testing. The experimental results showed that the tagging and classification of datasets in a way that is as detailed as possible can effectively improve the detection accuracy of the network model. Xiong et al. [9] developed a faster and more accurate system for the real-time vision detection, tracking, and locating of strawberries by combining YOLOv4, DeepSORT, and color threshold. In view of the low accuracy and poor robustness of the traditional green pepper detection methods, Li et al. [10] proposed an improved green pepper target detection algorithm based on Yolov4_tiny. Based on the backbone network of the classical target detection model, the algorithm introduces adaptive feature fusion and feature attention mechanism to improve the accuracy of the small-target recognition of green pepper while ensuring the accuracy of classification. Aiming at the characteristics of the small size and dense growth of plums, Wang [11] proposed a lightweight model named improved YOLOv4, based on YOLOv4. The experiments showed that the improved YOLOv4 model had higher average accuracy (mAP); in addition, compared with YOLOv4, the size was compressed by 77.85%, and the detection speed was accelerated by 112%. Yan et al. [12] proposed an apple detection algorithm based on improved YOLOv5s that can effectively identify graspable apples and ungraspable apples, and the average detection time for a single image was found to be only 0.015 s. In order to meet the requirements of accuracy, lightweight model, and fast response during picking in a trellised pear orchard, on the basis of optimizing the YOLO fruit detection algorithm, a pear detection algorithm based on improved YOLOv5s was developed, combined with a depth camera.

The terminal execution modes of agricultural picking robots usually include negative pressure adsorption, shear, mold cavity-sleeve, and flexible grasping. The Xiong team [13,14] of the Norwegian University of Life Sciences is devoted to the research of strawberry picking robots. The end-effector of the picking robot developed is a new cable-driven non-contact picking fixture with sensing function, which is composed of three active fingers, three passive covering fingers, and a cutter mechanism. The robot uses a collision-free motion planning algorithm to make picking safer and more convenient. For the picking of the cluster-shaped fruit of litchi, the Zhou team [15] of South China Agricultural University developed a picking robot whose end effector is mainly composed of an end-holder and a rotating cutter head. The robot uses a collision-free motion planning algorithm to make picking safer and more convenient. For the picking of cherry tomato, which is also a cluster fruit, Feng et al. [16] developed a scissor-like end-effector, with two cutters used to cut the stalk. By closing or opening the grip fixed to the cutter, the fruit can be cut and processed reliably.

For tomato picking, Yurni Oktarina et al. [17] in Indonesia designed a tomato picking robot. The end effector is a pair of scissors, sharp and flexible, which is driven by a servo motor. The strawberry-picking robot developed by Xiong's team [14] opens its mold cavity to "swallow" the fruit when picking, and it cuts the fruit stem with a blade to complete strawberry picking. For apple mold cavity-sleeve picking, the team designed a spherical two-finger structure gripper that can effectively reduce the fruit damage rate [18]. The team also developed a flexible gripper composed of two curved flexible fingers, and they improved and optimized it so that it could not only pick apples, but also pick pomegranates, grapefruits, and other fruits [19]. In order to further reduce the damage rate of apple picking, the team studied a bionic three-finger flexible gripper inspired by the octopus tentacle [20]. The end-effector of the pneumatic finger clip structure developed by Hohimer et al. [21] can pick apples flexibly and with high precision. Yu et al. [22] designed three-finger grippers made of flexible materials from an ergonomic point of view, being beneficial to protect apples from damage and achieve non-destructive picking. Due to the large difference in agronomic characteristics between fruits, the use outside the scope of the application scenario will greatly affect the integrity and picking efficiency of the fruit. Therefore, based on the above fruit pickers such as strawberry and tomato, an integrated end-picker for the trellised pear garden scene was developed.

The development of multiple robotic arms provides new ideas for picking research. The cherry-picking robot based on multi-joint robot arms developed by the ArimaS team utilizes a visual system to identify obstacles in the environment and perform path planning for a single robot arm [23,24]. Wageningen University developed a six-degree-of-freedom cucumber-picking robot that uses an identification device on the end effector to identify fruit stems and picks cucumbers by clamping the fruit stems and cutting them at high temperatures [25]. DanSteere developed an apple-picking robot in 2015 that is fast, efficient, and has a wide working range. It uses a four-degree-of-freedom robot arm and an air-suction end-effector [26]. Due to the limitation of picking efficiency of a single robot arm, the advantages of collaborative picking by multiple robotic arms are particularly prominent.

In order to improve the efficiency of apple picking, FFRobotics developed a parallel multiple robot arm apple-picking platform that adopts the way of grouping and dividing the working area [27]. In order to improve the efficiency of apple picking, FFRobotics developed an apple-picking platform based on a parallel multiple robot arm, which adopts the working mode of grouping and work area division [27]. This method can improve the picking efficiency and avoid the interference and collision between robot arms. Williams et al. [28] explored a kiwifruit-picking robot with four three-degree-of-freedom serial robot arms. The robot can collaboratively pick while effectively avoiding collisions between the robot arms. On the other hand, Fu et al. [29] developed a system containing four three-degree-of-freedom rectangular coordinate robots and applied it to the collaborative picking of kiwifruit. For strawberries, a fruit that is difficult to harvest, Harvest CROO developed a strawberry-harvesting robot for high-ridge cultivation that uses four parallel picking units. Each ridge of strawberries is equipped with an independent picking unit, thus improving work efficiency [30]. AGROBOT Robotics developed a robot suitable for elevated cultivation, using a solution of 24 parallel robot arms. The linear module utilizes mechanical isolation in the forward direction to avoid interference between robot arms, while robot arms on the same unit adopt a control strategy to achieve isolation [31].

As the process of population aging intensifies, a labor-intensive industry like picking urgently needs to be replaced by more efficient multiple robot arm systems. This paper focuses on the collaborative picking of multiple robotic arms in a trellised pear orchard. Taking the trellised pear garden as the object of research, the environment, agronomy, and physical characteristics of the fruit and the identification and positioning method of the pear were investigated. Based on the physical characteristics of pears, a pick-place integrated end-picker was designed. Based on kinematic analysis of robot arms and the construction of a private dataset, the YOLOv5s object detection algorithm was utilized in conjunction with a depth camera to achieve fruit positioning, and the hand–eye system calibration was carried out. It meets the research and development needs of an efficient picking robot in a pear orchard. In order to solve the optimal picking sequence, as well as reduce the scheduling time and the energy consumption, a task allocation method for a multiple robotic arms system was proposed, and the picking sequence was optimized through the simulated annealing algorithm, and finally, the experimental environment was set up for the robot arm picking experiment. The comparison experiment between different end-pickers and the comparison experiment before and after picking sequence optimization were conducted to test the efficiency of different end-pickers. On this basis, the task allocation method and optimal configuration were analyzed and verified by simulation experiments. This research can be utilized to advance robotic pear-picking research and development.

2. Pick-Place Integrated End-Picker

Kinematic analysis of the picking robot arm is the basis for trajectory planning and motion control in picking actions. Picking kinematics analysis mainly includes forward kinematics analysis and inverse kinematics analysis. Kinematics analysis describes the mapping relationship between the robot arm joint coordinate system and Cartesian space. Forward kinematics analysis is based on the rotation angle of each joint of a robot arm combined with the adjacent connecting rod coordinate system transformation matrix to obtain the mapping relationship between the robot arm picking end and base coordinate system. Inverse kinematics analysis is used to calculate the angle of each joint of a robot arm based on the mapping relationship between the picking end of the robot arm and the base coordinate system, and this solution often has several groups. At the same time, for the trellised pear garden, a pick-collect integrated end-picker was designed, the 3D model was established, and the physical processing was completed by 3D printing technology.

2.1. Kinematics of Robot Arm

2.1.1. Kinematic Analysis of the Robot Arm

In this paper, the improved D-H modeling method was utilized in kinematic modeling. The implementation steps are as follows:

To determine the coordinate system position between the connecting rods of a robot arm, the coordinate system was established according to the following criteria:

1. The Z_i axis coincides with the axis of the joint i; if the Z_i axis and Z_{i+1} axis intersect, the intersection point of the two axes is the origin of coordinate system, and if not, the origin is the intersection of the common perpendicular of the two axes and Z_i axis. The X_i axis is perpendicular to the Z_i axis and Z_{i+1} axis. If the Z_i axis and Z_{i+1} axis do not intersect, X_i axis points from the Z_i axis to the Z_{i+1} axis. After the Z_i axis and X_i axis are determined, the Y_i axis can be determined according to the right-hand rule.

The connecting rod coordinate systems of a robot arm can be determined through the above steps. To simplify the transformation of a coordinate system, the third and fourth coordinate systems are offset (Figure 1).

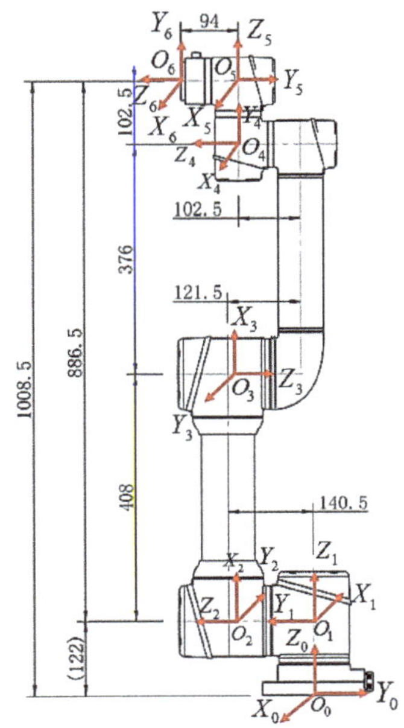

Figure 1. The connecting rod coordinate system.

2. Transforming the coordinate system of two adjacent connecting rods by translation and rotation. The implementation steps in the improved D-H modeling method are as follows: (1) Rotating the coordinate system $(X_{i-1}, Y_{i-1}, Z_{i-1})$ around the X_{i-1} axis so that the Z_{i-1} axis is parallel to the Z_i axis. (2) Translating the coordinate system $(X_{i-1}, Y_{i-1}, Z_{i-1})$ along the direction of the X_{i-1} axis until the Z_{i-1} axis coincides with the Z_i axis. (3) Rotating the coordinate system $(X_{i-1}, Y_{i-1}, Z_{i-1})$ around the Z_i axis so that the X_{i-1} axis is parallel to the X_i axis. (4) Translating the coordinate system $(X_{i-1}, Y_{i-1}, Z_{i-1})$ d_i along the Z_i axis so that the X_{i-1} axis coincides with the X_i axis. After the coordinate systems coincide, we determined the D-H parameters (Table 1).

Table 1. D-H parameter table.

Connecting Rod Number	Torsion Angle (°)	Length of Connecting Rod (m)	Joint Angle (°)	Setover of Connecting Rod (m)
1	0	0	180	0.122
2	−90	0	90	0.1215
3	180	0.408	0	0
4	180	0.376	−90	0
5	−90	0	0	0.1025
6	90	0	0	0.094

The calculating formula for the transformation matrix between the adjacent connecting rod coordinate systems is expressed in Equation (1):

$$\begin{aligned} {}^{i-1}_{i}T &= R(X_{i-1}, \alpha_{i-1})T(X_{i-1}, \alpha_{i-1})R(Z_i, \theta_i)T(Z_i, d_i) \\ &= \begin{bmatrix} \cos(\theta_i) & -\sin(\theta_i) & 0 & a_{i-1} \\ \cos(a_{i-1})\sin(\theta_i) & \cos(a_{i-1})\cos(\theta_i) & -\sin(a_{i-1}) & -d_i\sin(a_{i-1}) \\ \sin(a_{i-1})\sin(\theta_i) & \cos(\theta_i)\sin(a_{i-1}) & \cos(a_{i-1}) & d_i\cos(a_{i-1}) \\ 0 & 0 & 0 & 1 \end{bmatrix} \end{aligned} \tag{1}$$

3. Obtaining the transformation matrix ${}^0_1T, {}^1_2T, {}^2_3T, {}^3_4T, {}^4_5T, {}^5_6T$ between the connecting rods from the D-H parameter table.
4. Obtaining the transformation matrix between any two connecting rods by continuous multiplication from the transformation matrix between the adjacent connecting rods.

$$ {}^n_mT = {}^n_{n+1}T\,{}^{n+1}_{n+2}T\,{}^{n+2}_{n+3}T \ldots {}^{m-1}_{m}T, \text{ where } n < m. $$

2.1.2. Analysis of forward and Inverse Kinematics

The forward kinematics of a robot arm can calculate the position and attitude of the end-picker relative to the base coordinate system through the known angle information of each joint. The process is obtained by successive multiplication the transformation matrices of adjacent connecting rod coordinate systems:

$$ {}^0_6T = {}^0_1T\,{}^1_2T\,{}^2_3T\,{}^3_4T\,{}^4_5T\,{}^5_6T = \begin{bmatrix} r_{11} & r_{12} & r_{13} & p_x \\ r_{21} & r_{22} & r_{23} & p_y \\ r_{31} & r_{32} & r_{33} & p_z \\ 0 & 0 & 0 & 1 \end{bmatrix} \quad (2) $$

The inverse kinematics of a robot arm means obtaining the rotation angle of each robot arm joint based on the given transformation matrix of the end-picker relative to the base coordinate system. That is to say, on the premise of known ${}^0_6T$, seeking the solution of angles $\theta_1 - \theta_2$. In order to solve the six angles, we multiply the above equations on the left by ${}^0_1T^{-1}$:

$$ {}^1_6T = {}^0_1T^{-1}\,{}^0_6T = \begin{bmatrix} \cos(\theta_1) & \sin(\theta_1) & 0 & 0 \\ -\sin(\theta_1) & \cos(\theta_1) & 0 & 0 \\ 0 & 0 & 1 & -d_1 \\ 0 & 0 & 0 & 1 \end{bmatrix} \begin{bmatrix} r_{11} & r_{12} & r_{13} & p_x \\ r_{21} & r_{22} & r_{23} & p_y \\ r_{31} & r_{32} & r_{33} & p_z \\ 0 & 0 & 0 & 1 \end{bmatrix} = \begin{bmatrix} R_{11}' & R_{12}' & R_{13}' & P_x' \\ R_{21}' & R_{22}' & R_{23}' & P_y' \\ R_{31}' & R_{32}' & R_{33}' & P_z' \\ 0 & 0 & 0 & 1 \end{bmatrix} \quad (3) $$

Among them, each parameter can be calculated as

$$ \begin{array}{ll} R_{11}' = r_{11}\cos(\theta_1) - r_{21}\sin(\theta_1) & R_{13}' = r_{13}\cos(\theta_1) + r_{23}\sin(\theta_1) \\ R_{21}' = r_{21}\cos(\theta_1) - r_{11}\sin(\theta_1) & R_{23}' = r_{23}\cos(\theta_1) - r_{13}\sin(\theta_1) \\ R_{31}' = r_{31} & R_{33}' = r_{33} \\ R_{12}' = r_{12}\cos(\theta_1) + r_{22}\sin(\theta_1) & P_x' = p_x\cos(\theta_1) + p_y\sin(\theta_1) \\ R_{22}' = r_{22}\cos(\theta_1) - r_{12}\sin(\theta_1) & P_y' = p_y\cos(\theta_1) - p_x\sin(\theta_1) \\ R_{32}' = r_{32} & P_z' = p_z - d_1 \end{array} \quad (4) $$

In addition, it can be obtained by direct successive multiplication of the transformation matrices of the adjacent connecting rod coordinate systems:

$$ {}^1_6T = {}^1_2T\,{}^2_3T\,{}^3_4T\,{}^4_5T\,{}^5_6T = \begin{bmatrix} R_{11} & R_{12} & R_{13} & P_x \\ R_{21} & R_{22} & R_{23} & P_y \\ R_{31} & R_{32} & R_{33} & P_z \\ 0 & 0 & 0 & 1 \end{bmatrix} \quad (5) $$

According to the principle that corresponding elements of the matrix are equal, six angle values can be calculated:

Solving θ_1:

Let matrix elements R_{23}' and R_{23} be equal, and elements R_{24}' and R_{24} be equal, as follows:

$$ \begin{array}{l} r_{23}\cos(\theta_1) - r_{13}\sin(\theta_1) = \cos(\theta_5) \\ p_y\cos(\theta_1) - p_x\sin(\theta_1) = d_2 + d_6\cos(\theta_5) \end{array} \quad (6) $$

The above results imply that

$$ (p_y - d_6 r_{23}) \times \cos(\theta_1) + (d_6 r_{13} - p_x) \times \sin(\theta_1) = d_2 \quad (7) $$

Let $p_y - d_6 r_{23} = r\sin(\varphi) = B$, $d_6 r_{13} - p_x = r\cos(\varphi) = A$, $r = \sqrt{A^2+B^2}$, and then we have from a triangular replacement:

$$\begin{array}{l} \sin(\theta_1+\varphi) = d_2/r \\ \cos(\theta_1+\varphi) = \pm\sqrt{1-d^2/r^2} \end{array} \tag{8}$$

When $1 - d^2/r^2 \geq 0$, θ_1 has a solution, which is

$$\theta_1 = a\tan 2(d_2, \pm\sqrt{r^2+d_2{}^2}) - a\tan 2(B, A) \tag{9}$$

Solving θ_5:

After θ_1 is calculated, we obtain the following according to Formula (6):

$$\begin{array}{l} \sin(\theta_5) = \pm\sqrt{1-(r_{23}\cos(\theta_1)-r_{13}\sin(\theta_1))^2} \\ \theta_5 = a\tan 2\left(\pm\sqrt{1-(r_{23}\cos(\theta_1)-r_{13}\sin(\theta_1))^2}, r_{23}\cos(\theta_1)-r_{13}\sin(\theta_1)\right) \end{array} \tag{10}$$

Solving angle θ_6:

Let matrix elements $R_{21}{}'$ and R_{21} be equal, and elements $R_{22}{}'$ and R_{22} be equal, as follows:

$$\begin{array}{l} r_{21}\cos(\theta_1) - r_{11}\sin(\theta_1) = -\sin(\theta_5)\times\cos(\theta_6) \\ r_{22}\cos(\theta_1) - r_{12}\sin(\theta_1) = \sin(\theta_5)\times\sin(\theta_6) \end{array} \tag{11}$$

The above results imply that

$$\begin{array}{l} \cos(\theta_6) = \frac{r_{11}\sin(\theta_1)-r_{21}\cos(\theta_1)}{\sin(\theta_5)} \\ \sin(\theta_6) = \frac{r_{22}\cos(\theta_1)-r_{12}\sin(\theta_1)}{\sin(\theta_5)} \end{array} \tag{12}$$

When $\sin(\theta_5) \neq 0$, θ_6 has a solution, which is

$$\theta_6 = a\tan 2\left(\frac{r_{22}\cos(\theta_1)-r_{12}\sin(\theta_1)}{\sin(\theta_5)}, \frac{r_{11}\sin(\theta_1)-r_{21}\cos(\theta_1)}{\sin(\theta_5)}\right) \tag{13}$$

Solving θ_2:

Let matrix elements $R_{13}{}'$ and R_{13} be equal, and elements $R_{33}{}'$ and R_{33} be equal, as follows:

$$\begin{array}{l} r_{13}\cos(\theta_1) + r_{23}\sin(\theta_1) = \cos(\theta_2-\theta_3+\theta_4)\times\sin(\theta_5) \\ r_{33} = -\sin(\theta_2-\theta_3+\theta_4)\times\sin(\theta_5) \end{array} \tag{14}$$

The above results imply that

$$\begin{array}{l} \cos(\theta_2-\theta_3+\theta_4) = \frac{r_{13}\cos(\theta_1)-r_{23}\sin(\theta_1)}{\sin(\theta_5)} \\ \sin(\theta_2-\theta_3+\theta_4) = \frac{-r_{33}}{\sin(\theta_5)} \\ \theta_2-\theta_3+\theta_4 = a\tan 2\left(\frac{-r_{33}}{\sin(\theta_5)}, \frac{r_{13}\cos(\theta_1)-r_{23}\sin(\theta_1)}{\sin(\theta_5)}\right) \end{array} \tag{15}$$

Let matrix elements $P_x{}'$ and P_x be equal, and elements $P_z{}'$ and P_z be equal, as follows:

$$\begin{array}{l} p_x\cos(\theta_1) + p_y\sin(\theta_1) = a_2\cos(\theta_2) - d_5\sin(\theta_2-\theta_3+\theta_4) + a_3\cos(\theta_2-\theta_3) + d_6\sin(\theta_5)\cos(\theta_2-\theta_3+\theta_4) \\ p_z - d_1 = -a_2\sin(\theta_2) - d_5\cos(\theta_2-\theta_3+\theta_4) - a_3\sin(\theta_2-\theta_3) - d_6\sin(\theta_5)\sin(\theta_2-\theta_3+\theta_4) \end{array} \tag{16}$$

By transposition, it follows that

$$\begin{array}{l} a_2\cos(\theta_2) + a_3\sin(\theta_2-\theta_3) = p_x\cos(\theta_1) + p_y\sin(\theta_1) + d_5\sin(\theta_2-\theta_3+\theta_4) - d_6\sin(\theta_5)\cos(\theta_2-\theta_3+\theta_4) \\ a_2\sin(\theta_2) + a_3\sin(\theta_2-\theta_3) = d_1 - d_6\sin(\theta_5)\sin(\theta_2-\theta_3+\theta_4) - d_5\cos(\theta_2-\theta_3+\theta_4) - p_z \end{array} \tag{17}$$

Given

$$\begin{aligned} &p_x\cos(\theta_1)+p_y\sin(\theta_1)+d_5\sin(\theta_2-\theta_3+\theta_4)-d_6\sin(\theta_5)\cos(\theta_2-\theta_3+\theta_4)=C \\ &d_1-d_6\sin(\theta_5)\sin(\theta_2-\theta_3+\theta_4)-d_5\cos(\theta_2-\theta_3+\theta_4)-p_z=D \end{aligned} \tag{18}$$

we obtain

$$C\cos(\theta_2)+D\sin(\theta_2)=\frac{C^2+D^2+{a_2}^2-{a_3}^2}{2a_2}=E \tag{19}$$

Let

$$\sin\beta=\frac{C}{\sqrt{C^2+D^2}},\cos\beta=\frac{D}{\sqrt{C^2+D^2}}, \tag{20}$$

and then we have from triangular replacement the following:

$$\begin{aligned} &\sin(\beta+\theta_2)=\tfrac{E}{\sqrt{C^2+D^2}} \\ &\cos(\beta+\theta_2)=\pm\sqrt{1-E^2/(C^2+D^2)} \\ &\theta_2=a\tan 2(E,\pm\sqrt{C^2+D^2-E^2})-a\tan 2(C,D) \end{aligned} \tag{21}$$

When $C^2+D^2-E^2\geq 0$, θ_2 has a solution.
Compute θ_3:
After solving θ_3, compute $\theta_2-\theta_3$:

$$\begin{aligned} &\theta_2-\theta_3=a\tan 2\left(\tfrac{D-a_2\sin(\theta_2)}{a_3},\tfrac{C-a_2\cos(\theta_2)}{a_3}\right) \\ &\theta_3=\theta_2-(\theta_2-\theta_3) \end{aligned} \tag{22}$$

Solving θ_4:

$$\theta_4=(\theta_2-\theta_3+\theta_4)-(\theta_2-\theta_3) \tag{23}$$

All joint angles have been solved. A total of eight groups of inverse solutions can be solved, and a group of the most suitable solutions can be selected according to the size of the angle spinor.

2.2. The Design of the Executive Mechanism

In this study, the mold cavity-sleeve structure was used for pear picking. The active cutting mode will greatly increase the structural complexity of the end actuator, and the mold cavity sleeve itself will occupy a larger volume. If the cutting device is added, the volume and weight of the actuator will be increased. Therefore, a passive cutting mode for which the end actuator cooperates with the robot arm to complete the picking action was adopted to realize the miniaturization of the end-picker. By rational design of the opening and closing mechanism above the end-picker and cooperating with the robot arm, the crown pear stems can be cut. Figure 2 shows the overall assembly diagram of the end-picker.

Figure 3a shows the three-dimensional structure of the blade. The cutting mechanism of the end-picker designed for mold cavity picking is the key to separating pears from stems. There are two connecting holes on both sides of the structure that are connected to main body of the mold cavity. Two blades are arranged symmetrically, wherein opening and closing are controlled by the steering engine. After the fruit enters the mold cavity, the steering engine drives blades to clamp the stem and reset after picking is completed.

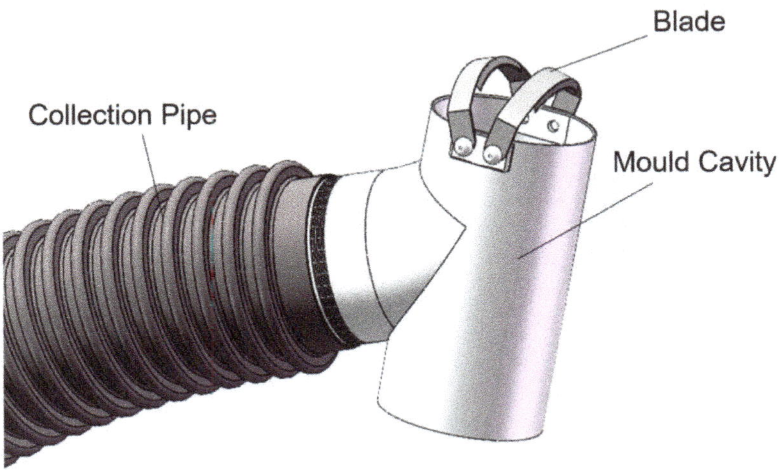

Figure 2. End-picker assembly diagram.

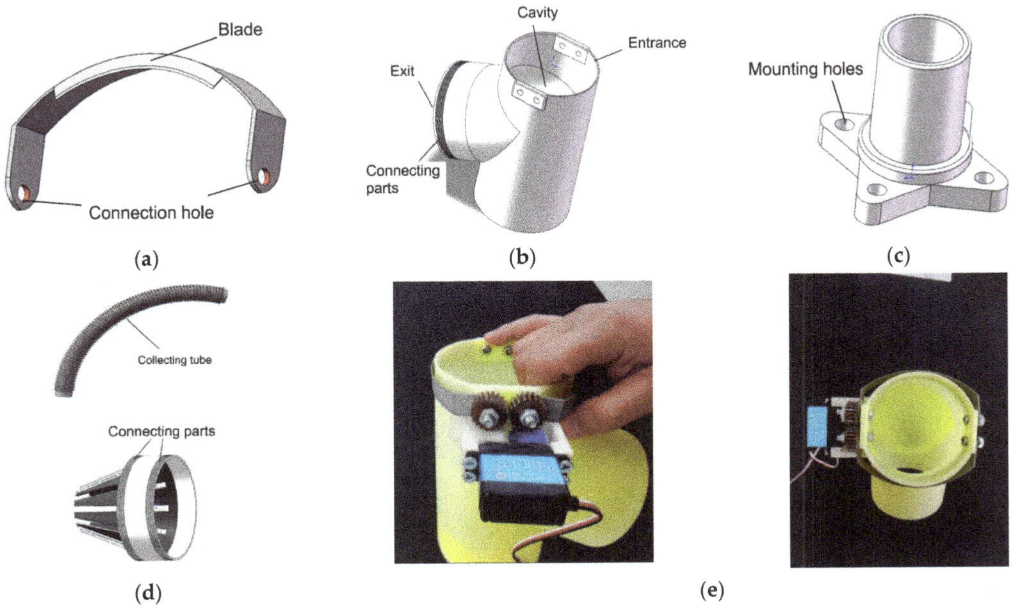

Figure 3. End-picker assembly diagram: (**a**) blade; (**b**) mold cavity; (**c**) connecting mechanism; (**d**) fruit-collecting mechanism; (**e**) transmission mechanism and assembly figure.

Figure 3b shows the mold cavity of the picker. The main body is cylindrical. The fruit entrance and exit of the mold cavity is a connecting tube with an inner diameter of 100 mm. After fruit enters from the inlet, it will directly enter the fruit-collecting tube from the outlet. As long as the fruit enters the mold cavity, there will be no damage to the fruit due to the robot arm not grasping it firmly. The bottom of the cavity of the picker is closed and exists as a buffer structure in the internal structure. A slope channel is formed in the cavity with foam and other buffering materials so that the fruit can be buffered and enter the collection tube along the ramp. Therefore, the mold cavity-sleeve method can protect the fruit well and avoid damage to the fruit during the picking process.

Figure 3c shows the connection mechanism, which is used to connect the end-picker and the end flange of the robot arm. The four mounting holes correspond to the four threaded holes of the end flange of the robot arm.

Figure 3d shows the collecting mechanism. The diameter of the collecting tube is 100 mm. And the tube is made of rubber to prevent pears from being damaged by friction with the tube inner wall during the collecting process. The deceleration mechanism is set in the middle of the collecting tube. When the fruit passes through, its kinetic energy will be reduced, thereby reducing the speed in the tube and the loss of fruit.

Figure 3e shows the transmission mechanism and photos illustrating the installation of each element. The transmission mechanism mainly has three gears, for which the module is 1, the number of teeth is 20, and the inner diameter of the hole is 6 mm. Two metal gears are fixed at the two blade mounting holes and mesh with each other. The driving mode of the steering gear is PWM control, which can accurately control the predetermined rotation angle. It is mainly used to drive the meshing gear fixed on the blade and then drive the opening and closing of the two blades for picking.

3. Target Detection and 3D Location

3.1. Dataset Construction

At present, there are few open-source pear datasets in the field of agricultural picking, and especially the data of the natural unpicked state and fruit under a trellised orchard are very scarce. The experimental dataset was collected by the orchard and the trellised pear orchard environment built in the laboratory. The image format was JPEG with a resolution of 3024 × 4032, and the video format was MP4 with a resolution of 4K. And the diversity of the dataset was improved by collecting datasets in two scenes of natural light during the day and filling light at night.

The visual equipment used for target detection and calibration in this paper was an Intel RealSense D435i depth camera, which was composed of an rgb camera, infrared camera, infrared transmitter, and IMU (depth resolution: 1280 × 720; RGB sensor resolution: 1920 × 1080).

The performance of deep learning models strongly depends on the quantity and quality of input data. Many deep learning projects fail to obtain enough data in practical applications, resulting in the underfitting of the model. In order to improve the model generalization ability, that is, the ability to predict new data, data augmentation techniques are widely used. Currently, supervised data augmentation schemes are commonly used, which operate the dataset through known image transformation rules to make the model learn more features and increase the richness of the dataset, thereby improving the performance and stability of the model. The data augmentation methods used in this article include (1) random cropping; (2) mirror image; (3) Gaussian noise; (4) grayscale image; and (5) rotating 90°.

Each image collected was manually screened and classificatorily annotated by a label using LabelImg software (v1.8.1) as shown in Figure 4. In the annotation process, a rectangle annotation method is used to draw the minimum bounding rectangle of the pear object in the image, and the categorical attribute of the rectangle frame is set to "pear". After the annotation is completed, the generated xml format label file is automatically saved, and the xml format text contains the height, width, and categorical information of the rectangular box. According to the VOC2007 dataset format, the trellised pear orchard dataset was made, wherein the dataset has a total of 2574 images, and the ratio of the training set to validation set was set to 7:3.

Figure 4. Dataset annotation.

3.2. Yolo-v5s Detection Model

Experimental running environment: CPU: i7-12700H, 16GB running memory; GPU: GTX 3060, 6 GB graphic memory; running system: Windows11; depth framework: PyTorch, GPU1.10.1, with CUDA11.6 and cudnn7.6.5. Running software: Pycharm2021.3 community edition, Python3.8.16.

Considering the experimental hardware platform conditions and model detection accuracy, the choice of network depth and width is very important. Under the condition of the minimum depth and width of the model, the YOLOv5s model is a very excellent object detection model that can achieve high-precision object detection with a smaller memory footprint and computing resources. The input layer of the YOLOv5s model is used to receive the original image data, and the Backbone layer can extract feature information from the original image, with the Neck layer being able to further compress the feature information. Finally, object detection and classification are performed in the Detect layer. The structure of the Yolov5s model used in this project is shown in Figure 5.

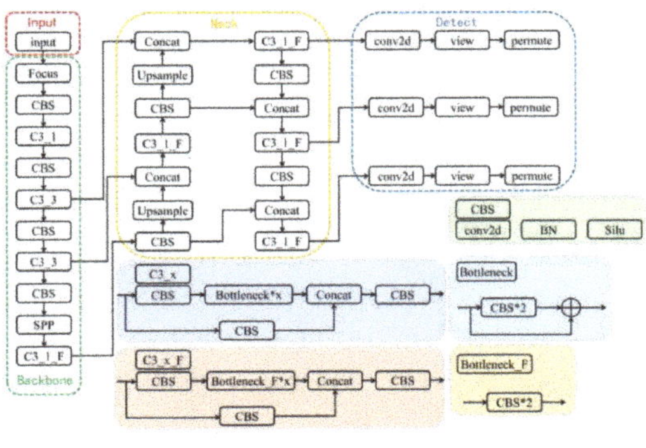

Figure 5. The Yolov5s model structure.

Based on the collected dataset above and the model, the results of network training by setting parameters are shown in the figure. Figure 6a shows the confusion matrix, which is a commonly used method to evaluate the performance of classification models and is usually used to calculate precision, recall, F1 score, and other indicators of classification algorithms. The horizontal axis of the confusion matrix represents the real label, the vertical axis represents the predicted label, and the value of each cell represents the number of samples that the real label matches to the predicted label. The probability of the pears in Figure 6a being correctly classified is 0.99. In Figure 6b, there is shown a certain relationship between F1 score and confidence thresholds. F1 score can help to comprehensively consider the performance of the model in the classification task, with neither too much emphasis on precision, nor too much emphasis on recall, so that the detection results of models are more balanced and reliable. Typically, the top of the curve is close to 1, indicating that the model performs well on the training dataset. Figure 6c shows the precision–confidence threshold curve. The higher confidence is, the higher the precision of the classification detection is. But, it is possible to miss some real samples with low probability of judgment. Figure 6d shows the precision–recall curve (PR curve), which is also a commonly used performance evaluation tool for evaluating the classification performance of the model in multiple categories. As can be seen from the figure, the higher precision is, the lower recall is. It is hoped that all categories can be detected as far as possible under the premise of high precision. Therefore, the area of the mAP curve should be as close to 1 as possible.

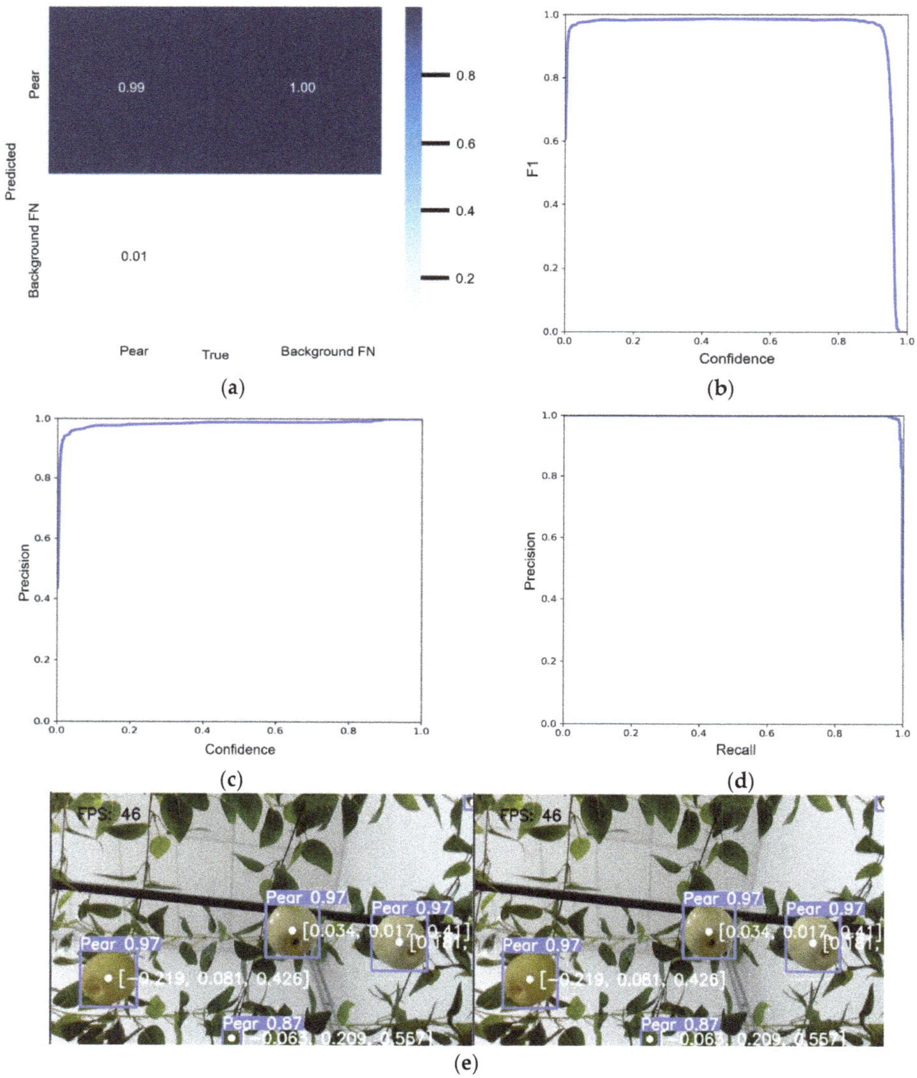

Figure 6. Performance evaluation. (**a**) Confusion matrix; (**b**) F1–confidence curve; (**c**) precision–confidence curve; (**d**) PR curve; (**e**) object detect results.

3.3. Hand–Eye System Calibration

In the process of camera calibration and hand–eye calibration, the transformation relationship of four coordinate systems is involved, which are the pixel coordinate system, image coordinate system, camera coordinate system, and world coordinate system, as shown in Figure 7. For this study, the world coordinate system was the robot arm base coordinate system.

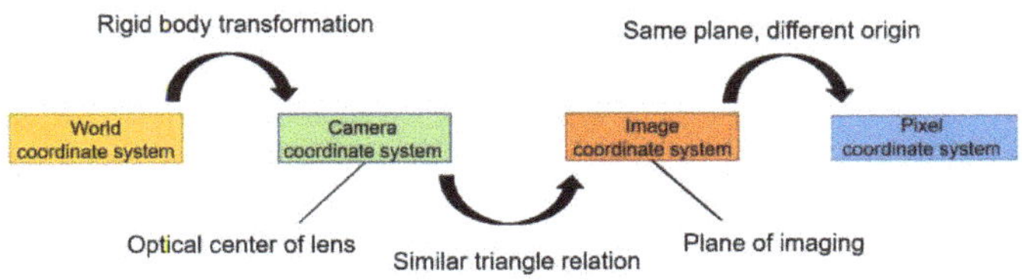

Figure 7. Coordinate transformation.

Hand–eye calibration mainly finds the relationship between the camera coordinate system and the robot arm base coordinate system. By deriving the transformation matrix between these two coordinate systems, the 3D coordinates of the detected target in the camera coordinate system can be converted to the robot arm base coordinate system. There are mainly two hand–eye calibration methods, as shown in Figure 8, which can be divided into "eyes in hand" and "eye to hand", depending on the setting position of the camera.

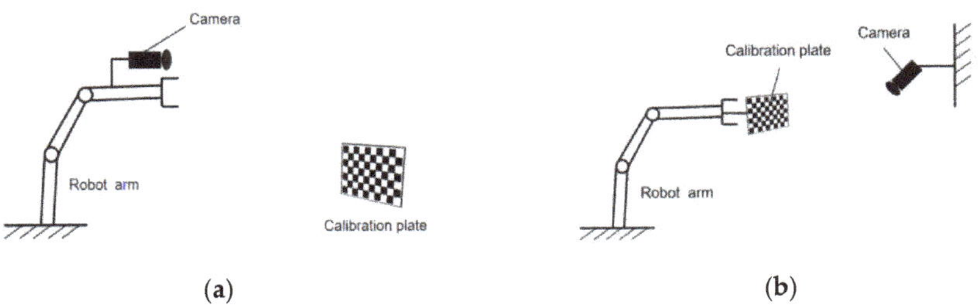

Figure 8. Setting position of the camera. (**a**) Eyes in hand; (**b**) eye to hand.

For the "eyes to hand" method of hand-eye calibration, during the movement of the robot arm, the camera coordinate system and the robot arm base coordinate system are fixed, and the transformation matrix between them is also unchanged. Hand–eye calibration is used to solve the transformation relationship between these two coordinate systems, as shown in Figure 9.

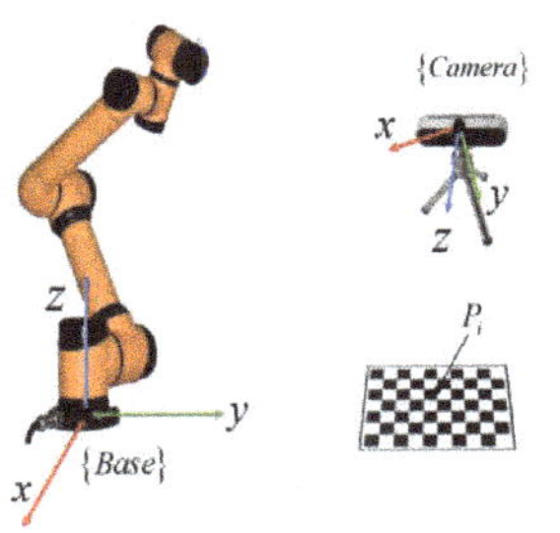

Figure 9. Calibration diagram.

We set the robot arm base coordinate system to {Base} and the camera coordinate system to {Camera}. If the coordinates of several fixed points P_i under these two coordinate systems are known, the corresponding transformation matrix of these two coordinate systems can be derived according to the coordinate transformation formula. The implementation steps are as follows:

(1) We set the coordinate of a spatial point P_i under {Base} to $P_i(x_i', y_i', z_i')$ and the coordinate under {Camera} to $P_i(x_i', y_i', z_i')$. On the basis of the relationship between these two coordinates and the homogeneous matrix form, the following formula can be derived:

$$\begin{cases} r_{11}x_i + r_{12}y_i + r_{13}z_i + a = x_i' \\ r_{21}x_i + r_{22}y_i + r_{33}z_i + b = y_i' \\ r_{31}x_i + r_{32}y_i + r_{33}z_i + c = z_i' \\ 1 = 1 \end{cases} \tag{24}$$

(2) We derive the pixel coordinates of the spatial point P_i by the object detection algorithm, and the specific coordinates of the point in the camera coordinate system by combining the camera internal parameters and the depth information of the depth camera. The coordinates in the robot arm base coordinate system could be read out directly by the robot arm teach pendant.

(3) We select twenty different points to fit the optimal results, solving the following parameters: $r_{11}, r_{12}, r_{13}, r_{21}, r_{22}, r_{23}, r_{31}, r_{32}, r_{33}, a, b, c$, as well as listing the overdetermined linear equations of these points in the camera coordinate system.

(4) We solve the overdetermined equations by the least squares method. We obtain the rotation matrix and translation vector between the camera coordinate system and the robot arm base coordinate system.

4. The Picking Task Planning of Multiple Robot Arms

4.1. The Task Allocation Method of Multiple Robot Arms

In this paper, the Monte Carlo method based on random probability was utilized to solve the working space of a robot arm. Figure 10 shows the working space of a single AUBO-i5 robot arm and the projection of the working space on xoz. The working space is approximately a sphere, and the picking area is divided based on this.

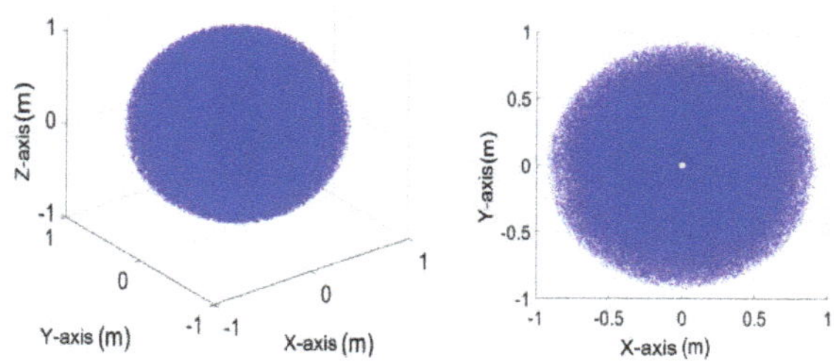

Figure 10. The working space of AUBO-i5.

It can be seen from Figure 11 that when the picking plane was 0.4009 m away from the unmanned vehicle plane, the radius of the picking plane was 0.8703 m; when the picking plane was 0.923 m away from the unmanned vehicle plane, the radius of the picking plane was 0.4696 m. As picking height increases, the picking range will decrease. The projection of the workspace on xoy, the circular area, decreased as the Z-axis absolute value increased.

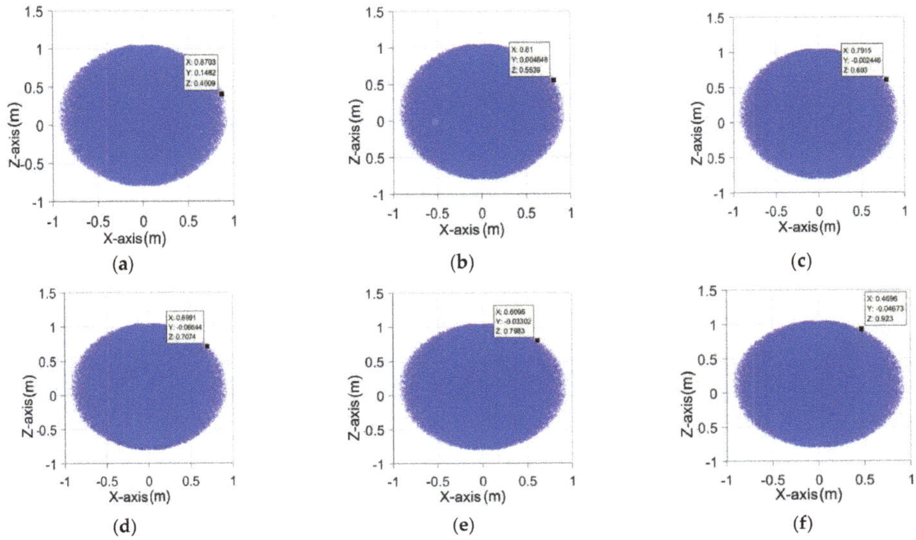

Figure 11. The analysis diagram of a picking area with a single robot arm. (**a**) z = 0.4 m, x = 0.87 m; (**b**) z = 0.55 m, x = 0.81 m; (**c**) z = 0.60 m, x = 0.80 m; (**d**) z = 0.71 m, x = 0.70 m; (**e**) z = 0.80 m, x = 0.61 m; (**f**) z = 0.92 m, x = 0.47 m.

The distance between robot arms was set to 1 m. The projection of a multiple robot arm workspace on xoy can be obtained using the Monte Carlo method (Figure 12).

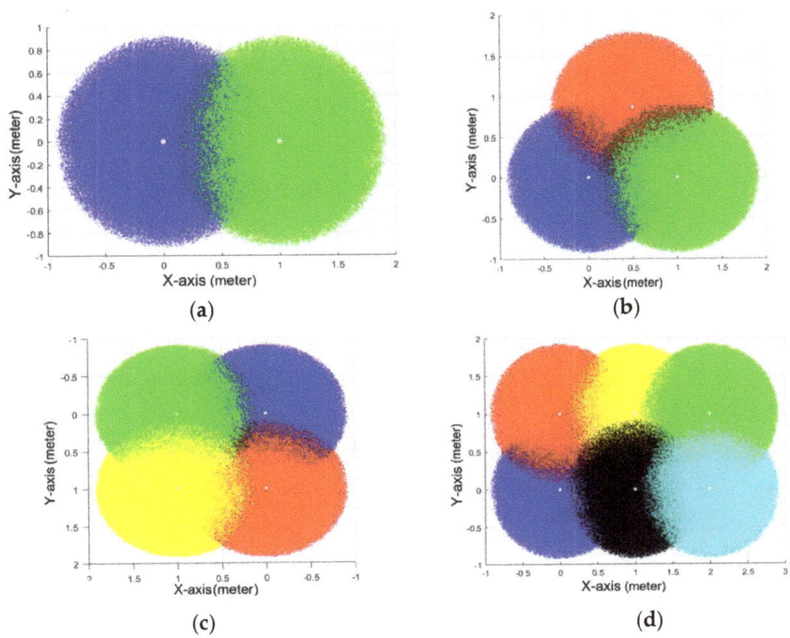

Figure 12. The analysis diagram of a picking range with multiple robot arms (Different colors represent different work areas). (**a**) the projection of a two-robot-arm workspace; (**b**) the projection of a three-robot-arm workspace; (**c**) the projection of a four-robot-arm workspace; (**d**) the projection of a six-robot-arm workspace.

It can be seen from Figure 12 that for the trellised pear orchard environment, at different heights from the unmanned vehicle plane, the picking area of each robot arm was approximately a circle. As picking height increased, the overlap of the picking area between adjacent robot arms was approximately an ellipse, wherein the area decreased as picking height increased.

The height of an unmanned vehicle is 1.1 m. The robot arm performs picking operations on a picking plane 0.6 m away from the unmanned vehicle plane. The operating area of the robotic arm is approximately a circle with a radius of 0.8 m. On the basis of the solution of multiple robot arm workspace in the previous section, the picking area of a robot arm was divided (Figure 13). Collaborative picking task allocation of multiple robotic arms can be described as the problem of multiple robotic arms cooperating to complete the picking task. The ultimate goal is to increase picking efficiency, and at the same time, tending to divide the task volume of each robot arm evenly and improve the utilization rate of each robot arm.

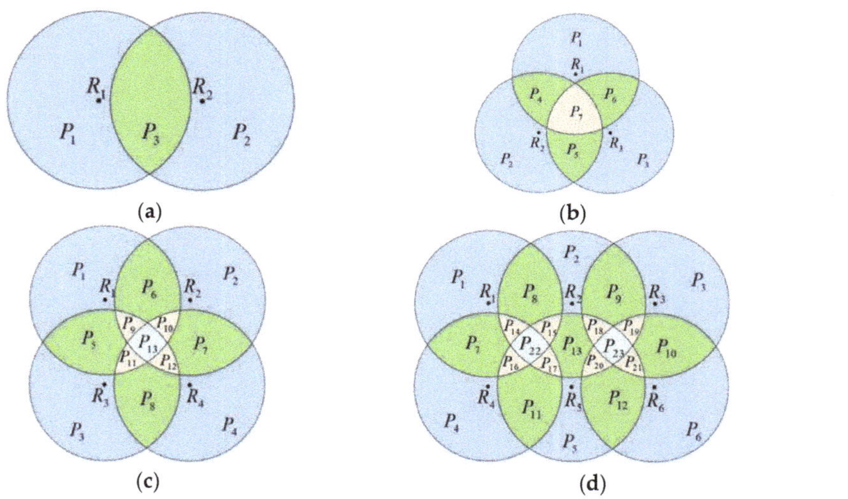

Figure 13. The diagram of a picking area with multiple robot arms. (**a**) The picking area of a two-robot-arm workspace; (**b**) the picking area of a three-robot-arm workspace; (**c**) the picking area of a four-robot-arm workspace; (**d**) the picking area of a six-robot-arm workspace.

The collaborative picking task allocation model of multiple robotic arms is specifically described as follows:

(1) A set of multiple robotic arms is represented by $R = \{R_1, R_2, R_3, R_4, \ldots, R_a\}$, where a represents the number of robot arms and R_i represents the robot arm, $1 \leq i \leq a$. (2) The picking area is represented by $P = \{P_1, P_2, P_3, P_4, \ldots, P_b\}$, where b represents the number of picking areas and P_j represents the picking sub-regions, $1 \leq j \leq b$. (3) The task volume in each sub-regions is represented by $G = \{G_1, G_2, G_3, G_4, \ldots, G_b\}$, where G_k represents the picking task volume in a sub-picking area k, $1 \leq k \leq b$. (4) Each robot arm is set in different positions, and the picking sub-regions are also different. The set of tasks that the robot arm can pick is $T_{Ri} = \{T_{R1}, T_{R2}, T_{R3}, T_{R4}, \ldots, T_{Ra}\}$, where i represents the i-th robot arm, $T_{Ri} \in G$. The task process is shown in Figure 14.

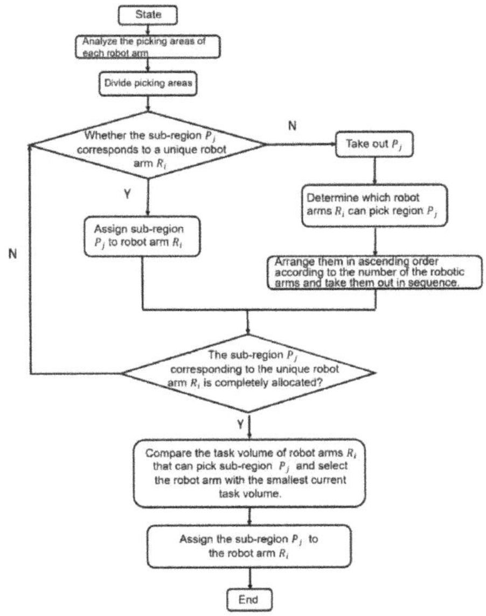

Figure 14. Task allocation flow chart.

Firstly, the number of robot arms, the spacing between the robot arms, and the height of the picking plane are determined. Then, the Monte Carlo method is utilized to analyze the workspace of multiple robot arms and divide the picking area.

After the picking area is divided, the picking area that can only be picked by the only robot arm is first matched with its corresponding one. Then, for other areas that can be picked by two or more robot arms, first, we arrange them from small to large according to the number of robot arms that can pick in this area. Then, we take them out in the order from small to large and select the robot arm with the least task volume until all the areas are allocated. This allocation strategy is mainly to distribute tasks evenly while avoiding collision interference between multiple robotic arms. The purpose of the picking robot arms in this area being taken out in order from small to large is to better balance the task volume of robot arms. This is because as the allocation process gradually proceeds, the number of robot arms corresponding to the picking area increases, and the algorithm can further balance the robot arms with a large difference of task volume assigned at the last time. At the same time, the picking area becomes smaller, and the probability of few fruits in the area becomes greater. So, it will not cause the task volume of robot arm with the fewest tasks to become much higher than that of other robot arms after the allocation. Taking two robot arms as an example, due to uneven fruit distribution in most cases, when the difference of the fruit number between P_1 and P_2 is large, according to the task allocation strategy, the public picking area P_3 will be assigned to the robot arm with a smaller task volume, so that task volume of the two robot arms tends to be evenly allocated, with little difference. The same method will be utilized in the common picking area of three, four, and six robot arms.

4.2. Picking Sequence Optimization

In this paper, in a trellised pear orchard environment, the heights of pears were basically the same. The Z-axis coordinate was ignored, and the multi-fruit picking path optimization problem was converted into a two-dimensional traveling salesman problem. The coordinates of each pear can be given by the depth camera, and then the distance between each two pears can be obtained. The picking path optimization problem is to find the shortest path that can traverse all the fruits and does not repeat [32]. The TSP model of picking path optimization can be expressed by a mathematical model.

The requirement of the picker design is to complete the most picking in the shortest path. If the simple algorithm is used to pick across the region, the picking path will have great redundancy, and it is more time-consuming to schedule the manipulator back and forth. Therefore, we used the simulated annealing algorithm for optimization, to solve the optimal picking sequence, to reduce the scheduling time of the robot arm, and to reduce energy consumption at the same time.

We set the weighted graph as $G = (V, E)$, and all pear coordinate-sets as $V = \{1, 2, 3, \ldots, n-1, n\}$. E is a set of path weights or lengths, and D_{ij} represents the distance between each pear, where $D_{ij} > 0, i, j \in V$.

$$x_{ij} = \begin{cases} 1, (i,j) \in L \\ 0, (i,j) \notin L \end{cases} \tag{25}$$

L is the solution sequence. The TSP mathematical model can be expressed in the following form:

$$\begin{aligned} & \text{mins} = \sum_{i=1}^{n} \sum_{j=1}^{n} d_{ij} x_{ij} \\ & s.t. \begin{cases} \sum\limits_{i=1}^{n} x_{ij} = 1, j \in V \\ \sum\limits_{j=1}^{n} x_{ij} = 1, i \in V \\ \sum\limits_{i,j \in S}^{n} x_{ij} \leq |S| - 1, \forall S \subseteq V, 1 < |S| < n \end{cases} \end{aligned} \tag{26}$$

The distance between adjacent fruits is set as d_{ij}. S is a non-empty subset of the vertex set V, and $|S|$ is the number of vertices of the set S in the weighted graph G.

This project utilized a simulated annealing algorithm that can avoid falling into local optimality to a certain extent, in order to optimize the picking sequence. The algorithm mainly consists of two parts: one is the Metropolis algorithm, and the other is the annealing cooling process, which corresponds to the internal cycle and the external cycle, respectively.

The algorithm steps are as follows (Figure 15):

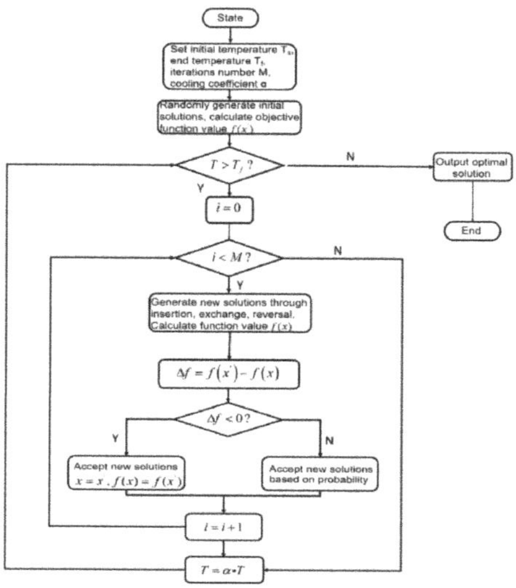

Figure 15. Simulated annealing algorithm flow chart.

(1) Initialize the temperature, and set the current temperature $T = T_S$, the termination temperature T_f, and the maximum number of iterations M. Randomly generate an initial solution x and then calculate the objective function value $f(x) = \sum_{i=1}^{n}\sum_{j=1}^{n} d_{ij}$. (2) At a certain temperature, the current solution is perturbed by insertion, exchange, reversal, and other ways to generate a new solution. (3) Solve the objective function value $f(x')$, and calculate $\Delta f = f(x') - f(x)$; if $\Delta f < 0$, then the new solution x' is accepted. If not, the new solution x' is accepted according to the probability. (4) At a certain temperature T, the perturbation and acceptance process is repeated for a certain number of times, that is, steps 2 and 3 are repeated. (5) Lower the temperature T. (6) Repeat steps 2–5 until convergence conditions are met.

The simulated annealing algorithm has three important parameters: initial temperature T_s, internal loop iteration number M, and cooling coefficient α, which determine the optimization capacity and comprehensive performance of the simulated annealing algorithm, that is, finding a better solution quickly and accurately with less time cost. (The above initial temperature does not represent the physical temperature in the real environment, but instead a parameter that needs to be optimized in the simulated annealing algorithm, which represents the base temperature in the optimization algorithm.)

In theory, the larger the initial temperature, the better. A low temperature causes the problem of the algorithm not searching enough in the solution space, resulting in missing the optimal solution in the optimizing process. But if the temperature is too high, it will take a lot of time. The number of inner loop iterations also affects the search ability of the algorithm in the solution space. The greater the number of iterations at the same temperature, the greater the chance of finding the optimal solution, but the time cost is often greatly increased. Then, a too large cooling coefficient easily leads to the loss of the optimal solution, while a too small cooling coefficient greatly increases the algorithm's time cost. The termination temperature determines when the algorithm ends. Setting the termination temperature too high will result in an insufficient search, so the termination temperature is generally set to a smaller value.

(1) The adjustment of initial temperature T_s

Only the initial temperature was changed, and other parameters were kept constant and assigned values. The number of iterations was set to 100, the cooling coefficient was 0.9, and the end temperature was 0.02. Twelve groups of temperatures were selected, and the simulated annealing program was run 10 times at each temperature to reduce the accidental error. Table 2 shows the data of the optimization process.

Table 2. Program optimization at 1000 degrees centigrade.

T_s = 1000	Traversal Distance Optimized (m)	Running Time (s)	Cooling Iteration Times
1	18.7174	4.1342	32.0000
2	20.6193	3.8558	21.0000
3	20.9443	3.4763	9.0000
4	19.1814	4.1061	29.0000
5	19.9797	3.9653	23.0000
6	20.3760	4.2626	32.0000
7	20.4530	4.3658	36.0000
8	21.6306	3.8912	21.0000
9	20.0824	3.9527	23.0000
10	19.6808	4.2147	31.0000
MIN	18.7174	3.4763	9.0000
AVG	20.1665	4.0225	25.7000
Standard deviation	0.8451	0.2546	7.9029

The average traversal distance and the minimum value at each temperature, as well as the standard deviation, were counted to generate the line chart (Figure 16). The blue line is the average traversal distance at each temperature, which was utilized to evaluate the overall optimization performance of the algorithm. The smaller the average traversal distance, the better the performance of the algorithm at that temperature. The yellow line represents the optimized minimum value at each temperature, which represents the ability to find the optimal solution at each set of temperatures. The smaller the minimum value, the stronger mining ability of the algorithm at that temperature. The green line represents the standard deviation of each set of data, which was utilized to judge the degree of discreteness of a set of data. If the standard deviation was larger, it means that the set of data fluctuated greatly, and the performance of the algorithm was unstable. The abscissa represents the temperature of each group, and the ordinate represents the corresponding distance at a certain temperature. According to Figure 16, when the initial temperature was 100 °C, the average traversal distance and the minimum distance were the largest, indicating that the performance of the algorithm was poor. As the temperature increased, the average traversal distance and the minimum distance decreased. After the initial temperature reached 1500 °C, the decrease in amplitude of both slowed down as the initial temperature increased. After the initial temperature reached 3000 °C, the average traversal distance and minimum distance remained basically unchanged, and the algorithm performance tended to be stable. At the same time, it can be seen that the standard deviation fluctuated up and down at the beginning and then decreased to zero at the end, indicating that the algorithm optimization performance became more stable. In summary, 3000 °C was selected as the initial temperature of the algorithm to ensure the performance and reduce the waste of time.

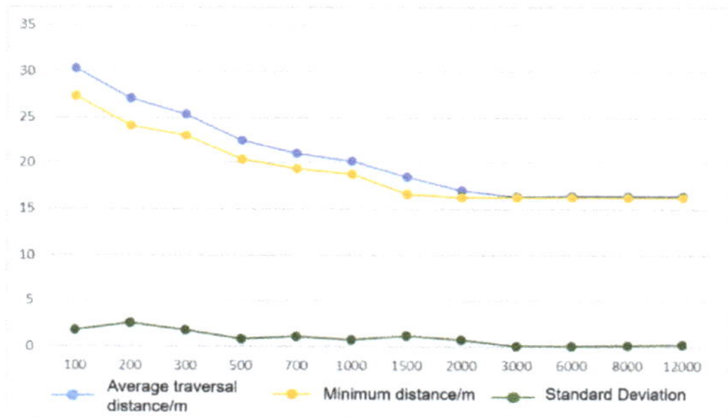

Figure 16. Optimization effect at different temperatures.

(2) The adjustment of the number of inner loop iteration M

Only the iteration number of inner loop was changed, and other parameters were kept constant and assigned. The initial temperature was set to 3000, the cooling coefficient was 0.9, and the end temperature was 0.02. Twelve groups of iteration numbers were selected, and the simulated annealing program was run 10 times at each iteration number to reduce the accidental error (Table 3).

Table 3. Program optimization under 100 loop times.

M = 100	Traversal Distance Optimized (m)	Running Time (s)	Cooling Iteration Times
1	16.3104	5.8098	83
2	17.5659	7.2418	119
3	16.2374	4.2802	32
4	16.3471	5.5327	60
5	16.8163	7.1329	122
6	16.5062	8.8083	171
7	16.334	6.6753	108
8	17.3419	7.0707	122
9	16.7095	8.8845	178
10	16.8789	8.5171	166
MIN	16.2374	4.2802	32
AVG	16.70476	6.99533	116.1
Standard deviation	0.4557	1.5009	48.0681

The average traversal distance, minimum value, and standard deviation under each internal iteration number were counted, and the trend line chart was drawn. When the number of internal iterations was too small, the optimization ability of the algorithm was limited, and the optimization effect of the average traversal distance was not obvious (Figure 17). With the increase in the number of internal iterations, the average traversal distance and the optimized minimum distance both decreased, indicating that the optimization ability of the algorithm was enhanced with the increase in the number of internal iterations. As the number of internal iterations increased, both the average traversal distance and the optimal minimum distance decreased, indicating that the optimization ability of the algorithm increased as the number of internal iterations increased. When the number of iterations reached 150, the performance of the algorithm almost did not improve with the increase in the iteration numbers. Therefore, the algorithm was already at a relatively optimal level at this time, and further increasing the number of internal iterations would only increase the time cost. In summary, the number of inner iterations was chosen to be 150.

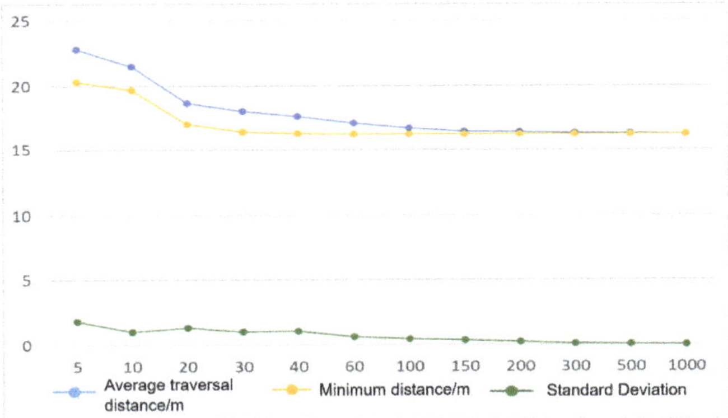

Figure 17. Optimization effect under different iteration numbers.

(3) The adjustment of cooling coefficient α

Only the cooling coefficient was changed, and other parameters were kept constant and assigned. The initial temperature was set to 3000, the iteration number of each temperature was 150, and the end temperature was 0.02. Twelve groups of cooling coefficient were selected, and the simulated annealing program was run 10 times at each cooling coefficient to reduce the accidental error (Table 4).

Table 4. Program optimization under 0.99 cooling coefficient.

$\alpha = 0.99$	Traversal Distance Optimized (m)	Running Time (s)	Cooling Iteration Times
1	16.2242	8.1109	107
2	16.2374	5.6527	55
3	16.2242	5.0363	43
4	16.3104	9.1028	133
5	16.2910	8.3292	113
6	16.2242	6.4719	71
7	16.2374	5.3622	51
8	16.2374	4.8088	37
9	16.2374	3.575	10
10	16.4018	7.1986	89
MIN	16.2242	3.575	10
AVG	16.26254	6.36484	70.9
Standard deviation	0.0571	1.7823	38.7941

The average traversal distance and the minimum value, as well as the standard deviation, at each iteration number were counted to generate the trend line chart (Figure 18). The blue line is the average traversal distance at each temperature. The yellow line represents the optimized minimum value at each temperature. The green line represents the standard deviation of each set of data. The abscissa represents the cooling coefficient of each group, and the ordinate represents the corresponding distance at a certain cooling coefficient. According to Figure 18, with the increase in the cooling coefficient, the optimization ability of the algorithm became better. When the cooling coefficient reached 0.99, with the increase in the cooling coefficient, the optimized effect of the simulated annealing algorithm was not obvious, and the optimized average traversal distance and the optimized minimum value of each group tended to be consistent. Therefore, 0.99 was an appropriate parameter value for this algorithm. In summary, the cooling coefficient of the algorithm was 0.99.

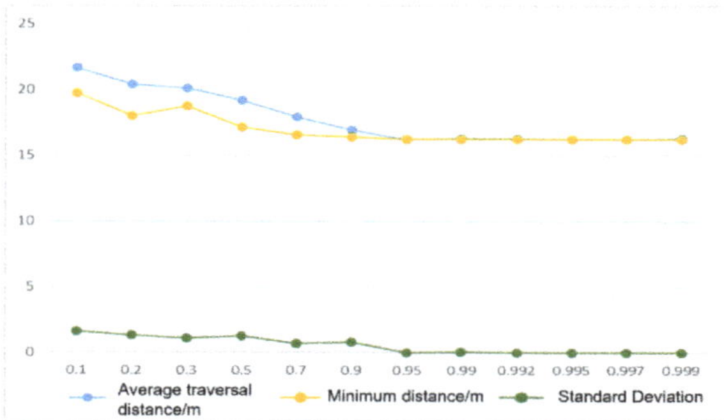

Figure 18. Optimization effect under different cooling coefficients.

Taking an area as an example, 20 points were randomly generated within the picking range of a robot arm to simulate the distribution of pears on a trellis. Since the impact in the vertical direction is small, only the coordinate information in X and Y directions was considered. Figure 19 separately shows that the randomly traversed path before optimization was disorderly and the picking path after optimization formed a neat loop, for which the optimization effect was obvious (Figure 19).

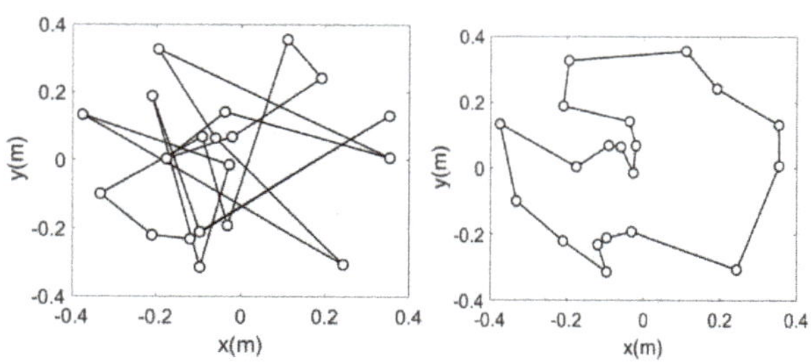

Figure 19. Comparison before and after the optimization of the simulated annealing algorithm.

The total number of iterations was 300. As the number of iterations increased, the total path length gradually decreased. In the range of 0–50 times, the curve was steeper, and the convergence speed was faster. At this time, the optimization effect was better. In the range of 50–120 times, the curve was a straight line, and the algorithm fell into a local optimum. As the number of iterations increased, the perturbation caused the solution to jump out of the current local optimum. After the number of iterations reached 240, a better result was achieved (Figure 20).

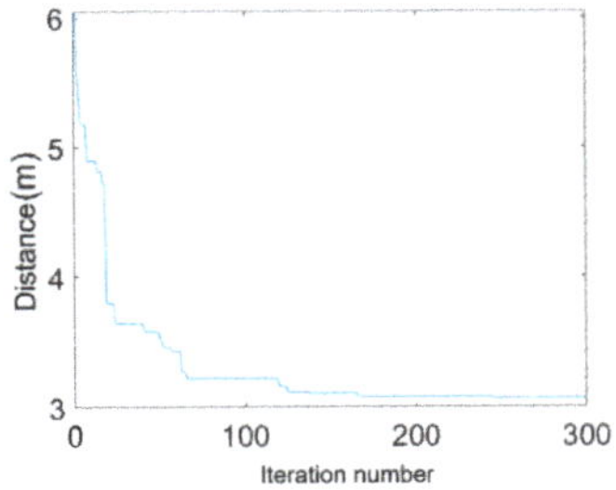

Figure 20. Iteration optimization of a simulated annealing algorithm.

5. Picking Experiment

5.1. Picking Experiment with Different End-Pickers

In this experiment, different numbers of fruits were set to test the picking effect of different end-pickers. Each set of fruits had the same arrangement position, the same picking sequence, and the same number. The picking experiment was carried out using the gripper and the pick-place integrated end-picker. In addition, all the fruits in this experiment were models, so all the cutting action of fruit stems was a simulation. The picking pose of the end-picker was vertically up (Figure 21b).

(a) (b)

Figure 21. Picking diagram. (**a**) Initial pose; (**b**) target pose.

The binocular camera at the bottom of the robot arm obtained the three-dimensional spatial coordinates of the fruit. After the coordinate transformation, the recognized fruit coordinates were transmitted to the robot arm. The flexible gripper picking process is shown in Figure 22. The pick-place integrated end-picker moved from the initial pose (Figure 23a) to the lower part of target position (Figure 23b), and it continued to move to the target position (Figure 23c). The mechanical arm carried the end-picker close to the fruit, setting the fruit into the cavity. At this time, the steering engine drove the cutting mechanism to close through the reduction gear and clamp the pear stem (Figure 23d). Then, the robot arm drove the cutting mechanism to move down and cut off the pear stem, and the fruit from the cavity fell into the collection tube in the direction of the arrow, and through the collection tube entered the storage basket, which completed the picking and collection (Figure 23e). Finally, after completing a picking task, the cutting mechanism was driven to open by the steering engine (Figure 23f).

Figure 22. The process of flexible grippers from picking fruits to placing them in the basket. (**a**) Reaching below the target position; (**b**) gripper open; (**c**) reaching the target location; (**d**) gripper closed; (**e**) putting the pear in the basket.

Figure 23. The integration process of the pick-place integrated end-picker from the initial pose to the completion of picking. (**a**) Initial pose; (**b**) reaching below the target position; (**c**) reaching the target location; (**d**) cutting mechanism closed; (**e**) cutting completed; (**f**) cutting mechanism open.

Different numbers of pears were randomly placed within the picking range of the robot arm. The picking experiments of two different end-pickers are shown in Table 5 below:

Table 5. Picking experimental data statistics.

Fruit Number	Picking Time (s)		Percent (%)	Picking Failure Number	
	Grippers	Pick-Place Integrated		Grippers	Pick-Place Integrated
2	35.32	21.96	37.83	0	0
4	69.65	45.59	34.54	1	1
6	97.65	61.28	37.24	2	1
8	138.35	81.09	41.39	2	0
10	165.94	111.63	32.73	3	2

It can be seen from the table that when picking with flexible claws, the picking time of each fruit was about 16 s, while the pick-place integrated end-picker took about 10 s to pick one fruit. The efficiency of the pick-place integrated end-picker was increased by approximately 37%. At the same time, the probability of picking failure using a flexible gripper was higher than that of the pick-place integrated end-picker. When the number of fruits was ten, the number of failed gripper pickings was three, for which the failure probability was 30%. Conversely, the failure probability of the integrated end-picker was only 20%. In several sets of experiments, the overall success probability of the integrated end-picker was about 86.67%, while that of the gripper was only 73.33%.

5.2. Comparison before and after Picking Sequence Optimization

In this experiment, different numbers of fruits were set to test the picking effect of the single robot arm system, and the picking sequence was optimized by the simulated annealing algorithm. The experiment used the integrated end-picker, for which each set of fruits was randomly distributed within the picking range of the robot arm, and the fruit distribution position was consistent before and after optimization. The picking experiment result was as shown in Table 6 below:

Table 6. Experiment result before and after optimization.

Fruit Number	Picking Time (s)		Percent (%)	Picking Failure Number	
	Random Traversal	Simulated Annealing		Random Traversal	Simulated Annealing
2	11.94	12.14	−1.68	0	0
4	28.65	23.51	17.94	0	1
6	41.59	31.24	24.89	1	0
8	50.98	39.61	22.30	2	2
10	63.43	46.58	26.56	2	1

It can be seen from the table that picking efficiency was improved by about 20% after the optimization of the simulated annealing algorithm, for which the improvement efficiency also increased with the number of fruits. The picking route was messy and lengthy before optimization, and the traversal distance was greatly reduced after optimization by the simulated annealing algorithm (Figure 24).

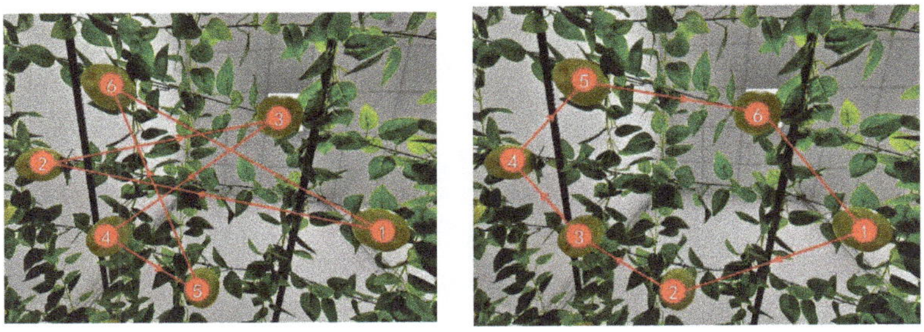

(**a**) (**b**)

Figure 24. Comparison before and after optimization. (**a**) Before optimization; (**b**) after optimization.

5.3. Simulation of Task Allocation for Multiple Robotic Arms

We randomly generated 5, 10, 20, 30, 40, and 50 points in the picking plane area of the two robot arms to simulate the growth distribution position of pears, and we counted the number of fruits falling into the three areas and then followed the proposed task allocation method of multi robotic arms allocating picking tasks to corresponding robot arms. Finally, we counted the task volume of each robot arm. From Table 7, it can be seen from the six sets of data that the task allocation method proposed in this paper was able to effectively improve the utilization rate of a robot arm and make the task volume of the multi-machine system tend to be evenly distributed.

Table 7. Task allocation for two robot arms.

Group	Fruits Number in Region			Task Volume of the Robotic Arm	
	P_1	P_2	P_3	R_1	R_2
5	2	0	3	2	3
10	5	3	2	5	5
20	5	12	3	8	12
30	17	9	4	17	13
40	21	14	5	21	19
50	26	16	8	26	24

We randomly generated 10, 20, 30, 40, 50, and 60 points in the working area of three robot arms to simulate the growth distribution position of pears, and then we counted the number of fruits falling into the seven areas, and finally, we allocated tasks. It can be seen from Table 8 that for the three-robot-arm system, taking the number of fruits as 40 as an example, after using the task allocation method proposed in this article, the task volume of the three robot arms were 12, 14, and 14 each. The minimum and maximum of robot arm task volume differed by two fruits.

Table 8. Task allocation for three robot arms.

Group	Fruit Number in the Region							Task Volume of the Robotic Arm		
	P_1	P_2	P_3	P_4	P_5	P_6	P_7	R_1	R_2	R_3
10	3	1	0	1	2	1	2	3	4	3
20	4	3	5	1	2	2	3	6	6	8
30	4	6	7	4	3	2	4	12	9	9
40	8	9	7	4	3	4	5	12	14	14
50	9	7	11	8	5	7	3	16	18	16
60	11	13	10	6	6	7	7	17	20	23

In summary, the task allocation method for the multiple robotic arm system proposed in this paper is effective and avoids the unreasonable task allocation that causes some of the robot arms to have a particularly large task volume and work all the time, while other robot arms have a particularly small task volume and a short working time, improving the utilization rate of each robot arm and the overall picking efficiency.

5.4. Simulation of Two Robot Arms

We exported the URDF file of the robot arm and the STL model in SOLIDWORKS, and we imported the file into Simulink. The two robot arms were 100 cm apart, and the visualization is shown in Figure 25a.

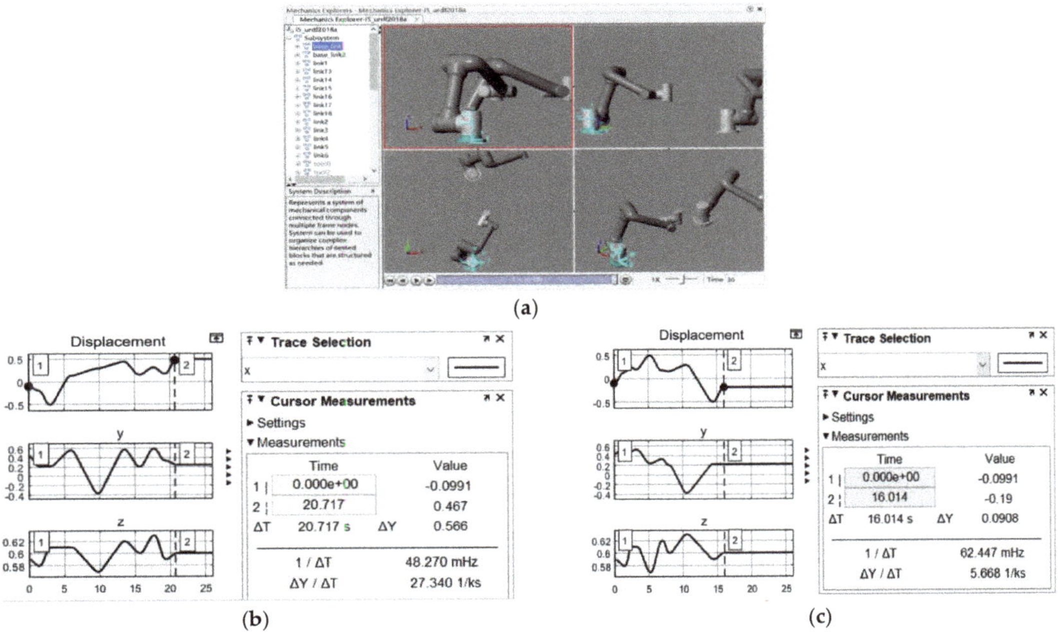

Figure 25. Time record with different algorithms. (**a**) Simulation visualization diagram; (**b**) random traversal; (**c**) simulated annealing.

A total of 4, 6, 8, and 10 points were randomly generated in the picking range of the two robot arms to simulate the position of the pear, and the proposed task allocation method was used to allocate the tasks of the two robot arms. The manipulator directly moved to the next target point after reaching a target point from the initial pose, so only the movement process time of the manipulator was recorded. The movement time was read from the oscilloscope as shown in Figure 25b,c, and the time data are shown in Table 9.

Table 9. Traverse time of two robot arms.

Group	Random Traversal (s)			Simulated Annealing (s)		
	R_1	R_2	Overall System	R_1	R_2	Overall System
4	13.854	11.651	13.854	10.533	9.032	10.533
6	15.217	18.965	18.965	10.657	14.367	14.367
8	21.717	25.361	25.361	16.014	18.549	18.549
10	30.941	25.758	30.941	23.187	19.552	23.187

As can be seen from Table 9, there was little difference in picking time between the two robot arms in each group after task assignment, and the picking time of the overall system was determined by the robot arm that spent the longest time. The time of traversing all points was reduced by about 20% by simulated annealing.

6. Conclusions

In view of the current inefficient work of picking robots, this paper studied the collaborative picking of multiple robotic arms in a trellised pear orchard environment to improve the picking efficiency of the system. The physical characteristics of pears were studied through experiments, and the structural design of the end-picker was carried out based on this. We constructed the private dataset of pears, and finally, we output the three-dimensional coordinates of the target fruit through the Yolo-v5s detection model combined with the depth camera and carried out the object detection through the trained weight, and following this, we completed the object detection task accurately. The camera calibration and hand–eye calibration were completed, and the images of the calibration plate at different angles and distances were collected with the camera. After inputting into the Matlab calibration toolbox, the camera's distortion coefficient and internal parameter matrix were obtained, for which the camera calibration was completed. Through the relationship between two fixed coordinate systems, twenty points were selected to form a hyperparameter equation, and finally, the hand–eye calibration matrix was obtained by fitting the results using the least squares method. The conversion from the pixel coordinate system to the robot arm base coordinate system was completed with good accuracy.

A task allocation method for a multiple robotic arm system was proposed, and the picking sequence was optimized through the simulated annealing algorithm. The several key parameters of the simulated annealing algorithm on the algorithm performance were studied, and the optimal parameter values were selected. After optimization, the picking efficiency was clearly improved. The final result was as follows: initial temperature T_s = 3000 °C, internal loop iteration number M = 150, and cooling coefficient α = 0.99.

The experimental results showed that the picking efficiency of the designed pick-place integrated end-picker is higher than that of the traditional gripper. The success rate of the picking mechanism designed in this paper was 86.67%, which is about 30% higher than that of the claw-gripper. The task allocation method proposed in this paper can make the task volume of a multiple-robotic-arm system tend to be evenly divided, and it obviously improved the utilization rate of each robot arm. Through the simulated annealing algorithm, compared with random traversal, the efficiency of the optimized picking path was increased by about 20%. Moreover, with the increase in fruit number, the efficiency showed an increasing trend.

However, this paper only focused on the structural design of the end-picker, and it did not conduct real experiments under different lighting conditions. In the next step, we will further develop the vision system suitable for a wider range of application scenarios and develop supporting devices that can reduce recognition interference such as light source occlusion, so as to further improve the accuracy rate of the target detection system and the success rate of the picking system. In addition, on the basis of studying the agronomic characteristics and the damage mechanism of fruits, we will further improve the mechanical structure and matching algorithm of the picker to achieve a better picking effect, which is also the next goal of our research.

Author Contributions: Conceptualization, H.Z. and S.W.; methodology, H.Z. and S.W.; software, H.Z. and S.W.; validation, H.Z., D.L. and S.W.; formal analysis, H.Z. and S.W.; investigation, H.Z.; resources, S.W.; data curation, D.L. and S.W.; writing—original draft preparation, H.Z., L.W. and S.W.; writing—review and editing, X.L. and S.W.; visualization, X.L. and S.W.; supervision, L.W.; project administration, L.W.; funding acquisition, S.W. All authors have read and agreed to the published version of the manuscript.

Funding: This work was funded by the National Natural Science Foundation of China under grant 32372592; CCF- Baidu Apollo Joint Development Project Fund.

Data Availability Statement: Data are contained within the article.

Conflicts of Interest: The authors declare no conflicts of interest.

References

1. Mu, L.; Cui, G.; Liu, Y.; Cui, Y.; Fu, L.; Gejima, Y. Design and simulation of an integrated end-effector for picking kiwifruit by robot. *Inf. Process. Agric.* **2020**, *7*, 58–71. [CrossRef]
2. Yang, C.H.; Xiong, L.Y.; Wang, Z.; Wang, Y.; Shi, G.; Kuremot, T.; Zhao, W.H.; Yang, Y. Integrated detection of citrus fruits and branches using a convolutional neural network. *Comput. Electron. Agric.* **2020**, *174*, 105469. [CrossRef]
3. Yang, Q.; Rong, J.; Wang, P.; Yang, Z.; Geng, C. Real-time detection and localization using SSD method for oyster mushroom picking robot. In Proceedings of the 2020 IEEE International Conference on Real-Time Computing and Robotics (RCAR), Asahikawa, Japan, 28–29 September 2020; pp. 158–163.

4. Kang, H.; Zhou, H.; Chen, C. Visual perception and modeling for autonomous apple harvesting. *IEEE Access* **2020**, *8*, 62151–62163. [CrossRef]
5. Peng, H.; Xue, C.; Shao, Y.; Chen, K.; Xiong, J.; Xie, Z.; Zhang, L. Semantic segmentation of litchi branches using DeepLabV3+ model. *IEEE Access* **2020**, *8*, 164546–164555. [CrossRef]
6. Li, J.; Tang, Y.; Zou, X.; Lin, G.; Wang, H. Detection of fruit-bearing branches and localization of litchi clusters for vision-based harvesting robots. *IEEE Access* **2020**, *8*, 117746–117758. [CrossRef]
7. Liang, C.; Xiong, J.; Zheng, Z.; Zhong, Z.; Li, Z.; Chen, S.; Yang, Z. A visual detection method for nighttime litchi fruits and fruiting stems. *Comput. Electron. Agric.* **2020**, *169*, 105192. [CrossRef]
8. Suo, R.; Gao, F.; Zhou, Z.; Fu, L.; Song, Z.; Dhupia, J.; Li, R.; Cui, Y. Improved multi-classes kiwifruit detection in orchard to avoid collisions during robotic picking. *Comput. Electron. Agric.* **2021**, *182*, 106052. [CrossRef]
9. Xiong, Y.; Ge, Y.; From, P.J. An improved obstacle separation method using deep learning for object detection and tracking in a hybrid visual control loop for fruit picking in clusters. *Comput. Electron. Agric.* **2021**, *191*, 106508. [CrossRef]
10. Li, X.; Pan, J.; Xie, F.; Zeng, J.; Li, Q.; Huang, X.; Liu, D.; Wang, X. Fast and accurate green pepper detection in complex backgrounds via an improved Yolov4-tiny model. *Comput. Electron. Agric.* **2021**, *191*, 106503. [CrossRef]
11. Wang, L.; Zhao, Y.; Liu, S.; Li, Y.; Chen, S.; Lan, Y. Precision detection of dense plums in orchards using the improved YOLOv4 model. *Front. Plant Sci.* **2022**, *13*, 839269. [CrossRef]
12. Yan, B.; Fan, P.; Lei, X.; Liu, Z.; Yang, F. A real-time apple targets detection method for picking robot based on improved YOLOv5. *Remote Sens.* **2021**, *13*, 1619. [CrossRef]
13. Xiong, Y.; Ge, Y.; From, P.J. An obstacle separation method for robotic picking of fruits in clusters. *Comput. Electron. Agric.* **2020**, *175*, 105397. [CrossRef]
14. Xiong, Y.; Peng, C.; Grimstad, L.; From, P.J.; Isler, V. Development and field evaluation of a strawberry harvesting robot with a cable-driven gripper. *Comput. Electron. Agric.* **2019**, *157*, 392–402. [CrossRef]
15. Ye, L.; Duan, J.; Yang, Z.; Zou, X.; Chen, M.; Zhang, S. Collision-free motion planning for the litchi-picking robot. *Comput. Electron. Agric.* **2021**, *185*, 106151. [CrossRef]
16. Feng, Q.; Zou, W.; Fan, P.; Zhang, C.; Wang, X. Design and test of robotic harvesting system for cherry tomato. *Int. J. Agric. Biol. Eng.* **2018**, *11*, 96–100. [CrossRef]
17. Oktarina, Y.; Dewi, T.; Risma, P.; Nawawi, M. Tomato harvesting arm robot manipulator; a pilot project. *J. Phys. Conf. Series.* **2020**, *1500*, 012003. [CrossRef]
18. Wei, J.; Yi, D.; Bo, X.; Guangyu, C.; Dean, Z. Adaptive variable parameter impedance control for apple harvesting robot compliant picking. *Complexity* **2020**, *2020*, 1–15. [CrossRef]
19. Liu, C.-H.; Chiu, C.-H.; Chen, T.-L.; Pai, T.-Y.; Chen, Y.; Hsu, M.-C. A soft robotic gripper module with 3D printed compliant fingers for grasping fruits. In Proceedings of the 2018 IEEE/ASME International Conference on Advanced Intelligent Mechatronics (AIM), Auckland, New Zealand, 9–12 July 2018; pp. 736–741.
20. Pi, J.; Liu, J.; Zhou, K.; Qian, M. An octopus-inspired bionic flexible gripper for apple grasping. *Agriculture* **2021**, *11*, 1014. [CrossRef]
21. Hohimer, C.J.; Wang, H.; Bhusal, S.; Miller, J.; Mo, C.; Karkee, M. Design and field evaluation of a robotic apple harvesting system with a 3D-printed soft-robotic end-effector. *Trans. ASABE* **2019**, *62*, 405–414. [CrossRef]
22. Yu, X.; Fan, Z.; Wang, X.; Wan, H.; Wang, P.; Zeng, X.; Jia, F. A lab-customized autonomous humanoid apple harvesting robot. *Comput. Electr. Eng.* **2021**, *96*, 107459. [CrossRef]
23. Arima, S.; Kondo, N.; Yagi, Y.; Monta, M.; Yoshida, Y. Harvesting robot for strawberry grown on table top culture, 1: Harvesting robot using 5 DOF manipulator. *J. Soc. High Technol. Agric.* **2001**, *13*, 159–166. [CrossRef]
24. Arima, S.; Monta, M.; Namba, K.; Yoshida, Y.; Kondo, N. Harvesting robot for strawberry grown on table top culture (part 2) harvesting robot with a suspended manipulator under cultivation bed. *Shokubutsu Kojo Gakkaishi* **2003**, *15*, 162–168. [CrossRef]
25. Van Henten, E.J.; Schenk, E.J.; Van Willigenburg, L.G.; Meuleman, J.; Barreiro, P. Collision-free inverse kinematics of the redundant seven-link manipulator used in a cucumber picking robot. *Biosyst. Eng.* **2010**, *106*, 112–124. [CrossRef]
26. Bontsema, J. Picking robot for peppers (interview met Jan Bontsema). *WageningenWorld* **2011**, *2011*, 6.
27. Kahani, A. Multi-Robot Crop Harvesting Machine. U.S. Patent 9,475,189, 25 October 2016.
28. Williams, H.A.; Jones, M.H.; Nejati, M.; Seabright, M.J.; Bell, J.; Penhall, N.D.; Barnett, J.J.; Duke, M.D.; Scarfe, A.J.; Ahn, H.S.; et al. Robotic kiwifruit harvesting using machine vision, convolutional neural networks, and robotic arms. *Biosyst. Eng.* **2019**, *181*, 140–156. [CrossRef]
29. Fu, L.; Tola, E.; Al-Mallahi, A.; Li, R.; Cui, Y. A novel image processing algorithm to separate linearly clustered kiwifruits. *Biosyst. Eng.* **2019**, *183*, 184–195. [CrossRef]
30. Trinidad, J.B. Machine for Automatically Harvesting Fruit Cultivated in Rows. U.S. Patent 13/124,577, 20 October 2011.

31. Pitzer, R. Automated Selective Harvesting of Crops with CONTINUOUS offload. U.S. Patent 9,554,513, 31 January 2017.
32. Guillén-Navarro, M.A.; Martínez-España, R.; Bueno-Crespo, A.; Morales-García, J.; Ayuso, B.; Cecilia, J.M. A decision support system for water optimization in anti-frost techniques by sprinklers. *Sensors* **2020**, *20*, 7129. [CrossRef]

agronomy

MDPI

Article

Precision Inter-Row Relative Positioning Method by Using 3D LiDAR in Planted Forests and Orchards

Limin Liu [1], Dong Ji [1], Fandi Zeng [1], Zhihuan Zhao [1,*] and Shubo Wang [2,*]

1 College of Mechanical and Electronic Engineering, Shandong Agriculture and Engineering University, Jinan 250100, China; liulimsy2882@163.com (L.L.); 13325259481@163.com (D.J.); zfd19508@163.com (F.Z.)
2 School of Automation, Institute of Intelligent Unmanned System, Qingdao University, Qingdao 266071, China
* Correspondence: zhaozhihuan@sdaeu.edu.cn (Z.Z.); shubowang@qdu.edu.cn (S.W.)

Abstract: Accurate positioning at the inter-row canopy can provide data support for precision variable-rate spraying. Therefore, there is an urgent need to design a reliable positioning method for the inter-row canopy of closed orchards (planted forests). In the study, the Extended Kalman Filter (EKF) fusion positioning method (method C) was first constructed by calibrating the IMU and encoder with errors. Meanwhile, 3D Light Detection and Ranging (LiDAR) observations were introduced to be fused into Method C. An EKF fusion positioning method (method D) based on 3D LiDAR corrected detection was designed. The method starts or closes method C by the presence or absence of the canopy. The vertically installed 3D LiDAR detected the canopy body center, providing the vehicle with inter-row vertical distance and heading. They were obtained through the distance between the center of the body and fixed row spacing. This can provide an accurate initial position for method C and correct the positioning trajectory. Finally, the positioning and canopy length measurement experiments were designed using a GPS positioning system. The results show that the method proposed in this study can significantly improve the accuracy of length measurement and positioning at the inter-row canopy, which does not significantly change with the distance traveled. In the orchard experiment, the average positioning deviations of the lateral and vertical distances at the inter-row canopy are 0.1 m and 0.2 m, respectively, with an average heading deviation of 6.75°, and the average relative error of canopy length measurement was 4.35%. The method can provide a simple and reliable inter-row positioning method for current remote-controlled and manned agricultural machinery when working in standardized 3D crops. This can modify the above-mentioned machinery to improve its automation level.

Keywords: 3D LiDAR; orchard; positioning; canopy length measurement

Citation: Liu, L.; Ji, D.; Zeng, F.; Zhao, Z.; Wang, S. Precision Inter-Row Relative Positioning Method by Using 3D LiDAR in Planted Forests and Orchards. *Agronomy* **2024**, *14*, 1279. https://doi.org/10.3390/agronomy14061279

Academic Editor: Mario Cunha

Received: 6 May 2024
Revised: 7 June 2024
Accepted: 11 June 2024
Published: 13 June 2024

1. Introduction

Orchards and planted forests are widely distributed worldwide [1]. Recently, precision variable-rate spraying has received much study in orchards and planted forests [2–4]. It requires precise canopy characteristics to provide data [5,6]. The geometric characteristics of plants are directly related to their growth and productivity. They can be used as indicators for estimating plant biomass and growth, yield, water consumption, health assessment, and long-term productivity testing [7,8]. Among them, canopy volume is the most commonly used canopy characteristic [9]. The canopy volume requires sensors to provide length, width and height [10]. The width and height are provided by distance sensors, and accurate canopy length requires highly accurate positioning sensors and methods [11].

At present, there are three main positioning methods: absolute, relative and fusion positioning [12]. The absolute positioning method actively or passively senses the surrounding information through its sensors, thereby achieving absolute positioning based on the external environment [13]. This is mainly the Global Navigation Satellite System

(GNSS) positioning method. Because of its high positioning accuracy, it has been widely used in agriculture [14]. Many studies have shown that the positioning accuracy of GNSS in field crops is more than 95% [15]. However, it is susceptible to weather and cannot be used in rainy, snowy or foggy weather [13]. Due to tree occlusion, the GNSS signal in orchards and planted forests is weak even with no signal [16]. Hence, the mounted position of the GNSS antenna is raised to receive GNSS signals [13]. Nevertheless, its installation is too high, and is highly susceptible to vibration and terrain changes. This will greatly affect the positioning accuracy of GNSS. The relative positioning method mainly uses sensors such as the Inertial Measurement Unit (IMU) and encoder [12]. It requires an accurate initial position and motion estimation method, and the obtained position is relative. It does not need to perceive the external environment, but only its own motion state [17]. Therefore, it is not easily affected by the weather and the surroundings. It has been widely used in fields. There are inherent errors in each sensor so they cannot be eliminated and accumulate with time [13]. It has high positioning accuracy for a short time, but is not suitable for long time precise positioning [17–19].

The fusion positioning method combines the advantages of the absolute positioning method, which has high accuracy in perceiving the surroundings, and the relative positioning method, which does not rely on the external environment. [12,20,21]. Therefore, it has been widely used in recent years [22]. Among them, the most typical method is the Simultaneous Localization and Mapping (SLAM) [23]. The method obtains high positioning accuracy, but the algorithm is complex and requires high computing [24,25]. The sensors of the method to perceive the surroundings are mainly visual sensors and Light Detection and Ranging (LiDAR) [25–27]. The LiDAR has the advantages of high positioning accuracy, large data, and less susceptibility to weather [28]. This makes LiDAR widely used in crop detection and navigation [21,29,30]. Among them, horizontally mounted LiDAR is used for navigation and vertically mounted for crop detection [31,32]. The horizontal installation of 3D LiDAR has a narrow vertical field of view so that the entire plant canopy can be measured far away from the canopy [33]. While the vertical installation can obtain abundant plant information on both sides, providing massive data support for variable-rate spraying, the narrow horizontal view cannot be used for navigation.

In summary, absolute positioning methods are highly susceptible to surroundings. Relative positioning methods have high positioning accuracy in a short time and low positioning accuracy in a long time. The fusion positioning methods are complex, with high computing demand, and the LiDAR needs to be installed horizontally. To address the above issues, we use a single vertically installed 3D LiDAR and relative positioning sensors (encoder and IMU) to achieve high-precision positioning and canopy length measurement at the inter-row canopy. Thus, 3D LiDAR is used to detect the presence or absence of the canopy and determine the absolute inter-row positioning (heading and inter-row vertical coordinates) of the vehicle. This can correct the data acquired by the positioning method (method C) based on the Extended Kalman Filter (EKF) at the canopy. The method C and the sensor starts or closes according to the presence or absence of the canopy. When restarting, the inter-row lateral coordinates (0) between rows are acquired. Meanwhile, the advantages of the relative positioning method with high positioning accuracy for a short time are utilized. When the method C is restarted, it can provide high-precision positioning and canopy length measurement at the inter-row canopy. The above is the EKF fusion positioning method (method D) based on 3D LiDAR corrected detection. In this study, we make the following hypotheses:

(1) Mechanical vibration has no effect on sensor accuracy.
(2) There is no significant magnetic field change in the surroundings.
(3) Temperature changes have no effect on the sensor.

It is limited because the method is mainly for modern orchards (planted forests) with fixed row spacing. Yet, it provides new methods and new ideas for the inter-row canopy positioning in modern orchards (planted forests) and can be extended to other 3D crops that are planted in a standardized manner. The method has a wide range of uses in 3D crops grown in fixed row spacing with weak and no GNSS signals. It is simple and reliable, and can be used to modify existing agricultural machinery into intelligent machinery.

2. Materials and Methods

2.1. Composition and Electronic Hardware System of Information Collection Vehicle

The hereby designed information collection vehicle (ICV), with dimensions of 1.2 m × 0.45 m × 1.5 m in length, width and height, respectively, and a maximum load of 200 kg, is shown in Figure 1a. The ICV mainly consists of an electronic hardware system, a chassis walking system, and a positioning system. The electronic hardware system is integrated, which can sense the surroundings in real-time, and obtain information about the ICV (Figure 1b). It is divided into a sensor module, processing module and power module in accordance with the functions. The chassis walking system is the vehicle shown in Figure 1a, which provides support for other systems. Meanwhile, the positioning system is the GPS shown in Figure 1b and can obtain the ICV trajectory.

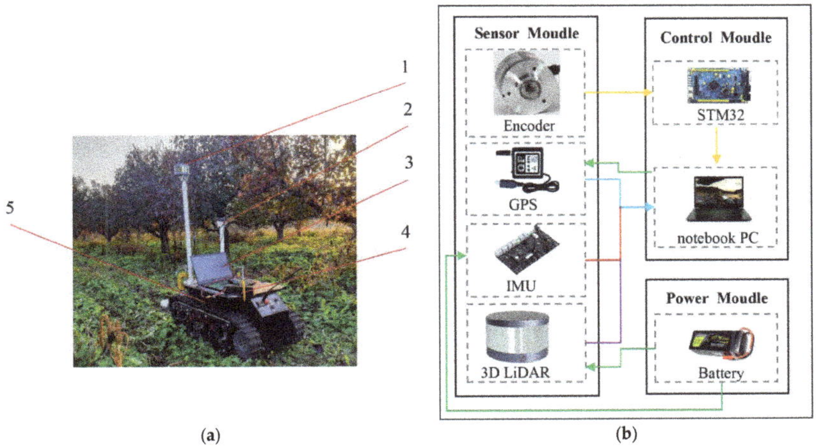

Figure 1. Composition and electronic hardware system of ICV. (**a**) ICV, and (**b**) Electronic hardware system design. 1. 3D LiDAR, 2. GPS, 3. Notebook PC, 4. Battery, 5. Chassis walking system. The green line in (**b**) indicates the power supply, while the others represent the information transmission.

2.1.1. Chassis Walking System

The chassis walking system is a Bunker chassis produced by AgileX Robotics Co., Ltd. (Shenzhen, China), with dimensions of 1.2 m × 0.4 m × 0.45 m in length, width and height, respectively, powered by 48 V. The active wheels drive the rubber tracks that move the ICV forward, the eight loading wheels support the body weight, and four guiding wheels are used to guide and support the tracks and adjust the tightness of the tracks. The ICV achieves differential steering and 360° in-situ steering by means of different speeds of the left and right active wheels. The spring suspension was designed to give the ICV high ground clearance and good ground adaptation. When walking, the maximum moving speed of ICV is 3.0 m/s, the maximum climbing angle is 30°, and the minimum distance between the frame and the ground is 20 cm, which allows it to move flexibly in the orchard. We added a controller to its interior to realize the function of traveling at a certain speed. In this study, the ICV is implemented at a speed of 1 m/s. Operators only need to control the left and right directions of the ICV and prevent collisions and bumps.

2.1.2. Sensor Module

The sensor module mainly consists of 3D LiDAR, an encoder, an IMU, etc. LiDAR senses the surroundings of the ICV and provides data support for auxiliary positioning.

The 16-wire mechanical LiDAR produced by RoboSense Technology Co., Ltd. in Shenzhen, China, can uninterruptedly scan the surrounding trees at 360° horizontally and 30° vertically (15° above and below the LiDAR level), and provide up to 300,000 points per second, with a maximum detection distance of 150 m, a detection accuracy of ±2 cm, and a vertical angle resolution of 2°. The horizontal angle resolution is 0.09°, 0.18° and 0.38° while working at 5 Hz, 10 Hz and 20 Hz, respectively (10 Hz is hereby adopted), with DC 12 V power supply and 100 M Ethernet communication with the notebook PC.

The encoder and IMU acquire the vehicle speed, heading angle and position information. The encoder is an E6B2-CWZ6C encoder (Omron, Osaka, Japan) with a resolution of 1000 P/R (pulse/ring), and is co-axially connected to the active wheel (diameter 22 cm) of the ICV via a connecting shaft. The IMU is a WT61C-RS485 IMU manufactured by Dongguan Weite Intelligent Technology Co., Ltd., Guangzhou, China, with a Kalman filter program, and provides stable and accurate data. The static accuracy is 0.05°/s; the dynamic accuracy is 0.2°/s in X and Y directions; the Z-axis accuracy is 1°/s without magnetic field interference; the acceleration accuracy is 0.0005 g; the gyroscope accuracy is 0.061°/s; and the maximum data output frequency is 200 Hz (100 Hz is hereby adopted).

2.1.3. Processing Module

With a data volume of 300,000 point clouds per second, the central processing unit (CPU) must be extremely powerful. For this reason, we chose a notebook PC equipped with an i7 8750H processor, 16 G of RAM, 128 G of SSD, NVIDIA GeForceGTX1050Ti GPU, 1 T of Mechanical hard drive, Windows 10 pre-installed, RS232, Ethernet, USB, and RS485 communication interfaces. This study requires MCU to take the encoder information and upload it to the notebook PC. In this paper, the M3S type STM32 MCU produced by QiXingChong Company in Dongguan, China is hereby selected, which has a chip of stm32f103zet6, cortex-M3 protocol, 144 pins, 512k flash memory, 72 MHz main frequency, and multiple communication interfaces such as CAN, USB, RS232 and RS485. The GPS module, MCU, and IMU are connected to the notebook PC through USB, RS232, and RS485 communication interfaces, respectively.

2.1.4. Power Module

In addition to the functional modules mentioned above, a power module is required to provide the power supply for each module. In this case, a 12 V ternary lithium battery (6S-12000mAh, Shenzhen, China) with a battery capacity of 12,000 mAh and a full charge voltage of about 22.2 V is selected. A voltage regulation module and a power display module are built-in, which can continuously supply power for 2–4.5 h under normal conditions and make automatic alarms when the voltage is lower than 11.5 V using a buzzer. It takes about 1.5 h to be fully charged. Meanwhile, there are voltage regulators installed to provide stable voltage for different functional modules (5 and 12 V power supply).

2.1.5. Positioning System

To verify the positioning performance of the ICV, the mobile trajectory measurement system needs to have a high accuracy so that it can be directly used as the reference true value. The "QFQ3" RTK GNSS positioning system produced by China Quanfang Navigation Company (Shenzhen, China), which is compatible with GLONASS, Galileo, QZSS, SBAS, BDS, and GPS, is hereby adopted. The positioning system incorporates real-time differential algorithms to provide centimeter positioning accuracy, ±1 cm horizontal positioning accuracy, initialization time < 5 s, data output frequency up to 20 Hz (20 Hz is used), and a power supply range of 5 to 24 V DC. The RTK GNSS mobile station (±2.5 cm movement accuracy) on the robot has a linear distance of 0.5 m from the center of the body and a vertical distance of 0.2 m.

2.2. The Fusion Positioning Method Based on EFK

There are inherent measurement errors in the encoder and IMU due to factors such as manufacturing and installation. The measurement error of the encoder was corrected by STM32 MCU encoder mode, whereas the IMU was corrected by the ellipsoid correction method (Appendix A).

2.2.1. Kinematic Model

The ICV kinematic model was constructed from kinematic and relative positioning sensor (encoder and IMU) data (Figure 2).

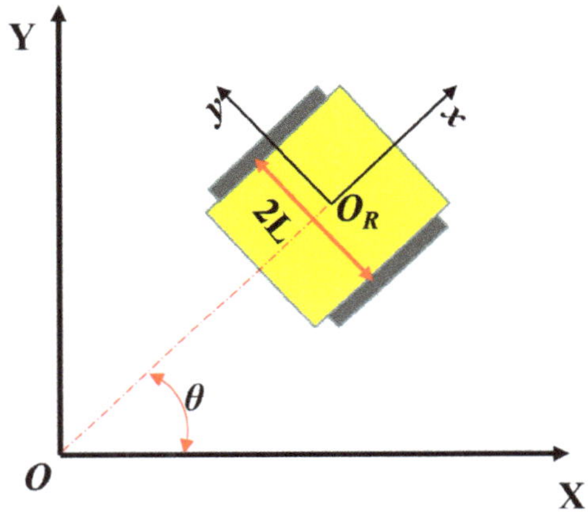

Figure 2. Kinematic model. Two coordinate systems are included in the figure, one is the world coordinate system and the other is the body coordinate system. The heading angle (θ), wheelbase (2L), and body center (O_R) of the vehicle are also labeled in the figure.

The kinematic model of the relative positioning method is constructed from the initial position and the data provided by the sensors (Equation (1)).

$$\overline{\mathbf{T}_t} = \mathbf{T}_{(t-1)} + \Delta\mathbf{T} = \begin{bmatrix} x_{(t-1)} + \Delta x \\ y_{(t-1)} + \Delta y \\ \theta_{(t-1)} + \omega_{(t)}\Delta t \\ v_{x(t-1)} + a_x\Delta t \\ v_{y(t-1)} + a_y\Delta t \end{bmatrix} \tag{1}$$

where, $\overline{\mathbf{T}_t}$ is the current motion state of the ICV; $\mathbf{T}_{(t-1)}$ is the previous motion state of the ICV; $\Delta\mathbf{T}$ is the change of the motion state of the ICV within the moment; $x_{(t-1)}$, $y_{(t-1)}$, $\theta_{(t-1)}$, $v_{x(t-1)}$, and $v_{y(t-1)}$ are the X-axis displacement, Y-axis displacement, heading angle, X-axis velocity and Y-axis velocity of the ICV at the previous moment, respectively; Δx, Δy, $\omega_{(t)}\Delta t$, $a_x\Delta t$ and $a_y\Delta t$ are the change of X-axis displacement, Y-axis displacement, heading angle, X-axis velocity and Y-axis velocity of the ICV within the moment, respectively; $\omega_{(t)}$, a_x and a_y are the current angular velocity, X-axis acceleration and Y-axis acceleration of the ICV, respectively.

Meanwhile, the expression form of Δx and Δy is as follows:

$$\begin{aligned} \Delta x &= v_{x(t-1)}\Delta t\cos\theta_{(t-1)} + v_{y(t-1)}\Delta t\sin\theta_{(t-1)} + 0.5a_x\Delta t^2\cos\theta_{(t-1)} + 0.5a_y\Delta t^2\sin\theta_{(t-1)} \\ \Delta y &= v_{x(t-1)}\Delta t\sin\theta_{(t-1)} - v_{y(t-1)}\Delta t\cos\theta_{(t-1)} + 0.5a_x\Delta t^2\sin\theta_{(t-1)} - 0.5a_y\Delta t^2\cos\theta_{(t-1)} \end{aligned} \tag{2}$$

2.2.2. Measurement Model

After the ICV kinematic model is determined, its measurement model needs to be constructed. The model should clearly represent the process of and the change of ICV motion through the encoder (Figure 3).

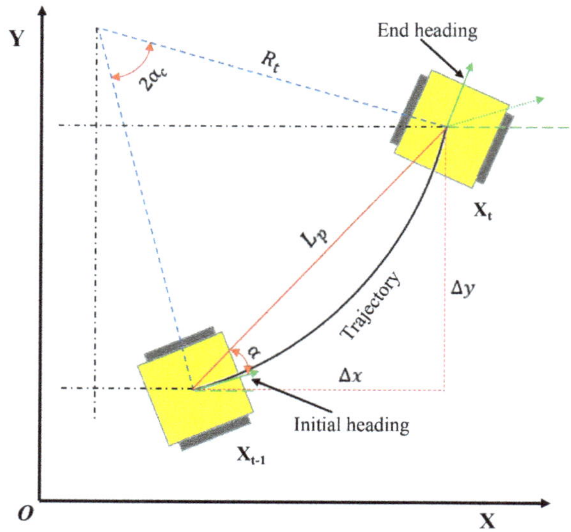

Figure 3. Encoder measurement model based on the sine theorem. The ICV heading, trajectory, displacement (L_P), displacement change (Δx and Δy) and heading deviation ($2\alpha_c$) are clearly shown. α is the chord tangent angle, so $\alpha = \alpha_c$, which is half of the heading deviation.

According to the geometric relationship and the sine theorem (Figure 3), the displacement increment of the ICV in the world coordinate can be obtained. Therefore, the increment (Δx) of the X-axis in world coordinates is:

$$\begin{gathered}\Delta x = L_p\cos(\beta_{(t-1)} + 2\alpha_c) = \frac{2(N_L+N_R)\mathrm{L}\sin\left(\frac{\pi(N_L-N_R)\Delta t}{2\mathrm{nL}}\right)}{N_L-N_R}\times \\ (\cos(\beta_{(t-1)})\cos(\frac{\pi(N_L-N_R)\Delta t}{2\mathrm{nL}}) - \sin(\beta_{(t-1)})\sin(\frac{\pi(N_L-N_R)\Delta t}{2\mathrm{nL}}))\end{gathered} \tag{3}$$

where, L_p is the ICV displacement; $\beta_{(t-1)}$ is the initial heading angle; $2\alpha_c$ is the heading deviation; N_L is the encoder pulses number for the left side of the ICV's active wheel; N_R is the encoder pulse number for the right side of the ICV's active wheel; L is half of the active wheel distance on both sides; and n is the pulse in one rotation.

So, the increment (Δy) of the Y-axis in world coordinates is:

$$\begin{gathered}\Delta y = L_p\sin(\beta_{(t-1)} + 2\alpha_c) = \frac{2(N_L+N_R)\mathrm{L}\sin\left(\frac{\pi(N_L-N_R)\Delta t}{2\mathrm{nL}}\right)}{N_L-N_R}\times \\ (\sin(\beta_{(t-1)})\cos(\frac{\pi(N_L-N_R)\Delta t}{2\mathrm{nL}}) + \cos(\beta_{(t-1)})\sin(\frac{\pi(N_L-N_R)\Delta t}{2\mathrm{nL}}))\end{gathered} \tag{4}$$

From this, the position and attitude of the ICV at any moment (t) is obtained as:

$$\mathbf{T}_t = \mathbf{T}_{(t-1)} = \begin{bmatrix} x_{(t-1)} \\ y_{(t-1)} \\ \beta_{(t-1)} \end{bmatrix} + \begin{bmatrix} \Delta x \\ \Delta y \\ \omega_t \Delta t \end{bmatrix} = \begin{bmatrix} x_{(t-1)} + \Delta x \\ y_{(t-1)} + \Delta y \\ \beta_{(t-1)} + \frac{\pi(N_L-N_R)\Delta t}{2\mathrm{nL}} \end{bmatrix} \tag{5}$$

It is incomplete to construct the measurement model only by the encoder, and the measurement model constructed by the IMU is also required. When the ICV is moving, the measurement model of the ICV at the moment obtained by the IMU is:

$$\mathbf{T}_t = \left(S_{x,t}, S_{y,t}, \theta_{ya,t}, v_{x,t}, v_{y,t}, \omega_{xa,t}, \omega_{ya,t}, a_x, a_y\right) \tag{6}$$

where, $S_{x,t}$ and $S_{y,t}$ are the X-axis and Y-axis displacements at the moment t, respectively, and obtained by integrating the acceleration twice; $\theta_{ya,t}$ is the heading angle; $v_{x,t}$ and $v_{y,t}$ are the X-axis and Y-axis velocities at the moment t, respectively, and obtained by integrating the acceleration; $\omega_{xa,t}$ and $\omega_{ya,t}$ are the X-axis and Y-axis angular velocities at the moment t, respectively, and obtained by integrating the acceleration, respectively; a_x and a_y are the X-axis and Y-axis accelerations at the moment t, respectively.

2.2.3. EKF Fusion Positioning Method Based on Encoder and IMU Data

Multi-sensor fusion positioning needs to minimize the effect of sensor noise. The data from the encoder and IMU are fused, which are used to reduce errors and improve positioning accuracy. The encoder has high accuracy in measuring speed and determining displacement, whereas the IMU has high accuracy in angular velocity and acceleration measurement. Therefore, the measured data between the encoder and IMU can be complementary. In this study, the EKF algorithm is used to fuse the two measured data to realize the estimation, prediction and measurement of the real-time position of the ICV (Appendix B) [34]. The flowchart of the method C is shown in Figure 4.

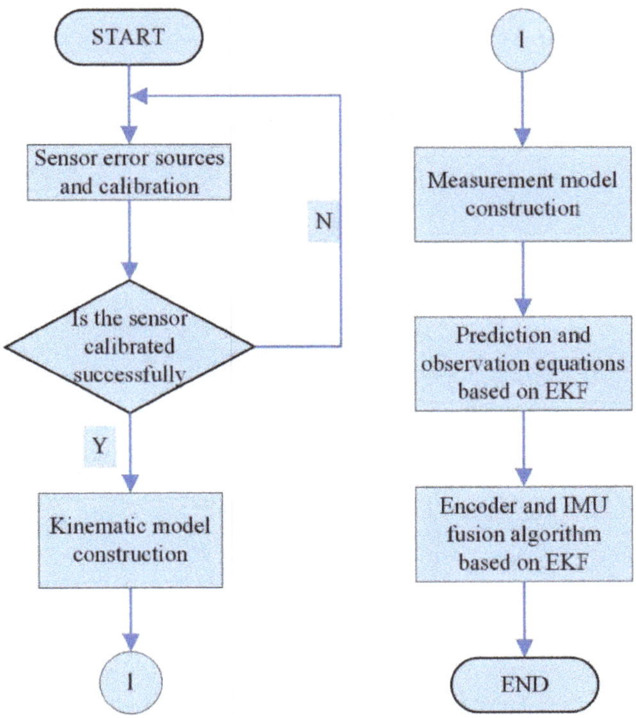

Figure 4. Flowchart of method C.

2.3. Methods for Obtaining the Vertical Position of the Canopy Body Center

The tree canopy is normally distributed and its body center is a stump. Because the LiDAR in this study is installed vertically, only part of the canopy can be detected. The canopy body centers cannot be obtained in real time through the whole tree point cloud. In this study, a threshold (0.75 times the row spacing) was used to determine whether there was a canopy on both sides of the ICV. If not, the point cloud was removed. If there is, the point cloud within one frame (0.1 s) was sparsely processed to reduce computational difficulty. Then, the point cloud was transformed into a voxel point with a length of 2 cm. And the mean of the point cloud is obtained by Singular Value Decomposition (SVD). At the mean position, the plane was constructed perpendicular to the tree rows. The point cloud on the plane is fitted to a circle by the Least Squares method. The circle center is the canopy body center of the point cloud for that frame. The vertical coordinate of the circle center is the vertical distance of the body center.

2.4. Principle of Accurate Positioning at the Inter-Row Canopy

The relative positioning method has high positioning accuracy in a short time. There are gaps between canopies in some orchards and planted forests. According to the above characteristics, method D is designed (Figure 5). After the electronic hardware system is powered on, 3D LiDAR senses the surrounding information and processes it. It first determines whether the point cloud is a gap or not. If there is a gap, the IMU, encoder and the method C will stop. If there is a canopy, start. When restarting, the heading and inter-row vertical coordinates of the initial position are obtained (Figure 6). Additionally, the inter-row lateral coordinates of the initial position are 0. At the inter-row canopy, the heading and vertical coordinates are obtained by 3D LiDAR in real time. It can be corrected for the heading and vertical coordinates of the method C.

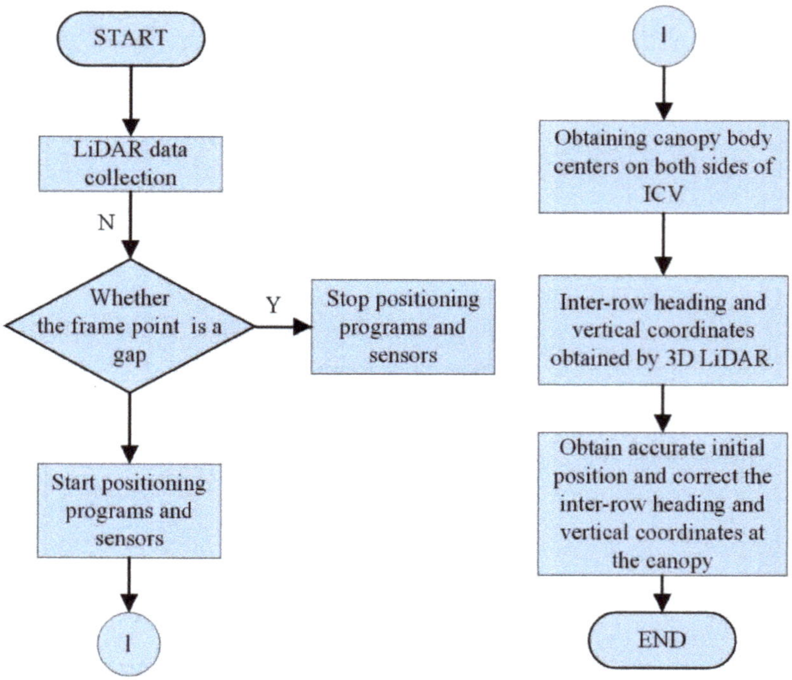

Figure 5. Flowchart of method D.

When the ICV is driving in orchards (planted forests) with fixed row spacing, the gaps and canopies on both sides can be detected by 3D LiDAR (Figure 6). When the ICV travels to any position, its heading (Equation (7)) and the vertical coordinates (Equation (8)) at the inter-row canopy can be obtained. They are calculated by the distance of inter-row canopy body-centered vertical coordinates on both sides and the row spacing. The above is the method D designed in this study. As a result, the purpose of high-precision positioning of the ICV can be obtained at the inter-row canopy.

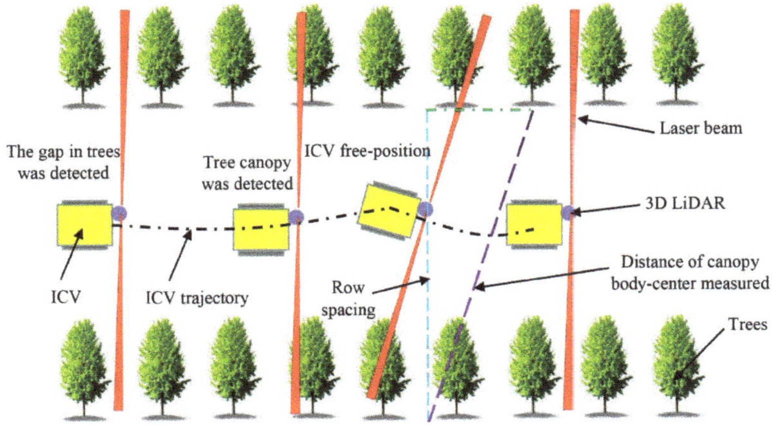

Figure 6. Principle diagram of method D. The laser beam detects tree canopies or gaps on both sides when the ICV is traveling between rows.

From the geometric relationship in Figure 6, we can get

$$\alpha_L = \text{atccos}\left(\frac{D_{tree}}{L_{lidar}}\right) \tag{7}$$

where, α_L is the heading angle; D_{tree} is the row spacing; L_{lidar} is the vertical distance of the canopy body centers on both sides detected by 3D LiDAR.

Based on the heading angle, row spacing and the vertical distance detected by 3D LiDAR, the inter-row vertical coordinate (Equation (8)) is obtained.

$$Y_w = \left(\frac{L_{lidarl}}{2} + d_L\right)\cos\alpha_L - \frac{D_{tree}}{2} \tag{8}$$

where, Y_w is the inter-row vertical coordinate; L_{lidarl} is the vertical distance detected by LiDAR from the body center to the left canopy; d_L is the vertical installation distance between LiDAR and ICV axis.

2.5. Experimental Design

To verify the positioning and canopy length measurement accuracy of method D at the inter-row canopy, campus and orchard comparison experiments were designed. The encoder positioning method (method A), the weighted average fusion positioning method (method B) of the encoder and IMU data (the weighted proportion of encoder is 0.4, and IMU is 0.6), and method C are compared with method D. In this case, the method D uses GPS positioning at the gap to ensure the positioning integrity. On 12 October 2021, ginkgo trees on both sides of the road (Figure 7a) were selected for the experiment on the sidewalk of China Agricultural University (116.2915° E, 40.0357° N) with good GPS signals. The average tree height of the trees is 10.56 m, with a plant spacing of 7.5 m, a row spacing of 15 m, and an average maximum length of 3.58 m. The length of the test area is 40 m, with a total of 8 ginkgo trees on both sides, which are divided into 4 tree groups.

After each experiment, a digital inclinometer (PT180, produced by R&D Instruments, Shenzhen, China, with a measurement range of −90°~90° and an accuracy of 0.01°) is used to correct the 3D LiDAR installation angle (Figure 7b).

Figure 7. Situation of the experimental campus and orchard. (**a**) Ginkgo trees on both sides of the sidewalk; (**b**) 3D LiDAR installation angle correction; (**c**) experimental orchard situation; (**d**) experimental site. The figures include both campus and orchard experimental site conditions. The two topographies are obviously different.

2.5.1. Positioning Accuracy

The GPS base station is placed in the middle of the tree row. Moreover, the GPS module is used to determine the latitude and longitude of the stumps and the boundary of the test area. The GPS mobile station is installed on the ICV to record its real-time trajectory. The ICV is controlled to move forward using a remote control. When it reaches the end point, the positioning algorithm program is stopped. The experiment is repeated 5 times.

The heading angle of the ICV is obtained by real-time GPS trajectory and the center line of the tree row (Figure 8b), and the equation for the center line of the tree row can be obtained as follows:

$$\frac{2x - x_{e1} - x_{e2}}{x_{e3} + x_{e4} - x_{e1} - x_{e2}} = \frac{2y - y_{e1} - y_{e2}}{y_{e3} + y_{e4} - y_{e1} - y_{e2}} \quad (9)$$

where, e1(x_{e1}, y_{e1}), e2(x_{e2}, y_{e2}) are the initial coordinates of the boundaries of the boundary of the experimental area on both sides, respectively; e3(x_{e3}, y_{e3}), e4(x_{e4}, y_{e4}) are the end coordinates, respectively.

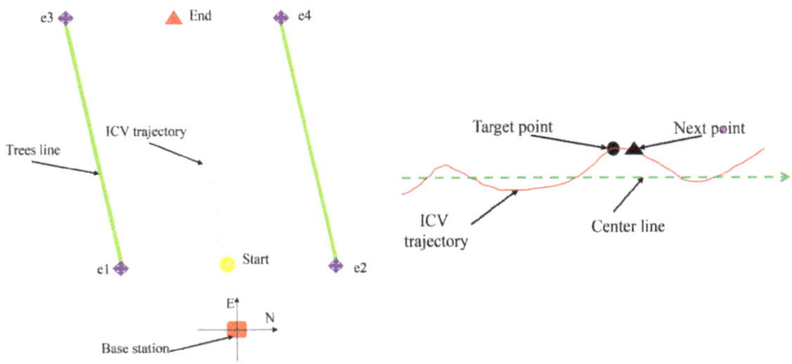

Figure 8. Schematic diagram of positioning and heading angle measurement methods. (**a**) Schematic diagram of the positioning test. (**b**) Calculation method of heading angle.

Further simplification can be obtained:

$$Ax + By + C = 0 \quad (10)$$

From this, it can be seen that the slope of the center line of the tree row ($K_1 = B/A$). We assume the current coordinates of the ICV are (x_0, y_0) and the coordinates of the next moment are (x_{0n}, y_{0n}). Consequently, the slope obtained from the two points should be $K_2 = (y_{0n} - y_0)/(x_{0n} - x_0)$, and it can be seen that the heading angle (α_{hg}) of the current ICV is:

$$\alpha_{hg} = |arctan(\frac{K_2 - K_1}{1 + K_1 * K_2})| \quad (11)$$

2.5.2. Vertical Coordinates of Canopy Body Centers

To verify the accuracy of obtaining the vertical coordinates of the canopy body center, the vertical position of the canopy body center was compared with the value (stump positions) obtained by GPS. The experiment is repeated 5 times.

2.5.3. Canopy Length

A meter ruler with an accuracy of ±1 cm was used to measure the maximum length of the canopy of each tree on both sides of the test area. The data were measured 5 times and recorded in a notebook. The maximum length of the canopy was obtained by 4 positioning methods and 3D LiDAR calculations. The acquired lengths were compared with the lengths obtained from manual measurements to obtain the relative error of the canopy maximum length measurements. The experiment was repeated 5 times. And the canopy point cloud of group 4 with significant errors was selected for analysis.

2.5.4. Orchard Validation

To verify the high-precision positioning of method D at the inter-row canopy, the experiment was carried out in a 4-year-old cherry orchard (Figure 7c, 0.91 ha) at Shangzhuang Experimental Station of China Agricultural University (116.1923° E, 40.1478° N) in Haidian District, Beijing, with good GPS signal. The height of cherry trees is 2.95 m, with an average canopy length of 1.12 m, a plant spacing of 1.5 m, an average canopy length of 1.16 m, and a row spacing of 4 m. The selected test area is 30 m and contains 40 cherry trees. Six cherry trees at the end point are selected for the data analysis. The positioning and canopy length measurement accuracy of the 4 positioning methods at the inter-row canopy is obtained, and the experiment is repeated 5 times. The experiment was conducted on 25 October 2021. During the experiment, the weather was continuously rainless, the daytime temperature was kept at 20.6–30.4 °C, and the wind speed was lower than 1.2 m/s.

2.6. Data Processing

To facilitate the processing of the positioning data, the positioning errors of method D are compared directly using combined positioning (GPS positioning and self-positioning) and GPS positioning. Therefore, the maximum lateral positioning deviation at the inter-row single canopy for method D is roughly defined:

$$E_{\text{tree}} = \frac{2E_{\text{pr}}}{n_{\text{tree}}} \tag{12}$$

where, E_{tree} is the maximum positioning deviation at the inter-row single canopy; E_{pr} is maximum lateral positioning deviation between rows; and n_{tree} is the number of trees in the experimental area.

In the study, the deviation between measured data and actual data is represented by error (Equation (13)). In this case, both positioning deviation and heading deviation are obtained through the Equation.

$$E_{\text{rr}} = \frac{\sum_{i=1}^{n}(\text{M}_{\text{ev}} - \text{A}_{\text{cv}})}{n} \tag{13}$$

where, E_{rr} is the error; M_{ev} is the measured value; A_{cv} is the actual value; n is the number of calculations, which is taken as 25.

Given the variation of the measured positioning, the relative error (Equation (14)) can better represent the accuracy of the measured results.

$$R_{\text{ee}} = \frac{\sum_{i=1}^{n}\frac{(\text{M}_{\text{ev}} - \text{A}_{\text{cv}})}{\text{A}_{\text{cv}}}}{n} \times 100\% \tag{14}$$

where, R_{ee} is the relative error, %.

The canopy point clouds are displayed using Open3D 0.11.0 (Intel Inc., Santa Clara, CA, USA) by Python 3.8. The collected data were analyzed using SPSS Statistics Version 20 (IBM Inc., Armonk, NY, USA) for Windows, and plotted using OriginPro Version 2020 (OriginLab Inc., Northampton, MA, USA). All the test results were tested for normal distribution using SPSS and conformed to normal distribution.

Duncan's post-hoc test was performed on the measured results using SPSS one-way analysis of variance (ANOVA) with a significance level of 0.05. In all cases, Duncan's post-hoc test was used to compare the positioning deviation, positioning relative error and the average value of the maximum canopy length of the 4 positioning methods at the 0.05 significance level.

3. Results

3.1. Motion Trajectories Obtained by the Four Positioning Methods

There are differences between the motion trajectories obtained by the four methods and the actual trajectory of the ICV obtained by GPS (Figure 9). The actual motion trajectory of the ICV is not a straight line due to the level of the operator and the ground conditions. The motion trajectories obtained by the four positioning methods are inferred relative positioning trajectories. Among them, method A only uses an encoder for position estimation of the motion trajectory, resulting in significant errors. This makes its trajectory differ from the actual trajectory. Methods B and C fuse the data from IMU and encoder for position estimation of motion trajectory, with relatively small errors. This makes its trajectory similar to the actual trajectory. Method D only performs position estimation and pose correction at the inter-row canopy, so its trajectory has a high similarity to the actual trajectory. There were significant differences between the actual trajectories of the ICV and methods A, B and C, but the differences with the method D were not significant. The actual trajectory of the ICV within the inter-row lateral distance (*X*-axis) of 2 m is almost the same as its motion trajectory obtained by the four positioning methods. It indicates that the positioning accuracy of the relative positioning sensor is high in a short time. As the inter-row lateral distance further increases, methods A, B, and C exhibit significant positioning deviations compared to the actual trajectory. Among them, method A showed significant positioning deviation in both inter-row lateral and vertical positioning (*Y*-axis). It has the largest positioning deviation. In addition, method C has a high positioning accuracy. It indicates that the EKF fusion positioning algorithm can maintain high positioning accuracy for a long time. Method D has the smallest positioning error. It indicates that method D has the highest positioning accuracy.

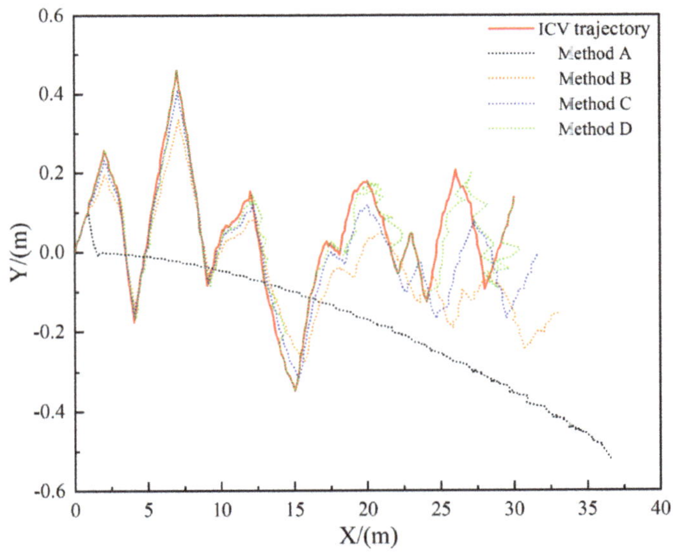

Figure 9. Trajectories of the ICV obtained by the four methods. The figure clearly illustrates the actual trajectory of the ICV and the measured trajectories of the four methods. The *X*-axis represents the inter-row lateral distance and the *Y*-axis represents the inter-row vertical distance.

3.2. Deviation and Relative Error of Positioning

Although the ICV-measured trajectories of the four methods can be clearly obtained in Figure 9, the positioning deviations and relative errors between them and the actual trajectories cannot be quantitatively described. As far as the deviation and relative error of lateral distance positioning are concerned, there were significant differences among the four methods (Figure 10a,c). The mean positioning deviation and relative positioning error of method A (2.19 m, 10.58%), B (0.91 m, 4.80%), C (0.44 m, 2.32%), and D (0.15 m, 0.68%) decreased sequentially (Figure 10a,c). The maximum positioning deviation and relative error of methods A (6.65 m, 18.12%), B (3.04 m, 9.39%), C (1.64 m, 5.39%), and D (0.88 m, 5.02%) also decrease sequentially (Figure 10a,c). The maximum lateral positioning deviation at the inter-row canopy was calculated to be 0.22 m for method D ((Equation (12)). In terms of inter-row vertical positioning deviation, there is a significant difference between method D and the other three methods (Figure 10b). The average positioning deviation of method D is the smallest (0.05 m), with lower significance than methods A (0.66 m), B (0.29 m), and C (0.14 m). In terms of the relative error of inter-row vertical positioning, there is a significant difference between method A and the other three methods (Figure 10d). The average relative error of method A (298.35%) was the largest, which was significantly higher than that of method B (47.69%), C (39.27%) and D (−2.08%). After removing outliers, the average relative errors of methods A, B, C, and D were 40.37%, 24.45%, 16.86%, and −1.82%, respectively.

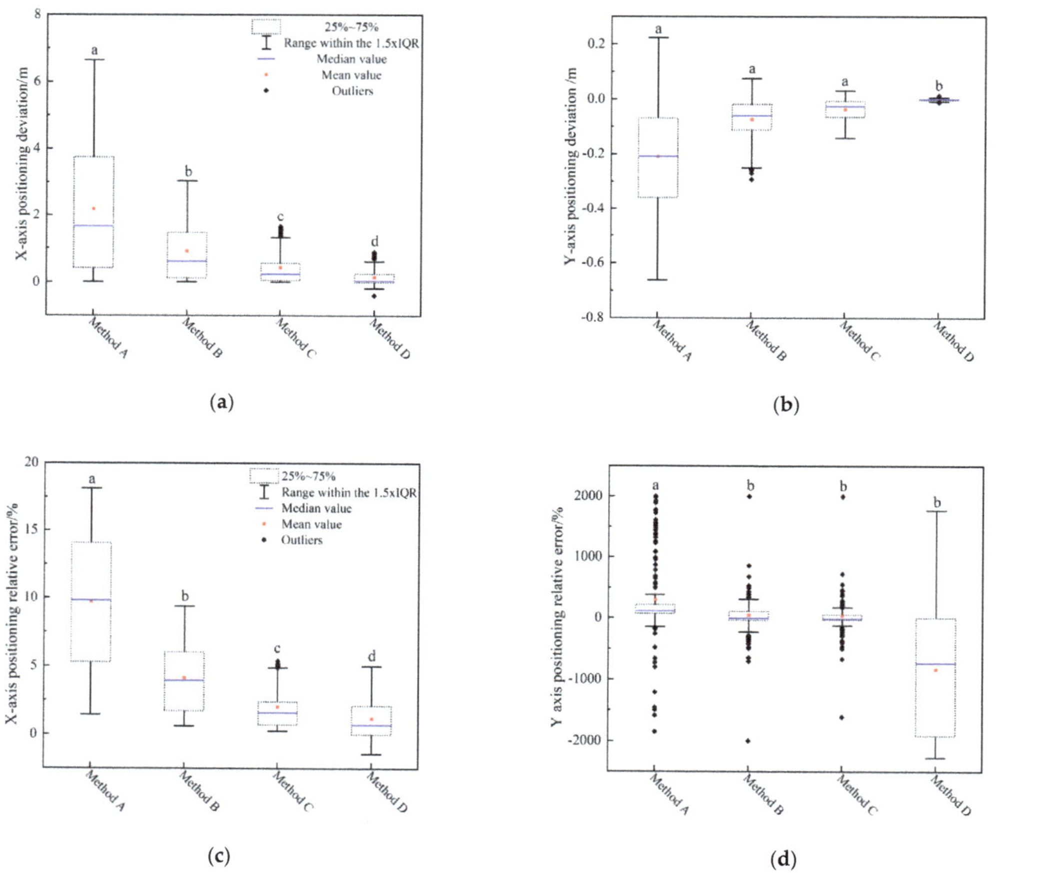

Figure 10. Plots of the ICV positioning errors obtained by the four methods. (**a**) Inter-row lateral positioning deviation; (**b**) Inter-row vertical positioning deviation; (**c**) Relative error of inter-row lateral positioning; (**d**) Relative error of inter-row vertical positioning. For convenience of drawing, the values of method D in (**d**) were enlarged by 400 times. If the absolute relative error of methods A, B, and C is greater than 2000%, it should be equal to ±2000%. Different letters indicate significant differences (Duncan test, α = 0.05).

3.3. Heading Deviation

The maximum heading deviation for methods A, B, C and D were 25.4°, 66.4°, 32.75° and 4.35°, respectively. Among them, method D has the smallest heading deviation and the standard deviation (Table 1). It indicates that the method D has a significant correction for heading deviation.

Table 1. Statistical data of heading deviation of the four positioning methods/°.

Positioning Method	Maximum Deviation	Minimum Deviation	Average	Upper Deviation	Lower Deviation
Method A	25.46	−12.87	14.78	10.45	−22.65
Method B	16.43	−7.98	9.86	6.42	−14.84
Method C	12.75	−6.23	7.42	5.31	−9.65
Method D	4.35	−3.08	2.94	1.41	−4.02

3.4. Body Center Vertical Coordinates

As shown in Figure 11, the canopy body centers (red points) of a certain plant were obtained, whereas green points represent the canopy point cloud. From the figure, it can be seen that the obtained partition canopy body centers are basically on the same plane and distributed relatively evenly. The relative error of the measured body center vertical distance is 2.36 ± 0.57%, indicating that this algorithm has high accuracy.

Figure 11. Acquisition of body center locations. Green points in the figure represent the canopy and red points indicate the partitioned canopy body centers.

3.5. Relative Error of Maximum Canopy Length Measurement

3.5.1. Relative Error of Maximum Canopy Length Measurements for the Four Positioning Methods

The relative measurement error of canopy length for methods A, B, and C increases with the increase of inter-row lateral distance, whereas the change was not significant for method D (Figure 12). In canopy group 1, the mean relative error of maximum canopy length measured by method A (4.63%) was significantly higher than that of methods B (2.57%), C (2.10%) and D (2.08%). The maximum relative measurement errors for methods A, B, C, and D were 6.24%, 4.42%, 4.89%, and 4.59%, respectively, and the difference was small. At canopy groups 2, 3, and 4, the relative errors (2.23, 2.15, and 2.36%) of canopy length measured by method D were significantly lower than those of methods A (8.37, 13.40, and 17.36%), B (5.30, 7.95, and 10.67%), and C (3.85, 5.76, and 7.57%) (Figure 12b–d). In canopy group 2, the maximum relative measurement errors of the four methods were 9.85%, 6.21%, 5.59%, and 4.26%, respectively. At this time, the relative measurement errors of method A are relatively large, while the relative measurement errors of methods B and C increase compared to method D (Figure 12b). With the further increase in traveling distance, when the ICV traveled to canopy group 3, their maximum measurement errors were 15.39%, 11.13%, 8.89% and 4.36%, respectively. At this time, the four methods are significantly different (Figure 12c). In canopy group 4, the maximum relative measurement errors were 19.54%, 12.49%, 8.84%, and 5.68%, respectively (Figure 12d).

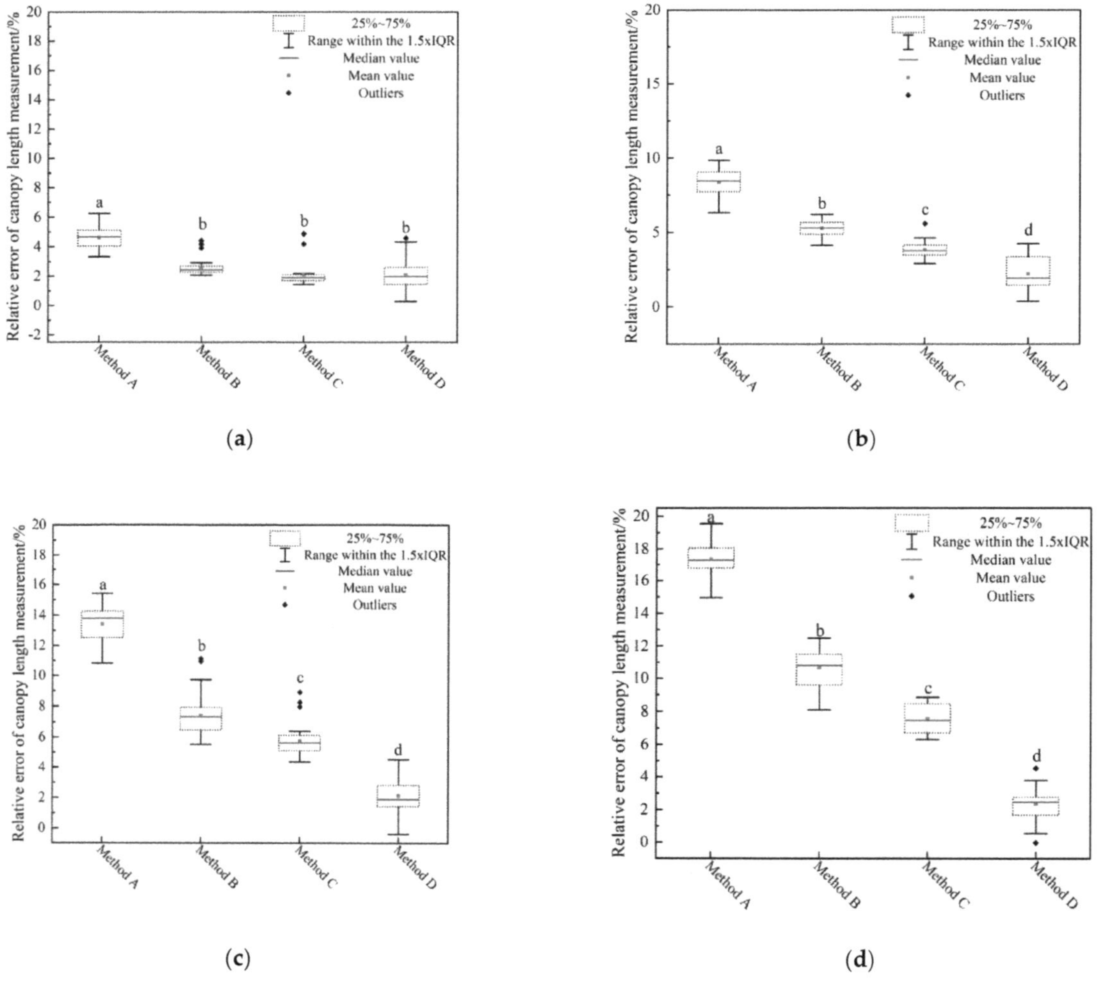

Figure 12. Relative measurement errors of canopy length obtained by the four methods. (**a**) Group 1; (**b**) Group 2; (**c**) Group 3; (**d**) Group 4. The nearest distance from the initial point is the group 1, and the nearest distance from the end point is the group 4. Different letters indicate significant differences (Duncan test, $\alpha = 0.05$).

3.5.2. Relative Errors of Maximum Canopy Length Measurements for the Four Positioning Methods

The maximum canopy lengths measured by the four positioning methods were not consistent (Figure 13). Among them, the maximum (Figure 13a, 4.37 m) was obtained by method A and the minimum (Figure 13d, 3.79 m) by method D. The maximum lengths of the canopy obtained by methods B and C were 4.11 m (Figure 13c) and 3.98 m (Figure 13d), respectively. The relative errors of the maximum length of the canopy measured by methods A, B, C and D were calculated to be 18.75%, 11.68%, 8.15% and 2.99%, respectively. It indicates that method D does not increase with increasing distance traveled in terms of measuring canopy length.

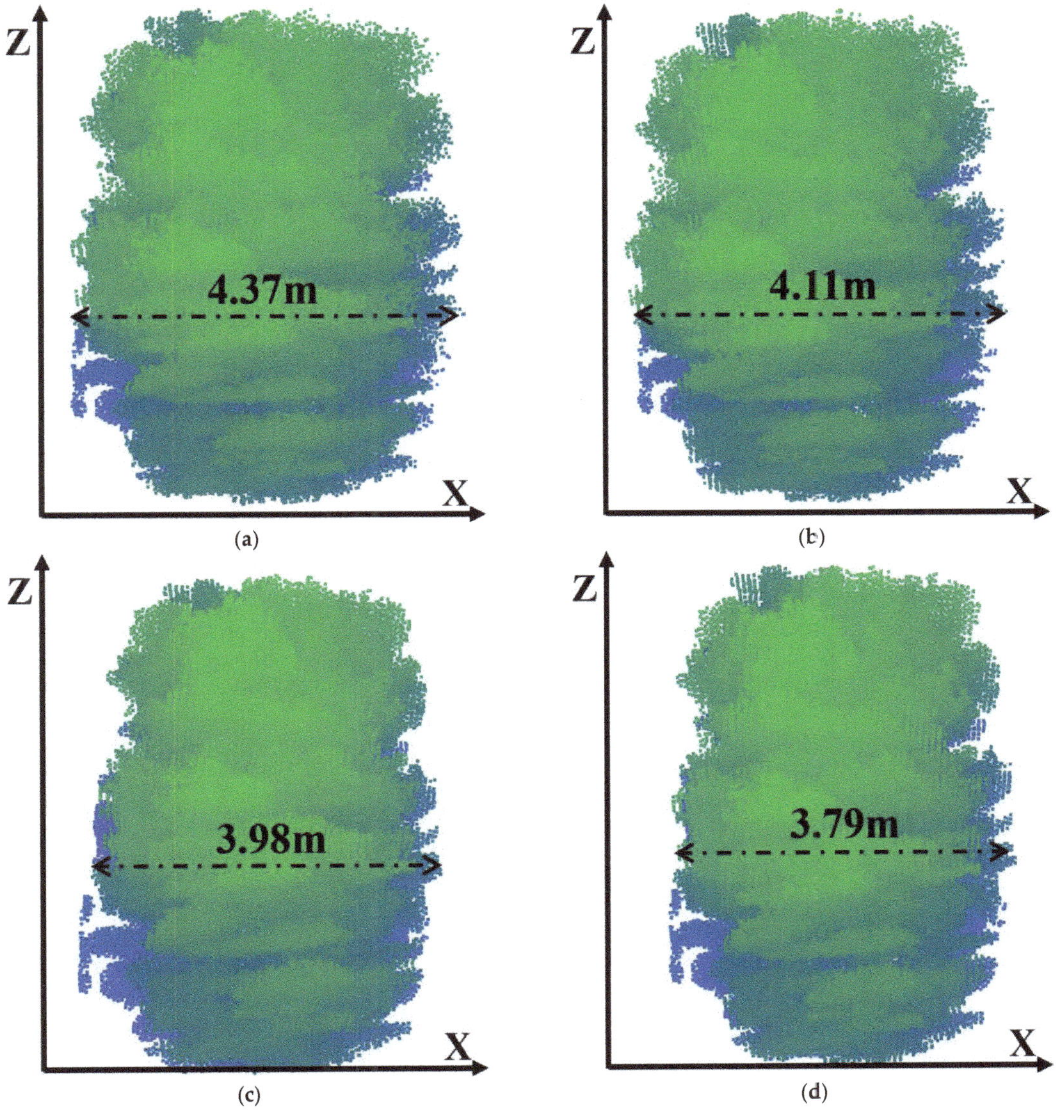

Figure 13. Point clouds of group four ginkgo tree canopy obtained by the four methods. (**a**) Method A; (**b**) Method B; (**c**) Method C; (**d**) Method D. The *X*-axis represents the direction of the ICV, and the Z-axis represents vertical ground upward. The actual measured maximum canopy of the ginkgo tree was 3.69 m.

3.6. Orchard Experiment

The relative error of the positioning and the canopy measurement were calculated by removing outliers (Table 2). The positioning and maximum length of the canopy measured by the four methods were significantly different. The inter-row lateral and vertical positioning deviations (8.73 ± 2.53 m, 2.45 ± 0.91 m) and relative positioning errors (32.55 ± 9.43%, 52.35 ± 19.55%) were significantly higher in method A than those in methods B, C, and D. Among them, the inter-row lateral and vertical distances positioning deviations (0.31 ± 0.11 m, 0.15 ± 0.08 m) and relative positioning errors (4.07 ± 1.81%, 3.27 ± 1.27%) of method D were the lowest. Accordingly, method A has the worst positioning effect, whereas method D has the best positioning effect. The lateral positioning relative error of method D is lower than the vertical positioning relative error, which is different from other methods. Besides, the heading deviation (6.75 ± 1.89°) and the relative error (4.35 ± 1.09%) of canopy length measurement of method D were significantly lower than those of methods A, B and C.

Table 2. Positioning statistics of the four positioning methods. The table includes the relative errors of the ICV's positioning, heading and canopy measurements in the orchard.

Positioning Method	*X*-axis Positioning Error/m	*X*-axis Relative Positioning Error/%	*Y*-axis Positioning Error/m	*Y*-axis Relative Positioning Error/%	Heading Deviation/°	Canopy Length Error/%
Method A	8.73 ± 2.53 a	32.55 ± 9.43 a	2.45 ± 0.91 a	52.35 ± 19.55 a	30.54 ± 12.31 a	37.34 ± 12.01 a
Method B	4.52 ± 1.35 b	21.32 ± 5.77 b	1.52 ± 0.63 b	32.43 ± 13.45 b	21.65 ± 7.54 b	23.77 ± 8.32 b
Method C	3.56 ± 0.82 c	16.03 ± 3.06 c	0.99 ± 0.37 c	21.05 ± 7.76 c	14.38 ± 3.18 c	18.68 ± 5.05 c
Method D	0.31 ± 0.11 d	4.07 ± 1.81 d	0.15 ± 0.05 d	3.27 ± 1.27 d	6.75 ± 1.89 d	4.35 ± 1.09 d

Note: Different letters indicate significant differences (Duncan test, α = 0.05).

4. Discussion

In this study, an EKF fusion positioning method (method D) based on 3D LiDAR detection correction is presented. The vertically installed 3D LiDAR (Figure 7b,d) was used to detect the presence or absence of the canopy. Based on this, it (method D) starts or closes the EKF fusion positioning method (method C) and relative positioning sensors (encoder and IMU). The inter-row initial heading and vertical position of method D were obtained through fixed row spacing and body canopy center (Figures 5 and 6). The restarted method C provides the value of 0 for the initial lateral coordinates (Figures 5 and 6). Consequently, the initial position of the method D is provided. It can take advantage of the short time positioning accuracy of relative positioning methods and sensors [12,35,36]. This characteristic was also demonstrated in this study (Figure 9). Furthermore, the lateral distance is corrected by the inter-row heading and vertical position acquired by 3D LiDAR. Hence, method D can accurately obtain the positioning at the inter-row canopy and the maximum length of the canopy (Figures 10, 12 and 13).

Since the IMU and encoder have inherent errors, the error of positioning methods cannot be eliminated and accumulates with time (driving distance) [36,37]. This is the reason why the deviations of positioning methods A, B, C and D gradually increase with time (Figure 9). Methods C and D utilize the advantageous information of encoders and IMU, which have high positioning accuracy for a longer time (Figures 9 and 10) [34]. This is consistent with the findings of Cui et al. [38]. Although the sensor signals of the relative positioning method are less susceptible to receiving interference, they cannot sense the surroundings to correct the ICV position and pose [39]. The method requires a precise initial position, and usually only one [12,37]. With the increase in driving time, the error gradually accumulates, which will cause significant positioning errors [12]. Therefore, researchers have integrated positioning sensors such as machine vision, ultrasonic and sensors LiDAR into relative positioning methods [12,40,41]. It will provide fusion positioning for relative positioning methods, thereby improving positioning accuracy. However, such methods require large computational power [22].

On the basis of vertically installing 3D LiDAR to detect plant information, we also use the single 3D LiDAR to assist in positioning. It not only requires determining gaps, but also obtaining canopy information. The heading and vertical positioning at the inter-row canopy are obtained. As a result, the accuracy of inter-row heading and vertical positioning can be significantly improved (Table 1, Figure 10b). The acquisition of accurate inter-row vertical distance is a prerequisite for correcting the above parameters. The real-time canopy body center acquisition method proposed in this study has high accuracy (Figure 11). The inter-row vertical distance obtained by 3D LiDAR and lateral distance obtained by method C are both implemented on a plane. The ICV tilts or vibrates when the terrain is rough and bumpy. This can cause large errors in the calibration of method D and the lateral positioning of method C. This is the reason why the relative error of inter-row vertical positioning is negative for method D and positive for the others (Figure 10d). It is also the reason why the relative positioning error of the X-axis and Y-axis in the orchard experiment is larger than that of the sidewalk (Table 2, the length of the experimental area is consistent). Even so, method D had high positioning accuracy and canopy measurement accuracy in the orchard experiments.

Method D can provide multiple accurate initial positions within the experimental area. As a consequence, the canopy length measurements relative errors for the four groups' canopy of method D are not significantly different. whereas those of methods A, B, and C increased with increasing inter-row lateral distance (Figures 12 and 13). The maximum lateral positioning deviation at the inter-row canopy of a single tree calculated by Equation (12) is 0.22 m. The maximum canopy length obtained by method D is 0.1 m larger than the actual crown length (Figure 13d). It indicates that the calculation method for the maximum inter-row lateral positioning deviation at the canopy of a single tree is feasible. In addition, the maximum canopy lengths of the ginkgo trees (average maximum canopy length of 3.52 m) and the three cherry trees (average maximum canopy length of 1.16 m) on campus were essentially the same. But the lateral positioning error at the inter-row canopy of the ginkgo trees on campus was smaller than cherry trees. It indicates that the topography of the orchard has a significant impact on both positioning and canopy measurement accuracy.

In summary, it can be seen that high-precision positioning at the inter-row canopy is a prerequisite to ensure the accuracy of canopy length measurement. Yet, method D has significant limitations. It can only be used in orchards (planted forests) with fixed row spacing and large gaps and young or sparse orchards (planted forests). In the future, the gap above a standardized planted 3D crop could be utilized to close or start method C. Fixed row spacing is utilized through method D to provide inter-row heading and vertical positioning to extend the method use. Meanwhile, it is necessary to introduce the ICV roll angle to correct the point cloud and further improve positioning accuracy. Where the GNSS signal is good, its signal is incorporated into the method to provide accomplished positioning information. Moreover, the method is used to modify the existing manned and remote-controlled sprayers into precision variable-rate sprayers, providing new vitality for the old machines. Although it has achieved good positioning and canopy length measurement accuracy, it can only be used as an auxiliary positioning rather than a conventional positioning method. It can provide new ideas for inter-row vehicle positioning and canopy length measurement methods at the canopy for variable-rate spraying in planted forests and orchards. Additionally, the method is drawn upon to solve the problem of low positioning and canopy length measurement accuracy in agricultural surroundings with poor GPS signals.

5. Conclusions

The kinematics and measurement models of the ICV were constructed by IMU and encoder. The EKF fusion positioning method (method C) is designed by fusing the sensor data using the EKF filtering algorithm. The 3D LiDAR is fused into method C to achieve the EKF fusion positioning method (method D) based on 3D LiDAR detected correction. Method D eliminates the cumulative errors of the IMU and the encoder by closing gaps or starting method C at inter-row canopies. Furthermore, 3D LiDAR provides initial heading and inter-row vertical through body center and fixed row spacing and assists in positioning.

As a result, the accuracy of inter-row positioning and canopy length measurement can be greatly improved. From the campus and orchard experiments and discussions, it can be concluded that:

(1) In the campus test area, the positioning at the inter-row canopy for method D is relatively small and does not increase with inter-row travel distance. The lateral positioning at the individual tree canopies increased with the increase in driving distance. Among them, the deviation of inter-row lateral and vertical positioning at the canopy was less than 0.22 m and 0.15 m, respectively, the heading deviation was less than 4.35°, and the relative error of canopy length measurement was less than 5.68%. It indicates that the method has high positioning accuracy on flat terrain.

(2) The orchard experiments showed that the positioning deviations were larger than those on the campus due to the influence of orchard terrain or mechanical vibration. Hence, the average inter-row lateral and vertical positioning deviations at the canopy were 0.1 m and 0.2 m, the average heading deviation was 6.75°, and the average relative error of canopy length measurement was 4.35%. It indicates that the positioning accuracy of the method is still very high on rugged terrain.

(3) The method is suitable for 3D crops with standardized planting and gaps between canopies, which has significant limitations. The method can solve the problem of low accuracy of positioning and canopy length measurement in 3D crops with poor GPS signals. It has great potential for application.

Although the method proposed in this study solves the problem of inter-row canopy positioning and canopy measurement, there are still many problems. We will use IMU to acquire the rolling angle of the vehicle and improve the accuracy of 3D LiDAR to acquire the vertical coordinates of the body center. Moreover, the accuracy of method C in different terrains will be improved based on the acquired rolling angle. In addition, method D is extended from 2D to 3D to address many problems in actual operations. Furthermore, the method is applied to other crops with fixed row spacing, and gaps in the upper canopy, such as tomatoes in greenhouses.

Author Contributions: Conceptualization, L.L., S.W. and Z.Z.; methodology, L.L. and D.J.; software, L.L. and D.J.; validation, S.W. and Z.Z.; formal analysis, L.L., S.W. and Z.Z.; investigation, L.L., D.J., F.Z. and Z.Z.; resources, S.W. and Z.Z.; data curation, L.L., D.J., F.Z. and Z.Z.; writing—original draft, L.L., D.J., S.W. and F.Z.; visualization, L.L. and F.Z.; and funding Z.Z., S.W. and L.L.; acquisition, Z.Z. All authors have read and agreed to the published version of the manuscript.

Funding: This work was supported by Program for Innovative Research Team in SDAEU (sgykycxtd2020-03), and CCF- Baidu Apollo Joint Development Project Fund.

Data Availability Statement: The original contributions presented in the study are included in the article, further inquiries can be directed to the corresponding author.

Acknowledgments: The authors would like to give special thanks to Tian Zhong for providing the test orchard.

Conflicts of Interest: The authors declare no conflict of interest.

Appendix A

Appendix A.1 Error Sources and Calibration of Encoder

A rotary encoder is a sensor used to measure rotation angle, speed, and direction, which converts angular displacement or angular velocity into a series of electrical digital pulses. The incremental encoder is used in this study. Although incremental rotary encoders have many advantages, they are driven by mechanical transmission to rotate the encoder disk. It makes the encoder highly susceptible to mechanical vibration, resulting in the generation of burrs that affect the number of pulses generated. This ultimately leads to significant measurement errors. To ensure the computing power of the microcontroller, the use of software methods to improve the measurement accuracy of the encoder is abandoned. On the contrary, we used the encoder mode designed by ST company for the STM32 microcontroller. The method utilizes hardware to improve the measurement accuracy of the encoder used in this study.

The counting methods of the encoder mode are shown in Table A1, which are divided into three modes: counting only at TI1 (mode 1), counting only at TI2 (mode 2), and counting both at TI1 and TI2 (mode 3). When the encoder rotates forward, regardless of the mode, the count value increases. Similarly, when reversing, the count values are decreasing.

Table A1. Relationship between counting direction and signal.

Active Edge	Level of Opposite Signals *	TI1FP1 Signal		TI2FP2 Signal	
		Increase	Decrease	Increase	Decrease
Count only at TI1	High	—	+	\	\
	Lower	+	—	\	\
Count only at TI2	High	\	\	+	—
	Lower	\	\	—	+
Count at both TI1 and TI2	High	—	+	+	—
	Lower	+	—	—	+

Note: * TI1FP1 corresponds to TI2, and TI2FP2 corresponds to TI1. "+" Indicates an increasing count; "—" Indicates a decrease in count; "\" Indicates no counting.

From the three counting modes mentioned above, it can be seen that mode 1 and mode 2 are counting at 2 times the frequency, while mode 3 is counting at 4 times the frequency. As shown in Figure A1, the encoder generates four counts for every pulse counter generated, while effectively removing the impact caused by dithering. Compared to the other modes, mode 3 can better remove burrs and interference. Therefore this study chooses mode 3.

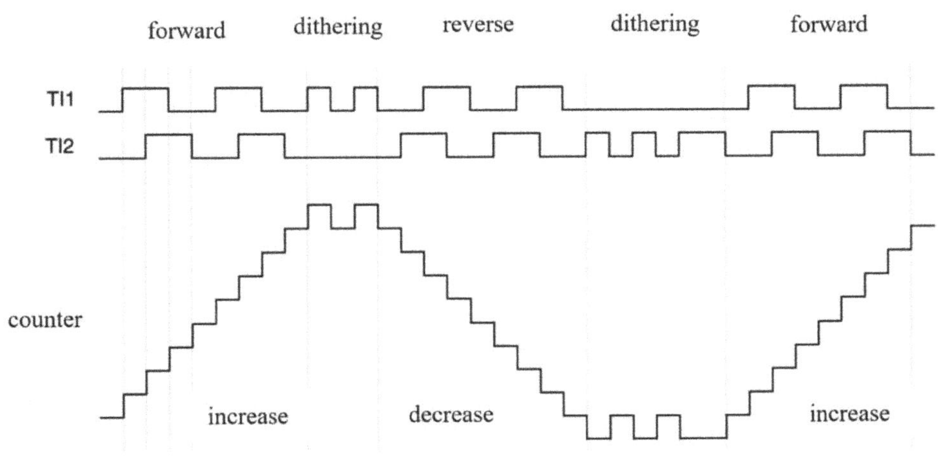

Figure A1. Counter schematic for encoder mode 3.

A comparison test was conducted using the number of pulses captured by A-phase and microcontroller external interrupts alone (method 1), and the number of pulses captured by the encoder software 4× technique (method 2) with the present method (method 3). This experiment was conducted five times, with eight measurements taken each time. One-way ANOVA analysis (Wall Duncan's test) was performed using SPSS 26.0, and the results are shown in Table A2. The relative errors of the three methods were significantly differences, with method 1 having a relative error of approximately 4 times that of method 2. Compared to method 2, method 3 has a smaller relative error, indicating that encoder mode 3 has a good burr removal effect.

Table A2. Relative Error of Three Measurement Methods.

Measuring Method	Method 1	Method 2	Method 3
Data	8.56 ± 0.73 a	2.23 ± 0.21 b	0.19 ± 0.03 c

Note: The data in the table are described using mean and standard deviation, with different letters indicating significant differences at the 0.05 level.

Appendix A.2 Error Sources and Calibration of IMU

Due to production and processing reasons, IMUs are highly susceptible to factors such as installation, zero drift, and random noise. In addition to the aforementioned sources of error, scaling errors may also occur due to inconsistent physical parameters of components on magnetometers, accelerometers, and gyroscopes. In addition, sensors may experience certain drift and accumulate errors during long-term use. To improve measurement accuracy, it is necessary to calibrate the parameters of the geomagnetic pole and gyroscope before using them.

The accelerometer is in a stationary state, and each attitude is only affected by gravity. In three-dimensional space, the gravity points are all on a spherical surface. However, there may be deviations in the measurement units between the axes, and the gravity points of each attitude fall on an ellipsoid. The center of the ellipsoid is the offset of the acceleration, which is the calibration value. The magnetometer senses the same magnetic field intensity for each attitude in a constant environment, so there is no need to measure the magnetic field intensity in a stationary state. Instead, it is slowly rotated along the three axes, which ensures a more uniform distribution of the collected spatial attitude data from the magnetometer.

Due to the presence of errors, the measurement data of accelerometers and magnetometers will be distributed on an ellipsoid that is not centered around the zero point. The calibration process is the process of fitting the ellipsoid, and the model of the ellipsoid is as follows:

Because of the presence of errors, the measurement data of accelerometers and magnetometers will be distributed on an ellipsoid that is not centered around the zero point. The calibration process is the process of fitting the ellipsoid, and the model of the ellipsoid is as follows:

$$\left(\frac{x-x_0}{A_0}\right)^2+\left(\frac{y-y_0}{B_0}\right)^2+\left(\frac{z-z_0}{C_0}\right)^2=1 \tag{A1}$$

According to the ellipsoid fitting equation, there are six unknown variables, where (x_0, y_0, z_0) is the fitted ellipsoid center and also the zero bias on the accelerometer or gyroscope. (A_0, B_0, C_0) is the scaling factor on the three axes, and the above equation can be simplified into:

$$x^2+a_0y^2+b_0z^2+c_0x+d_0x+e_0x+f_0=0 \tag{A2}$$

The center of the ellipsoid and the scale factor of the required solution, as well as a_0, b_0, c_0, d_0, e_0 and f_0 have the following relationship:

$$\begin{aligned} x_0=-\frac{{c_0}^2}{2}, y_0=-\frac{d_0}{2a_0}, z_0=-\frac{e_0}{2b_0}, A_0=\frac{\sqrt{{c_0}^2+\frac{{d_0}^2}{a_0}+\frac{{e_0}^2}{b_0}+4f_0}}{2} \\ B_0=\frac{1}{2}\sqrt{\frac{{c_0}^2+4f_0}{a_0}+\frac{{d_0}^2}{{a_0}^2}+\frac{{e_0}^2}{a_0b_0}}, C_0=\frac{1}{2}\sqrt{\frac{{c_0}^2+4f_0}{b_0}+\frac{{d_0}^2}{a_0b_0}+\frac{{e_0}^2}{{b_0}^2}} \end{aligned} \tag{A3}$$

To obtain the values of a_0, b_0, c_0, d_0, e_0 and f_0, samples of 7 or more magnetometers and accelerometers need to be measured. By inputting their values into Equation (A3), the calibrated error parameters of the magnetometer and accelerometer can be obtained.

This study uses PyCharm 2022 and Python 3.8 programming to collect data from accelerometers in a stationary state and magnetometers in a moving state. We use the above equation to solve for their ellipsoidal center coordinates and ellipsoidal radius lengths. As shown in Table A3, the center and half-axis lengths of the ellipsoids of the accelerometer and magnetometers obtained are similar. Among them, the half-axis lengths of the three ellipsoids of the magnetometer are not significantly different, and the fitted ellipsoids obtained are basically consistent with those of the sphere (Figure A2b). The length of the *a* and *b* axes of the accelerometer in this IMU is basically the same. The length of the *c*-axis is about 10 times that of the first two. The fitted shape is similar to a rugby (Figure A2a). From this, it can be seen that due to the presence of bias and error, the ellipsoidal centers of the accelerometer and magnetometers are not on the original (0 g, 0 g, 0 g) and (0, 0, 0). It is necessary to transfer the centers of each ellipsoid to the original center for correction, normalize the magnetometer, and convert the coordinates of all axes to a sphere with a radius length of 72.1410. Meanwhile, normalize the a and b axes of the accelerometer while keeping the c axis unchanged.

Table A3. Ellipsoid calibration parameters.

Type	Elliptical Center Coordinates			Elliptical Half-Axis Length		
	x_0	y_0	z_0	a	b	c
Accelerometer	0.0092 g	−0.0158 g	1.0035 g	0.0017	0.0019	0.0117
Magnetometer	−2.4917	1.2728	0.6507	74.9957	72.1410	72.5667

As shown in Figure A2, the results of ellipsoidal fitting for accelerometer and magnetometers are shown. As mentioned earlier, one is similar to rugby, whereas the other is similar to a sphere. The ellipsoidal half-axis of the accelerometer is represented by a green line, whereas the ellipsoidal half-axis of the magnetometer is represented by a yellow line.

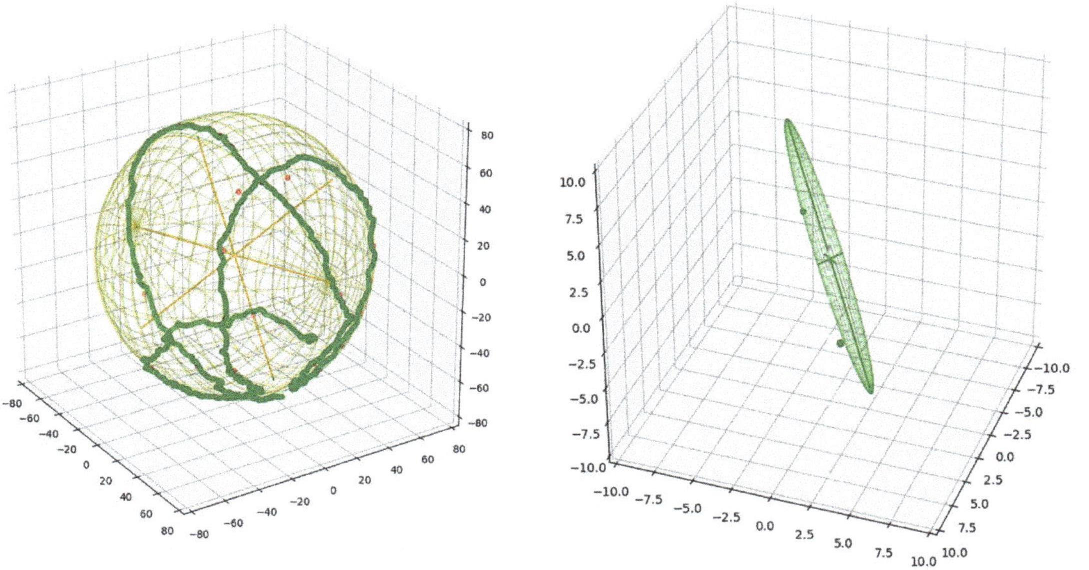

(a) (b)

Figure A2. Ellipsoid fitting results. (**a**) Ellipsoidal fitting results of the accelerometer. (**b**) Ellipsoidal fitting results of the magnetometer.

The measurement data on the three axes of the gyroscope will have zero bias due to the influence of power supply voltage and temperature. The existence of zero bias will lead to cumulative errors. This will result in a decrease in measurement accuracy, so it is necessary to calibrate the gyroscope. Place the IMU stationary on a horizontal plane, sample it, and take the average value of each axis to obtain the zero deviation on its axis.

$$\Delta\omega_x = \sum_{i=0}^{k} \frac{\omega_x{}^i}{k}, \Delta\omega_y = \sum_{i=0}^{k} \frac{\omega_y{}^i}{k}, \Delta\omega_z = \sum_{i=0}^{k} \frac{\omega_z{}^i}{k} \tag{A4}$$

By standing the IMU on flat outdoor ground for 1 min and using Python 3.8 programming to collect data, the entire process is repeated five times. The obtained values of $\Delta\omega_x$, $\Delta\omega_y$ and $\Delta\omega_z$ are -1.0139, 1.2950, and 0.000030371 rad/s, respectively.

Appendix B

Appendix B.1 A Positioning Method Based on EKF

We use the EKF algorithm to fuse the data measured by IMU and encoder, and obtain real-time position estimation, prediction, and measurement.

Appendix B.2 Prediction and Observation Equations Based on EKF

EKF is an optimal estimate for linear Gaussian systems. However, in many positioning problems, state transformation and measurement are both nonlinear. EKF takes a first-order approximation for this nonlinear system using Taylor expansion. Then they use EKF to complete state estimation. Assuming that the prediction and observation equations of ICV are both nonlinear systems, the EKF algorithm is as follows:

$$\begin{aligned} \mathbf{T}_t &= g(\boldsymbol{U}_t, \mathbf{T}_{t-1}) + \varepsilon_t \\ \mathbf{Z}_t &= h(\mathbf{T}_t) + \sigma_t \end{aligned} \tag{A5}$$

where, $\mathbf{T}_t$ and $\mathbf{T}_{t-1}$ represent the states of the ICV at t moment and $t-1$ moment, respectively; $\mathbf{Z}_t$ is the observed values of the ICV at t moment; ε_t and σ_t are respectively state (process) noise and observation (measurement) noise; Both of them follow a Gaussian distribution or normal distribution; $\boldsymbol{U}_t$ is the input control variables within $t-1$ moment to t moment; $g(.)$ and $h(.)$ are the state transition matrix and observation matrix, respectively.

Expanding $g(.)$ in Equation (A5) using Taylor equation at the mean $\boldsymbol{\lambda}_{t-1}$ of $t-1$ moment, taking a first-order approximation, we can obtain:

$$g(\boldsymbol{U}_t, \mathbf{T}_{t-1}) = g(\boldsymbol{U}_t, \boldsymbol{\lambda}_{t-1}) + \mathbf{G}_t(\mathbf{T}_{t-1} - \boldsymbol{\lambda}_{t-1}) \tag{A6}$$

Meanwhile, the $h(.)$ in Equation (A6) is expanded using the Taylor equation at the predicted mean $\overline{\boldsymbol{\lambda}}$, and taking a first-order approximation can obtain:

$$h(\mathbf{T}_t) = h(\overline{\boldsymbol{\lambda}}) + \mathbf{H}_t(\mathbf{T}_t - \overline{\boldsymbol{\lambda}}) \tag{A7}$$

In Equations (A6) and (A7), $\mathbf{G}_t$ and $\mathbf{H}_t$ are Jacobian matrices:

$$\mathbf{G}_t = \frac{\partial g(\boldsymbol{U}_t, \boldsymbol{\lambda}_{t-1})}{\partial \mathbf{T}_{t-1}} |\boldsymbol{U}_t, \boldsymbol{\lambda}_{t-1} \tag{A8}$$

$$\mathbf{H}_t = \frac{\partial h(\mathbf{T}_t)}{\partial \mathbf{T}_t} |\overline{\boldsymbol{\lambda}} \tag{A9}$$

Appendix B.3 Encoder and IMU Fusion Positioning Method Based on EKF

Based on the prediction and observation equations mentioned above and the principle of the EKF algorithm, the steps of the EKF algorithm using IMU and encoder are further obtained. The specific details are as follows:

The first step is to predict the motion state of ICV:

$$\overline{\boldsymbol{\lambda}}_t = \mathbf{g}(\mathbf{U}_t, \boldsymbol{\lambda}_{t-1}) \tag{A10}$$

$$\overline{\mathbf{F}}_t = \mathbf{G}_t\mathbf{F}_t\mathbf{G}_t{}^{\mathrm{T}} + \mathbf{Q}_t \tag{A11}$$

where, $\overline{\lambda}_t$ and $\overline{\mathbf{F}}_t$ represent the mean and variance of the state prediction; $\mathbf{Q}_t$ is process noise, it conforms to a normal distribution.

Simultaneously update the measured ICV motion status:

$$\mathbf{K}_t = \overline{\mathbf{F}}_t\mathbf{H}_t{}^{\mathrm{T}}\left(\mathbf{H}_t\overline{\mathbf{F}}_t\mathbf{H}_t{}^{\mathrm{T}} + \mathbf{R}_t\right)^{-1} \tag{A12}$$

$$\lambda_t = \overline{\lambda}_t + \mathbf{K}_t(\mathbf{Z}_t - \mathbf{h}(\overline{\lambda}_t)) \tag{A13}$$

$$\mathbf{F}_t = (\mathbf{I} - \mathbf{K}_t\mathbf{H}_t)\overline{\mathbf{F}}_t \tag{A14}$$

where, $\overline{\lambda}_t$ and $\overline{\mathbf{F}}_t$ represent the mean and variance of the state prediction; $\mathbf{Q}_t$ is process noise, it conforms to a normal distribution. $\mathbf{K}_t$ is the Kalman gain; $\mathbf{R}_t$ is measure noise; λ_t and $\mathbf{F}_t$ are the mean and variance of the EKF, respectively; $\mathbf{I}$ is unit matrices

In this study, the specific process of the measurement data fusion algorithm for two positioning sensors based on EKF will be described in the following. Assuming the motion state of ICV at t moment is $\boldsymbol{U}_t = (\omega, \mathrm{a_x}, \mathrm{a_y})$. The mean state at $t-1$ moment is λ_{t-1}, its state vector is Equation (1) in the formal manuscript. By incorporating the motion model Equation (2) into the ICV, the mean predicted state in Equation (A10) can be obtained as:

$$\begin{aligned}\overline{\lambda}_t = &\begin{bmatrix} 1 & 0 & 0 & \Delta t\cos\theta_{(t-1)} & \Delta t\sin\theta_{(t-1)} \\ 0 & 1 & 0 & \Delta t\sin\theta_{(t-1)} & -\Delta t\cos\theta_{(t-1)} \\ 0 & 0 & 1 & 0 & 0 \\ 0 & 0 & 0 & 1 & 0 \\ 0 & 0 & 0 & 0 & 1 \end{bmatrix}\lambda_{t-1} \\ &+ \begin{bmatrix} 0 & 0.5\Delta t^2\cos\theta_{(t-1)} & 0.5\Delta t^2\sin\theta_{(t-1)} \\ 0 & 0.5\Delta t^2\sin\theta_{(t-1)} & -0.5\Delta t^2\cos\theta_{(t-1)} \\ \Delta t & 0 & 0 \\ 0 & \Delta t & 0 \\ 0 & 0 & \Delta t \end{bmatrix}\boldsymbol{U}_t\end{aligned} \tag{A15}$$

By taking the derivative of Equation (2) in the formal manuscript over λ_{t-1} in sequence, the Jacobian matrix can be obtained as:

$$\mathbf{G}_t = \begin{bmatrix} 1 & 0 & \partial\theta_1 & \Delta t\cos\theta_{(t-1)} & \Delta t\sin\theta_{(t-1)} \\ 0 & 1 & \partial\theta_2 & \Delta t\sin\theta_{(t-1)} & -\Delta t\cos\theta_{(t-1)} \\ 0 & 0 & 1 & 0 & 0 \\ 0 & 0 & 0 & 1 & 0 \\ 0 & 0 & 0 & 0 & 1 \end{bmatrix} \tag{A16}$$

The expression for $\partial\theta_1$ and $\partial\theta_2$ are as follows:

$$\begin{aligned}\partial\theta_1 &= -v_{x(t-1)}\Delta t\sin\theta_{(t-1)} + v_{y(t-1)}\Delta t\cos\theta_{(t-1)} - 0.5a_x\Delta t^2\sin\theta_{(t-1)} + 0.5a_y\Delta t^2\cos\theta_{(t-1)} \\ \partial\theta_2 &= v_{x(t-1)}\Delta t\cos\theta_{(t-1)} + v_{y(t-1)}\Delta t\sin\theta_{(t-1)} + 0.5a_x\Delta t^2\cos\theta_{(t-1)} + 0.5a_y\Delta t^2\sin\theta_{(t-1)}\end{aligned} \tag{A17}$$

At this time, according to the EKF algorithm, the covariance ($\mathbf{F}_t$) of the predicted state of ICV can be obtained by incorporating Equation (2) and process noise into Equation (1). However, the measurement error of the two positioning sensors is relatively large. If one is used for prediction and the other is used for correction, there will still be significant deviation in the position and attitude of the fused positioning. Therefore, combining the motion model of ICV with its motion state as prediction, and using the data from both sensors (the average of both) as measurement models. When the measurement error of one sensor is large, choose other measurement data.

References

1. Chowdhury, M.; Thomas, E.V.; Jha, A.; Kushwah, A.; Kurmi, R.; Khura, T.K.; Sarkar, P.; Patra, K. An Automatic Pressure Control System for Precise Spray Pattern Analysis on Spray Patternator. *Comput. Electron. Agric.* **2023**, *214*, 108287. [CrossRef]
2. Rogers, H.; De La Iglesia, B.; Zebin, T.; Cielniak, G.; Magri, B. An Automated Precision Spraying Evaluation System. In Proceedings of the Lecture Notes in Computer Science (Including Subseries Lecture Notes in Artificial Intelligence and Lecture Notes in Bioinformatics), Cambridge, UK, 5–8 September 2023; Volume 14136 LNAI.
3. Kolendo, Ł.; Kozniewski, M.; Ksepko, M.; Chmur, S.; Neroj, B. Parameterization of the Individual Tree Detection Method Using Large Dataset from Ground Sample Plots and Airborne Laser Scanning for Stands Inventory in Coniferous Forest. *Remote Sens.* **2021**, *13*, 2753. [CrossRef]
4. Wu, H.; Wang, S. Design and Optimization of Intelligent Orchard Frost Prevention Machine under Low-Carbon Emission Reduction. *J. Clean. Prod.* **2023**, *433*, 139808. [CrossRef]
5. Li, L.; He, X.; Song, J.; Wang, X.; Jia, X.; Liu, C. Design and Experiment of Automatic Profiling Orchard Sprayer Based on Variable Air Volume and Flow Rate. *Nongye Gongcheng Xuebao/Trans. Chin. Soc. Agric. Eng.* **2017**, *33*, 70–76.
6. Qin, J.; Wang, W.; Mao, W.; Yuan, M.; Liu, H.; Ren, Z.; Shi, S.; Yang, F. Research on a Map-Based Cooperative Navigation System for Spraying–Dosing Robot Group. *Agronomy* **2022**, *12*, 3114. [CrossRef]
7. Gao, Z.-Q.; Zhao, C.-X.; Cheng, J.-J.; Zhang, X.-C. Tree Structure and 3-D Distribution of Radiation in Canopy of Apple Trees with Different Canopy Structures in China. *Chin. J. Eco-Agric.* **2012**, *20*, 63–68. [CrossRef]
8. Khemira, H.; Schrader, L.E.; Peryea, F.J.; Kammereck, R.; Burrows, R. Effect of Rootstock on Nitrogen and Water Use in Apple Trees. *HortScience* **2019**, *32*, 486A-486. [CrossRef]
9. Liu, L.; Wang, J.; Mao, W.; Shi, G.; Zhang, X.; Jiang, H. Canopy Information Acquisition Method of Fruit Trees Based on Fused Sensor Array. *Nongye Jixie Xuebao/Trans. Chin. Soc. Agric. Mach.* **2018**, *49*, 359.
10. Jiang, H.; Liu, L.; Liu, P.; Wang, J.; Zhang, X.; Gao, D. Online Calculation Method of Fruit Trees Canopy Volume for Precision Spray. *Nongye Jixie Xuebao/Trans. Chin. Soc. Agric. Mach.* **2019**, *50*, 120–129.
11. Lan, Y.; Yan, Y.; Wang, B.; Song, C.; Wang, G. Current Status and Future Development of the Key Technologies for Intelligent Pesticide Spraying Robots. *Nongye Gongcheng Xuebao/Trans. Chin. Soc. Agric. Eng.* **2022**, *38*, 30–40.
12. Saqib, N. Positioning—A Literature Review. *PSU Res. Rev.* **2021**, *5*, 141–169. [CrossRef]
13. Yao, H.; Liang, X.; Chen, R.; Wang, X.; Qi, H.; Chen, L.; Wang, Y. A Benchmark of Absolute and Relative Positioning Solutions in GNSS Denied Environments. *IEEE Internet Things J.* **2024**, *11*, 4243–4273. [CrossRef]
14. Perez-Ruiz, M.; Upadhyaya, S. GNSS in Precision Agricultural Operations. In *New Approach of Indoor and Outdoor Localization Systems*; Intech: Houston, TX, USA, 2012.
15. Jin, S.; Wang, Q.; Dardanelli, G. A Review on Multi-GNSS for Earth Observation and Emerging Applications. *Remote Sens.* **2022**, *14*, 3930. [CrossRef]
16. Perez-Ruiz, M.; Martínez-Guanter, J.; Upadhyaya, S.K. High-Precision GNSS for Agricultural Operations. In *GPS and GNSS Technology in Geosciences*; Elsevier: Amsterdam, The Netherlands, 2021.
17. de Ponte Müller, F. Survey on Ranging Sensors and Cooperative Techniques for Relative Positioning of Vehicles. *Sensors* **2017**, *17*, 271. [CrossRef]
18. Mohanty, A.; Wu, A.; Bhamidipati, S.; Gao, G. Precise Relative Positioning via Tight-Coupling of GPS Carrier Phase and Multiple UWBs. *IEEE Robot. Autom. Lett.* **2022**, *7*, 5757–5762. [CrossRef]
19. Teso-Fz-Betoño, D.; Zulueta, E.; Sanchez-Chica, A.; Fernandez-Gamiz, U.; Teso-Fz-Betoño, A.; Lopez-Guede, J.M. Neural Architecture Search for the Estimation of Relative Positioning of the Autonomous Mobile Robot. *Log. J. IGPL* **2023**, *31*, 634–647. [CrossRef]
20. Xia, Y.; Lei, X.; Pan, J.; Chen, L.W.; Zhang, Z.; Lyu, X. Research on Orchard Navigation Method Based on Fusion of 3D SLAM and Point Cloud Positioning. *Front. Plant Sci.* **2023**, *14*, 1207742. [CrossRef]
21. Chen, B.; Zhao, H.; Zhu, R.; Hu, Y. Marked-LIEO: Visual Marker-Aided LiDAR/IMU/Encoder Integrated Odometry. *Sensors* **2022**, *22*, 4749. [CrossRef]
22. Liu, H.; Duan, Y.; Shen, Y. Real-Time Navigation Method of Orchard Mobile Robot Based on Laser Radar Dual Source Information Fusion. *Nongye Jixie Xuebao/Trans. Chin. Soc. Agric. Mach.* **2023**, *54*, 249–258.
23. Chen, M.; Tang, Y.; Zou, X.; Huang, Z.; Zhou, H.; Chen, S. 3D Global Mapping of Large-Scale Unstructured Orchard Integrating Eye-in-Hand Stereo Vision and SLAM. *Comput. Electron. Agric.* **2021**, *187*, 106237. [CrossRef]
24. Liu, W.; He, X.; Liu, Y.; Wu, Z.; Yuan, C.; Liu, L.; Qi, P.; Li, T. Navigation Method between Rows for Orchard Based on 3D LiDAR. *Nongye Gongcheng Xuebao/Trans. Chin. Soc. Agric. Eng.* **2021**, *37*, 165–174.
25. Jiang, S.; Qi, P.; Han, L.; Liu, L.; Li, Y.; Huang, Z.; Liu, Y.; He, X. Navigation System for Orchard Spraying Robot Based on 3D LiDAR SLAM with NDT_ICP Point Cloud Registration. *Comput. Electron. Agric.* **2024**, *220*, 108870. [CrossRef]
26. Jiang, A.; Ahamed, T. Navigation of an Autonomous Spraying Robot for Orchard Operations Using LiDAR for Tree Trunk Detection. *Sensors* **2023**, *23*, 4808. [CrossRef] [PubMed]

27. Zhang, H.; Li, X.; Wang, L.; Liu, D.; Wang, S. Construction and Optimization of a Collaborative Harvesting System for Multiple Robotic Arms and an End-Picker in a Trellised Pear Orchard Environment. *Agronomy* **2024**, *14*, 80. [CrossRef]
28. Guevara, J.; Auat Cheein, F.A.; Gené-Mola, J.; Rosell-Polo, J.R.; Gregorio, E. Analyzing and Overcoming the Effects of GNSS Error on LiDAR Based Orchard Parameters Estimation. *Comput. Electron. Agric.* **2020**, *170*, 105255. [CrossRef]
29. Baltazar, A.R.; Santos, F.N.D.; De Sousa, M.L.; Moreira, A.P.; Cunha, J.B. 2D LiDAR-Based System for Canopy Sensing in Smart Spraying Applications. *IEEE Access* **2023**, *11*, 43583–43591. [CrossRef]
30. Jiang, S.; Wang, S.; Yi, Z.; Zhang, M.; Lv, X. Autonomous Navigation System of Greenhouse Mobile Robot Based on 3D Lidar and 2D Lidar SLAM. *Front. Plant Sci.* **2022**, *13*, 5281. [CrossRef] [PubMed]
31. Liu, L.; Liu, Y.; He, X.; Liu, W. Precision Variable-Rate Spraying Robot by Using Single 3D LIDAR in Orchards. *Agronomy* **2022**, *12*, 2509. [CrossRef]
32. Liu, L.; He, X.; Liu, W.; Liu, Z.; Han, H.; Li, Y. Autonomous Navigation and Automatic Target Spraying Robot for Orchards. *Smart Agric.* **2022**, *4*, 63.
33. Rivera, G.; Porras, R.; Florencia, R.; Sánchez-Solís, J.P. LiDAR Applications in Precision Agriculture for Cultivating Crops: A Review of Recent Advances. *Comput. Electron. Agric.* **2023**, *207*, 107737. [CrossRef]
34. Hu, F.; Wu, G. Distributed Error Correction of EKF Algorithm in Multi-Sensor Fusion Localization Model. *IEEE Access* **2020**, *8*, 93211–93218. [CrossRef]
35. Zheng, W.; Gong, G.; Tian, J.; Lu, S.; Wang, R.; Yin, Z.; Li, X.; Yin, L. Design of a Modified Transformer Architecture Based on Relative Position Coding. *Int. J. Comput. Intell. Syst.* **2023**, *16*, 168. [CrossRef]
36. Sagar, A.S.M.S.; Kim, T.; Park, S.; Lee, H.S.; Kim, H.S. Relative-Position Estimation Based on Loosely Coupled UWB–IMU Fusion for Wearable IoT Devices. *Comput. Mater. Contin.* **2023**, *75*, 1941–1961. [CrossRef]
37. Shaw, P.; Uszkoreit, J.; Vaswani, A. Self-Attention with Relative Position Representations. In Proceedings of the NAACL HLT 2018–2018 Conference of the North American Chapter of the Association for Computational Linguistics: Human Language Technologies—Proceedings of the Conference, New Orleans, LA, USA, 1–6 June 2018; Volume 2.
38. Cui, Y.; Zhang, Y.; Huang, Y.; Wang, Z.; Fu, H. Novel WiFi/MEMS Integrated Indoor Navigation System Based on Two-Stage EKF. *Micromachines* **2019**, *10*, 198. [CrossRef] [PubMed]
39. He, Y.; Wu, J.; Xie, G.; Hong, X.; Zhang, Y. Data-Driven Relative Position Detection Technology for High-Speed Maglev Train. *Measurement* **2021**, *180*, 109468. [CrossRef]
40. Xue, H.; Fu, H.; Dai, B. IMU-Aided High-Frequency Lidar Odometry for Autonomous Driving. *Appl. Sci.* **2019**, *9*, 1506. [CrossRef]
41. Petrović, I.; Sečnik, M.; Hočevar, M.; Berk, P. Vine Canopy Reconstruction and Assessment with Terrestrial Lidar and Aerial Imaging. *Remote Sens.* **2022**, *14*, 5894. [CrossRef]

agronomy

Article

Enhanced Estimation of Rice Leaf Nitrogen Content via the Integration of Hybrid Preferred Features and Deep Learning Methodologies

Yiping Peng [1,2,3,†], Wenliang Zhong [1,2,3,†], Zhiping Peng [1,2,3], Yuting Tu [1,2,3], Yanggui Xu [1,2,3], Zhuxian Li [1,2,3], Jianyi Liang [1,2,3], Jichuan Huang [1,2,3,*], Xu Liu [4] and Youqiang Fu [5]

1 Institute of Agricultural Resources and Environment, Guangdong Academy of Agricultural Sciences, Guangzhou 510640, China; pengyiping@gdaas.cn (Y.P.)
2 Key Laboratory of Plant Nutrition and Fertilizer in South Region, Ministry of Agriculture, Guangzhou 510640, China
3 Guangdong Key Laboratory of Nutrient Cycling and Farmland Conservation, Guangzhou 510640, China
4 Institute of Agricultural Economics and Information, Guangdong Academy of Agricultural Sciences, Guangzhou 510640, China
5 Rice Research Institute, Guangdong Academy of Agricultural Sciences, Guangzhou 510640, China
* Correspondence: huangjichuan@gdaas.cn
† These authors contributed equally to this work.

Citation: Peng, Y.; Zhong, W.; Peng, Z.; Tu, Y.; Xu, Y.; Li, Z.; Liang, J.; Huang, J.; Liu, X.; Fu, Y. Enhanced Estimation of Rice Leaf Nitrogen Content via the Integration of Hybrid Preferred Features and Deep Learning Methodologies. *Agronomy* **2024**, *14*, 1248. https://doi.org/10.3390/agronomy14061248

Academic Editor: Gniewko Niedbała

Received: 8 May 2024
Revised: 27 May 2024
Accepted: 6 June 2024
Published: 9 June 2024

Abstract: Efficiently obtaining leaf nitrogen content (LNC) in rice to monitor the nutritional health status is crucial in achieving precision fertilization on demand. Unmanned aerial vehicle (UAV)-based hyperspectral technology is an important tool for determining LNC. However, the intricate coupling between spectral information and nitrogen remains elusive. To address this, this study proposed an estimation method for LNC that integrates hybrid preferred features with deep learning modeling algorithms based on UAV hyperspectral imagery. The proposed approach leverages XGBoost, Pearson correlation coefficient (PCC), and a synergistic combination of both to identify the characteristic variables for LNC estimation. We then construct estimation models of LNC using statistical regression methods (partial least-squares regression (PLSR)) and machine learning algorithms (random forest (RF); deep neural networks (DNN)). The optimal model is utilized to map the spatial distribution of LNC at the field scale. The study was conducted at the National Agricultural Science and Technology Park, Guangzhou, located in Baiyun District of Guangdong, China. The results reveal that the combined PCC-XGBoost algorithm significantly enhances the accuracy of rice nitrogen inversion compared to the standalone screening approach. Notably, the model built with the DNN algorithm exhibits the highest predictive performance and demonstrates great potential in mapping the spatial distribution of LNC. This indicates the potential role of the proposed model in precision fertilization and the enhancement of nitrogen utilization efficiency in rice cultivation. The outcomes of this study offer a valuable reference for enhancing agricultural practices and sustainable crop management.

Keywords: UAV hyperspectral; leaf nitrogen content (LNC); feature optimization; deep learning

1. Introduction

Nitrogen serves as a vital nutrient in fostering the growth and maturation of rice plants. The suitable application of nitrogen fertilizer stands as a pivotal strategy in attaining both abundant yields and superior quality in rice crops [1]. According to statistics, approximately 200 million tons of nitrogen are applied globally to farmland, yet its utilization efficiency remains below 50%. The resultant nitrogen loss results in problems such as atmospheric and water contamination, biodiversity depletion, and the exacerbation of climate fluctuations [2,3]. Accurate estimations of nitrogen levels in rice leaves can be used to overcome these challenges, facilitating precise fertilization tailored to the plant's nutritional requirements. This can consequently reduce the volume of chemical fertilizers used, enhance their efficacy, and mitigate nitrogen loss [4]. However, conventional chemical testing methods are costly, time-consuming, and inherently lag behind real-time agricultural demands [5]. Unmanned aerial vehicle (UAV)-based hyperspectral technology can overcome these bottlenecks due to its exceptional spectral resolution, spatial precision, and detection sensitivity. Leveraging these attributes, UAV-based monitoring has emerged as a potent tool for gauging the nitrogen status of rice paddies. It could invert nitrogen information based on UAV spectral data and calculate the rice nitrogen deficit in combination with critical nitrogen concentration curve, effectively supporting dynamic nitrogen monitoring and optimizing the management of nitrogen fertilizers [6,7]. This significantly contributes to the management of nitrogen pollution in farmland, promoting a more sustainable and environmentally friendly agricultural practice.

Currently, estimations of rice nitrogen content based on UAV hyperspectral technology primarily concentrate on two key steps: the screening of characteristic variables and the optimization of estimation models. The former, which is crucial for accurately estimating leaf nitrogen content (LNC), can be categorized into two groups based on the methodological approach. The first approach screens the characteristic variables using the correlation between the variables and LNC. Methods such as the maximum information coefficient, Pearson correlation coefficient (PCC), and uninformative variables elimination are commonly employed for this approach. However, due to the inherent weakness of spectral information and the challenges in its acquisition, relying solely on statistical analysis methods can often result in an incomplete determination of optimal feature bands. The second approach centers on feature-importance-based variable screening. Algorithms such as random forest, gradient boosting decision tree, and extreme gradient boosting are examples of this category. These methods iteratively select a subset of features from the initial feature set, train the learner, evaluate the subset based on the learner's performance, and ultimately identify features with higher importance [8,9]. However, scholars have identified inconsistencies in selected wavebands when using various methods to estimate crop nitrogen content. The multi-method ensemble for wavebands selection may improve the definition of particular spectral regions in relation to specific absorption features, thereby increasing the reliability of the results, surpassing the capabilities of a single method [10,11]. In addition, given the intricate relationship between spectra and rice nitrogen, employing just one screening method may result in the omission of crucial information, thereby compromising inversion accuracy.

At present, the direct inversion method and vegetation index construction method are the two primary approaches used for rice nitrogen estimation based on UAV hyperspectral technology. The direct inversion method determines nitrogen information by establishing a relationship model between nitrogen content and spectral reflectance. For example, Yang et al. [12] utilized characteristic variables identified through principal component analysis (PCA) and employed the support vector machine (SVM) algorithm to construct a relationship model between nitrogen content and these variables. Their findings demonstrate the efficacy of the PCA-SVM method in assessing LNC. However, combining bands has been reported to provide more comprehensive information and enhance the accuracy of nitrogen estimations compared to the use of a single band. Consequently, scholars have established nitrogen estimation models through the construction of vegetation indices. This method is currently the most widely adopted approach. For example, Yu et al. [13] employed the successive projections algorithm to extract characteristic bands and proposed the nitrogen characteristic transfer index (NCTI) composed of these bands. The authors then constructed an LNC estimation model using the linear regression method, achieving an R^2 value of 0.774. Similarly, Wang et al. [8] integrated vegetation, color and texture indices, hyperspectral parameters, and machine learning algorithms to construct an estimation model of nitrogen concentration in rice stems and leaves, with R^2 values of 0.7 and 0.8, respectively. These studies offer valuable insights for nitrogen estimation in rice, yet the complex relationship between spectra and nitrogen requires the introduction of novel algorithms to enhance model accuracy.

The contribution of this work was to develop a novel method for LNC estimation in rice that combines the screening of characteristic variables with model optimization. In this method, spectral variables were screened using a hybrid method involving PCC and the extreme gradient boosting (XGBoost) algorithm. A deep neural network algorithm was subsequently employed to establish a model capturing the relationship between characteristic variables and LNC. This model was then compared against commonly employed methods such as partial least squares regression (PLSR), and random forest (RF). The optimal model was subsequently utilized to map the spatial distribution of LNC.

2. Materials and Methods

2.1. Field Experimental Design

The experiment was performed at the National Agricultural Science and Technology Park, Guangzhou, located in Baiyun District of Guangdong, China (23°23′38″ N, 113°25′37″ E). This area has a subtropical monsoon climate, with an average annual temperature of approximately 24 °C. We selected the conventional rice variety Meixiangzhan No. 2 as the experimental material, with an average plant spacing of 25 cm. The experiment comprised six nitrogen fertilizer gradients, namely, N1 (0 kg/hm^2), N2 (37.5 kg/hm^2), N3 (75.0 kg/hm^2), N4 (112.5 kg/hm^2), N5 (150.0 kg/hm^2), and N6 (187.5 kg/hm^2). Each gradient was replicated five times, resulting in a total of 30 experimental blocks. The phosphorus and potassium fertilizer application followed the local standard rates. Figure 1 presents the specific experimental layout. The area of each block was approximately 19.0 m^2 for No. 1–6 and 22.5 m^2 for No. 7–30.

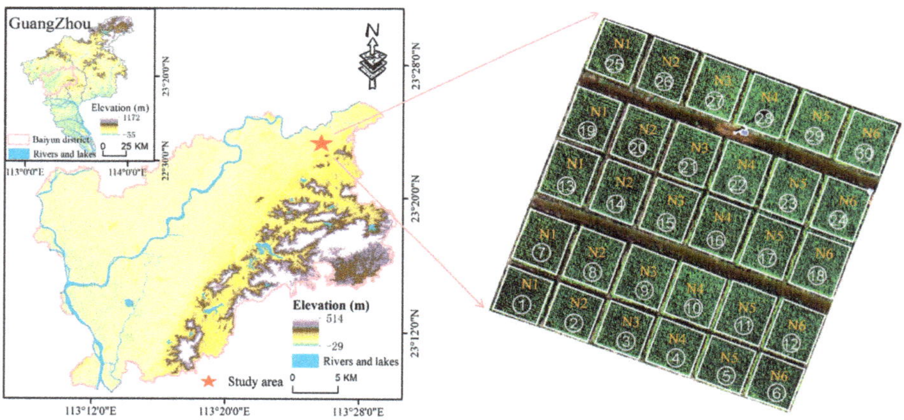

Figure 1. Layout of the experimental blocks.

2.2. Data and Pre-Processing

2.2.1. Collection and Pre-Processing of UAV Hyperspectral Imagery

Hyperspectral images were collected during the tillering (3 September 2022), jointing-to-booting (27 September 2022), and heading-to-flowering (21 October 2022) stages of rice using a UAV-based hyperspectral imaging system (Gaiasky mini3-VN, Dualix Spectral Image Technology Co. Ltd., Wuxi, China). This system offers a spectral range spanning from 400 to 1000 nm, with a spectral resolution of 5 nm and a total of 224 bands (Figure 2). Data collection took place on days of clear weather, with the flight altitude maintained at 50 m. To ensure data accuracy, calibration was conducted with a standard white plate and standard gray cloth prior to flights. The acquired images were pre-processed using SpecView 2.9.3.10, HySpectralStitcher 1.0.1, and ENVI 5.6. Pre-processing steps included geometric correction, radiometric correction, image stitching, clipping, etc. We computed the average of the spectra in the region of interest (ROI) for each block. Gauss filtering was then employed to denoise the spectral data, yielding canopy spectral reflectance data for each block (Table 1).

Parameters	Values
Wavelength	400-1000nm
Channels	448
Lens	16mm
Sampling rate	1.35nm
Spectral resolution	5nm @32μm slit
Pixel pitch	16(μm)×8(μm)
Weight	1.2kg
Power	DC 12, 45W
Carrying platform	DJI M350
Digitization	12bit
Field of view	35°@16mm
Spatial resolution	0.187(@16mm, height=300m)

Figure 2. UAV hyperspectral imaging system and parameters used in the experiment.

Table 1. Basic information of UAV hyperspectral images.

Instrumentation	Flight Altitude	Block Size	Spectral Range	Forward Overlap	Side Overlap
Gaiasky mini 3-VN	50 m	No.1–6: 19.0 m^2 No. 7–30: 22.5 m^2	400–1000 nm	80%	65%

2.2.2. Acquisition of Rice Agronomic Parameters

Rice plant samples were gathered from five points in an "X" shape (Figure 3) at each block and the acquisition time coincided with the collection of the UAV hyperspectral images. The collected samples were then washed and the fresh leaves of each rice plant were weighed, labeled, and placed in bags for oven-drying at 105 °C for 30 min. The temperature was then adjusted to 80 °C for drying until a constant weight was achieved. After drying, the samples were crushed using a pulverizer through a 0.25 mm sieve. The Kjeldahl method was subsequently employed to determine the LNC. Table 2 reports the statistical results. The LNC was observed to progressively decrease with rice growth. The mean LNC during the tillering, jointing-to-booting, and heading-to-flowering stages was 2.63%, 2.03%, and 1.69%, respectively. Moreover, LNC exhibited the highest standard deviation (0.44) and coefficient of variation (16.90%) at the tillering stage, indicating greater data variability.

Table 2. Statistical parameters of leaf nitrogen contents from the rice plant samples (unit: %).

Growth Stages	Minimum	Maximum	Mean	SD	CV (%)
tillering	2.02	3.66	2.63	0.44	16.90
jointing-to-booting	1.69	2.55	2.03	0.23	11.56
heading-to-flowering	1.42	2.06	1.69	0.18	10.38

Note: SD, standard deviation; CV, coefficient of variation.

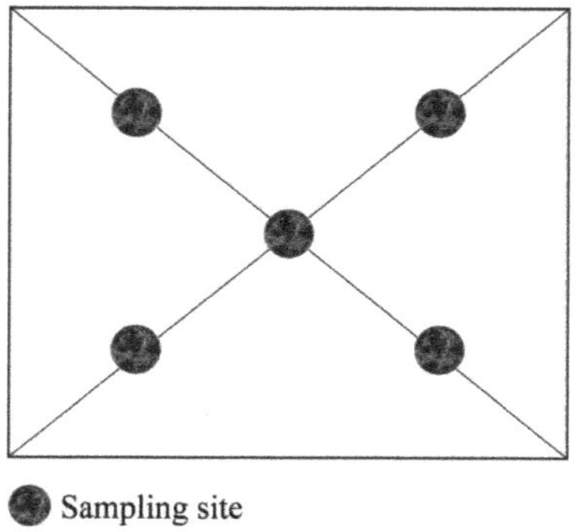

Figure 3. Sampling site schematic.

2.3. *Vegetation Index Construction*

Vegetation indices (VIs) serve as valuable indicators of crop health and growth, reflecting the nutritional status of crops [14]. Drawing from the relevant studies [15–19], 20 VIs were selected as variables for estimating the LNC (Table 3).

Table 3. Definitions and formulae of the selected vegetation indices.

VI	Formula	Reference	VI	Formula	Reference
NPCI	$(R_{670} - R_{460})/(R_{670} + R_{460})$	[15]	SIPI	$(R_{800} - R_{445})/(R_{800} - R_{680})$	[16]
SR	R_{750}/R_{550}	[15]	PSRI	$(R_{680} - R_{500})/R_{750}$	[17]
MSR	$(R_{800}/R_{760} - 1)/(R_{800}/R_{670} + 1)^{0.5}$	[15]	GI	R_{554}/R_{677}	[17]
PBI	R_{810}/R_{560}	[15]	PSND	$(R_{800} - R_{470})/(R_{800} + R_{470})$	[17]
LCI	$(R_{850} - R_{710})/(R_{850} + R_{680})$		PSSR	R_{800}/R_{500}	[17]
GNDVI	$(R_{750} - R_{550})/(R_{750} + R_{550})$	[16]	RARS	R_{760}/R_{500}	[17]
SRPI	R_{430}/R_{680}	[16]	OSAVI	$1.16 \times (R_{800} - R_{670})/(R_{800} + R_{670} + 0.16)$	[17]
PRI	$(R_{570} - R_{531})/(R_{570} + R_{531})$	[16]	RENDVI	$(R_{750} - R_{705})/(R_{750} + R_{705})$	[18]
MTCI	$(R_{750} - R_{710})/(R_{710} - R_{680})$	[16]	DCNI	$(R_{720} - R_{700})/(R_{700} - R_{670})/(R_{720} - R_{670} + 0.03)$	[19]
NDRE	$(R_{790} - R_{720})/(R_{790} + R_{720})$	[16]	NDVI	$(R_{800} - R_{670})/(R_{800} + R_{670})$	[19]

Note: Rx denotes the spectral reflectance at wavelength x nm; NPCI, normalized pigment chlorophyll ratio index; SR, simple ratio vegetation index; MSR, modified simple ratio; PBI, plant biochemical index; LCI, leaf chlorophyll index; GNDVI, green normalized difference vegetation index; SRPI, simple ratio pigment index; PRI, photochemical reflectance index; MTCI, MERIS terrestrial chlorophyll index; NDRE, normalized difference red edge index; SIPI, structure intensive pigment index; PSRI, plant senescence reflectance index; GI, greenness index; PSND, pigment-specific normalized difference; PSSR, pigment-specific simple ratio; RARS, ratio analysis of reflectance spectra; OSAVI, optimized soil adjusted vegetation index; NDVI, normalized difference vegetation index; RENDVI, red edge NDVI; DCNI, double-peak canopy nitrogen index.

2.4. *Identification of Characteristic Variables for LNC Estimation*

The determination of the characteristic variables is a critical step in the development of the LNC estimation method [6]. However, conventional statistical analysis methods often struggle to precisely determine the optimal characteristic variables due to the inherent challenges in capturing weak spectral information. Moreover, the relationship between spectral data and agronomic parameters is typically complex, encompassing a combination of linear and nonlinear associations [20]. In this study, we employ a hybrid approach integrating correlation analysis with XGBoost to screen characteristic variables for estimating LNC and compare it with the equivalent singular method. These methods are briefly described in the following sections:

- Pearson correlation coefficient

A significant correlation has been reported between the spectral reflectance of certain bands and the LNC [6]. In this study, bands exhibiting high correlation coefficients with statistical significance at $p \leq 0.01$ were identified as characteristic variables using the Pearson correlation coefficient, r_i, determined as [21]:

$$r_i = \frac{\sum_{n=1}^{N}\left(R_{ni} - \overline{R_i}\right)(y_n - \overline{y})}{\sqrt{\sum_{n=1}^{N}\left(R_{ni} - \overline{R_i}\right)^2 \sum_{n=1}^{N}(y_n - \overline{y})^2}} \quad (1)$$

where r_i is the correlation coefficient between LNC and spectral reflectance; R_{ni} is the spectral reflectance of the n-th sample in the i-th band; $\overline{R_i}$ is the mean reflectance of samples in the i-th band; y_n is the LNC of n-th sample; and $\overline{y}$ is the mean value of LNC.

- Extreme gradient boosting

We employed the XGBoost algorithm to select the optimal characteristic variables for LNC estimation. XGBoost is a modified gradient-boosting algorithm that averages and subsequently ranks the feature importance (FI) of each tree. It employs the following calculation method [22]:

$$\mathrm{FI(T,F)} = \mathrm{H(T)} - \mathrm{H(T|F)} = -\sum_{i=1}^{j} \mathrm{p}_i \log_2 \mathrm{p}_i - \sum_{\mathrm{F}} \mathrm{p(F)} \times \sum_{i=1}^{j} \mathrm{p}(i|\mathrm{F}) \log_2 \mathrm{p}(i|\mathrm{F}), \quad (2)$$

where H(T) and H(T|F) refer to the entropy of the parent and child nodes based on the F-feature segmentation, respectively, and p_i represents the score of the labeled samples at the i-th node.

2.5. Model Construction and Validation

Three algorithms, namely partial least squares regression (PLSR), random forest (RF), and deep neural network (DNN), were employed to develop the LNC estimation model using the selected characteristic variables. These algorithms were created using Python software (version 3.10). In the following, we present a summary of each algorithm.

- Partial least squares regression

Partial least-squares regression (PLSR) integrates the strengths of three analytical techniques: principal component analysis, canonical correlation analysis, and multiple linear regression. It leverages all available data to construct a model, extracting maximal information that reflects data variation. PLSR demonstrates a strong predictive capability and is particularly effective at handling datasets with strong linear correlations among variables [23]. This method includes a principal component analysis of both the spectral data matrix and the matrix of rice leaf content during modeling, resulting in a regression model based on the contributions of the derived variables.

- Random forest

Random forest (RF) is an ensemble learning algorithm proficient in effectively modeling the nonlinear relationship between the characteristic and explanatory variables. It boasts a strong generalization ability and robustness and a rapid training speed, and only requires a minimal number of tuning parameters. The model employs a bootstrap strategy to generate new samples of equal size from the original data. Two-thirds of the original sample are typically utilized to construct decision trees. The remaining one-third serves as out-of-bag data (OOB) for inner cross-validation to evaluate the estimated accuracy of RF. The results from all decision trees are aggregated and their average serves as the final prediction outcome [24].

- Deep neural network

Deep neural networks (DNNs) are advanced neural network architectures characterized by multi-layered structures comprising input layers, multiple hidden layers, and output layers. Each layer is interconnected, with connections existing between nodes in adjacent layers. However, nodes within the same layer and across layers are not interconnected. Through fully connected layers and combinations of activation functions, various nodes are linked to construct a sophisticated multi-layer neural network model [25]. Figure 4 depicts the structural layout of DNNs.

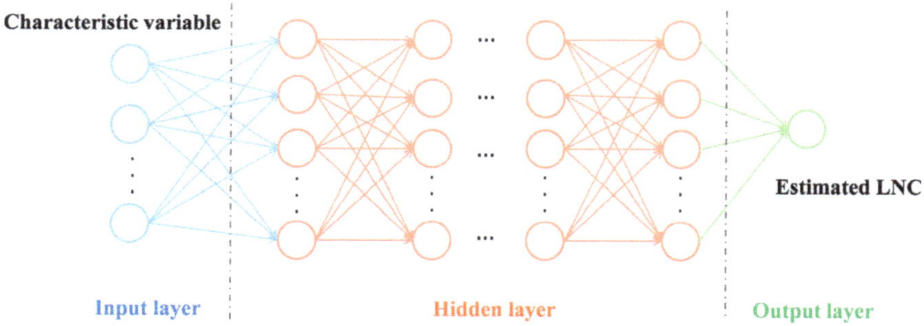

Figure 4. Structure of deep neural networks.

DNNs undergo training via forward and backward propagation algorithms, continually updating each weight (w) and bias (b) within the network. The backward propagation algorithm, pivotal to DNN functionality, retroactively propagates the error between the estimated and measured output values. It iteratively adjusts the w and b across each layer of the network model, aiming to minimize the error until it aligns with specified accuracy criteria [26]. The weight and bias are calculated as follows [25]:

$$w^* = w - \alpha \frac{\alpha j(w,b)}{\alpha w} \tag{3}$$

$$b^* = b - \alpha \frac{\alpha j(w,b)}{\alpha b} \tag{4}$$

where w^* and w denote the weights after and before the update, respectively; b^* and b represent the biases after and before the update, respectively; α is the learning rate; and $j(w,b)$ is the loss function of the model. The optimal parameters w^* and w are attained through iterative gradient descent. In this study, a DNN comprising three hidden layers was constructed. The number of nodes in the input layer corresponded to the number of characteristic variables. We adopted tanh as the activation function. The network was trained for 3000 iterations and the mean squared error (MSE) was adopted as the Loss function to assess the performance.

- Accuracy verification

The collected data (90 samples) were divided into a training set (60 samples) and a test set (30 samples). The test set was utilized to evaluate the performance of the LNC estimation models by computing the coefficient of determination (R^2) and root mean squared error (RMSE). Models with a higher R^2 and lower RMSE are considered to have a better estimation effect [27].

$$R^2 = 1 - \frac{\sum_{i=1}^{n} (y_i - \hat{y}_i)^2}{\sum_{i=1}^{n} (y_i - \bar{y})^2} \tag{5}$$

$$RMSE = \sqrt{\frac{\sum_{i=1}^{n} (\hat{y}_i - y_i)^2}{n}} \tag{6}$$

where y_i and $\hat{y}_i$ are the measured and estimated values, respectively; $\bar{y}$ is the mean of the measured values; and n signifies the sample number.

3. Results

3.1. Determining the Characteristic Variables for the Estimation of LNC

3.1.1. Spectral Characteristics of Rice at Different Growth Stages

Figure 5 presents the spectral curves of rice at various growth stages under different nitrogen (N) application levels. The rice canopy spectra exhibit low reflectance ($r < 0.1$) within the 400–670 nm range, followed by a sharp increase in reflectance within the red-edge range (670–760 nm). Notably, marked differences in spectra are observed across varying nitrogen levels within the 750–1000 nm range, with nitrogen fertilizer application demonstrating a positive correlation with rice spectral reflectance. This phenomenon may arise from an optimal nitrogen supply, which enhances rice growth and thereby results in increased reflectance levels.

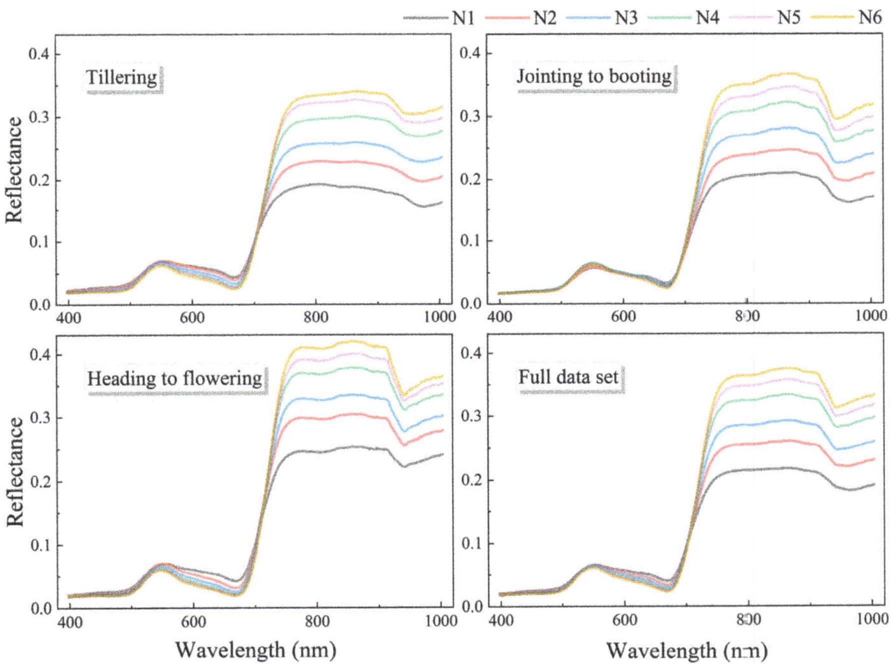

Figure 5. Spectral curves of rice at different growth stages.

3.1.2. Screening of Characteristic Variables for LNC Estimation

The PCC and XGBoost algorithms were employed to identify the characteristic variables for estimating LNC. Following numerous experiments, the screening criteria for PCC and XGBoost were established as $|r| > 0.76$ and $FI > 0.01$, respectively. Stepwise regression was then utilized to mitigate multicollinearity among the selected characteristic variables (Figure 6). Note that the characteristic variables identified by both algorithms exhibited some degree of overlap ($band_{750.92}$, PSSR, and RARS).

3.2. Estimation and Accuracy Assessment of LNC

The characteristic variables obtained solely from PCC (Figure 6b), solely from XGBoost (Figure 6a), and from their combination ($band_{756.40}$, $band_{770.11}$, $band_{742.71}$, $band_{569.61}$, $band_{731.79}$, $band_{671.94}$, $band_{463.34}$, $band_{495.06}$, $band_{680.08}$, $band_{880.59}$, $band_{500.37}$, $band_{750.92}$, PBI, LCI, PSSR, and RARS) were utilized as independent variables, respectively, with LNC serving as the dependent variable to construct the estimation models of LNC using PLSR, RF, and DNN algorithms (Figure 7). For the linear models (PLSR), the model based on the characteristic variables derived from the linear screening algorithm (PCC) exhibits a scatter distribution closer to the 1:1 line compared to models built with characteristic variables from the XGBoost algorithm, indicating a superior estimation performance of the former. Conversely, for nonlinear models (RF, DNN), the model constructed using the characteristic variables from XGBoost demonstrates a better performance. Furthermore, models based on the characteristic variables obtained from the combination of PCC and XGBoost consistently yield improved estimation results ($R^2 > 0.8$). Among these, the DNN model exhibits the best performance for estimating the LNC, with a validation R^2 of 0.89 and an RMSE of 0.17%.

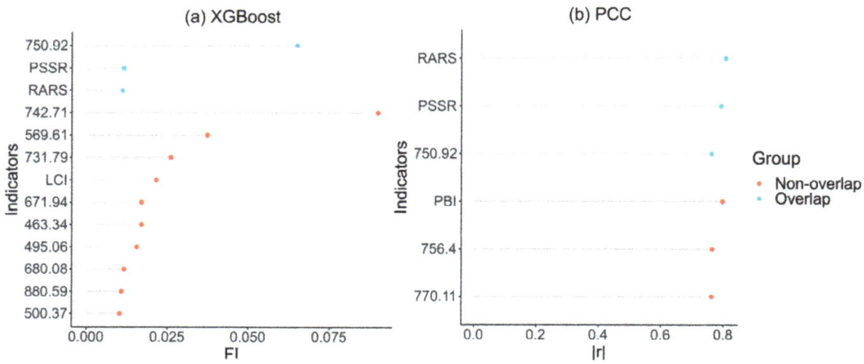

Figure 6. FI and | r | of the selected characteristic variables for estimating LNC.

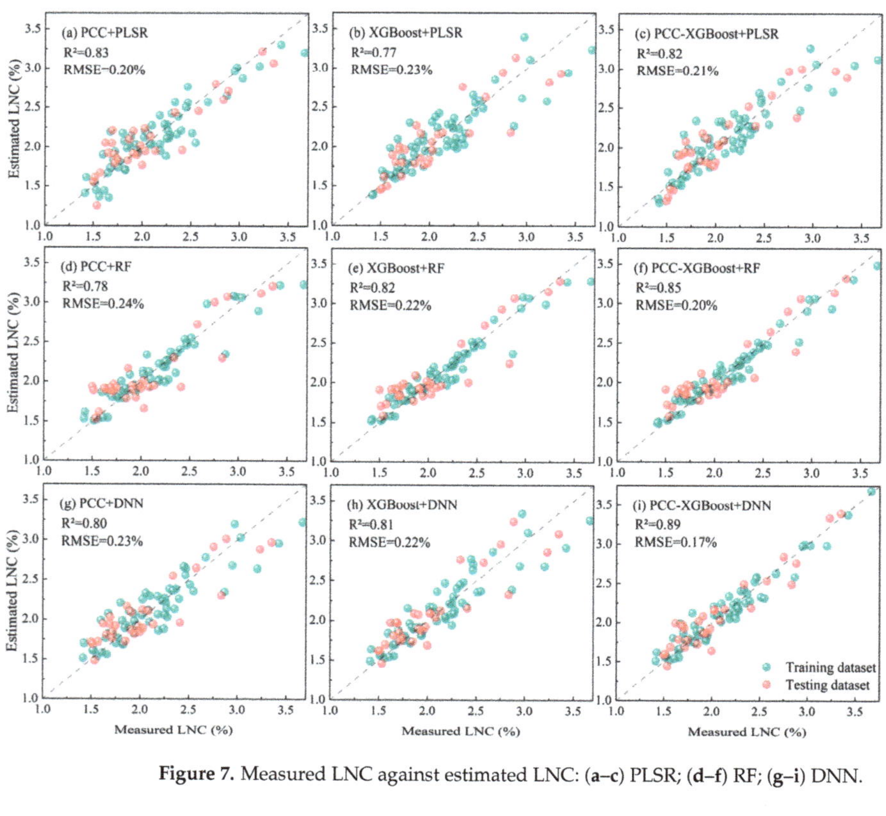

Figure 7. Measured LNC against estimated LNC: (**a**–**c**) PLSR; (**d**–**f**) RF; (**g**–**i**) DNN.

3.3. Mapping LNC at the Field Scale

Previous research has indicated that nitrogen-monitoring models based on the entire growth period are applicable for nitrogen monitoring across all growth stages [8]. Therefore, we adopted the optimal model (PCC-XGBoost-DNN) to map the spatial distribution of LNC during various growth stages at the field scale (Figure 8). LNC was higher during the tillering stage compared to the jointing-to-booting and heading-to-flowering stages. Moreover, LNC exhibited an increasing trend corresponding to fertilizer application, which is consistent with the distribution pattern of the sample data. These findings offer valuable insights for subsequent precision fertilization strategies in rice cultivation.

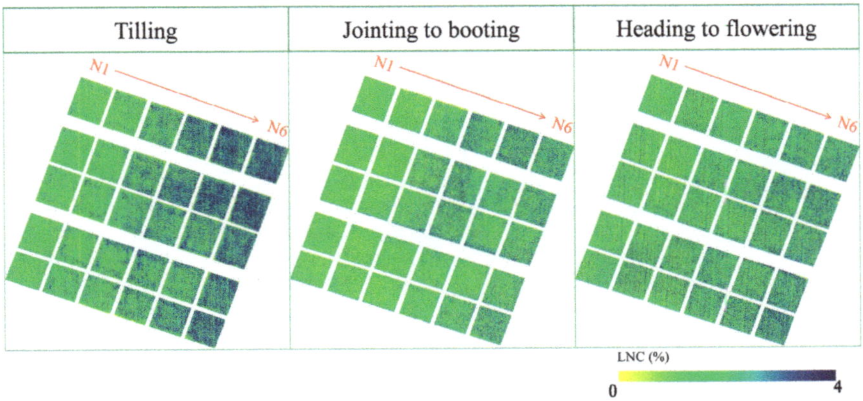

Figure 8. Spatial distribution of LNC.

4. Discussion

As the global population continues to expand, the world faces significant challenges in mitigating the environmental repercussions of excessive nitrogen inputs while striving to maintain high crop yields [28]. China's annual nitrogen fertilizer input for rice cultivation is reported to reach 6.3 million tons, representing approximately one-third of the world's total rice nitrogen fertilizer consumption. However, nitrogen utilization efficiency stands at less than 50%, resulting in a loss of over half of the applied nitrogen to the environment. This phenomenon adversely impacts air and water quality, posing threats to environmental safety and public health [29,30]. Given the pivotal role of nitrogen information in assessing rice growth status and yield, real-time access to such data serves as a crucial reference for optimizing nitrogen fertilizer application and enhancing nitrogen utilization efficiency [31]. UAV hyperspectral technology has proven to be a vital tool for obtaining this information [6].

Scholars have achieved successful LNC estimations using both linear [32] and nonlinear [20,33] screening algorithms. This prompted us to speculate on the existence of a blending relationship (linear and nonlinear) between nitrogen and the corresponding spectral information. Thus, we conducted an experiment to validate this hypothesis. We combined a correlation analysis with XGBoost to screen characteristic variables and compared the results with those obtained using a single-method approach. Our findings indicate that the feature variables identified through the combined correlation analysis and XGBoost algorithm significantly enhance rice nitrogen inversion, offering a novel approach to bolstering the accuracy of rapid crop nutrition diagnosis. This hints at the potential of the multi-method ensemble selection of characteristic variables in regression problems, which is consistent with former research [10,11]. Furthermore, among all the tested models, the DNN model emerged as the most effective in terms of LNC estimation. This may be attributed to the DNN's inherent nonlinear mapping ability and its resilience to noisy data, enabling it to handle complex datasets better than shallow modeling approaches. The study offers a valuable reference for diagnosing the nutritional status and optimizing nitrogen fertilizer management in rice, which plays a crucial role in ensuring the sustainability of crop production.

In this study, we only utilized the cultivar 'Meixiangzhan 2' as the test material. Thus, further work is required to ascertain the method's applicability to other rice varieties. In addition, South China belongs to the typical oceanic subtropical monsoon climate influence zone, with frequent cloudy and rainy weather. Due to constraints such as weather conditions and collection costs, the dataset utilized in this study remains relatively limited, potentially introducing some uncertainty into the results. To address this limitation, future research will explore the integration of data generation algorithms such as generative adversarial networks (GAN) to augment the dataset and enhance model robustness. Furthermore, this study solely focuses on estimating LNC. Subsequent research will integrate spatial nitrogen distribution data, expert insights, and regional soil characteristics to formulate a prescription map for rice fertilizer application. This holistic approach aims to facilitate precise fertilizer application, thereby enhancing fertilizer reductions and efficiency.

5. Conclusions

The precise determination of pertinent characteristic variables is key to developing accurate estimation models for LNC. In this study, we propose a novel approach that combines the strengths of PCC and the XGBoost algorithm to comprehensively screen for the optimal characteristic variables for LNC estimation. Utilizing these selected variables and field observations of LNC, we subsequently developed LNC estimation models. The most accurate estimation model was then adopted to explore the possibility of spatially mapping the LNC at the field scale. The results reveal that the integration of PCC and XGBoost enables a more meticulous screening of characteristic variables than solely adopting single-class methods. Furthermore, based on the RMSE values derived from test datasets, the DNN was determined as the most accurate model at the sample-point level. The proposed method offers the potential to map the LNC at the field scale using UAV hyperspectral imagery. The estimation accuracy of LNC was enhanced by combining PCC and XGBoost for characteristic variables screening, offering insights into optimizing nitrogen fertilizer's application in rice cultivation.

Author Contributions: Conceptualization, Y.P. and Z.P.; methodology, Y.P. and W.Z.; software, Y.P. and W.Z.; validation, Y.P., Z.L. and J.L.; investigation, Z.P., Y.T., X.L. and Y.X.; resources, Z.P. and J.H.; writing—original draft preparation, Y.P. and W.Z.; writing—review and editing, Y.P., Y.T., Z.L., Y.X., J.L., Y.F., X.L., J.H. and W.Z.; funding acquisition, J.H. All authors have read and agreed to the published version of the manuscript.

Funding: This research was funded by The Collaborative Innovation Center Project of Guangdong Academy of Agricultural Science (XTXM202201), Key Technologies R&D Program of Guangdong Province (2023B0202010027) and Guangdong Rural Science and Technology Commissioner Project (KTP20210417, KTP20210335).

Data Availability Statement: Data are contained within the article.

Conflicts of Interest: The authors declare no conflicts of interest.

References

1. Li, D.; Chen, J.M.; Yan, Y.; Zheng, H.B.; Yao, X.; Zhu, Y.; Cao, W.X.; Cheng, T. Estimating leaf nitrogen content by coupling a nitrogen allocation model with canopy reflectance. *Remote Sens. Environ.* **2022**, *283*, 113314. [CrossRef]
2. Zhang, X.; Zou, T.; Lassaletta, L.; Mueller, N.D.; Tubiello, F.N.; Lisk, M.D.; Lu, C.Q.; Conant, R.T.; Dorich, C.D.; Gerber, J.; et al. Quantification of global and national nitrogen budgets for crop production. *Nat. Food* **2021**, *2*, 529–540. [CrossRef] [PubMed]
3. Gu, B.J.; Zhang, X.M.; Lam, S.K.; Yu, Y.L.; Grinsven, H.J.M.; Zhang, S.H.; Wang, X.X.; Bodirsky, B.L.; Wang, S.T.; Duan, J.K.; et al. Cost-effective mitigation of nitrogen pollution from global croplands. *Nature* **2023**, *613*, 77–84. [CrossRef] [PubMed]
4. Xu, T.Y.; Jin, Z.Y.; Guo, Z.H.; Yang, L.; Bai, J.C.; Feng, S.; Yu, F.H. Simultaneous inversion method of nitrogen and phosphorus contents in rice leaves using CARS-RUN-ELM algorithm. *Trans. CSAE* **2022**, *38*, 148–155.

5. Liu, H.Y.; Zhu, H.C.; Li, Z.H.; Yang, G.J. Quantitative analysis and hyperspectral remote sensing of the nitrogen nutrition index in winter wheat. *Int. J. Remote Sens.* **2020**, *41*, 858–881. [CrossRef]
6. Fu, Y.Y.; Yang, G.J.; Pu, R.L.; Li, Z.H.; Li, H.L.; Xu, X.G.; Song, X.Y.; Yang, X.D.; Zhao, C.J. An overview of crop nitrogen status assessment using hyperspectral remote sensing: Current status and perspectives. *Eur. J. Agron.* **2021**, *124*, 126241. [CrossRef]
7. Yu, F.H.; Bai, J.C.; Jin, Z.Y.; Guo, Z.H.; Yang, J.X.; Chen, C.L. Combining the critical nitrogen concentration and machine learning algorithms to estimate nitrogen deficiency in rice from UAV hyperspectral data. *J. Integr. Agric.* **2023**, *22*, 1216–1229. [CrossRef]
8. Wang, D.L.; Li, R.; Liu, T.; Liu, S.P.; Sun, C.M.; Guo, W.S. Combining vegetation, color, and texture indices with hyperspectral parameters using machine-learning methods to estimate nitrogen concentration in rice stems and leaves. *Field Crops Res.* **2023**, *304*, 109175. [CrossRef]
9. Feng, S.; Xu, T.Y.; Yu, F.H.; Chen, C.L.; Yang, X.; Wang, N.Y. Research of method for inverting nitrogen content in canopy leaves of japonica rice in northeastern China based on hyperspectral remote sensing of unmanned aerial vehicle. *Spectrosc. Spectr. Anal.* **2019**, *39*, 3281–3287.
10. Feilhauer, H.; Asner, G.P.; Martin, R.E. Multi-method ensemble selection of spectral bands related to leaf biochemistry. *Remote Sens. Environ.* **2015**, *164*, 57–65. [CrossRef]
11. Fu, Y.Y.; Yang, G.J.; Li, Z.H.; Li, H.L.; Li, Z.H.; Xu, X.G.; Song, X.Y.; Zhang, Y.H.; Duan, D.D.; Zhao, C.J.; et al. Progress of hyperspectral data processing and modelling for cereal crop nitrogen monitoring. *Comput. Electron. Agric.* **2020**, *172*, 105321. [CrossRef]
12. Yang, H.Y.; Zhou, Q.; Yang, J.; Sun, Y.T.; Lu, Y.; Yin, H. Study on nitrogen nutrition diagnosis of rice leaves based on hyperspectrum. *Acta Agric. Zhejiangensis* **2019**, *31*, 1575–1582.
13. Yu, F.H.; Xing, S.M.; Guo, Z.H.; Bai, J.C.; Xu, T.Y. Remote sensing inversion of the nitrogen content in rice leaves using character transfer vegetation index. *Trans. CSAE* **2022**, *38*, 175–182.
14. Zeng, Y.L.; Hao, D.L.; Huete, A.; Dechant, B.; Berry, J.; Chen, J.M.; Joiner, J.; Frankenberg, C.; Bond-Lamberty, B.; Ryu, Y.; et al. Optical vegetation indices for monitoring terrestrial ecosystems globally. *Nat. Rev. Earth Environ.* **2022**, *3*, 477–493. [CrossRef]
15. Feng, H.K.; Fan, Y.G.; Tao, H.L.; Yang, F.Q.; Yang, G.J.; Zhao, C.J. Monitoring of nitrogen content in winter wheat based on UAV hyperspectral imagery. *Spectrosc. Spectr. Anal.* **2023**, *43*, 3239–3246.
16. Tian, Y.C.; Yao, X.; Yang, J.; Cao, W.X.; Hannaway, D.B.; Zhu, Y. Assessing newly developed and published vegetation indices for estimating rice leaf nitrogen concentration with ground- and space-based hyperspectral reflectance. *Field Crops Res.* **2011**, *120*, 299–310. [CrossRef]
17. Tao, H.L.; Feng, H.K.; Xu, L.J.; Miao, M.K.; Yang, G.J.; Yang, X.D.; Fan, L.L. Estimation of the yield and plant height of winter wheat using UAV-Based hyperspectral images. *Sensors* **2020**, *20*, 1231. [CrossRef] [PubMed]
18. Zagajewski, B.; Kycko, M.; Tmmervik, H.; Bochenek, Z.; Wojtun, B.; Bjerke, J.; Klos, A. Feasibility of hyperspectral vegetation indices for the detection of chlorophyll concentration in three high arctic plants: Salix polaris, bistorta vivipara, and dryas octopetala. *Acta Soc. Bot. Pol.* **2018**, *87*, 3604. [CrossRef]
19. Chen, P.F.; Haboudane, D.; Tremblay, N.; Wang, J.H.; Vigneault, P.; Li, B.G. New spectral indicator assessing the efficiency of crop nitrogen treatment in corn and wheat. *Remote Sens. Environ.* **2010**, *114*, 1987–1997. [CrossRef]
20. Tan, K.Z.; Wang, S.W.; Song, Y.Z.; Liu, Y.; Gong, Z.P. Estimating nitrogen status of rice canopy using hyperspectral reflectance combined with BPSO-SVR in cold region. *Chemom. Intell. Lab. Syst.* **2018**, *172*, 68–79. [CrossRef]
21. Bu, Y.H.; Jiang, X.N.; Tian, J.P.; Hu, X.J.; Fei, X.; Huang, D.; Luo, H.B. Rapid and accurate detection of starch content in mixed sorghum by hyperspectral imaging combined with data fusion technology. *J. Food Process Eng.* **2022**, *45*, e14129. [CrossRef]
22. Zhao, L.; Zhou, W.; Peng, Y.P.; Hu, Y.M.; Ma, T.; Xie, Y.K.; Wang, L.Y.; Liu, J.C.; Liu, Z.H. A new AG-AGB estimation model based on MODIS and SRTM data in Qinghai Province, China. *Ecol. Indic.* **2021**, *133*, 108378. [CrossRef]
23. Wang, Q.; Chang, Q.R.; Li, K.; Chen, X.K.; Miao, H.L.; Shi, B.T.; Zeng, X.L.; Li, Z.F. Estimation of winter wheat canopy chlorophyll content based on principal component analysis and random forest regression. *J. Triticeae Crops* **2024**, *44*, 532–542.
24. Feng, Q.; Liu, J.; Gong, J. UAV Remote sensing for urban vegetation mapping using random forest and texture analysis. *Remote Sens.* **2015**, *7*, 1074–1094. [CrossRef]
25. Gao, M.; Huang, X.Y.; Wang, F.; Zhang, H.L.; Zhao, H.X.; Gao, X.Y. Sea surface salinity inversion based on DNN model. *Adv. Mar. Sci.* **2022**, *40*, 496–504.
26. Sze, V.; Chen, Y.H.; Yang, T.J.; Emer, J.S. Efficient processing of deep neural networks: A tutorial and survey. *Proc. IEEE* **2017**, *105*, 2295–2329. [CrossRef]
27. Xue, J.; Zhang, X.L.; Chen, S.C.; Lu, R.; Wang, Z.; Wang, N.; Hong, Y.S.; Chen, X.Y.; Xiao, Y.; Ma, Y.X.; et al. The validity domain of sensor fusion in sensing soil quality indicators. *Geoderma* **2023**, *438*, 116657. [CrossRef]
28. Shen, W.Z.; He, J.; Li, S.S.; Zhuang, Y.H.; Wang, H.Y.; Liu, H.B.; Zhang, L.; Kappler, A. Opportunity and shift of nitrogen use in China. *Geogr. Sustain.* **2024**, *5*, 33–40. [CrossRef]
29. Gu, J.F.; Yang, J.C. Nitrogen (N) transformation in paddy rice field: Its effect on N uptake and relation to improved N management. *Crop Environ.* **2022**, *1*, 7–14. [CrossRef]
30. Li, W.C.; Xu, W.; Yin, G.F.; Zhang, X.L.; Zhang, Z.H.; Xi, B.; Lei, Q.L.; Zhai, L.M.; Zhang, Q.; Yang, L.Z.; et al. Critical processes and major factors that drive nitrogen transport from farmland to surface water bodies. *Front. Agric. Sci. Eng.* **2023**, *10*, 541–552. [CrossRef]

31. Wang, W.H.; Zheng, H.B.; Wu, Y.P.; Yao, X.; Zhu, Y.; Cao, W.X.; Cheng, T. An assessment of background removal approaches for improved estimation of rice leaf nitrogen concentration with unmanned aerial vehicle multispectral imagery at various observation times. *Field Crops Res.* **2022**, *283*, 108543. [CrossRef]
32. Tian, Y.C.; Yang, J.; Yao, X.; Zhu, Y.; Cao, W.X. A newly developed blue nitrogen index for estimating canopy leaf nitrogen concentration of rice. *Chin. J. Appl. Ecol.* **2010**, *21*, 966–972.
33. Thorp, K.R.; Wang, G.; Bronson, K.F.; Badaruddin, M.; Mon, J. Hyperspectral data mining to identify relevant canopy spectral features for estimating durum wheat growth, nitrogen status, and grain yield. *Comput. Electron. Agric.* **2017**, *136*, 1–12. [CrossRef]

agronomy

MDPI

Article

Enhanced Neural Network for Rapid Identification of Crop Water and Nitrogen Content Using Multispectral Imaging

Yaoqi Peng [1,2,3], Mengzhu He [1,2], Zengwei Zheng [1,2,*] and Yong He [3,*]

1 School of Computer and Computing Science, Hangzhou City University, Hangzhou 310015, China; pengyaoqi@yeah.net (Y.P.); hemz@hzcu.edu.cn (M.H.)
2 Zhejiang Provincial Engineering Research Center for Intelligent Plant Factory, Hangzhou 310015, China
3 College of Biosystems Engineering and Food Science, Zhejiang University, Hangzhou 310015, China
* Correspondence: zhengzw@hzcu.edu.cn (Z.Z.); yhe@zju.edu.cn (Y.H.)

Abstract: Precision irrigation and fertilization in agriculture are vital for sustainable crop production, relying on accurate determination of the crop's nutritional status. However, there are challenges in optimizing traditional neural networks to achieve this accurately. This paper aims to propose a rapid identification method for crop water and nitrogen content using optimized neural networks. This method addresses the difficulty in optimizing the traditional backpropagation neural network (BPNN) structure. It uses 179 multi−spectral images of crops (such as maize) as samples for the neural network model. Particle swarm optimization (PSO) is applied to optimize the hidden layer nodes. Additionally, this paper proposes a double−hidden−layer network structure to improve the model's prediction accuracy. The proposed double−hidden−layer PSO−BPNN model showed a 9.87% improvement in prediction accuracy compared with the traditional BPNN model. The correlation coefficient R^2 for predicted crop nitrogen and water content was 0.9045 and 0.8734, respectively. The experimental results demonstrate high training efficiency and accuracy. This method lays a strong foundation for developing precision irrigation and fertilization plans for modern agriculture and holds promising prospects.

Keywords: nitrogen content; water content; double hidden layer; BP neural network; particle swarm optimization

Citation: Peng, Y.; He, M.; Zheng, Z.; He, Y. Enhanced Neural Network for Rapid Identification of Crop Water and Nitrogen Content Using Multispectral Imaging. *Agronomy* **2023**, *13*, 2464. https://doi.org/10.3390/agronomy13102464

Academic Editor: Shubo Wang

Received: 9 August 2023
Revised: 17 September 2023
Accepted: 19 September 2023
Published: 23 September 2023

1. Introduction

Irrigation and fertilization are essential key factors in the crop growth stage [1] To improve the current situation of overuse in traditional farmland production, it is necessary to obtain information on crop water and nitrogen in advance to realize precise irrigation and fertilization [2,3]. The demand for water and fertilizer of crops is influenced by several factors, including but not limited to sunshine, air temperature and humidity, soil temperature and humidity, and CO_2 concentration. These factors contribute to the complexity of the crop−growing environment, which can be characterized as a time−delay system with a multitude of parameters that exhibit nonlinearity and strong interdependence. An artificial neural network has powerful self−learning, self−organizing, and self−adapting ability, which makes it possible for the network to deal with uncertain or unknown complex nonlinear systems, and by optimizing the network structure it can fully approximate any complex nonlinear relations. In the development history of the artificial neural network, for a long time there was no effective algorithm to adjust the connection weight of hidden layers [4–6]. This lasted until the error backpropagation (BP) algorithm was proposed, the weight adjustment problem of a multilayer feedforward neural network for solving nonlinear continuous functions was successfully solved, and the BP neural network was used in many applications [7,8]. For example, the remotely sensed leaf area index (LAI) and vegetation temperature condition index (VTCI) are closely related to crop growth and crop water stress as two key variables for indicating crop growth conditions and estimating crop yields in the Guanzhong Plain, and the BP neural network and the IPSO–BP neural network were used to calculate the weight coefficients and thresholds of the VTCI and LAI at the four growth stages and to establish an integrated index, *I*, during the main growth period [9]. Three simulated tidal flow systems and a full system of continuous vertical flow of synthetic wastewater were treated by effluent removal with the help of a BP neural network; by comparing the influent and effluent concentrations, the results show that the ability of the BP artificial neural network model to predict nutrient concentrations in the effluent was good; there were only small errors when correlating the predicted values and the actual values [10]. Based on the monitoring data of soil moisture, soil electrical conductivity, air temperature, and light intensity, a prediction model of crop water demand based on the 4–8–1–structure BP neural network was established to guide water–saving irrigation in the crop production process [11]. Many studies have proved that a BPNN has a good effect on data prediction.

Although BPNNs have been widely used, they have some defects and deficiencies, including the following: (1) Because the learning rate is fixed, the convergence speed of the network is slow, and the network requires a long training time [12]. For some complex problems, the training time of a BP algorithm may be very long, which is mainly due to the long learning duration [13]. It can be improved by using a variable learning rate or an adaptive learning rate [14]. (2) A BP algorithm can make the weight converge to a certain value, but it does not guarantee that the value is the global minimum of the error plane because the gradient descent method may produce a local minimum value [15]. The additional momentum method can be used to solve this problem [16]. (3) BPNN learning and memory are unstable. In other words, if learning samples are added, the trained network must be trained from the beginning, and there is no memory for the previous weights and thresholds [17]. However, it can save the better weights of the prediction, classification, or clustering [18]. (4) There is no theoretical guidance for the selection of the number of layers and units of the network's hidden layer; these are generally determined by experience or by repeated experiments [19]. Motivated by these problems, we focused on determining the number of hidden layer nodes of a BP neural network based on particle swarm optimization (PSO).

In this study, field crops were taken as the research object. The spectral information of crop water and nitrogen was extracted by a multi–spectral camera, and the information of crop leaf water and nitrogen content was measured by a hand–held sensor. Considering that artificial neural networks can process nonlinear adaptive information well, by analyzing the correlation between crop spectrum characteristics and moisture nitrogen, an improved BPNN model using multispectral crop images to rapidly identify crop nitrogen and water contents was constructed. Finally, this study provides a theoretical basis for precision irrigation and fertilization of field crops. We also considered the influence of the number of hidden layers. The main contributions of this study are as follows:

- PSO was used to optimize the number of hidden layer nodes in a BP neural network, which improved the training efficiency and reduced the time and tedium of determining the number of hidden layer nodes by experience.
- In addition to increasing the number of hidden layer nodes and improving the prediction accuracy of a BP neural network, we found that a double–hidden–layer structure can effectively reduce the performance errors of the network and improve its performance.
- A prediction model of crop nitrogen and water contents based on PSO–BPNN with a double–hidden–layer structure was established. Experiments showed that the model was highly efficient at predicting the nitrogen and water contents of the crop.

The primary objectives of this study are as follows:

Section 2: to provide a comprehensive description of the data acquisition scheme, model descriptions, and mathematical preliminaries utilized in the study.

Section 3: to present and explain the optimization principle and modeling scheme employed for the model, specifically focusing on the application of particle swarm optimization (PSO) and the proposed double−hidden−layer network structure.

Section 4: to present and discuss the comparative experimental results obtained from the study, analyzing the performance and accuracy of the PSO−BPNN model in comparison to the traditional BPNN model.

Section 5: to draw conclusions based on the findings and implications of the study, summarizing the key outcomes, and discussing the potential prospects and applications of the proposed method.

2. Materials and Methods

2.1. Data Acquisition

This study aimed to determine the water and nitrogen content in a crop canopy by employing a plant nutrient analyzer(device model: YLS−D, Hubei, China) in the field. The collection methodology involved dividing the field into multiple rectangular grid areas, from which random samples were extracted. Each sample was meticulously obtained from distinct sections of the plant canopy, namely, the upper, middle, and lower regions. Subsequently, the measurements of water and nitrogen content at each sampling location were diligently recorded. To obtain accurate results, the average value of the three canopy regions was deemed representative of the overall water and nitrogen content of the crop canopy. Concurrently, a handheld multi−spectral camera(device model: RedEdge−M, Seattle, Washington, US) was employed to capture vertical images of the canopy at the sampling positions. The real−time previewing of the multi−spectral image position was enabled through Wi−Fi connectivity with a mobile phone. By maintaining a camera−lens−to−canopy distance of approximately 30 cm and ensuring proper alignment with the sensor's five channels, images were captured and saved in a 16−bit TIFF format. Notably, the comprehensive dataset encompassed 179 sets of crop canopy multispectral images, along with the corresponding water and nitrogen content data. A detailed depiction of the data acquisition process can be found in Figure 1.

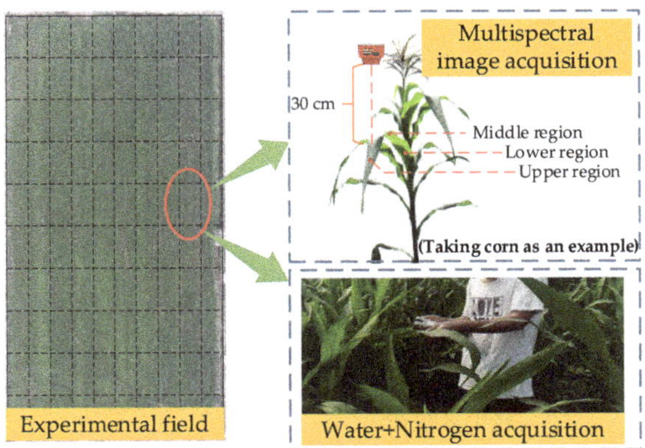

Figure 1. Crop data acquisition process.

2.2. BPNN Model Description

A BPNN is a kind of multilayer feedforward network trained by an error backpropagation algorithm [20]. It is one of the most widely used neural network models. The neurons in the input layer are responsible for receiving the input information from the outside and transmitting it to the neurons in the middle layer. The middle layer is the internal information processing layer, which is responsible for information transformation. Depending on the demands of information change, the middle layer can be designed as a single hidden layer or multiple hidden layers. The output hidden layer transmits the information to each neuron in the output layer. After further processing, it completes a learning forward propagation process and outputs the information processing results to the outside world from the output layer. A double–hidden–layer neural network structure was used in this study. Network precision was improved by increasing the number of hidden layers [21]. The topological structure of a BP neural network is shown in Figure 2.

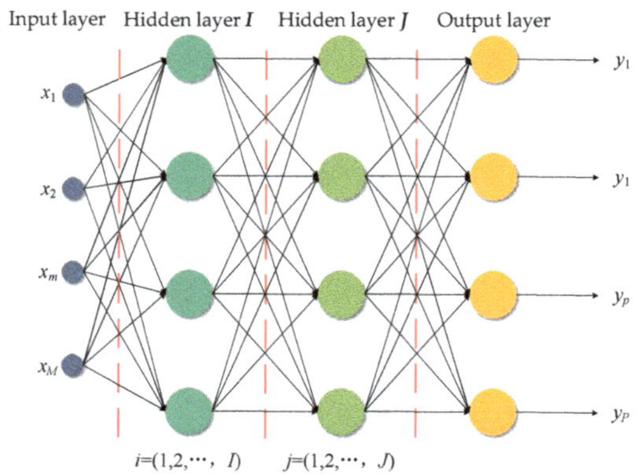

Figure 2. The topological structure of a BP neural network.

The learning process of a BP neural network comprises mainly the following parts:

(i) Setting variables and parameters. $X_k = [x_{k1}, x_{k2}, \ldots, x_{kM}]$, ($k = 1, 2, \ldots, N$) is the input variable, also known as training samples, and N is the number of training samples.

$W_{MI}(n) = \begin{bmatrix} w_{11}(n) & w_{12}(n) & \cdots & w_{1I}(n) \\ w_{21}(n) & w_{22}(n) & \cdots & w_{2I}(n) \\ \vdots & \vdots & \vdots & \vdots \\ w_{M1}(n) & w_{M2}(n) & \cdots & w_{MI}(n) \end{bmatrix}$ is the weight vector between the input layer and the hidden layer I in the nth iteration.

$W_{IJ}(n) = \begin{bmatrix} w_{11}(n) & w_{12}(n) & \cdots & w_{1J}(n) \\ w_{21}(n) & w_{22}(n) & \cdots & w_{2J}(n) \\ \vdots & \vdots & \vdots & \vdots \\ w_{I1}(n) & w_{I2}(n) & \cdots & w_{IJ}(n) \end{bmatrix}$ is the weight vector between the hidden layer I and the hidden layer J in the nth iteration.

$W_{JP}(n) = \begin{bmatrix} w_{11}(n) & w_{12}(n) & \cdots & w_{1P}(n) \\ w_{21}(n) & w_{22}(n) & \cdots & w_{2P}(n) \\ \vdots & \vdots & \vdots & \vdots \\ w_{J1}(n) & w_{J2}(n) & \cdots & w_{JP}(n) \end{bmatrix}$ is the weight vector between the hidden layer J and the output layer in the nth iteration.

$Y_k(n) = [y_{k1}(n), y_{k2}(n), \cdots, y_{kP}(n)]$, ($k = 1, 2, \ldots, N$) is the actual output of the network in the nth iteration. $d_k = [d_{k1}, d_{k2}, \cdots, d_{kP}]$, ($k = 1, 2, \ldots, N$) is the desired output.

η is the learning rate, and n is the number of iterations.

(ii) Initialization. Assign a smaller random nonzero value to W_{MI} (0), W_{IJ} (0), W_{JP} (0), and $n = 0$.

(iii) Random input sample X_k.

(iv) The input sample X_k, input signal u, and output signal v of each layer of the BPNN are calculated forward, where $v_p^P(n) = y_{kp}(n)$, $p = 1, 2, \ldots, P$.

(v) Calculate the error $E(n)$ from the expected output d_k and the actual output $Y_k(n)$ obtained in the previous step to judge whether it meets the requirements. If it meets the requirements, go to step viii; if not, go to step vi.

(vi) Determine whether the $n + 1$ is greater than the maximum number of iterations. If it is greater, go to step viii. If it is not greater, the local gradient δ of each layer of neurons is inversely calculated for the input sample X_k. The equations are

$$\delta_p^P(n) = y_p(n)(1 - y_p(n))(d_p(n) - y_p(n)),\ p = 1, 2, \ldots P \tag{1}$$

$$\delta_j^J(n) = f\prime(u_j^J(n))\sum_{p=1}^{P}\delta_p^P(n)w_{jp}(n),\ j = 1, 2, \ldots J \tag{2}$$

$$\delta_i^I(n) = f\prime(u_i^I(n))\sum_{j=1}^{J}\delta_j^J(n)w_{ij}(n),\ i = 1, 2, \ldots I \tag{3}$$

(vii) Calculate the weight correction Δw and correct the weight; if $n = n + 1$, go to step iv.

$$\Delta w_{jp}(n) = \eta\delta_p^P(n)v_j^J(n)\ w_{jp}(n+1) = w_{jp}(n) + \Delta w_{jp}(n)\ j = 1, 2, \ldots J;\ p = 1, 2, \ldots P \tag{4}$$

$$\Delta w_{ij}(n) = \eta\delta_j^J(n)v_i^I(n)\ w_{ij}(n+1) = w_{ij}(n) + \Delta w_{ij}(n)\ i = 1, 2, \ldots I;\ j = 1, 2, \ldots J \tag{5}$$

$$\Delta w_{mi}(n) = \eta\delta_i^I(n)x_{km}(n)\ w_{mi}(n+1) = w_{mi}(n) + \Delta w_{mi}(n)\ m = 1, 2, \ldots M;\ i = 1, 2, \ldots I \tag{6}$$

(viii) Judge whether all the training samples have been learned. If they have, the learning process is finished. If they have not, go to step iii.

2.3. Application Principle of the Particle Swarm Optimization Algorithm

PSO has the characteristics of evolutionary computation and swarm intelligence. Similar to other algorithms, PSO can search for the best solution in complex space through cooperation and competition among individuals [22].

In a PSO algorithm, the solution of each optimization problem is regarded as a "bird" or "particle" in the search space [23]. At the beginning of the algorithm, the initial solution is generated; that is, the population composed of m particles is randomly initialized in the feasible solution space, where the position $Z_i = \{z_{i1}, z_{i2}, \cdots z_{in}\}$ of each particle represents a solution to the problem, and a new solution is searched for according to the objective function calculation. In each iteration, the particle tracks two extrema to update itself; one is the best solution p_{id} found by the particle itself, and the other is the best solution p_{gd} found by the entire population, which is the global extremum. Also, each particle has a velocity of $V_i = \{v_{i1}, w_{i2}, \cdots w_{in}\}$. When the two optimal solutions are found, each particle updates its velocity according to Equation (7):

$$\begin{aligned} v_{id} &= wv_{id}(t) + \eta_1 rand()[p_{id} - z_{id}(t)] + \eta_2 rand()[p_{gd} - z_{id}(t)] \\ z_{id}(t+1) &= z_{id}(t) + v_{id}(t+1) \end{aligned} \tag{7}$$

where $Z_{id}(t + 1)$ is the velocity of the ith particle in the d dimension in the $t + 1$ iteration, w is the inertia weight, η_1 and η_2 are acceleration constants, and *rand*() is a random number between 0 and 1. Also, the upper limit of the velocity can be set to prevent the particle velocity from being too great; that is, when $v_{id}(t+1) > v_{\max}$, $v_{id}(t+1) = v_{\max}$; when $v_{id}(t+1) < -v_{\max}$, $v_{id}(t+1) = -v_{\max}$.

From the updated equation of particles, we can see that the moving direction of particles is determined by three parts: their original velocity v_{id}; the distance $p_{id} - z_{id}(t+1)$ from their best experience; and the distance $p_{gd} - z_{id}(t)$ from the best experience of the group, and their relative importance is determined by the weight coefficients w, η_1, and η_2. When the end condition of the algorithm is reached, that is, a sufficiently optimal solution is found or the maximum number of iterations is reached, the algorithm ends. The basic flow of PSO is shown in Figure 3.

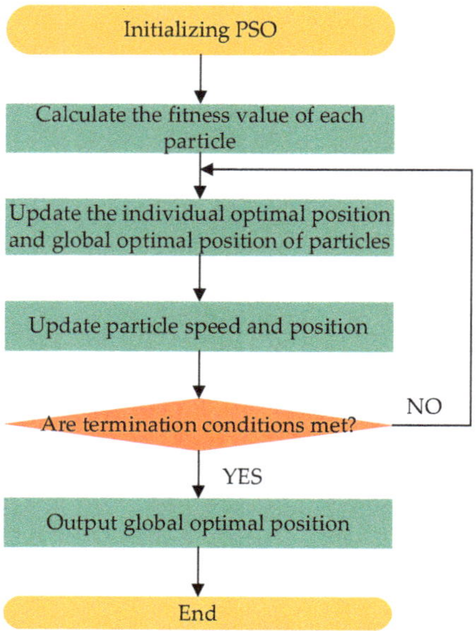

Figure 3. The basic flow of PSO.

3. Construction of BPNN Model Based on PSO Optimization

3.1. Optimization of BPNN by PSO Algorithm

To solve the defects of a BPNN, we combined a PSO algorithm with a BPNN algorithm and applied the PSO algorithm to optimize the initial weight and the threshold values of the neural network. The overall algorithm flow is shown in Figure 4.

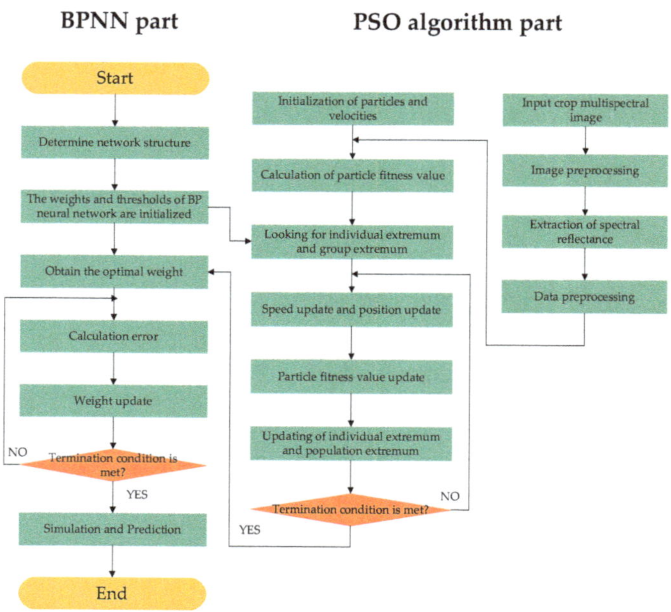

Figure 4. Combined PSO–BP algorithm flow chart.

The implementation steps of the PSO–BP algorithm are

***Step* 1.** The structure and parameters of the BP neural network are initialized.

***Step* 2.** Combined with the connection weights of the BP neural network, the network structure of the PSO is initialized. First, the weight vector $W = \{w_1, w_2, \ldots, w_n\}$ of the BP network is constructed as the space particle in the PSO optimization algorithm, and then the parameters of the PSO network are set, namely, inertia weight, the acceleration constant, particle speed, and position in Equations (8) and (9).

***Step* 3.** The speed and position of the weighted particles are updated. The particles start from an initial position X_{id} in space with a certain initial velocity V_{id}, where i is the number and d is the dimension. In the process of particle motion, the velocity and position will change constantly, and the update formulas are Equations (8) and (9), respectively:

$$V_{id}^{k+1} = \omega V_{id}^{k} + c_1 r_1 (P_{id}^{k} - X_{id}^{k}) + c_2 r_2 (P_{gd}^{k} - X_{gd}^{k}) \tag{8}$$

$$X_{id}^{k+1} = X_{id}^{k} + V_{id}^{k+1} \tag{9}$$

where Equation (8) is the velocity of the particle and Equation (9) is the position of the particle.

***Step* 4.** Find the global optimal extremum. First, the fitness function value of each particle in the space is calculated. When the particle is iterated many times, a new fitness function value is calculated. If the fitness function value of the new particle is better than the current value, the individual extreme value p_{best} and the population extreme value g_{best} are updated until the best extreme value is found. The mean square error of the BP neural network on the training set is taken as the fitness function, and the calculation method is Equation (10).

$$E(x_p) = \frac{1}{N}\sum_{p=1}^{n}\sum_{k=0}^{m}(y_{pk}(x_p) - t_{pk})^2 \tag{10}$$

where x_p is the input sample of group p, $p = 1, 2, \ldots, n$; y_{pk} is the kth output of input x_p sample; and t_{pk} is the expected value of the kth output of input x_p sample, $k = 1, 2, \ldots, m$.

***Step* 5.** Weight optimization is achieved. Compare the best fitness function value obtained in ***Step* 4** with the preset objective, or judge whether the maximum iteration times have been reached. If the requirements are met, it indicates that the global best weight has been found and the operation is finished.

***Step* 6.** In the PSO--BP network, the outputs of the hidden and output layers are calculated as follows:

$$x\prime = f(\sum_{i=0}^{2} w_{ij} - \theta_1) d = f(\sum_{j=0}^{2} w_{jk} x\prime - \theta_2) \tag{11}$$

$$f(u) = \frac{1}{1 + e^{-u_j}} = \frac{1}{1 + e^{-(\sum w_j x_j - \theta_j)}} \tag{12}$$

***Step* 7.** Error judgment. Calculate whether the error function meets the expectation. If so, the network training ends and retains the trained weight; otherwise, the error of each layer is calculated layer by layer. The calculation of expectation and error is as follows:

$$e_p = \frac{1}{2}\sum_{k}\left(y_k - \hat{y}_k\right)^2 \tag{13}$$

$$\delta_{jk}^{p_1} = \left(t_k^{p_1} - d_k^{p_1}\right) d_k^{p_1}\left(1 - d_k^{p_1}\right) \delta_{ij}^{p_1} = \sum_{k=0}^{p_1} \delta_{jk}^{p_1} w_{jk} x\prime^{p_1} (1 - x\prime^{p_1}) \tag{14}$$

***Step* 8.** Adjust the weight of each layer of the network; the specific calculation is

$$w_{jk}(n_0+1) = w_{jk}(n_0) + \eta \sum_{p_1=1}^{p} \delta_{jk}^{p_1} x I^{p1} w_{ij}(n_0+1) = w_{ij}(n_0) + \eta \sum_{p_1=1}^{p} \delta_{ij}^{p_1} x I^{p1} \quad (15)$$

***Step* 9.** After the adjustment, continue to input the sample and repeat the calculation process in ***Step* 6** with the new weight. Once the error meets the requirements of Equation (14), the training is stopped.

***Step* 10.** Save the trained neural network models and predict the crop nitrogen and moisture contents of the trained neural network.

3.2. Construction of Prediction Model of Crop Water and Nitrogen Contents Based on BPNN with Double Hidden Layers

The performance of the PSO algorithm is affected by various interaction parameters, such as group size N, inertia factor ω, learning factors c_1 and c_2, maximum speed $v_{\max}$, and maximum iteration number G_k. The range of the population size is generally from 20 to 50. If $\omega = 0$, then the adjustment of particle velocity is related only to the current position and the historical best position, and the change in velocity size has nothing to do with it [24]. When the inertia factor ω is added, particles can effectively explore other parts of space, not only in the vicinity of the individual and global best positions but also in the function of global and local exploration. When ω is large, the global search ability is strong; when ω is small, the local search ability is strong [25]. Generally, $c_1 = c_2 = 2$. Ambroziak's research shows that if we want to obtain good results and simplify the operation, c_1 and c_1 are better as constants [26]. These parameters must be given artificially at the beginning of training. The specific parameter values of this study are shown in Table 1.

Table 1. Parameter selection of PSO algorithm.

N	ω	c_1	c_2	$v_{\max}$	G_k
50	0.1	2	2	0.5	200

In this study, 179 sets of multispectral images of crop leaves were collected to extract the reflectance value, and the nitrogen and water contents of the leaves were measured by a plant nutrient analyzer. Finally, the reflectance of the blue, green, red, near−infrared (NIR), and RedEdge bands, as well as crop nitrogen and water contents, were obtained—a total of seven variables. A correlation analysis of the seven variables is shown in Figure 5. The correlation between the reflectance of each spectral band and the nitrogen and water contents of the crops was not strong, showing a nonlinear mapping relation. It is therefore more suitable to use a neural network to build a prediction model.

Five neurons were in the input layer of the BPNN—that is, the reflectance of five spectral bands—and two neurons were in the output layer; namely, the nitrogen and water contents of the crops. Because the units of input data and output data were different, to facilitate data training the data had to be normalized into a dimensionless form before modeling. The training data in this study were normalized to between 0.001 and 0.999. The naturalization equation is given by Equation (16).

$$P = \frac{p - p_{min}}{p_{max} - p_{min}} \times 0.998 + 0.001 \quad (16)$$

where p is the original data, p_{min} is the minimum value of the data with the same dimension, p_{max} is the maximum value of the data with the same dimension, and P is the normalized data.

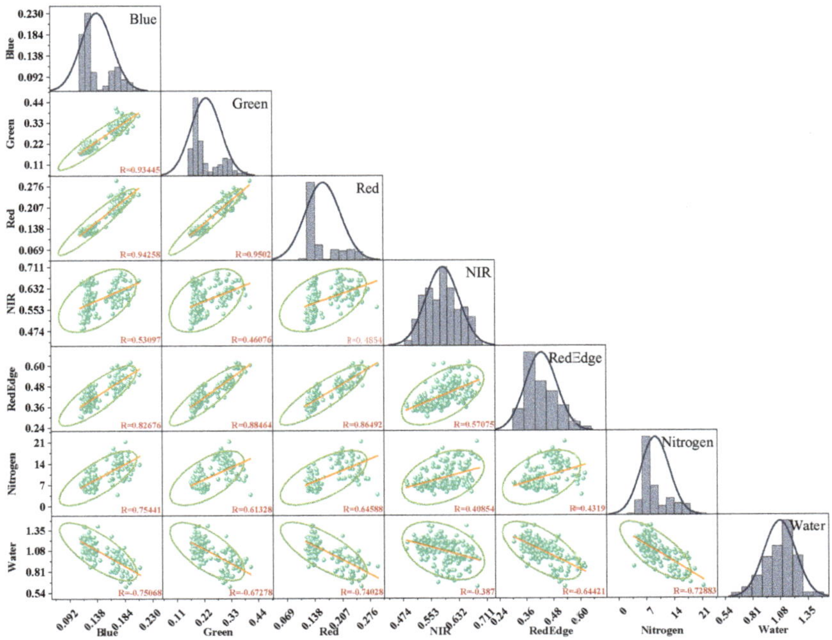

Figure 5. Correlation analysis of seven variables.

In the process of modeling with the BPNN, the number of hidden layer neurons in the network structure could not be determined. Generally, the greater the number of hidden layers in a neural network, the stronger its nonlinear mapping capability. Upon reaching an appropriate number of neurons, further increasing the number would not help much to improve the accuracy of the network but would increase the amount of calculation [27]. Currently, the number of neurons is determined mainly by trial calculation via an empirical equation; the general empirical formulas are given by Equations (17) and (18).

$$l < \sqrt{(m+n)} + a \tag{17}$$

$$l < 2m + 1 \tag{18}$$

l is the number of hidden neurons, m is the number of neurons in the input layer, n is the number of neurons in the output layer, and a is a constant from 0 to 10 [28].

According to the empirical formulas of Equations (17) and (18), it was determined that the number of neurons in the first hidden layer ranged from 2 to 13. The networks with different hidden layers were trained 10 times, and the numbers of epochs and mean squared error (MSE) values were recorded. The recorded results are shown in Figure 6. In Figure 6, the color of the 3D sphere represents the number of neurons in the hidden layer, and the size of the sphere represents the MSE. After the trial calculation, when the number of neurons in the hidden layer was 12, the correlation coefficient $R-value$ of the trained BPNN was the largest, while the number of iterations and the MSE value were low.

This study aimed to establish a BPNN with double hidden layers. Through trial calculation, we determined the number of nodes in the first hidden layer to be 12. It was found that the best ratio of the number of nodes in the first hidden layer and the second hidden layer was 3:1 in the case of high−dimensional input [29]. Therefore, the number of nodes in the second hidden layer of the neural network was determined to be 4. Finally, a BPNN with a 5−12−4−2 structure was established. The structure diagram is shown in Figure 7.

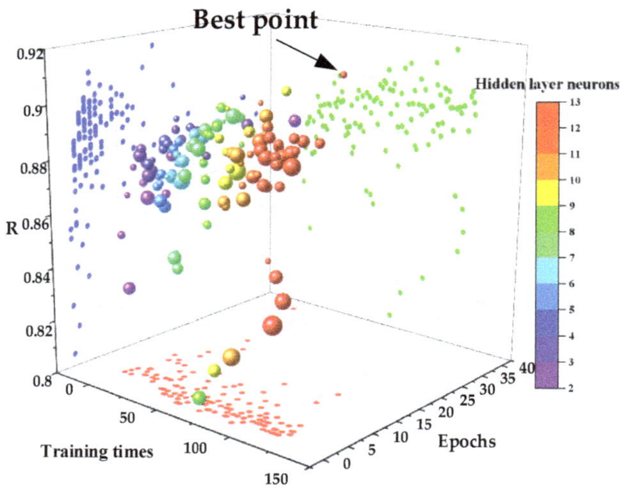

Figure 6. First hidden −layer trial results.

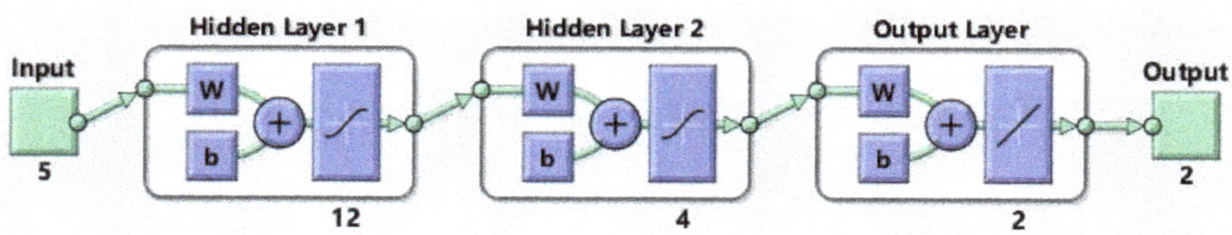

Figure 7. BPNN with a 5−12−4−2 structure (It contains 5 input neurons, with 12 neurons in the first hidden layer, 4 neurons in the second hidden layer, and 2 neurons in the output layer).

Below is the pseudo−code for optimizing a BP neural network using PSO:

1. Initialization:
    - Define the population size and the maximum number of iterations.
    - Randomly initialize the position and velocity of particles in the search space.
    - Initialize the global best position and fitness value.
2. Particle Movement and Update:
    - For each iteration, update the velocity and position of each particle using the PSO equations.
    - Apply velocity and position limits if necessary.
    - Evaluate the fitness of each particle using the BP neural network with the current position as the weights.
    - Update the personal and global best positions if a particle finds better solutions.
3. Termination Condition:
    - Stop the process when either the maximum number of iterations is reached, or a desired fitness value is achieved.
4. Return the Best Solution:
    - After the termination condition is met, return the position of the particle with the best fitness value as the optimized weights for the BP neural network.

4. Comparative Experimental Results and Analysis

Next, we trained 150 groups of samples, and the remaining 29 groups were used for verification. We compared and analyzed the performances of a traditional BPNN, a single−hidden−layer BP neural network optimized by PSO (PSO−1H−BPNN), and a double−hidden−layer BP neural network optimized by PSO (PSO−2H−BPNN), and also determined the prediction accuracies of the three neural networks.

4.1. Performance Analysis of the Neural Networks

Three kinds of neural networks were trained by 150 groups of samples. The convergence and network performances of the training are shown in Figures 8 and 9. In Figure 8, we can see that the MSE of the BPNN was the largest and that of PSO−2H−BPNN was the smallest. This shows that a BP neural network with a double−hidden−layer structure optimized by PSO can effectively reduce the number of training errors. Also, increasing the number of hidden layers increases the number of iterations, but the increase is not obvious. This also confirms the results of a previous study that increasing the number of hidden layers will increase the training time [30].

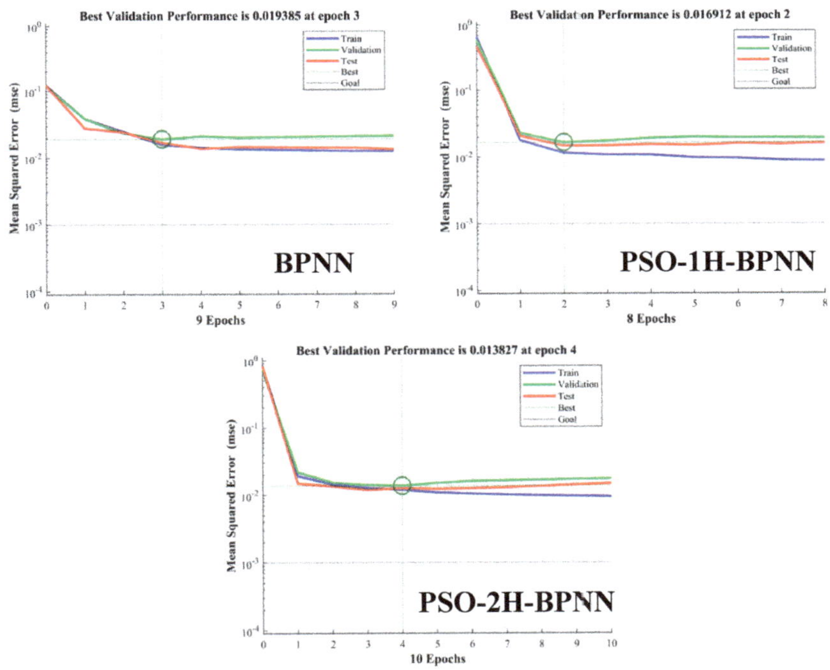

Figure 8. Convergence process of neural network training.

In Figure 9, we can see that the decision coefficient R of the three neural networks gradually increased after training, and the decision coefficient R of PSO−2H−BPNN reached 0.92978. Compared with a conventional BPNN, the network performances of PSO−1H−BPNN and PSO−2H−BPNN improved by 3.97% and 9.87%, respectively, which indicates that PSO-2H-BPNN had the best performance of the three kinds of neural networks.

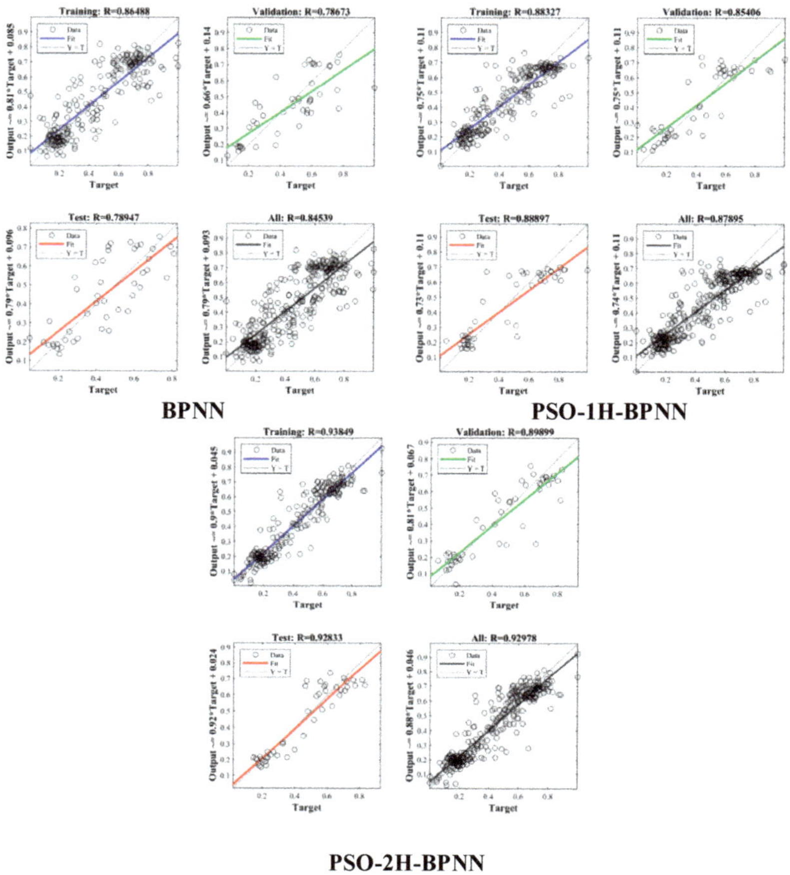

Figure 9. Neural network performance.

4.2. Prediction Accuracy Analysis of Three Kinds of Neural Networks

We used three kinds of neural networks to simulate the remaining 29 groups of samples. The results of the predicted and expected outputs are shown in Figure 10. From Figure 10, it can be seen that in the simulation of either nitrogen content or water content, the predicted output and the expected output had the strongest consistency in PSO−2H−BPNN. Also, the deviation degree of the PSO−2H−BPNN prediction error was the smallest of the three kinds of neural networks.

Finally, a linear correlation analysis was performed between the predicted output and the expected output of the sample, as shown in Figure 11. From Figure 11, it can be seen that for the predicted nitrogen content of the crop the correlation coefficient R^2 of the BPNN, PSO−1H−BPNN, and PSO−2H−BPNN reached 0.7995, 0.8352, and 0.9045, respectively, and the correlation coefficient R^2 of PSO−2H−BPNN was the highest. On the predicted water content of the crop, the correlation coefficient R^2 of the BPNN, PSO−1H−BPNN, and PSO−2H−BPNN reached 0.7533, 0.88099, and 0.8734, respectively, and the correlation coefficient R^2 of PSO−2H−BPNN was also the highest. In the same kind of neural network prediction, the accuracy of predicting crop nitrogen content was higher than that of predicting crop water content.

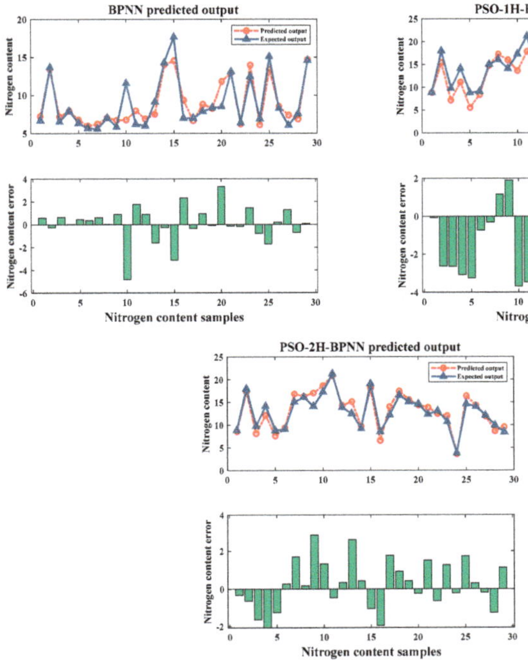

(**a**) Simulation verification of crop nitrogen content.

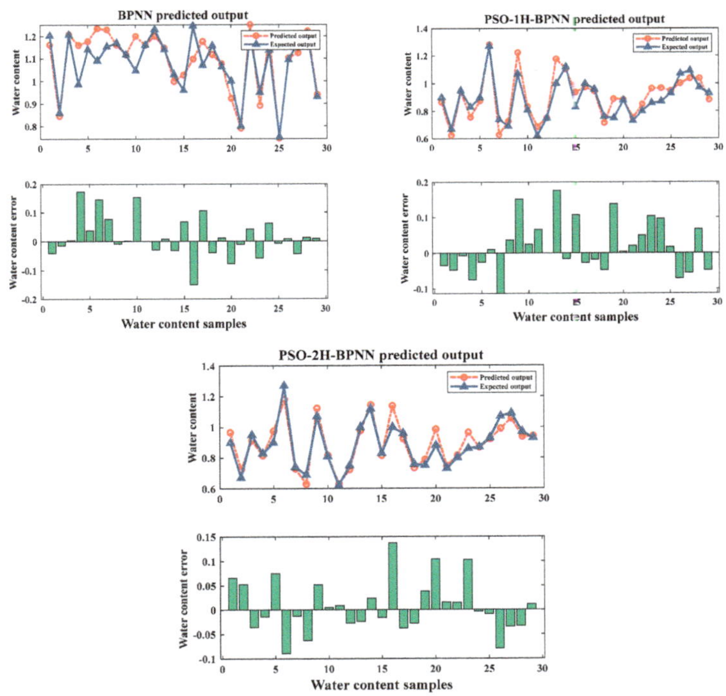

(**b**) Simulation verification of crop water content.

Figure 10. Simulation results of three kinds of neural networks.

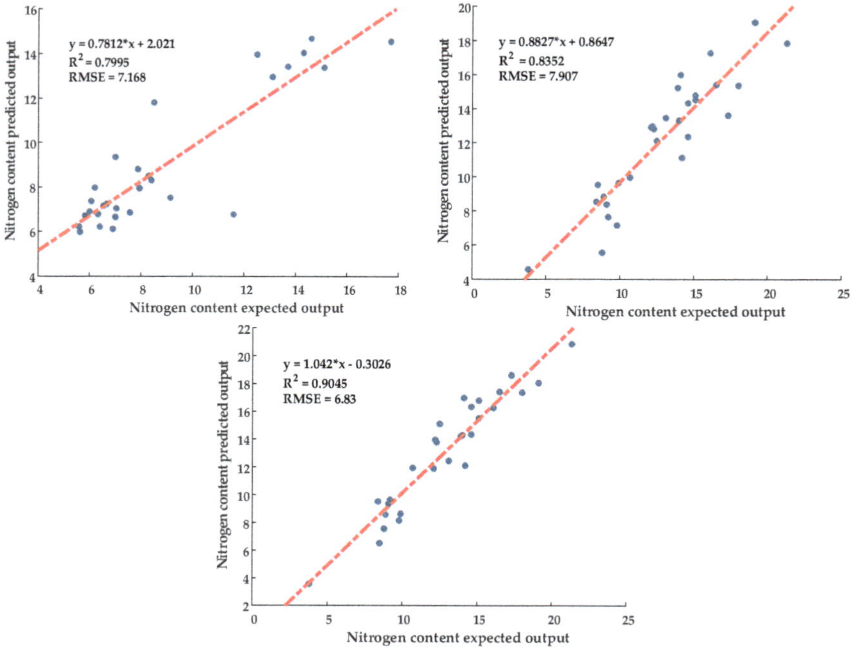

(**a**) Linear correlation analysis of crop nitrogen content.

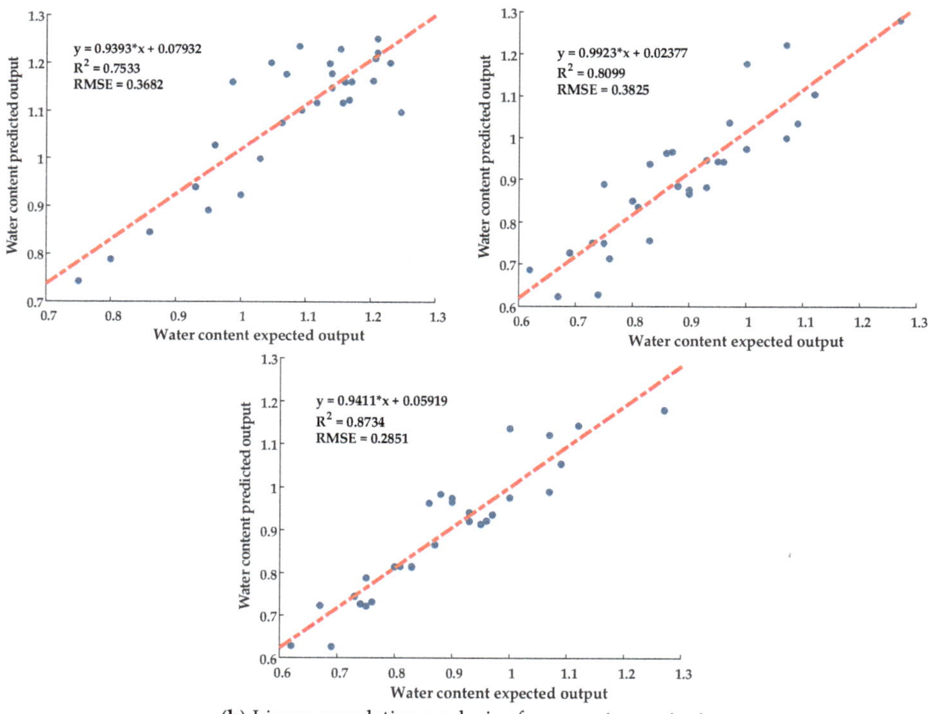

(**b**) Linear correlation analysis of crop water content.

Figure 11. Linear correlation analysis of three kinds of neural networks.

5. Conclusions

In this study, we designed a double−hidden−layer BP neural network optimized by a PSO algorithm. Compared with a conventional BPNN, the network performance of PSO−2H−BPNN was improved by 9.87%. On the predicted nitrogen content of the crop, the correlation coefficient R^2 of PSO−2H−BPNN reached 0.9045, and on the predicted water content of the crop, the correlation coefficient R^2 of PSO−2H−BPNN reached 0.8734. Both had the highest determination coefficient. For the same kind of neural network prediction, the accuracy of predicting the crop nitrogen content was higher than that of the crop water content.

Although the network can quickly identify crop moisture and nitrogen content, it has a high requirement for collecting multi−spectral images of crop canopy. For canopy multispectral image data collected under sufficient lighting conditions, high accuracy can be easily achieved, but canopy multispectral image data collected under cloudy or evening conditions may reduce recognition accuracy. In future studies, we will try using a different algorithm to replace PSO, continue to optimize BP neural networks, and continually improve the prediction accuracy of the network.

Author Contributions: Writing—original draft preparation, Y.P.; writing—review and editing, M.H.; project administration, Z.Z.; supervision, Y.H. All authors have read and agreed to the published version of the manuscript.

Funding: This research was supported in part by the Zhejiang Provincial Natural Science Foundation of China (Grant No. LGN21F020002).

Data Availability Statement: Not applicable.

Acknowledgments: The authors are thankful for the Zhejiang Provincial Natural Science Foundation for funding this work through the Grant No. (LGN21F020002).

Conflicts of Interest: The authors declare no conflict of interest.

References

1. Alcaide Zaragoza, C.; González Perea, R.; Fernández García, I.; Camacho Poyato, E.; Rodríguez Díaz, J.A. Open source application for optimum irrigation and fertilization using reclaimed water in olive orchards. *Comput. Electron. Agric.* **2020**, *173*, 105407. [CrossRef]
2. Perich, G.; Aasen, H.; Verrelst, J.; Argento, F.; Walter, A.; Liebisch, F. Crop Nitrogen Retrieval Methods for Simulated Sentinel−2 Data Using In−Field Spectrometer Data. *Remote Sens.* **2021**, *13*, 2404. [CrossRef] [PubMed]
3. Rattalino Edreira, J.I.; Guilpart, N.; Sadras, V.; Cassman, K.G.; van Ittersum, M.K.; Schils, R.L.M.; Grassini, P. Water productivity of rainfed maize and wheat: A local to global perspective. *Agric. For. Meteorol.* **2018**, *259*, 364–373. [CrossRef]
4. Heydarizad, M.; Gimeno, L.; Minaei, M.; Shahsavan Gharehghouni, M. Stable Isotope Signatures in Tehran's Precipitation: Insights from Artificial Neural Networks, Stepwise Regression, Wavelet Coherence, and Ensemble Machine Learning Approaches. *Water* **2023**, *15*, 2357. [CrossRef]
5. Kamal, S.; Prajapati, H.S.; Cahill, N.D.; Hailstone, R.K. Probe Aberration Correction in Scanning Electron Microscopy Using Artificial Neural Networks. *Microsc. Microanal.* **2023**, *29*, 739–740. [CrossRef]
6. Li, B. A Productivity Prediction Method Based on Artificial Neural Networks and Particle Swarm Optimization for Shale−Gas Horizontal Wells. *Fluid. Dyn. Mater. Process.* **2023**, *19*, 2729–2748. [CrossRef]
7. Wei, L.; Xv, S.; Li, B. Short−term wind power prediction using an improved grey wolf optimization algorithm with back−propagation neural network. *Clean. Energy* **2022**, *6*, 288–296. [CrossRef]
8. Xu, S.; Wan, H.; Zhao, X.; Zhang, Y.; Yang, J.; Jin, W.; He, Y. Optimization of extraction and purification processes of six flavonoid components from Radix Astragali using BP neural network combined with particle swarm optimization and genetic algorithm. *Ind. Crops Prod.* **2022**, *178*, 114556. [CrossRef]
9. Tian, H.; Wang, P.; Tansey, K.; Zhang, S.; Zhang, J.; Li, H. An IPSO−BP neural network for estimating wheat yield using two remotely sensed variables in the Guanzhong Plain, PR China. *Comput. Electron. Agric.* **2020**, *169*, 105180. [CrossRef]
10. Li, W.; Cui, L.; Zhang, Y.; Cai, Z.; Zhang, M.; Xu, W.; Zhao, X.; Lei, Y.; Pan, X.; Li, J.; et al. Using a Backpropagation Artificial Neural Network to Predict Nutrient Removal in Tidal Flow Constructed Wetlands. *Water* **2018**, *10*, 83. [CrossRef]
11. Peng, Y.; Xiao, Y.; Fu, Z.; Dong, Y.; Zheng, Y.; Yan, H.; Li, X. Precision irrigation perspectives on the sustainable water−saving of field crop production in China: Water demand prediction and irrigation scheme optimization. *J. Clean. Prod.* **2019**, *230*, 365–377. [CrossRef]

12. Wan, T.; Bai, Y.; Wang, T.; Wei, Z. BPNN–based optimal strategy for dynamic energy optimization with providing proper thermal comfort under the different outdoor air temperatures. *Appl. Energy* **2022**, *313*, 118899. [CrossRef]
13. Zhang, D.; Lou, S. The application research of neural network and BP algorithm in stock price pattern classification and prediction. *Future Gener. Comput. Syst.* **2021**, *115*, 872–879. [CrossRef]
14. Takase, T.; Oyama, S.; Kurihara, M. Effective neural network training with adaptive learning rate based on training loss. *Neural Netw.* **2018**, *101*, 68–78. [CrossRef] [PubMed]
15. Knoll, C.; Mehta, D.; Chen, T.; Pernkopf, F. Fixed Points of Belief Propagation—An Analysis via Polynomial Homotopy Continuation. *IEEE Trans. Pattern Anal. Mach. Intell.* **2018**, *40*, 2124–2136. [CrossRef] [PubMed]
16. Aprilia, B.; Marzuki; Taufiq, I. Performance of backpropagation artificial neural network to predict el nino southern oscillation using several indexes as onset indicators. *J. Phys. Conf. Ser.* **2021**, *1876*, 12004. [CrossRef]
17. Wei, M.; Hu, X.; Yuan, H. Residual displacement estimation of the bilinear SDOF systems under the near–fault ground motions using the BP neural network. *Adv. Struct. Eng.* **2022**, *25*, 552–571. [CrossRef]
18. Zhou, C.; Gui, S.; Liu, Y.; Ma, J.; Wang, H. Fault Location of Distribution Network Based on Back Propagation Neural Network Optimization Algorithm. *Processes* **2023**, *11*, 1947. [CrossRef]
19. Zhang, J.; Gao, P.; Fang, F. An ATPSO–BP neural network modeling and its application in mechanical property prediction. *Comput. Mater. Sci.* **2019**, *163*, 262–266. [CrossRef]
20. Zhu, H.; Liu, J.; Yu, J.; Yang, P. Artificial neural network–based predictive model for supersonic ejector in refrigeration system. *Case Stud. Therm. Eng.* **2023**, *49*, 103313. [CrossRef]
21. Mahadeva, R.; Kumar, M.; Patole, S.P.; Manik, G. Employing artificial neural network for accurate modeling, simulation and performance analysis of an RO–based desalination process. *Sustain. Comput. Inform. Syst.* **2022**, *35*, 100735. [CrossRef]
22. Mokarram, V.; Banan, M.R. A new PSO–based algorithm for multi–objective optimization with continuous and discrete design variables. *Struct. Multidiscip. Optim.* **2018**, *57*, 509–533. [CrossRef]
23. Meng, Z.; Zhong, Y.; Mao, G.; Liang, Y. PSO–sono: A novel PSO variant for single–objective numerical optimization. *Inf. Sci.* **2022**, *586*, 176–191. [CrossRef]
24. Merugumalla, M.K.; Navuri, P.K. Chaotic inertia weight and constriction factor–based PSO algorithm for BLDC motor drive control. *Int. J. Process Syst. Eng.* **2019**, *5*, 30–52. [CrossRef]
25. Fan, Y.; Zhang, Y.; Guo, B.; Luo, X.; Peng, Q.; Jin, Z. A hybrid sparrow search algorithm of the hyperparameter optimization in deep learning. *Mathematics* **2022**, *10*, 3019. [CrossRef]
26. Ambroziak, A.; Chojecki, A. The PID controller optimisation module using Fuzzy Self–Tuning PSO for Air Handling Unit in continuous operation. *Eng. Appl. Artif. Intell.* **2023**, *117*, 105485. [CrossRef]
27. Wen, S.; Xiao, S.; Yang, Y.; Yan, Z.; Zeng, Z.; Huang, T. Adjusting learning rate of memristor–based multilayer neural networks via fuzzy method. *IEEE Trans. Comput. Aided Des. Integr. Circuits Syst.* **2018**, *38*, 1084–1094. [CrossRef]
28. Acheampong, A.O.; Boateng, E.B. Modelling carbon emission intensity: Application of artificial neural network. *J. Clean. Prod.* **2019**, *225*, 833–856. [CrossRef]
29. Majalca, R.; Acosta, P.R. Convex Hulls and the size and the Size of the Hidden Layer in a MLP Based Classifier. *IEEE Lat. Am. Trans.* **2019**, *17*, 991–999. [CrossRef]
30. Tian, J.; Liu, Y.; Zheng, W.; Yin, L. Smog prediction based on the deep belief–BP neural network model (DBN–BP). *Urban. Clim.* **2022**, *41*, 101078. [CrossRef]

agronomy

Article

The Medium-Blocking Discharge Vibration-Uniform Material Plasma Seed Treatment Device Based on EDEM

Yunting Hui, Chen Huang, Yangyang Liao, Decheng Wang, Yong You * and Xu Bai

College of Engineering, China Agricultural University, Beijing 100083, China; hyt@cau.edu.cn (Y.H.); hc@cau.edu.cn (C.H.); liaoyangyang@cau.edu.cn (Y.L.); wdc@cau.edu.cn (D.W.); baixu@midea.com (X.B.)
* Correspondence: youyong@cau.edu.cn

Abstract: Pre-sowing treatment of seeds by plasma can improve seed vigor and promote seed germination and growth. To solve the problems of low processing volume and uneven treatment in plasma seed treatment devices, according to the process scheme of medium-blocking discharge plasma seed treatment, a medium-blocking discharge vibration-uniform material plasma seed treatment device was designed, the structure and working principle of the vibration-uniform material device were systematically analyzed, and the mathematical model of seed force was established. According to electromagnetic vibration theory, the seed sorting and conveying principles were analyzed in the lower trough, and the relevant parameters were selected and calculated. Using EDEM discrete element simulation software, a numerical simulation of alfalfa seed feeding and vibration-uniform material process was carried out. A three-factor, three-level orthogonal test was established. The results showed that the vibration amplitude and groove shape significantly affected the coefficient of variation of seed uniformity on the groove during the seed feeding and vibration-uniform material processes, and the groove wheel speed had a certain effect on the coefficient of variation of uniformity. The main order of factors affecting the uniformity of seed spreading was vibration amplitude B > notch shape C > speed A. The optimal speed was 35 r/min, the optimal notch shape was circular, and the optimal vibration amplitude was 0.55 mm.

Keywords: alfalfa seeds; uniform fabric; discrete elements

Citation: Hui, Y.; Huang, C.; Liao, Y.; Wang, D.; You, Y.; Bai, X. The Medium-Blocking Discharge Vibration-Uniform Material Plasma Seed Treatment Device Based on EDEM. *Agronomy* **2023**, *13*, 2055. https://doi.org/10.3390/agronomy13082055

Academic Editors: Chao Chen and Roberto Marani

Received: 12 May 2023
Revised: 12 July 2023
Accepted: 20 July 2023
Published: 3 August 2023

1. Introduction

In recent years, with the improvement of people's living standards and the change of consumption concept, the demand for meat, eggs, and milk has significantly increased [1]. Alfalfa has a high yield and excellent grass quality and is loved by various livestock and poultry [2]. As the main source of protein feed, alfalfa plays an important role in the protein feed shortage in China's animal husbandry [3,4]. Therefore, researching technologies to improve alfalfa yield is of great practical significance.

Seeds are important agricultural production materials, and their vitality affects the entire life cycle of plants. However, the vitality of seeds decreases due to dormancy and physiological aging [5,6]. Effective pre-sowing treatment of seeds is an important way to improve seed viability and to achieve increased crop yields and incomes [7,8]. Currently, commonly used pre-sowing treatment methods for seeds include chemical, biological, and physical treatments [4]. Chemical treatment methods generally use some chemical agents, which can easily damage the environment and even harm human and animal health [9]. Biological treatment methods, such as biological control and plant hormones, are costly and have unclear effects and are also susceptible to environmental influences [10–12]. In recent years, physical treatment methods have become more and more popular due to their environmental protection and good effects [13]. Among them, plasma seed treatment technology is widely used due to its high efficiency, environmental friendliness, and strong applicability [14,15].

Foreign studies have determined that plasma is an effective pre-treatment method for promoting seed germination and growth [16]. Sera et al. [17] observed that, after plasma treatment, the growth of wheat and oat seeds was promoted, and the growth of young roots was accelerated. Moreover, cold plasma has a good sterilizing effect on seeds. Schnabel et al. [18] used medium-blocking discharge and microwave plasma to treat rapeseed seeds, both of which reduced the number of spore rods. Cold plasma can also improve the surface morphology and wetting properties of seeds. Nalwa et al. [19] used low-pressure glow discharge oxygen cold plasma to treat sweet pepper seeds. The surface morphology of the seeds changed and resulted in better seedling characteristics after planting. There have been many studies on plasma seed treatment technology in China. Meng Yiran found that using plasma to treat seeds can promote seed growth, increase yield, enhance seed stress resistance, and kill pathogenic bacteria on grain surfaces [20]. Wang Decheng et al. [21] designed a plasma seed treatment device with temperature control to address the issue of high temperatures caused by continuous treatment with low-pressure radio frequency plasma.

The transportation mode of seeds affects the effect of a plasma seed treatment. The commonly used transportation mode of the sample tray and conveyor belt will lead to an uneven seed treatment effect due to the relatively fixed seed position. Therefore, vibration transportation is widely used because of its simple structure and reliable operation [22]. Xing Jiejie et al. [23] used ADAMS and other software to establish a parameterized electromagnetic vibration orientation device and seed simulation model of corn seeds and verified that the vibration effect of the model was in line with reality from the three aspects of vibration analysis, contact force, and model vibration effect. Xia Hongmei et al. [24] designed a guided vibration seed supply device that can realize uniform seed transportation. Lim [25] pointed toward a possible methodology to model vibrating granular bed systems with inelastic bases using continuum theories. Stepanenko et al. [26] obtained the dependence of the function of the flow rate of grain material on the stepped surface of the vibrating feeder.

Currently, the application of plasma in agriculture, both domestically and internationally, is not comprehensive or well understood in terms of its mechanisms. Most of the developed processing equipment is experimental machines with limited processing capacity. For some batch processing machines that use a conveyor belt as the transportation method, the seeds can block each other, making it difficult to achieve uniform treatment [27–29]. This article aims to address the issues of insufficient processing capacity and uneven treatment of current plasma seed processing machines through motion and dynamic analysis of seeds on a vibrating device. Based on the process plan of medium-blocking discharge plasma seed treatment, we designed a uniform material flow vibration plasma seed treatment device to provide technical support and a theoretical basis for the industrialization of seed treatments.

2. Materials and Methods

2.1. Establishment of Alfalfa Seed Model

To ensure the accuracy of the simulation results, 100 alfalfa seeds were randomly selected from each, and the length, width, and thickness of the seeds were measured by digital vernier calipers (range 0~150 mm, accuracy 0.01 mm) [30], and the dimensions of the three axes of the seeds were 2.29 mm long, 1.38 mm wide, and 0.95 mm high. Alfalfa seeds are shown in Figure 1. The Poisson's ratio, shear modulus, density, and contact parameters of the alfalfa seeds were obtained through literature research and physical experiments [31,32]. The material properties and contact parameters of the seed discrete element model are shown in Tables 1 and 2 [33,34]. The discrete element model of alfalfa seeds was established using EDEM 2021 simulation software [35], as shown in Figure 2. As the alfalfa seeds are irregular in shape, their discrete element models are difficult to build by individual particles. Therefore, multiple spherical particles are aggregated to build models to simulate the actual seed properties more accurately. The Hertz–Mindlin nonslip model was selected as the particle contact model [36,37].

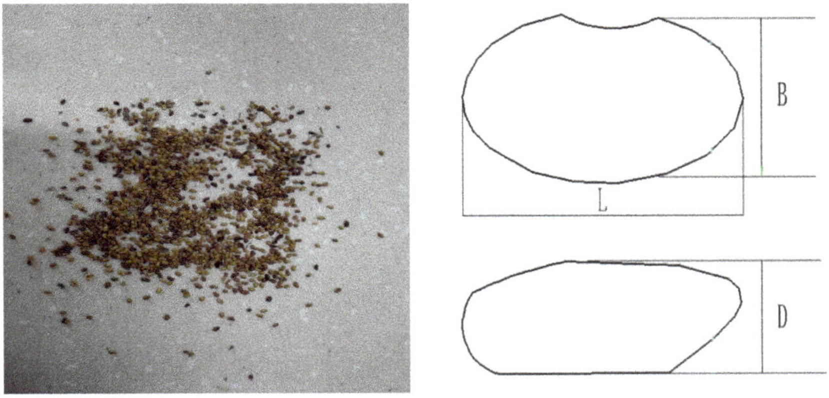

Figure 1. Alfalfa seeds. (**Left**): Alfalfa seeds physical picture. (**Right**): Alfalfa seed size diagram ("L" means "long", "B" means wide, and "D" means "high").

Table 1. Seed discrete element model and material parameters.

Parameters	Numerical Value
The Poisson's ratio of alfalfa seeds	0.4
The shear modulus of alfalfa seeds (MPa)	10
The density of alfalfa seeds (g/cm^{-3})	0.65
The Poisson's ratio of steel	0.3
The shear modulus of steel (MPa)	102
The density of steel (g/cm^{-3})	7850
The Poisson's ratio of nylon	0.4
The shear modulus of nylon (MPa)	90
The density of nylon (g/cm^{-3})	1200

Table 2. Contact parameters between seed and material.

Contact Parameters	Recovery Factor	Coefficient of Static Friction	Coefficient of Rolling Friction
alfalfa-alfalfa	0.21	0.191	0.005
alfalfa-nylon	0.47	0.500	0.010
alfalfa-steel	0.63	0.075	0.023

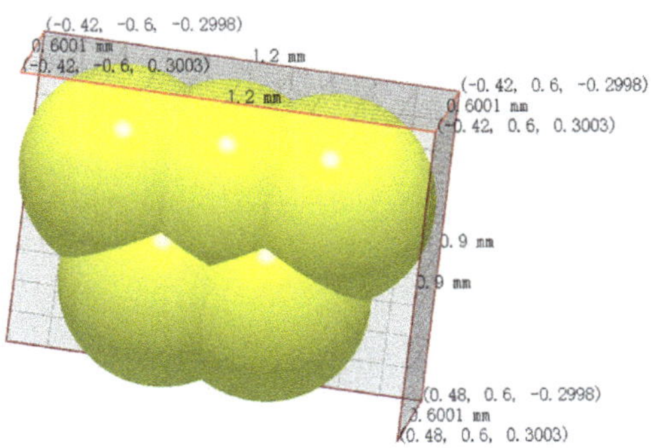

Figure 2. Discrete element model of alfalfa seeds.

2.2. Device Model

2.2.1. Medium-Blocking Discharge Plasma Seed Treatment Device

Technical requirements for the medium-blocking discharge plasma seed treatment device:

(I). General requirements

The medium-blocking discharge plasma seed treatment device operates under atmospheric pressure conditions, the batch processing volume reached up to 10 kg, the generation power of the medium-blocking discharge ranged from 0 to 500 W (adjustable), and the medium-blocking discharge treatment time for seeds ranged from 5 to 20 s (adjustable).

(II). Functional requirements

(a) The feeding amount was accurate and reliable, and the feeding regulation range was large. Under the condition of striving for precision, the feeding device was structurally designed so that the number of seeds fed each time was accurate, which was conducive to the seeds lying flat and evenly into the treatment bin.
(b) Feeding process achieved even spreading of seeds through the device's vibration conveyance. Whether the seeds could evenly enter between the polar plates for media-blocking discharge plasma treatment was one of the key technologies of this device.
(c) The transportation was stable, and the operation was reliable. The stable conveying structure prevented the seeds from being affected in the distribution state during the transmission process, thereby improving the stability of equipment performance.

In summary, the main technical parameters of the medium-blocking discharge vibration-uniform material plasma seed treatment device are shown in Table 3.

Table 3. The main technical parameters of the medium-blocking discharge vibration-uniform material plasma seed treatment device.

Name	Numerical Value
Plasma type	Medium-blocking discharge
Plasma generation power (W)	0~500
Seed processing time (s)	5~20 (adjustable)
Working air pressure (Pa)	10^5 (atmospheric pressure)
Treatment volume maximum (kg/batch)	10 (in alfalfa)
Temperature (°C)	15~20

According to the design requirements of the medium-blocking discharge plasma seed treatment device, a process plan for a medium-blocking discharge plasma seed treatment device with uniform material flow and vibration was proposed, as shown in Figure 3. The working process of the medium-blocking discharge plasma seed treatment device with uniform material flow and vibration consisted of four parts: the feeding device ensured accurate feeding of seeds; the uniform material distribution device ensured automatic uniform spreading of the seeds during transportation; the conveyor device realized smooth and batch transportation of seeds; and the medium-blocking discharge plasma generation system was used to provide a stable plasma environment for the seeds in the transportation device. The entire process was monitored and controlled in real time by the control system.

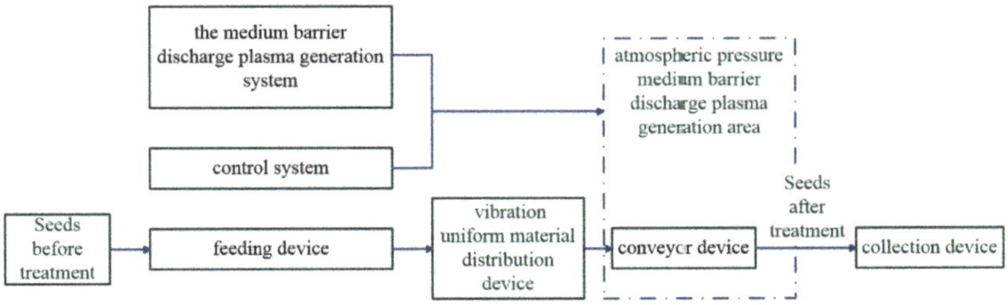

Figure 3. Process diagram of medium-blocking discharge plasma seed treatment.

Based on the process plan of the medium-blocking discharge plasma seed treatment device, a medium-blocking discharge plasma seed treatment device with uniform material flow and vibration was designed. The device mainly consisted of a feeding device, a vibration-uniform material distribution device, a conveyor device, a medium-blocking discharge plasma generation system, a collection device, etc. Among them, the feeding device included a feeding bin, seed dispenser body, outer groove wheel, connecting shaft, deceleration motor, etc. The vibration-uniform material distribution device included a groove body, groove body bracket, plate spring, vibrator, clamping iron, connecting fork, base, and other components. The conveyor device included a belt, roller, support frame, and so on. The medium-blocking discharge plasma generation system mainly included an RF power supply, voltage regulator, electrode plate, shielding layer, etc. The collection device mainly consisted of a guiding support frame, collection bin, etc. The structure schematic of the medium-blocking discharge plasma seed treatment device with a vibratory uniform transport function is shown in Figure 4.

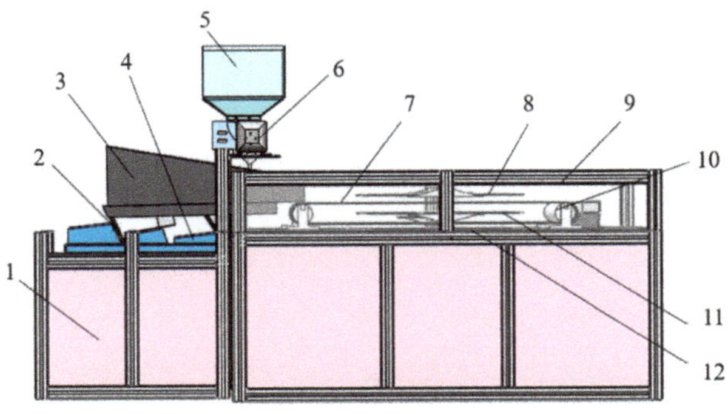

Figure 4. The structure schematic of the medium-blocking discharge plasma seed treatment device with a vibratory uniform transport function. 1. Rack; 2. Plate spring; 3. Groove body, Discharger body; 4. Groove body bracket; 5. Material box; 6. Motor; 7. Belt; 8. Upper level board; 9. Upper level board shield; 10. Roller; 11. Lower level board; 12. Lower level board shield.

During operation, the control system controlled the deceleration motor to feed the seeds through the feeding device. The seeds fell onto the vibration-uniform material distribution device, and a unidirectional half-wave rectified current was inputted to the vibrator, causing the vibrator to generate a periodic electromagnetic force that drove the groove body to move back and forth, achieving automatic and uniform spreading of seeds and forward transportation. Finally, the forward-moving seeds fell onto the conveyor belt, where they were transported by the motor-driven belt inside the drum to the processing area for plasma treatment. After treatment, the processed seeds were transported by the conveyor belt to the collection device for collection.

2.2.2. Kinematic Analysis of Seed Movement on the Material Groove

The vibration-uniform material distribution device used electromagnetic vibration, and the principle of electromagnetic vibration was to use the periodic electromagnetic force generated by the electromagnetic vibrator as the excitation force to maintain a persistent and stable vibration. The excitation force generated by the electromagnetic vibrator was related to the power supply mode. The power supply mode of the electromagnetic vibrator was mostly unidirectional half-wave rectification, as shown in Figure 5.

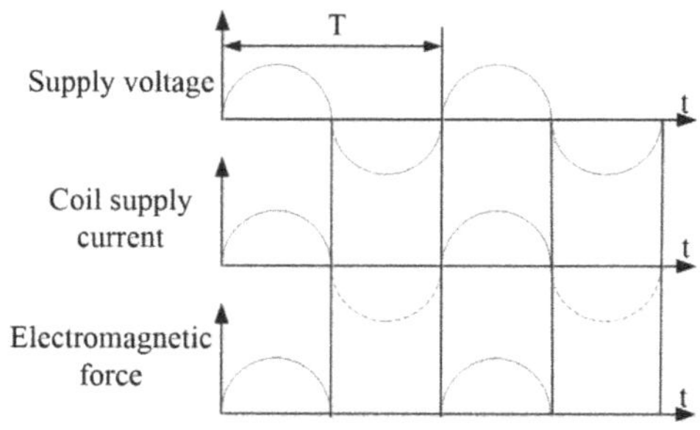

Figure 5. Graph of current vs. electromagnetic force.

In order to analyze the movement of seeds on the uniform material distribution device, a force analysis was conducted on the seeds on the material groove. A coordinate system was established with the surface direction of the material groove as the x-axis and the vertical direction of the material groove as the y-axis, as shown in Figure 6.

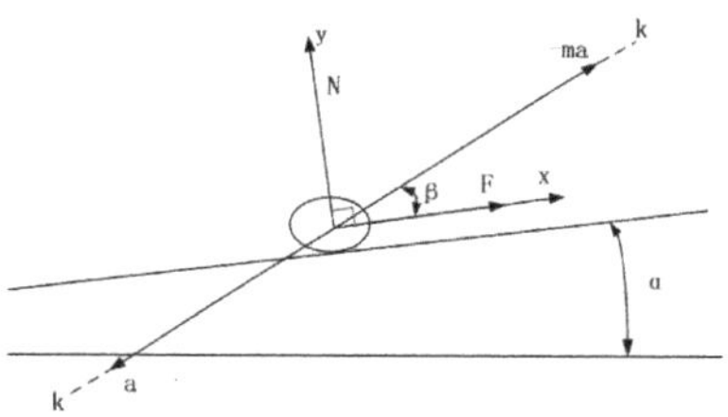

Figure 6. Force analysis diagram of seeds on the material groove.

Note: In the diagram, m—the mass of the seed; g—the gravitational acceleration; α—the installation angle of the groove body; β—the vibration direction angle; N—the positive pressure of the groove body on the seeds; F—the frictional force of the groove body on the seeds.

Due to the suction of the electromagnet, the material groove moved from the upper right corner to the lower left corner with the seeds at an acceleration of a, and the seeds were affected by the inertia force of ma. The vertical force to the surface of the material groove, ma sinβ, reduced the positive pressure on the seeds, thereby reducing the frictional force. As shown in Figure 7a, when the parallel force to the surface of the material groove, ma cosβ, was greater than the frictional force F, the seed slid upward along the x-axis; when the vertical force on the seed, ma sinβ, was greater than the gravity force of the seed itself, mg cosα, the seed began to jump and underwent projectile motion. If the throwing time was equal to the descending time of the material groove, the seed had the longest running time and the farthest distance traveled on the material groove when it contacted the groove, as shown in Figure 7b, from point B to C; if the throwing time was less than the descending time of the material groove, the seed would return to the groove earlier and follow the groove down, similar to taking "two steps forward and one step back" on the material groove, resulting in a smaller displacement of the seed's forward movement, as shown in Figure 7c. If the throwing time of the seed was greater than the descending time of the material groove, the seed would jump higher and return later, but land closer to the groove, as shown in Figure 7d.

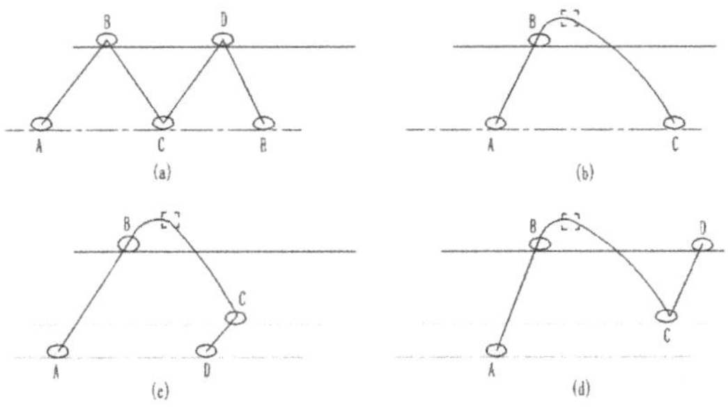

Figure 7. The motion states of seeds on the material groove. (**a**)The seed slid upward along the x-axis. (**b**) The throwing time was equal to the descending time of the material groove. (**c**) The throwing time was less than the descending time of the material groove. (**d**) he throwing time of the seed was greater than the descending time of the material groove. "A" indicates the initial position of the seed when the vibration device starts working; "B" indicates the position of the seed when the vibration device reaches the highest point; "C" indicates the location where the seed makes second contact with the material groove.

(1) Displacement, velocity, and acceleration of the material groove.

To further determine the motion of seeds on the groove, a mathematical model was established for the motion of seeds on the material groove. By performing dynamic analysis of the groove performing harmonic motion and a seed on its surface, the displacement formula of the working surface of the groove was derived, thereby determining the displacement, velocity, and acceleration of the groove.

$$S = \lambda sin\omega t \tag{1}$$

$$\omega t = \varphi \tag{2}$$

In the equation: λ—the single amplitude of the material groove vibrating along the k-k direction; ω—the circular frequency of the vibration; φ—the phase angle of the vibration; t—time.

Decomposing the vibration displacement along the x-axis and y-axis directions, the fractional displacements along the x-axis and y-axis are obtained, respectively, as shown in Equations (3) and (4):

$$S_y = \lambda sin\omega t sin\beta \tag{3}$$

$$S_x = \lambda sin\omega t cos\beta \tag{4}$$

Taking the first derivative and second derivative of time t for Formulas (1), (3) and (4), the velocities v_x and v_y along the x-axis and y-axis, as well as the accelerations a_x and a_y along the x-axis and y-axis were obtained as follows:

$$v_y = \lambda\omega cos\omega t sin\beta \tag{5}$$

$$v_x = \lambda\omega cos\omega t cos\beta \tag{6}$$

$$a_y = -\lambda\omega^2 sin\omega t sin\beta \tag{7}$$

$$a_x = -\lambda\omega^2 sin\omega t cos\beta \tag{8}$$

(2) The conditions for the seed to perform a throwing motion on the material groove.

First, the movement of a single seed on the groove was analyzed. Under this condition, the interaction forces between the seeds were completely neglected. It was assumed that the seed had relative motion with the surface of the material groove, with relative displacements $\Delta\chi$ and Δy in the x-axis and y-axis directions, relative velocities $\Delta\dot{\chi}$ and $\Delta\dot{y}$, and relative accelerations $\Delta\ddot{\chi}$ and $\Delta\ddot{y}$.

The force that promoted the sliding of the seed along the x-axis direction was the sum of the inertial and gravitational components of the seed parallel to the surface of the groove, as shown in Formula (9).

$$F_x = -m\left(a_x + \Delta\ddot{\chi}\right) + mgsin\alpha \tag{9}$$

The positive pressure of the groove surface on the seeds was given by Formula (10):

$$N = -m\left(a_y + \Delta\ddot{y}\right) + mgcos\alpha \tag{10}$$

In the formula: m—the mass of the seed; α—the inclination angle of the tank, when conveying downward take the "+" sign, when conveying upward take the "−" sign.

When the seed was in sliding motion relative to the trough surface, it contacted the surface of the trough, with positive pressure $N \geq 0$ and acceleration $\Delta\ddot{y} = 0$. When treated as a throwing motion $N = 0$ and $\Delta\ddot{y} \neq 0$. Formula (11) was obtained.

$$sin\varphi_d = gcos\alpha/\omega^2\lambda sin\beta \tag{11}$$

In the formula: φ_d—initial angle of the throwing motion, i.e., the phase angle at the beginning of the throwing motion instantaneously.

$$\varphi_d = arcsin\frac{1}{D} \tag{12}$$

$$D = Ksin\beta/cos\alpha \tag{13}$$

$$K = \omega^2\lambda/g \tag{14}$$

In the formula: D—throwing index, indicating the characteristics of the throwing motion; K—vibration intensity.

From the above formula, it can be seen that when $D > 1$, there was a solution to the initial throwing phase angle φ_d, and the seeds could have throwing motion on the surface of the groove; when $D < 1$, φ_d had no solution and the seeds could not perform throwing motion on the surface of the groove. Therefore, it could be concluded that the throwing index should not be less than 1 in order to make the seeds do throwing motion on the surface of the groove.

Since the vibration intensity $K = \omega^2\lambda/g$ and $\omega = \frac{2\pi n}{60}$, after choosing the amplitude λ, the vibration frequency was calculated according to Formula (15):

$$n = 30\sqrt{\frac{Dgcos\alpha}{\pi^2\lambda sin\beta}} \tag{15}$$

If the vibration frequency n was pre-selected, the amplitude was calculated according to Formula (16):

$$\lambda = \frac{900Dgcos\alpha}{\pi^2 n^2 sin\beta} \tag{16}$$

(3) Selection of throwing index D.

The movement state of the seed on the surface of the groove depended on the throwing index D. As can be seen from Formulas (12) and (13), the throwing index D should not be less than 1 in order to make the seeds do throwing movement on the surface of the groove.

The value of D normally ranged from 1 to 3, so that the time of the throwing process was less than the vibration cycle, thus improving the efficiency of the machine.

(4) The inclination angle α of the groove body and the vibration direction angle β.

When the seeds did throwing motion on the groove, from the perspective of improving the conveying speed, there was an optimal vibration direction angle at different trough inclination corresponding to each vibration intensity K. In order to improve the conveying ability of the vibration-uniform material distribution device for the seeds, the groove body was installed with a downward tilt because the inclination angle α was generally between 10° and 15° and was not too large. Therefore, this design selected an inclination angle of 10°. Typically, the value range of K was between 2 and 5, but due to the small size of the seed particles, the K value was relatively large, and the value of K was set to 8 in this design. Thus, the vibration direction angle β can be determined to be 21°.

(5) Vibration amplitude and vibration frequency

Generally, electromagnetic vibrating machines used high frequency and small amplitudes, with amplitudes usually around 3000 times/min and vibration amplitudes of 0.5–1 mm. Due to the small volume of material acted upon in this design, the vibration amplitude was not necessarily too large. Therefore, the vibration frequency was 60 Hz, and the vibration amplitude was 0.5 mm.

(6) Theoretical conveying speed.

The theoretical conveying speed of seeds on the material groove was calculated using Formula (17):

$$v_d = f(D)\omega a cos\beta \tag{17}$$

Since the throwing index D was known, $f(D)$ was determined as 0.92 from the dimensionless coefficient curve, and v_d was calculated to be 0.19 m/s.

(7) Actual conveying speed.

The actual conveying speed of seeds on the material groove was calculated using Formula (18):

$$v_m = \gamma_a C_h C_m C_w v_d \tag{18}$$

In the formula: γ_a—inclination correction coefficient, $\gamma_a = 0.9$; C_h—is the material layer thickness influence factor, which was taken as 0.9 here; C_m— the material property influence factor, and since alfalfa seeds belong to granular materials, C_m was selected as 0.9; C_w¯the sliding motion influence factor, since D = 2.5, the effect of sliding motion was ignored, so it was taken as 1 here.

Finally, the actual conveying speed of seeds on the material chute, v_m, was calculated as 0.14 m/s.

2.3. Experimental Design

Whether the seeds were spread evenly during the vibrating conveying process on the material groove was an evaluation index for the performance of the feeding device and the uniform vibration conveying device. A "Grid Bin Group" was set in the middle of the material groove to display the number of seed particles passing through the set area, as shown in Figure 8.

Three types of groove wheel shapes were selected, namely circular arc groove, conical arc groove, and right-angled groove. The rotational speeds of the groove wheels were set as 35 r/min, 45 r/min, and 55 r/min. The vibration amplitudes of the material groove were set as 0.45 mm, 0.5 mm, and 0.55 mm. The simulation was carried out under the above conditions, and a three-factor, three-level simulation experiment was conducted using the uniformity variation coefficient as the evaluation index. The factor level table is shown in Table 4.

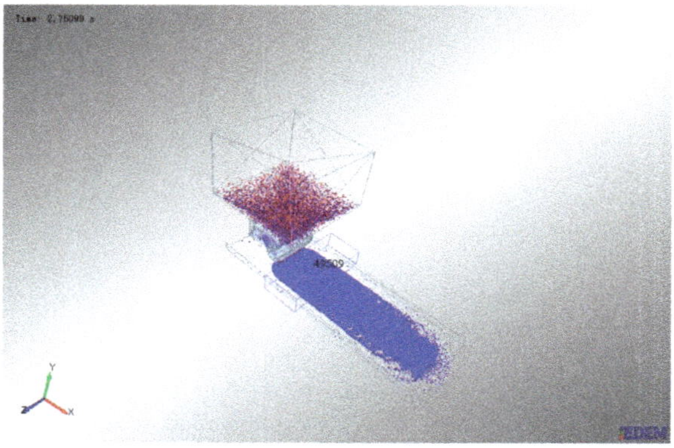

Figure 8. Setting of particle flow rate.

Table 4. Table of factor levels.

Level	Factor		
	A (r/min) Rotational Speed	B (mm) Vibration Amplitudes	C Groove Wheel Recess Shape
1	35	0.45	circular arc groove
2	45	0.5	conical arc groove
3	55	0.55	right-angled groove

The coefficient of variation was given by Formula (19):

$$Y_z = \sqrt{\frac{1}{n-1}\sum(X_i - x)^2} \times 100\% \tag{19}$$

In the formula: X—the average number of seeds distributed per measurement segment; n—the total number of samples in the experiment; X_i—the number of seeds per 100 mm measuring section.

3. Results and Discussion

The orthogonal table $L_9(3^4)$ was selected to arrange the orthogonal test scheme, and the fourth column was blanked as the error term in the variance treatment. The uniformity variation coefficient was used as the evaluation index to analyze and determine the optimal groove shape in the feeding device and the optimal amplitude parameter in the uniform vibration device. The experimental results are shown in Table 5.

Table 5. Orthogonal test results.

Serial Number	A (r/min) Rotational Speed	B (mm) Vibration Amplitudes	C Groove Wheel Recess Shape	The Coefficient of Variation (%)
1	1	1	1	36.18
2	1	2	2	42.72
3	1	3	3	36.06
4	2	1	2	42.72
5	2	2	3	33.91
6	2	3	1	35.49
7	3	1	3	33.84
8	3	2	1	35.51
9	3	3	2	31.48

3.1. Analysis of Extreme Differences

When using the uniformity variation coefficient as the evaluation index, the extreme difference analysis of the seed feeding vibration conveying simulation orthogonal test is shown in Table 6.

Table 6. Table of extreme difference analysis of the feeding vibration conveying simulation orthogonal test.

Serial Number	A (r/min) Rotational Speed	B (mm) Vibration Amplitudes	C Groove Wheel Recess Shape	The Coefficient of Variation (%)
1	1	1	1	37.30
2	1	2	2	27.10
3	1	3	3	28.50
4	2	1	2	45.30
5	2	2	3	37.80
6	2	3	1	21.20
7	3	1	3	60.40
8	3	2	1	28.50
9	3	3	2	24.50
$\overline{K}_{11}$	30.97	47.667	29.00	
$\overline{K}_{12}$	34.77	31.13	32.30	
$\overline{K}_{13}$	37.80	24.73	42.23	
R_j	6.83	3.24	13.23	
excellent level	A_1	B_3	C_1	
Primary and secondary factors	B > C > A			

When the uniformity variation coefficient was used as the evaluation index, the smaller its value, the better. From Table 6, it can be seen that the primary and secondary factors affecting the uniformity of seed spreading in the feeding and vibrating conveying process were in the following order: vibration amplitude B > groove shape C > rotational speed A. The optimal levels were: the optimal rotational speed was A_1 = 35 r/min; the optimal groove shape was C_1 = circular arc shape; and the optimal vibration amplitude was B_3 = 0.55 mm.

3.2. Analysis of Variance

The variance analysis of the seed feeding and vibrating conveying simulation orthogonal experiment is shown in Table 7.

Table 7. Analysis of variance for the seed feeding and vibrating conveying simulation orthogonal experiment.

Source	Sum of Squares Formula	Degree of Freedom	Mean Square	F	Significance-P
Modified model	1195.267 [a]	6	199.211	27.360	0.036
intercept distance	10,719.15	1	10,719.15	1472.186	0.001
Rotational Speed	70.336	2	35.168	4.830	0.172
Vibration amplitude	840.249	2	420.124	57.701	0.017
groove shape	284.682	2	142.341	19.549	0.049
Error	14.562	2	7.281		
Total	11,928.9	9			
Total after correction	1209.83	8			

Note: $p < 0.05$ indicates a significant effect; $p \geq 0.05$ indicates no effect, "a" indicates that the modified model is significant.

From the analysis of variance table, it can be seen that when using the uniformity variation coefficient as the evaluation index, $R^2 = 0.952$, indicating a good fitting effect. Under a confidence level of 95%, the vibration amplitude and groove shape had a significant impact on the uniformity variation coefficient of seeds on the groove body, while the rotational speed of the groove wheel had no influence on the uniformity variation coefficient.

3.3. Uniformity Test

The uniformity test was conducted in the Forage Machinery Laboratory of China Agricultural University. Seeds were fed from the feeding device. The rotational speed of the groove wheel of the feeding device was set at 35 r/min, and the vibration frequency of the vibration-uniform material distribution device was set at 60 Hz. The uniformity test of the plasma seed treatment device was carried out (Figure 9). The samples for the uniformity test were the seeds transported to the discharge area, and random samples were taken within the discharge area with a sampling area of 20 mm × 20 mm. Sampling was repeated 10 times, and each sample was tested twice. The weight of the obtained alfalfa seed samples was measured using an electronic balance (OHAUS Instruments Ltd. Shanghai, China, precision 0.0001 g).

Figure 9. The uniformity test of the plasma seed treatment device. **Left**: The uniformity test process. **Right**: The vibration-uniform material distribution device.

According to the uniformity evaluation method of CNAS-GL003:2018 "Guidelines for the Evaluation of Uniformity and Stability of Proficiency Testing Samples", one-way ANOVA was selected to evaluate the uniformity of the seeds when entering the discharge region [38]. The calculation formula is shown below:

$$M = \sum_{i=1}^{m} n_i \tag{20}$$

In the formula: n_i—the number of times the ith seed sample was tested; M—total number of tests for all seed samples; m—total number of seed samples to be tested.

The average value of the test for a single seed sample:

$$\overline{x_i} = \sum_{j=1}^{n} \frac{x_{ij}}{n_i} \tag{21}$$

In the formula: x_{ij}—the value of the ith seed sample tested at the jth time; $\overline{x_i}$—test averages for each seed sample.

Total average of all samples $\overline{\overline{x_i}}$:

$$\overline{\overline{x_i}} = \sum_{i=1}^{m} \frac{\overline{x_i}}{m} \tag{22}$$

Mean square value between samples:

$$MS_1 = \frac{1}{m-1}\sum_{i=1}^{m} (\overline{x_i} - \overline{x})^2 n_i \tag{23}$$

Mean square error within sample:

$$MS_2 = \frac{1}{M-m}\sum_{i=1}^{m}\sum_{j=1}^{n}\left(\overline{x_{ij}} - \overline{x}\right)^2 \tag{24}$$

Statistics used to characterize inter-sample variability in one-way ANOVA methods F:

$$F = \frac{MS_1}{MS_2} \tag{25}$$

The weighed alfalfa samples were subjected to one-way ANOVA, and the results are shown in Table 8. The F statistic was 0.652. By referring to GB/T 4086.4-1983, the critical value for $F_{0.05(9,10)}$ was found to be 3.02. Compared with the value of F of the statistic, it met the requirement that F was less than the critical value, which indicates that at the 0.05 level of significance, this uniformity of the tested alfalfa seed samples in the discharge region met the requirement of uniformity for proficiency testing.

Table 8. Results of one-way ANOVA for alfalfa samples.

Source of Variance	Degree of Freedom	Sum of Squares	Mean Square	F
Between samples	9	0.02821	0.003134	0.652
Within-sample	10	0.04805	0.004805	

4. Conclusions

(1) The key components of the medium-blocking discharge vibration-uniform conveying plasma seed treatment device were designed, and kinematic analysis of the seed on the groove was conducted to determine the relevant parameters. The operating frequency of the vibration-uniform feeding device was 60 Hz, the vibration amplitude was 0.5 mm, the vibration direction angle β was 21°, the trough inclination angle α was 10°, and the actual conveying speed of the seed on the trough v_m was 0.14 m/s.

(2) Using the discrete element method and based on EDEM software, the effects of the rotational speed of the groove wheel, vibration amplitude, and groove shape on the coefficient of variation of seed uniformity during feeding and vibrating conveying were analyzed. The results showed that the vibration amplitude and groove shape had a significant impact on the uniformity variation coefficient of seeds on the groove body, while the rotational speed of the wheel had no influence on the uniformity variation coefficient. The optimal rotational speed of the wheel was 35 r/min, the best groove shape was the circular arc groove, and the optimal vibration amplitude was 0.55 mm.

(3) Simulation experiments were conducted on the seed feeding and vibration conveyance processes, and the optimal rotational speed of the wheel, groove shape, and vibration amplitude were determined. The uniformity test of the plasma seed treatment device was conducted on a vibration-uniform material distribution device. It was demonstrated that at a significance level of 0.05, the uniformity of the tested alfalfa seed samples in the discharge area met the requirements of proficiency testing uniformity. In the future, prototype manufacturing and processing will be carried out based on these parameters, and the machined prototype will be used for testing to compare and verify the reliability of the simulation results.

Author Contributions: Conceptualization, Y.H.; Methodology, C.H.; Formal analysis, Y.L.; Resources, D.W.; Writing—review & editing, Y.Y.; Investigation, X.B. All authors have read and agreed to the published version of the manuscript.

Funding: This research was funded by the China Agriculture Research System (CARS-34) and the National Key R&D Program of China (2022YFD2001902).

Data Availability Statement: Not applicable.

Conflicts of Interest: The authors declare no conflict of interest.

References

1. Zhong, J.; Ni, K.; Yang, J.; Yu, Z.; Tao, Y. The present situation and prospect of the processing technology of forage grass in China. *Chin. Sci. Bull.* **2018**, *63*, 660–665. [CrossRef]
2. Pan, W.C. Main composition of alfalfa and its application in animal husbandry. *J. Tradit. Chin. Vet. Med.* **2015**, *34*, 73–77.
3. Bizzuti, B.E.; Pérez-Márquez, S.; van Cleef, F.D.O.S.; Ovani, V.S.; Costa, W.S.; Lima, P.M.T.; Louvandini, H.; Abdalla, A.L. In Vitro Degradability and Methane Production from By-Products Fed to Ruminants. *Agronomy* **2023**, *13*, 1043. [CrossRef]
4. El-Hifny, Z.; Bakheit, B.; Hassan, M.; El-Rady, W. Forage and Seed Yield Variation of Alfalfa Cultivars in Response to Planting Date. *SVU-Int. J. Agric. Sci.* **2019**, *1*, 21–33. [CrossRef]
5. Yan, X.; Wang, C.; Guo, Y. Storage, utilization and conservation of forage germplasm in China. *Pratacultural Sci.* **2008**, *25*, 85–92.
6. Aniszewski, T.; Haikonen, J.; Helwig, B.; Konert, G.; Oleksińska, Z.; Stenman, A.; Ylinampa, T. Vigor, vitality and seed dormancy of Avena sativa cultivars in a long-term experiment. *J. Appl. Bot. Food Qual.* **2012**, *85*, 150–158.
7. Kubeyev, E.I.; Antropov, B.S. Decomposition of Technological Processes for Evaluating the Performance of Production Line for Pre Sowing Treatment of Seeds. *Agric. Mach. Technol.* **2018**, *12*, 22–27. [CrossRef]
8. Kaitlynn, L.; Gautam, P.S.; Hansapani, R.; Rupesh, K.; Alexis, R. Pre-Sowing Treatments Improve Germinability of South Texas Native Plant Seeds. *Plants* **2021**, *10*, 2545.
9. Jha, A.C.; Rai, B.; Jha, M.M. Effect of Seed Treating Chemicals on Seed Germination, Root and Shoot Development and Health of Seedlings. *Ann. Biol.* **2004**, *20*, 199–201.
10. Sujanya, S.; Chandra, S. Effect of Biological Agents as Substitutes for Chemical Seed Treatment on Groundnut Cultivation. *Indian J. Innov. Dev.* **2012**, *1*, 47–53.
11. da Luz, D.C. Biological and chemical treatment combinations for corn seeds. *Fitopatol. Bras.* **2003**, *28*, 37–40.
12. Krupa, M.; Witkowicz, R. Biostimulants as a Response to the Negative Impact of Agricultural Chemicals on Vegetation Indices and Yield of Common Buckwheat (*Fagopyrum esculentum* Moench). *Agriculture* **2023**, *13*, 825. [CrossRef]
13. Erohin, A.I.; Tsukanova, Z.R. Physical methods of presowing cultivation of seeds and efficacy of their use. *Legum. Amp. Groat Crops.* **2014**, *7*, 646.
14. Thakur, N.; Vasudevan, S.N. Plasma Treatment and Seed Quality Advancement: A Review. *Agric. Rev.* **2021**, *42*, 197–202. [CrossRef]
15. Penado, K.N.M.; Mahinay, C.L.S.; Culaba, I.B. Effect of atmospheric plasma treatment on seed germination of rice (*Oryza sativa* L.). *Jpn. J. Appl. Phys.* **2018**, *57*, 01AG08. [CrossRef]
16. Ahmed, N.; Shahid, M.; Siow, K.; Razip, F.; Haron, F.; Patra, A.; Fazry, S. Germination and growth improvement of papaya utilizing oxygen (O_2) plasma treatment. *J. Phys. Appl. Phys.* **2022**, *55*, 255205. [CrossRef]
17. Sera, B.; Spatenka, P.; Sery, M.; Vrchotova, N.; Hruskova, I. Influence of Plasma Treatment on Wheat and Oat Germination and Early Growth. *Plasma Sci. IEEE Trans.* **2010**, *38*, 2963–2968. [CrossRef]
18. Schnabel, U.; Niquet, R.; Krohmann, U.; Winter, J.; Schlüter, O.; Weltmann, K.-D.; Ehlbeck, J. Decontamination of Microbiologically Contaminated Specimen by Direct and Indirect Plasma Treatment. *Plasma Process. Polym.* **2012**, *9*, 569–575. [CrossRef]
19. Nalwa, C.; Thakur, A.; Vikram, A.; Rane, R.; Vaid, A. Effect of Cold Plasma Treatment and Priming in Bell Pepper (*Capsicum annuum* L.). *Int. J. Bio-Resour. Stress Manag.* **2017**, *8*, 535–538. [CrossRef]
20. Meng, Y. Physiological and Biochemical Effects of Discharge Plasma on Seed Germination of Wheat. Master's Thesis, Northwest Agriculture and Forestry University of Science and Technology, Xianyang, China, 2018.
21. Feng, J. Optimization Design and Experiment on Plasma Seed Treatment Device. Ph.D. Thesis, China Agricultural University, Beijing, China, 2018.
22. Wen, B. *Vibration Design Theory and Dynamic Design Methods*; Machinery Industry Press: Beijing, China, 2002.
23. Xing, J.; Xu, L.; Shi, L.; Liu, W.; Gao, Z.; Yuan, Q. Establishment and verification for the simulation model of directional device based on electromagnetic vibration for corn seeds. *J. China Agric. Univ.* **2017**, *22*, 129–134.
24. Xia, H.; Zhou, S.; Liu, Y.; Zhao, K.; Li, Z. Design and Test of Directional Vibrating Seed-feeding Device for Flat Solanaceous Vegetable Seeds. *Trans. Chin. Soc. Agric. Mach.* **2020**, *51*, 82–88.
25. Lim, E.W.C. Vibrated granular bed on a bumpy surface. *Phys. Rev. E* **2009**, *79*, 041302. [CrossRef] [PubMed]
26. Stepanenko, S.; Kotov, B.; Rud, A.; Zamrii, M. Theoretical studies of the process of grain material movement on the surface of the stepped vibrating feeder. *Vib. Eng. Technol.* **2022**, *14*, 25–32. [CrossRef]
27. Munkhuu, N. Research on Forage Grass Seed Treatment by Optimizing Low Temperature Plasma Test Platform. Ph.D. Thesis, China Agricultural University, Beijing, China, 2015.
28. Jo, Y.K.; Cho, J.; Tsai, T.C.; Staack, D.; Kang, M.H.; Roh, J.H.; Shin, D.B.; Cromwell, W.; Gross, D. A Non-thermal Plasma Seed Treatment Method for Management of a Seedborne Fungal Pathogen on Rice Seed. *Crop. Sci.* **2014**, *54*, 796–803. [CrossRef]
29. Wang, D.; Shao, C.; Fang, X.; Zhang, D. Application Status and Development Trend of Low-temperature Plasma Equipment Used for Seed Treatment before Sowing. *Agric. Eng.* **2013**, *3*, 1–4+34.
30. Zhou, Z.; Wang, Y.; Cao, C.; Wang, P. Separation Characteristics of Seed by Size. *J. Beijing Agric. Eng. Univ.* **1988**, *8*, 61–75.
31. Zhao, M. *The Sorting Properties of Alfalfa Seed and the Optimizing Design of the Control System of the Sorting Machinery*; China Agricultural University: Beijing, China, 2015.
32. Li, S.; Wang, G.; Wang, D.; Wang, Z. The Model Analysis and Simulation of Gramineae-Leguminous Mixed Seeding Based on EDEM. *China Dairy Cattle* **2016**, *21*, 58–60. [CrossRef]

33. Zhang, Y.; Cui, Q.; Wang, F.; Li, H. Measurement of physical and mechanical parameters for the small seed seeding. *J. Shanxi Agric. Univ. Sci. Ed.* **2016**, *36*, 224–228. [CrossRef]
34. Liu, X. *Study and Application of Bulk Mechanical Properties of Small Seed Grains*; Shanxi Agricultural University: Jinzhong, China, 2015.
35. Hu, G. *Discrete Element Method Analytical Simulation of Particle Systems*; Wuhan University of Technology Press: Wuhan, China, 2010.
36. Leblicq, T.; Smeets, B.; Vanmaercke, S.; Ramon, H.; Saeys, W. A discrete element approach for modelling bendable crop stems. *Comput. Electron. Agric.* **2016**, *124*, 141–149. [CrossRef]
37. Lenaerts, B.; Aertsen, T.; Tijskens, E.; De Ketelaere, B.; Ramon, H.; De Baerdemaeker, J.; Saeys, W. Simulation of grain–straw separation by Discrete Element Modeling with bendable straw particles. *Comput. Electron. Agric.* **2014**, *101*, 24–33. [CrossRef]
38. Xu, W.; Li, W.; Yang, X. Evaluation of Homogeneity and Stability of Proficiency Testing Samples. *China Wood-Based Panels* **2020**, *27*, 33–36.

agronomy

MDPI

Article

Transfer Learning-Based Lightweight SSD Model for Detection of Pests in Citrus

Linhui Wang [1,2,3], Wangpeng Shi [1], Yonghong Tang [2], Zhizhuang Liu [2], Xiongkui He [1,3,*], Hongyan Xiao [4] and Yu Yang [2]

1 Sanya Institute of China Agricultural University, Sanya 572019, China; wlh3362@huse.edu.cn (L.W.); wpshi@cau.edu.cn (W.S.)
2 School of Intelligent Manufacturing, Hunan University of Science and Engineering, Yongzhou 425100, China; tyh2977@huse.edu.cn (Y.T.); liuzz168@huse.edu.cn (Z.L.); cindyyang@huse.edu.cn (Y.Y.)
3 College of Agricultural Unmanned Systems, China Agricultural University, Beijing 100091, China
4 Institute of Nanfan & Seed Industry, Guangdong Academy of Sciences, Sanya 572019, China; jetwang30@gmail.com
* Correspondence: xiongkui@cau.edu.cn; Tel.: +86-10-62731446

Abstract: In citrus cultivation, it is a difficult task for farmers to classify different pests correctly and make proper decisions to prevent citrus damage. This work proposes an efficient modified lightweight transfer learning model which combines the effectiveness and accuracy of citrus pest characterization with mobile terminal counting. Firstly, we utilized typical transfer learning feature extraction networks such as ResNet50, InceptionV3, VGG16, and MobileNetV3, and pre-trained the single-shot multibox detector (SSD) network to compare and analyze the classification accuracy and efficiency of each model. Then, to further reduce the amount of calculations needed, we miniaturized the prediction convolution kernel at the end of the model and added a residual block of a 1 × 1 convolution kernel to predict category scores and frame offsets. Finally, we transplanted the preferred lightweight SSD model into the mobile terminals developed by us to verify its usability. Compared to other transfer learning models, the modified MobileNetV3+RPBM can enable the SSD network to achieve accurate detection of Panonychus Citri Mcgregor and Aphids, with a mean average precision (mAP) up to 86.10% and the counting accuracy reaching 91.0% and 89.0%, respectively. In terms of speed, the mean latency of MobileNetV3+RPBM is as low as 185 ms. It was concluded that this novel and efficient modified MobileNetV3+RPBM+SSD model is effective at classifying citrus pests, and can be integrated into devices that are embedded for mobile rapid detection as well as for counting pests in citrus orchards. The work presented herein can help encourage farm managers to judge the degree of pest damage and make correct decisions regarding pesticide application in orchard management.

Keywords: transfer learning; MobileNetV3; pest detection; embedded system; citrus pest

Citation: Wang, L.; Shi, W.; Tang, Y.; Liu, Z.; He, X.; Xiao, H.; Yang, Y. Transfer Learning-Based Lightweight SSD Model for Detection of Pests in Citrus. *Agronomy* **2023**, *13*, 1710. https://doi.org/10.3390/agronomy13071710

Academic Editor: Gniewko Niedbała

Received: 24 May 2023
Revised: 21 June 2023
Accepted: 23 June 2023
Published: 26 June 2023

1. Introduction

Horticultural management faces the problem of identifying and classifying pests. Citrus yield and quality are adversely affected by pest damage, which occurs frequently [1]. Due to its complex structure and the high similarity in appearance between different species [2,3], pest classification is a challenging task. To prevent the transmission of pests that cause citrus diseases, it is necessary to identify and classify pests as soon as possible in crops and to select pesticides and biological prevention methods that are effective.

Manual capture and sampling methods for pest identification are ineffective, laborious, and labor-intensive. Commonly, methods for detecting pests that utilize physical information technology, including the acoustic detection method [4,5], electronic nose technology [6,7], and spectral imaging technology [8,9], have spatial and temporal limitations when used in complex orchard environments. As a result of the breakthrough development in target detection and recognition technology due to the upgrading of computer hardware and deep learning technologies, it is gradually becoming possible to use digital image technology to accurately analyze pest images and understand the degree of pest damage [10,11]. To overcome the limitations of traditional detection methods, the development of a visual computerized system using machine learning for image processing for the accurate classification and identification of pests will gradually become a trend in the field of agricultural research [12].

In recent years, the CNN-based deep learning model, as a class of powerful image classification tools, has been widely applied to various problems in the agricultural industry, including for the identification of plant diseases [13], the classification of fruits [14], weed identification [15], and the classification of pests in crops [16]. Li et al. [17] put forward a detection method for pests associated with rape that relied on a deep convolutional neural network, which achieved the rapid and accurate detection of five rape pests, including Aphids, cabbage caterpillar larvae, rape bugs, flea beetles, and Phaedon brassicae Baly, with an average accuracy of 94.12%. Kuzuhara et al. [18] proposed a two-stage pest detection and identification technique utilizing enhanced convolutional neural networks and using the Xception model to re-identify the CNN output.

The single-shot multibox detector (SSD) model prioritizes inference speed and can guarantee better accuracy. It has become the model of choice for many mobile target recognition researchers. He et al. [19] designed a rapeseed pest imaging system and developed a supporting Android application based on the SSD model, which can be combined with UAV and Internet of Things technology to monitor rapeseed pests. In addition, there are many cases of detection systems similar to those designed by He [20,21]. The accuracy of plant species identification was improved by using Google Net, AlexNet, and VGGNet models in [22]. Khan et al. [23] conducted an intensive pre-training SSD model to extract depth features to classify six kinds of ale and banana fruit diseases to improve classification accuracy and the degree of accuracy. In [24], an SSD network analyzed complex pest images to determine powerful local features. Therefore, the classification of 12 important rice field pests with a high average accuracy was achieved. However, the above cases of pest detection have only remained in the aspect of model building and have not realized specific applications. To this end, it is particularly urgent to design a mobile pest imaging system combined with the SSD model to realize intelligent pest monitoring.

In this paper, we take citrus trees as the research object and Panonychus citri Mcgregor (PCM) and Aphids as the detection objects. We combine several typical transfer learning models, pre-train the SSD network, compare and analyze the classification accuracy and efficiency of each model, and select the best model, which can effectively alleviate the problem of a poor model effect caused by insufficient training samples, and can be adapted to the detection of various diseases. The contribution of this paper is as follows:

- We compare the latest MobileNet, GoogLeNet, ReseNet and VGGNet feature extraction networks, and optimize the best feature extraction network to further improve the detection speed and accuracy of the SSD.
- Before the model performs prediction, we add a miniaturized residual block of a 1×1 convolution kernel to each feature map used for prediction to predict category scores and frame offsets.
- The effective modified MobileNetV3-SSD model is transplanted into the embedded terminal developed by us to realize the rapid monitoring of pests.

Citrus plantations are investigated using a rapid detection method to obtain pest information, which is of great significance for the identification of citrus pests and provides a theoretical basis for the precise management of the orchard and the design of accurate fertilization and application equipment.

2. Materials and Methods

2.1. Datasets of Pests

The dataset images were collected on 13 October 2021 at the selenium-enriched navel orange planting base in Yongzhou, Hunan Province, China (26°22′40″ N, 111°46′28″ E), as this is a high incidence period for PCM and Aphids. The weather was clear and sunny that day. During sampling, a CMOS camera (Sony ILCE-7RM4A, Sony Corporation, Tokyo, Japan) was used as the sampling tool to acquire image data with a single citrus tree as the basic unit, and a total of 100 citrus tree images were collected. In consideration of a sampling equilibrium, five locations in the middle of citrus trees were selected as sampling points: front (A), back (C), left (D), right (B), and top (E). The sampling diagram is shown in Figure 1. In total, 1000 images were collected for the two types of pests, and the size of the images was uniformly adjusted to 224 × 224 pixels after preprocessing. To enhance the network model's accuracy in detecting anomalies, clipping, rotation, translation, and other methods were adopted to enhance the dataset images, and the two pest datasets were amplified to a total of 5000 images.

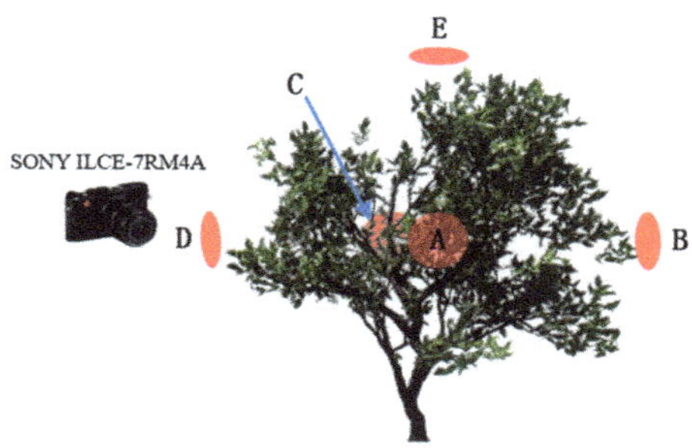

Figure 1. Sampling diagram. (A, B, C, D, and E represent the front, right, back, left, and top positions of the canopy of citrus trees, respectively, and are the positions for data collection).

LabelImg software "https://github.com/heartexlabs/labelImg (accessed on 25 June 2023)" was used to label the preprocessed pest images, and an xml data source file was generated. The labeled 10,000 sample images were split into three datasets, the training set, validation set, and test set, which were assigned randomly at a proportion of 8:1:1.

2.2. Modified SSD

The classic SSD network is a one-stage target detection network modeled after YOLO. It was proposed to improve the rough design of the anchor set by the YOLO network. It is designed using mainly multi-scale and multi-aspect ratio dense anchor points and a feature pyramid network. The SSD model is based on the base network of VGG16 and ends with several newly added layers. Instead of using the last feature map of ConvNet, it uses the multiple layers in a pyramidal feature hierarchy of ConvNet to predict objects with different scales, which is very beneficial for imaging PCM and aphid targets of various sizes as in our datasets. In this paper, under the premise of ensuring accuracy, to reduce the amount of model parameters and improve the detection speed, we will improve the SSD model from two aspects: feature extraction network optimization and prediction convolution kernel miniaturization.

2.2.1. Optimization of Feature Extraction Networks

The original SSD is based on VGG16 and is a feature to extract the backbone network. However, with the development of the transfer learning model structure, pre-training models are becoming more abundant. To optimize the best feature extraction network, we chose the ResNet50 model of the ResNet series and the InceptionV3 model of the GoogLeNet series for validation. However, the above three models have complex structures and parameters that are too large, making them difficult to apply in mobile or embedded devices. MobileNetV3 has outstanding structural lightweight features and has obvious advantages under the premise that embedded systems require model simplification. To this end, we chose the latest MobileNetV3 model as the fourth comparison network and trained it on the datasets in combination with the SSD framework for comparing the detection performance of each model.

1 VGG16

In the Network of Visual Geometry Groups [25], the structure depth is extended to 16 layers by using a very small (3 × 3) convolution filter. Two completely connected layers are formed by successive 3 × 3 convolutions and 2 × 2 maximum pool layers in the VGG model. The softmax output of is the last layer. The VGG16 model structure is displayed in Figure 2.

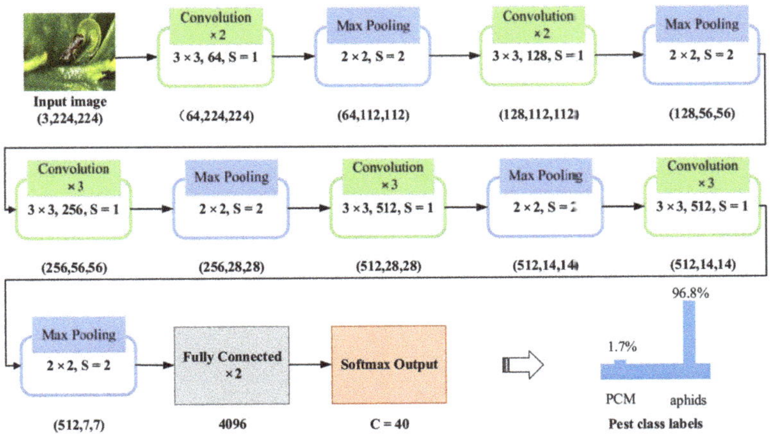

Figure 2. VGG16 model (S-Stride).

2 ResNet

The deep residual network, also known as ResNet, provides a direct path for information dissemination throughout the entire network by virtue of the deep structure of the system [26]. Due to improvements at the network level, the performance and accuracy of the deep network decline rapidly. The gradient disappears in RESNET during backpropagation. The parasitic neural network has a fast (or jumping) connection with a normal period, which allows it to understand global characteristics. Quick connections can be added immediately after multiple weight layers. During training, the network can skip unnecessary layers and optimize the number of correction layers to increase the speed of the process. Mathematically, conclusion H(x) can be expressed as follows:

$$H(x) = F(x) + x \tag{1}$$

Weights are learned from residual mappings, as shown in the following equation:

$$F(x) = H(x) - x \tag{2}$$

where layers of nonlinear weights are represented by F(x).

In this paper, we evaluated the ResNet50 residual network model for cropland pest classification. The Resnet50 model consists of 50 parameter layers on a deep spool network. As shown in Figure 3, these parameters are associated with the transition through the learning cycle. There are 7 × 7 volumes, 64 cores, 3 × 3 layers in the maximum pool, 2 steps, 16 remaining blocks, 7 × 7 layers in the intermediate pool, 7 × 7 layers in interval 7, and a softmax output layer before the connection is completed. A softmax output level of 2 represents the number of aggregated pests.

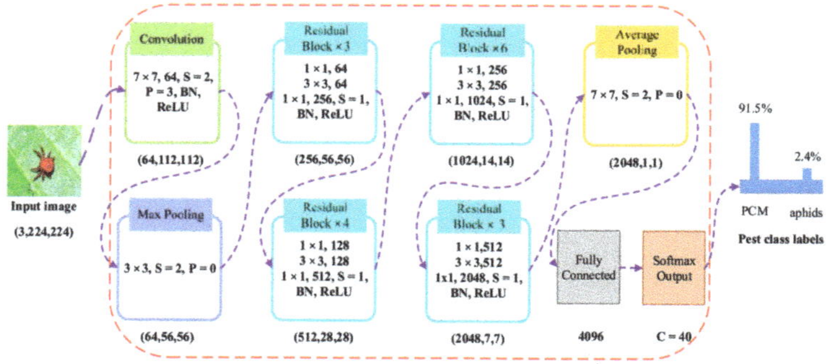

Figure 3. Transfer learning architecture with modified ResNet50 model (BN—batch normalization, S—stride, and P—padding).

3 GoogLeNet

GoogLeNet [27] is a depth model that uses a CNN. Using this model, knowledge transfer learning has improved computing efficiency and produced good results in a variety of fields [28,29]. In total, the GoogLeNet architecture comprises twenty-two deep elements, including two volumes, four largest pools, nine linear encapsulation modules, and an intermediate pool used in the final linear encapsulation module (Figure 4). Each inception module uses 1 × 1 convolution, and a dimensionality reduction operation is performed before the multi-dimensional 3 × 3 and 5 × 5 convolutions. In GoogLeNet, since an efficient inception model can be implemented with very small convolutions, the number of parameters is reduced. Compared to AlexNet, the computational cost is two times less.

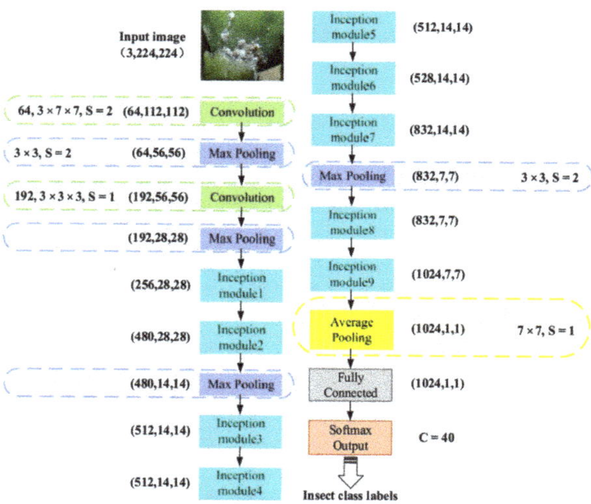

Figure 4. GoogLeNet architecture using transfer learning (S—stride).

4 Modified MobileNetV3

CNN networks for mobile terminals have experienced rapid growth over the past few years. From 2017 to 2019, three versions of a mobile network were continuously improved in terms of architecture. MobileNetV1 [30] was developed based on the conventional

VGG architecture combined with depth-separable convolutions. After one year since the introduction of MobileNetV2 [31], there has been a linear bottleneck and an inverse residual characteristic. As MobileNetV3 was developed with NAS and NetAdapt optimization, expensive layers were discarded, and the nonlinear functions of h-swish were replaced by ReLU, which is a modified version of swish nonlinearity with a faster calculation speed and better quantization effect. The main idea is to use the piecewise linear hard analog $\frac{\text{ReLU6}(x+3)}{6}$ to replace the sigmod, where the nuance occurs in the use of ReLU6 instead of regular clipping constants. Therefore, h-swish is defined as follows [32]:

$$\text{h} - \text{swish}[\text{x}] = \text{x}\frac{\text{ReLU 6}(\text{x}+3)}{6} \tag{3}$$

First of all, judging from the actual measurement results, using ReLU6 optimization is applicable to almost all software and hardware frameworks. Second, in quantized mode, it removes the potential loss of numerical precision caused by different implementations of the approximate sigmoid. Finally, in practice, h-swish can be implemented as a piecewise function that reduces the number of memory accesses, thus greatly reducing the latency cost.

In addition, MobileNetV3 introduces the SE module after the 3×3 depth-separable convolution in the inverse residual module, and firstly performs global pooling compression (Squeeze) to obtain a $1 \times 1 \times C$ vector. Then, after two "fully connected layer-activation (Excitation)" operations (in order to reduce the calculation time, the number of output channels of the first "fully connected layer-activation" operation is compressed to 1/4 of the original), it outputs the $1 \times 1 \times C$ vector. Finally, the obtained vector is bitwise multiplied with the result of the depthwise separable convolution to adjust the weight of each channel to improve the network accuracy. In the design of the overall network structure of MobileNetV3, firstly, the NAS algorithm is used to search and optimize the network structure (such as the arrangement and structure of blocks in the network) to obtain the general network structure, and finally, the NetAdapt algorithm is used to determine the number of channels for each filter. Figure 5 shows the MobileNetV3 architecture for feature extraction from citrus pest images.

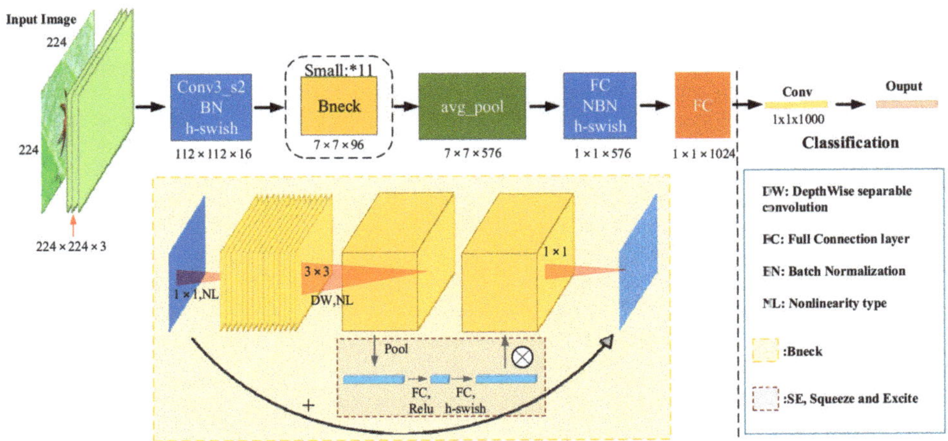

Figure 5. Modified MobileNetV3 architecture.

2.2.2. Predictive Convolution Kernel Miniaturization

Through the analysis of [33], it was found that the accuracy of the model using the 1×1 convolution kernel of the SSD network residual block is not much different from the accuracy of the model when using the 3×3 convolution kernel, but it can greatly reduce the calculation cost. Before the model performs prediction, we added a residual block of a 1×1 convolution kernel to each feature map used for prediction to predict category scores and frame offsets. We named it residual prediction block miniaturization (RPBM), and its structure is shown in Figure 6.

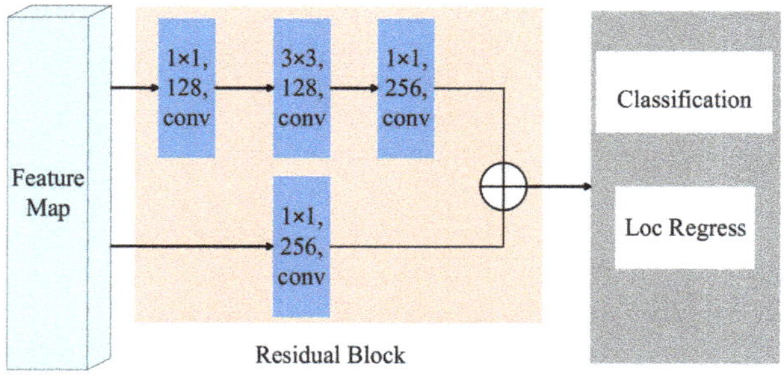

Figure 6. Miniaturized residual prediction network.

2.2.3. Model Training Environment

Hardware configuration: The TIntel E5-2660 processor is a single-core device, with a memory size of 32 GB. The graphics processor used was a 3070ti graphics card (NVIDIA Corporation, Santa Clara, CA, USA).

Software environment: The OS used was Version 16.04 of Ubuntu, the integrated development environment was Anaconda3, and TensorFlow was used as the deep learning framework.

2.2.4. Model Evaluation Indicators

To objectively evaluate the characteristics of a variety of feature networks, this article evaluates from four perspectives: algorithm running speed, the number of model parameters, accuracy rate (AR), and mean average precision (mAP).

Usually, the time required for a classifier to train and predict an image is the computational time, which is also called latency [34]. To minimize the error, the mean of latency (moL) for multiple images was used as the speed indicator in this manuscript.

The number of model parameters is one of the important indicators for embedded devices to run deep learning models [35]. The weight parameters of each layer were added to obtain the total number of parameters, in which the model parameters were stored as a floating point type and the model parameters were obtained by calculating with 4 bytes.

The correct rate of sample indicates how many samples were included correctly in the total number of samples. A classification is generally better if the correct rate is more high. Therefore, a true positive implies that the number of positive samples is predicted to be the number of true positive samples. The number of negative samples is predicted to be the number of true negative samples. In false positives, the number of negative samples is predicted to be the number of positive samples. When a false negative occurs, the number of positive samples is predicted to be the number of negative samples. As a result, the AR can be expressed as follows [36]:

$$\mathrm{AR} = \frac{\mathrm{TP} + \mathrm{TN}}{\mathrm{TP} + \mathrm{TN} + \mathrm{FP} + \mathrm{FN}} \tag{4}$$

There are some instances in which a high accuracy rate is not enough to indicate that the performance of the algorithm is excellent, so mAP is introduced as a measurement index. As a result of target recognition, each class can draw a recall line that is accurate and fast. AP represents the area under the curve, whereas the map represents the average of multiple test lines' AP, and its calculation formula is as follows [37]:

$$\text{mAP} = \frac{1}{|Q_R|} \sum_{q \in Q_R} \text{AP}(q) \tag{5}$$

where Q_R represents the number of validation sets and q represents a certain validation set.

The counting accuracy (CA) indicator is mainly used to record the correct identification and number of pest species counted by the detector. The correct identification of the pest species refers to the condition of predicting correct positive samples, and the calculation formula of its percentage in the actual total samples (true positive rate—TPR) is as follows:

$$\text{TPR} = \frac{\text{TP}}{\text{TP} + \text{TN} + \text{FP} + \text{FN}} \times 100\% \tag{6}$$

Correct counting means that the number of a certain type of pest in the statistical sample is consistent with the actual number of this type of pest in the sample. We assume that the percentage of the number of samples with correct identification and counting of the pest species in the total number of samples is PP, and the percentage of the number of samples with correct identification of pest species but incorrect counting in the total number of samples is PN, then the formula for the counting accuracy rate is as follows:

$$\text{CA} = \frac{\text{PP}}{\text{PP} + \text{PN}} \times 100\% \tag{7}$$

2.3. Pest Detection Embedded Mobile System

To verify the usability of the above transfer learning model in embedded devices, we developed a citrus pest detector (Figure 7). The detector includes a processor, a high-definition camera (resolution: 1280 × 960 pixels, model: RER-USB4KHDR01, manufacturer: REVISION), a Beidou (Beidou Navigation Satellite System, BDS) positioning module, a LoRa communication module and a display module. The processor is a K510 AI chip (Beijing Canaan Jess Information Technology Co., Ltd., Beijing, China). The trained SSD model stored data using the memory of the K510 processor.

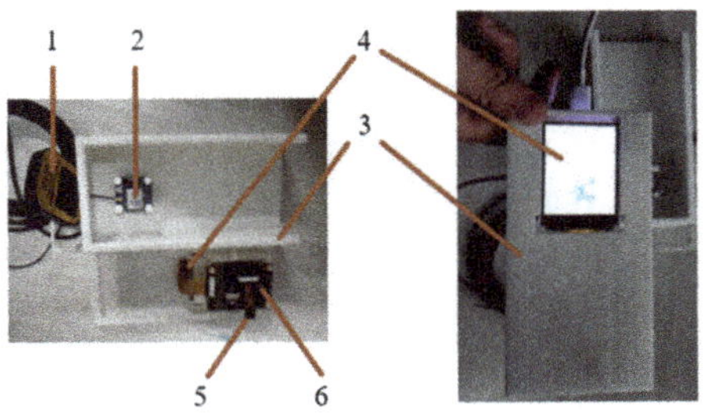

Figure 7. Citrus pest detector (1. BDS module; 2. LoRa; 3. package shell; 4. LCD screen; 5. camera; 6. K510 AI controller).

Regarding the model deployment process, first, we converted the PyTorch model into a model described by the Open Neural Network Exchange (ONNX) intermediate representation and formed an .onnx file, which contains the name and data type of the model input and output. Next, we used the inference engine-ONNX Runtime to run the .onnx file to complete model deployment.

The software workflow of the detector is shown in Figure 8. Firstly, the pest image information is obtained through the camera. After preprocessing, the trained modified SSD embedded in K510 is used to identify whether there are pests. For images with pests, feature extraction and classification are performed to identify the type and to count the pests. Then, the location information is obtained through the BDS module, and output is displayed with the pest identification results. Finally, the detection information is sent to the server through the LoRa communication module, which is convenient for users to predict the development trend of pests, and thus facilitate the overall control and prevention of pests.

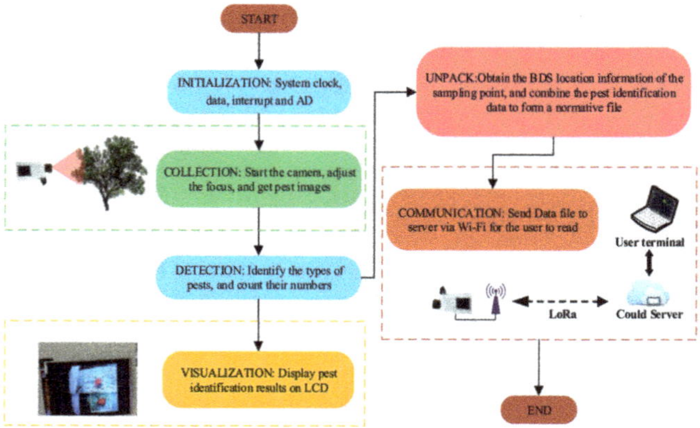

Figure 8. Detector software flow chart.

3. Results

We evaluated the modified SSD model on our PCM and aphid datasets. we analyzed the model training process and compared various feature extraction networks such as ResNet50, InceptionV3, VGG16, MobileNetV3, and modified MobileNetV3, and optimized the best feature extraction model. In order to verify the advantages of the modified SSD model in reducing the amount of computations needed and improving the accuracy, we compared the modified SSD model with other popular frameworks. Finally, we transplanted our model into the embedded detector that we developed, and analyzed its accuracy and efficiency.

3.1. Results of Citrus Pest Identification Model

3.1.1. Analysis of Model Training Process

Two kinds of pest training set and validation set sample data were selected for training, and changes to the loss function in the first 40 epochs were recorded, as shown in Figure 9. In the first 10 iterations, the loss of the training sets and the validation sets decreased rapidly, and then gradually stabilized between 10 and 25 iterations, indicating that no more features could be obtained. After 25 iterations, there was no change in the loss of training sets, but there was a slow and fluctuating increase in the loss of test sets, indicating that an overfitting phenomenon occurred; in other words, the model becomes too strict to ideally suit the features of the training sets, which causes the model to deviate from reality. In order to prevent the occurrence of over-fitting, an appropriate number of iterations should be selected, or a weight decay method should be used, that is, a small factor is used to reduce the weight during each epoch.

The sample images of the training sets were input into the SSD network model with epochs of 1, 2, 3, 4, 5, 10, 15, 20, 25, 50, 75, 100, 125, and 150. The accuracy curves of the detection and classification of the two pests are shown in Figure 10. From the perspective of the accuracy rate, after the network model is stabilized, the training effect of PCM is better as the accuracy rate reaches 91.7%, and the accuracy rate of Aphids is 91.1%. The main reason for this is that PCM is relatively fixed in its morphology, while Aphids are changeable in their morphology, and so the model generalization ability is slightly insufficient.

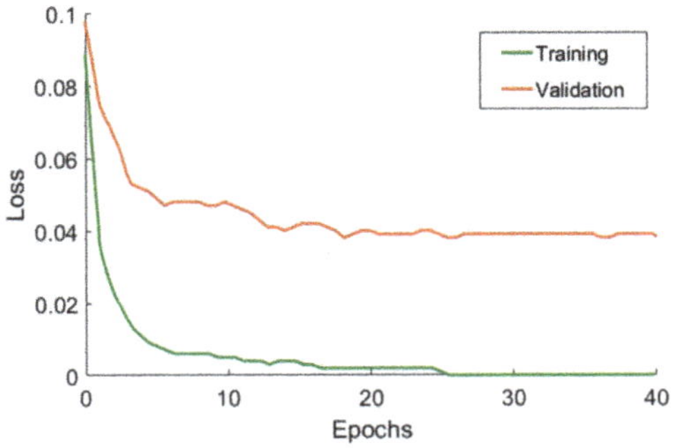

Figure 9. Convergence trend of loss during training.

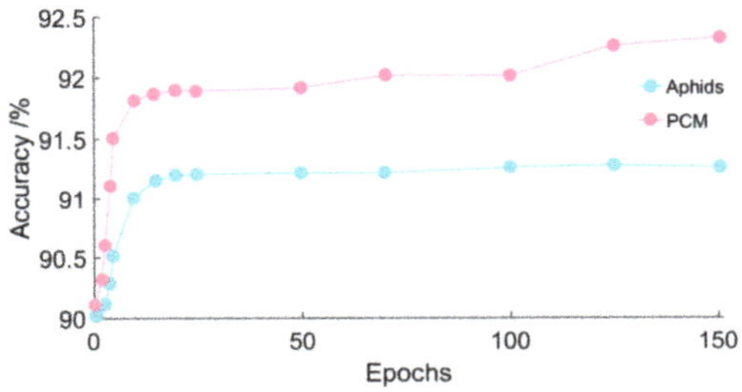

Figure 10. Accuracy of pest detection under different epochs.

3.1.2. Optimization Results and Analysis of Feature Network

The training set samples of two pests, using ResNet50, InceptionV3, VGG16, MobileNetV3 and MobileNetV3 +RPBM, were compared and analyzed to determine their detection effects. The curve changes in the loss of different feature extraction networks with the increase in the number of iterations are shown in Figure 11.

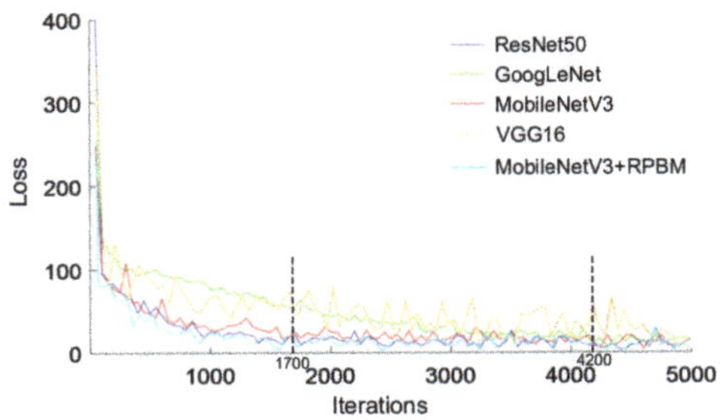

Figure 11. The loss value of SSD under different feature networks.

According to the loss curve from Figure 11, as the number of iterations increases, the loss of the five models may converge rapidly, but MobileNetV3+RPBM is the fastest. When the network iterates 1700 times, the loss of ResNet50, MobileNetV3, and MobileNetV3+RPBM gradually becomes stable, while the GoogLeNet model is stable after 4200 iterations, and the training time is relatively long. The loss of VGG16 converges the slowest.

As shown in Table 1, we observed that the Params of MobileNetV3 are only 15.147 M, which is approximately 1/38 of that of VGG16, and its mAP and AR scores are 86.40% and 91.07%, respectively, which are lower than VGG16 but higher than GoogleNet. The moL of VGG16 is 679 ms higher than that of GoogLeNet (459 ms), while that of MobileNetV3 is low at 286 ms. Although the AR and mAP of MobileNetV3 are lower than those of VGG16, the former has absolute advantages over MobileNetV3 in latency and efficiency. If we only compare whether to add the RPBM after MobileNetV3, we can find that the AR and mAP of MobileNetV3+RPBM and MobileNetV3 are not that different, but the Params of MobileNetV3+RPBM are much smaller than those of MobileNetV3, suggesting that MobileNetV3+RPBM is more suitable for use in our Citrus pest detector as a light feature extraction network.

Table 1. Comparison of performance parameters of different models.

Feature Extraction Network	mAP/%	AR/%	Params/M	moL/ms
VGG16	89.22	91.34	584.179	679
GoogLeNet	85.80	90.18	37.864	459
ResNet50	90.11	96.41	1478.179	1078
MobileNetV3	86.40	91.07	15.147	286
MobileNetV3+RPBM	86.10	91.00	10.025	185

3.1.3. Comparison with Other Frameworks

As can be seen from Table 2, the mAP of MobileNetV3+RPBM+SSD is higher than that of YOLOv7-tiny by 17.44%. It is even higher than that of FFSSD at 73.41% of the moL cost of the FFSSD. Although the Params of Pelee, proposed in [38], are lower than that of our model, it only achieves an 84.44% mAP when we take the model trained on our datasets.

Table 2. Comparison of performance parameters of different models.

Framework	mAP/%	Params/M	moL/ms
YOLOv7-tiny [39]	68.66	6.201	42
Pelee [38]	84.44	9.430	172
FFSSD [40]	78.98	13.550	252
MobileNetV3+RPBM+SSD (our)	86.10	10.025	185

3.2. *Analysis of the Practical Validation Results of Citrus Pest Identification Model*

The test sets of the two pests were collected, and the PP, PN, and NN values were recorded. The results are shown in Table 3.

Table 3. Detector experimental results.

Type	RRC [1]/%	RRWC [2]/%	TPR/%	CA/%
PCM	82.0	9.0	91.0	90.1
Aphids	39.0	50.0	89.0	43.8

[1] RRC, right recognition and counting; [2] RRWC, right recognition but wrong counting.

From the perspective of successful identification, the identification rates of PCM and Aphids are relatively high, as the sampling characteristics were correctly selected in 91.0% and 89.0% of cases. PCM has a CA value of 90.1%, while Aphids have a CA value of 43.8%. Essentially, Aphids densely inhabit areas, stay close to each other, and are more aggressive. Thus, an overlapping phenomenon occurs. It is difficult to mark all of the pests during marking, resulting in some aphid samples becoming negative samples during training, and the accuracy is reduced. The visual detection results of the LCD display are shown in Figure 12.

Figure 12. Mobile pest detector test results: (**a**) low-density detection results, (**b**) middle-density detection results, (**c**) high-density detection results (note: the number of pests less than 10 is low density, and between 11 and 50 is middle- density, otherwise, it is high density).

From the perspective of the detection characteristics of the number of pests, when the density of pests is small, the counting results are more accurate. When the density is high, there may be dozens of pests on leaf veins and stems, which leads to problems such as mutual covering and difficulty in focusing, resulting in a low counting accuracy.

4. Discussion

This paper takes Panonychus citri Mcgregor and Aphids as the research objects, and compares and analyzes the performance of the SSD model under the pre-training of various transfer learning feature extraction networks in citrus pest recognition. In order to further test the applicability of the model in a mobile embedded terminal, a prediction convolution kernel miniaturization method is proposed to improve the SSD model, and the advantages of this method in terms of computational cost and accuracy are compared and analyzed.

Through the comparative analysis of the parameters, the accuracy, and the execution efficiency of VGG16, GoogLeNet, ResNet50, MobileNetV3, and MobileNetV3+RPBM, we selected MobileNetV3+RPBM as the feature extraction network for citrus pest images as the mAP and AR reach up to 86.10% and 91.00%, respectively, and the moL is as low as 185 ms, which is invaluable for embedded devices with limited computing resources, and which is favorable for running the pest detection model in the embedded system. Although the moL of MobileNetV3+RPBM+SSD is relatively high, the mAP of it is highest when compared with YOLOv7-tiny, FFSSD, and Pelee. Therefore, we used MobileNetV3+RPBM+SSD as the detection model in the citrus pest detector.

The classification and counting of PCM and Aphids were achieved by using MobileNetV3+RPBM+ SSD, and the identification accuracy rate reached up to 91.0% and 89.0%, respectively. In terms of CA, PCM reaches 90.1%, while Aphids were only at 43.8%. The main reason for this is that the dense colony characteristics of Aphids lead to serious overlapping in the image data, which affects the counting accuracy.

The convolution kernel miniaturization method proposed in this paper only plays a role in improving the processing speed. The future optimization of our model may be necessary to further improve the accuracy of pest detection and counting.

5. Conclusions

In order to realize the rapid and accurate detection of pest information in citrus orchards and improve intelligent management in this field, this paper designed a portable intelligent detection system to obtain pest information by combining the advantages of deep learning technology and embedded devices. In the design of the detection model, reducing the amount of model parameters, improving the detection speed, and ensuring the accuracy were the comprehensive goals. To this end, this paper improved the SSD model from two aspects: feature extraction network optimization and prediction convolution kernel miniaturization. The parameters of the proposed novel MobileNetV3+RPBM model were reduced by 5.122 M compared with the optimal MobileNetV3 parameters, while the mAP and AR detection accuracy indicators for the two citrus pests were still maintained above 85%, which shows that our modified SSD model can indeed reduce the number of parameters and latency, which is of great significance for the intelligent target detection of mobile portable devices with limited computing power.

Author Contributions: Conceptualization, X.H. and Z.L.; methodology, L.W. and W.S.; software and algorithm, L.W.; validation, H.X.; formal analysis, H.X.; investigation, Y.Y. and L.W.; resources, Y.T.; data curation, H.X.; writing—original draft preparation, L.W.; writing—review and editing, X.H., W.S. and Y.T.; visualization, L.W.; supervision, Z.L.; project administration, X.H.; funding acquisition, X.H. and L.W. All authors have read and agreed to the published version of the manuscript.

Funding: This research was funded by the Scientific Research Project of Hunan Provincial Department of Education, grant number 21B0741. This research was also partly supported by the Sanya Institute of China Agricultural University Guiding Fund Project, grant number SYND-2021-06; the 2015 Talent Development Program of China Agricultural University; the Natural Science Foundation of Hunan Province, grant number 2020JJ2015; the Yongzhou Guiding Science and Technology Plan Project, grant number 2021-YZKJZD-007; and the stable support project of Guangdong Academy of Sciences, grant number 2022-SSGAS-05. Xiongkui He is the corresponding author.

Institutional Review Board Statement: Not applicable.

Informed Consent Statement: Not applicable.

Data Availability Statement: Not applicable.

Conflicts of Interest: The authors declare no conflict of interest.

References

1. Khanramaki, M.; Askari Asli-Ardeh, E.; Kozegar, E. Citrus pests classification using an ensemble of deep learning models. *Comput. Electron. Agric.* **2021**, *186*, 106192. [CrossRef]
2. He, X.; Bonds, J.; Herpbmt, A.; Langenakens, J. Recent development of unmanned aerial vehicle for plant protection in East Asia. *Int. J. Agric. Biol. Eng.* **2017**, *10*, 18–30.
3. Yang, J.; Guo, X.; Li, Y.; Marinello, F.; Ercisli, S.; Zhang, Z. A survey of few-shot learning in smart agriculture: Developments, applications, and challenges. *Plant Methods* **2022**, *18*, 28. [CrossRef] [PubMed]
4. Njoroge, A.W.; Mankin, R.W.; Smith, B.; Baributsa, D. Effects of hypoxia on acoustic activity of two stored-product pests, adult emergence, and grain quality. *J. Econ. Entomol.* **2019**, *112*, 1989–1996. [CrossRef] [PubMed]
5. Zhang, R.R. PEDS-AI: A Novel Unmanned Aerial Vehicle Based Artificial Intelligence Powered Visual-Acoustic Pest Early Detection and Identification System for Field Deployment and Surveillance. In Proceedings of the 2023 IEEE Conference on Technologies for Sustainability (SusTech), Portland, OR, USA, 23 May 2023.

6. Zhou, B.; Dai, Y.; Li, C.; Wang, J. Electronic nose for detection of cotton pests at the flowering stage. *Trans. Chin. Soc. Agric. Eng.* **2020**, *36*, 194–200.
7. Zheng, Z.; Zhang, C. Electronic noses based on metal oxide semiconductor sensors for detecting crop diseases and insect pests. *Comput. Electron. Agric.* **2022**, *197*, 106988. [CrossRef]
8. Wu, X.; Zhang, W.; Qiu, Z.; Cen, H.; He, Y. A novel method for detection of *Pieris rapae* larvae on cabbage leaves using NIR hyperspectral imaging. *Appl. Eng. Agric.* **2016**, *32*, 311–316.
9. Qin, Y.; Wu, Y.; Wang, Q.; Yu, S. Method for pests detecting in stored grain based on spectral residual saliency edge detection. *Grain Oil Sci. Technol.* **2019**, *2*, 33–38. [CrossRef]
10. Yang, X.; Liu, M.; Xu, J.; Zhao, L.; Wei, S.; Li, W.; Chen, M.; Chen, M.; Li, M. Image segmentation and recognition algorithm of greenhouse whitefly and thrip adults for automatic monitoring device. *Trans. Chin. Soc. Agric. Eng.* **2018**, *34*, 164–170.
11. Song, L. Recognition model of disease image based on discriminative deep belief networks. *Comput. Eng. Appl.* **2017**, *53*, 32–36.
12. Martineau, M.; Conte, D.; Raveaux, R.; Arnault, I.; Munier, D.; Venturini, G. A survey on image-based pest classification. *Pattern Recognit.* **2017**, *65*, 273–284. [CrossRef]
13. Deng, L.; Wang, Y.; Han, Z.; Yu, R. Research on pest pest image detection and recognition based on bio-inspired methods. *Biosyst. Eng.* **2018**, *169*, 139–148. [CrossRef]
14. Mukhiddinov, M.; Muminov, A.; Cho, J. Improved Classification Approach for Fruits and Vegetables Freshness Based on Deep Learning. *Sensors* **2022**, *22*, 8192. [CrossRef] [PubMed]
15. Lu, Y.; Yi, S.; Zeng, N.; Liu, Y.; Zhang, Y. Identification of rice diseases using deep convolutional neural networks. *Neurocomputing* **2017**, *267*, 378–384. [CrossRef]
16. Liu, B.; Zhang, Y.; He, D.J.; Li, Y.X. Identification of Apple Leaf Diseases Based on Deep Convolutional Neural Networks. *Symmetry* **2018**, *10*, 11. [CrossRef]
17. Li, H.; Long, C.; Zeng, M.; Shen, J. A detecting method for the rape pests based on deep convolutional neural network. *J. Hunan Agric. Univ. (Nat. Sci.)* **2019**, *45*, 560–564.
18. Kuzuhara, H.; Takimoto, H.; Sato, Y.; Kanagawa, A. pest detection and identification method based on deep learning for realizing a pest control system. In Proceedings of the 2020 59th Annual Conference of the Society of Instrument and Control Engineers of Japan (SICE), Chiang Mai, Thailand, 23–26 September 2020.
19. He, Y.; Zeng, H.; Fan, Y.Y.; Ji, S.; Wu, J. Application of deep learning in integrated pest management: A real-time system for detection and diagnosis of oilseed rape pests. *Mob. Inf. Syst.* **2019**, *2019*, 4570808. [CrossRef]
20. Qiu, G.; Tao, W.; Huang, Y.; Zhang, F. Research on farm pest monitoring and early warning system under the NB-IoT Framework. *J. Fuqing Branch Fujian Norm. Univ.* **2020**, *162*, 1–7.
21. Luo, Q.; Huang, R.; Zhu, Y. Real-time monitoring and prewarning system for grain storehouse pests based on deep learning. *J. Jiangsu Univ. (Nat. Sci. Ed.)* **2019**, *40*, 203–208.
22. Ghazi, M.M.; Yanikoglu, B.; Aptoula, E. Plant identification using deep neural networks via optimization of transfer learning parameters. *Neurocomputing* **2017**, *235*, 228–235. [CrossRef]
23. Khan, M.A.; Akram, T.; Sharif, M.; Awais, M.; Javed, K.; Ali, H.; Saba, T. CCDF: Automatic system for segmentation and recognition of fruit crops diseases based on correlation coefficient and deep CNN features. *Comput. Electron. Agric.* **2018**, *155*, 220–236. [CrossRef]
24. Liu, Z.; Gao, J.; Yang, G.; Zhang, H.; He, Y. Localization and Classification of Paddy Field Pests using a Saliency Map and Deep Convolutional Neural Network. *Sci. Rep.* **2016**, *6*, 20410. [CrossRef] [PubMed]
25. He, K.; Zhang, X.; Ren, S.; Sun, J. Deep residual learning for image recognition. In Proceedings of the IEEE Conference on Computer Vision and Pattern Recognition, Las Vegas, NV, USA, 27–30 June 2016.
26. Al-Qizwini, M.; Barjasteh, I.; Al-Qassab, H.; Radha, H. Deep learning algorithm for autonomous driving using GoogLeNet. In Proceedings of the 2017 IEEE Intelligent Vehicles Symposium (IV), Los Angeles, CA, USA, 11–14 June 2017.
27. Wang, D.; Deng, L.; Ni, J.; Gao, J.; Zhu, H.; Han, Z. Recognition pest by image-based transfer learning. *J. Sci. Food Agric.* **2019**, *99*, 4524–4531.
28. Suh, H.K.; Ijsselmuiden, J.; Hofstee, J.W.; Henten, E.J. Transfer learning for the classification of sugar beet and volunteer potato under field conditions. *Biosyst. Eng.* **2018**, *174*, 50–65. [CrossRef]
29. Simonyan, K.; Zisserman, A. Very deep convolutional networks for large-scale image recognition. In Proceedings of the International Conference on Learning Representations 2015, San Diego, CA, USA, 7–9 May 2015.
30. Gavai, N.R.; Jakhade, Y.A.; Tribhuvan, S.A.; Bhattad, R. MobileNets for flower classification using TensorFlow. In Proceedings of the 2017 International Conference on Big Data, IoT and Data Science (BID), Pune, India, 20–22 December 2017.
31. Sandler, M.; Howard, A.; Zhu, M.; Zhmoginov, A.; Chen, L.C. MobileNetv2: Inverted residuals and linear bottlenecks. In Proceedings of the IEEE Conference on Computer Vision and Pattern Recognition, Salt Lake City, UT, USA, 18–22 June 2018.
32. Howard, A.; Sandler, M.; Chu, G.; Chen, L.C.; Chen, B.; Tan, M.; Wang, W.; Zhu, Y.; Pang, R.; Vasudevan, V.; et al. Searching for MobileNetV3. In Proceedings of the IEEE International Conference on Computer Vision, Seoul, Korea, 27 October–2 November 2019.
33. Zou, Z.; Chen, K.; Shi, Z.; Guo, Y.; Ye, J. Object Detection in 20 Years: A Survey. *Proc. IEEE* **2023**, *111*, 257–276. [CrossRef]
34. Wang, L.; Lan, Y.; Liu, Z.; Yue, X.; Deng, S.; Guo, Y. Development and experiment of the portable real-time detection system for citrus pests. *Trans. Chin. Soc. Agric. Eng.* **2021**, *37*, 282–288.

35. Wang, S.; Han, Y.; Chen, J.; He, X.; He, X.; Zhang, Z.; Liu, X.; Zhang, K. Weed Density Extraction Based on Few-Shot Learning Through UAV Remote Sensing RGB and Multispectral Images in Ecological Irrigation Area. *Front. Plant Sci.* **2022**, *12*, 735230. [CrossRef]
36. Mohamed, E.K.; Fahad, A.; Sultan, A.; Sultan, A. A new mobile application of agricultural pests recognition using deep learning in cloud computing system. *Alex. Eng. J.* **2021**, *60*, 4423–4432.
37. Jin, X.; Sun, Y.; Che, J.; Bagavathiannan, M.; Yu, J.; Chen, Y. A novel deep learning-based method for detection of weeds in vegetables. *Pest Manag. Sci.* **2022**, *78*, 1861–1869. [CrossRef]
38. Robert, J.W.; Li, X.; Charles, X.L. Pelee: A Real-Time Object Detection System on Mobile Device. In Proceedings of the Advances in Neural Information Processing Systems 31 (NeurIPS 2018), Palais des Congrès de Montréal, Montréal, QC, Canada, 2–8 December 2018.
39. Wang, C.; Bochkovskiy, A.; Liao, H.M. YOLOv7: Trainable bag-of-freebies sets new state-of-the-art for real-time object detectors. In Proceedings of the IEEE/CVF Conference on Computer Vision and Pattern Recognition (CVPR), Vancouver, BC, Canada, 18–22 June 2023.
40. Cao, G.; Xie, X.; Yang, W.; Liao, Q.; Shi, G.; Wu, J. Feature-fused SSD: Fast detection for small objects. *Proc. SPIE* **2018**, *10615*, 381–388.

agronomy

MDPI

Article

Analysis of Near-Infrared Spectral Properties and Quantitative Detection of *Rose Oxide* in Wine

Xuebing Bai [1,2], Yaqiang Xu [1], Xinlong Chen [1], Binxiu Dai [1], Yongsheng Tao [1,2,*] and Xiaolin Xiong [3,*]

1 College of Enology, Northwest A&F University, Xianyang 712100, China; bxb@nwafu.edu.cn (X.B.); xuyaqiang@nwafu.edu.cn (Y.X.); daibx@nwafu.edu.cn (B.D.)
2 Ningxia Helan Mountain's East Foothill Wine Experiment and Demonstration Station of Northwest A&F University, Yongning 750104, China
3 Xue Lin Yuan (Shenzhen) Wine Culture Co., Ltd., Shenzhen 518000, China
* Correspondence: taoyongsheng@nwsuaf.edu.cn (Y.T.); somso99999@gmail.com (X.X.)

Abstract: This study aims to investigate the near-infrared spectral properties of *Rose Oxide* (4-Methyl-2-(2-methyl-1-propenyl) tetrahydropyran) in wine, establish a quantitative detection, and build relationships between the chemical groups of *Rose Oxide* and near-infrared characteristic bands, so as to provide ideas and references for the near-infrared detection of a low-content aroma substance in wine. In total, 133 samples with different wine matrices were analyzed using Fourier transform–near-infrared (FT-NIR) spectroscopy. Min–max normalization (MMN), principal component analysis (PCA), and synergy interval partial least squares regression (Si-PLSR) were used for pre-processing, outlier rejection, analysis of spectral properties, and modeling. Finally, the quantitative detection model was established using the PLSR method and the wine sample containing *Rose Oxide* was verified externally. Eight subintervals (4000–4400 cm^{-1}, 4400–4800 cm^{-1}, 5600–6000 cm^{-1}, 6000–6400 cm^{-1}, 6400–6800 cm^{-1}, 6800–7200 cm^{-1}, 7200–7600 cm^{-1}, 8400–8800 cm^{-1}) were determined as the characteristic band intervals of *Rose Oxide* in the NIR region. Among them, 5600–6000 cm^{-1} was assigned to the first overtone C–H stretching in tetrahydropyran ring and methyl as well as the combination C–H stretching of the CH_3 function groups, 6000–6400 cm^{-1} was assigned to the first overtone C–H stretching of the C–H=group and the combination C=C stretching in isobutyl, and 8400–8800 cm^{-1} was assigned to the second overtone C–H stretching and C–O stretching in tetrahydropyran ring as well as the C–H stretching vibration in methyl. In addition, 4000–4800 cm^{-1}, 6400–6800 cm^{-1}, and 7200–7600 cm^{-1} were assigned to the C–H stretching vibration, while 6400–7600 cm^{-1} was assigned to the C–O stretching vibration. The training result showed that the calibration model (r_{cv}^2 of 0.96 and $RMSE_{CV}$ of 2.33) and external validation model (r_{cv}^2 of 0.84 and $RMSE_{CV}$ of 2.72) of *Rose Oxide* in wine were acceptable, indicating a good predictive ability. The spectral assignment of *Rose Oxide* provides a new way for the NIR study of other terpenes in wine, and the use of the established Si-PLSR model for the rapid determination of *Rose Oxide* content in wine is feasible.

Keywords: *Rose Oxide* (4-Methyl-2-(2-methyl-1-propenyl) tetrahydropyran); de-aromatic wine; NIR spectroscopy; Si-PLSR; wavebands analysis

Citation: Bai, X.; Xu, Y.; Chen, X.; Dai, B.; Tao, Y.; Xiong, X. Analysis of Near-Infrared Spectral Properties and Quantitative Detection of *Rose Oxide* in Wine. *Agronomy* **2023**, *13*, 1123. https://doi.org/10.3390/agronomy13041123

Academic Editor: Shubo Wang

Received: 22 March 2023
Revised: 10 April 2023
Accepted: 12 April 2023
Published: 14 April 2023

1. Introduction

Rose Oxide (4-Methyl-2-(2-methyl-1-propenyl) tetrahydropyran), with strong fragrance of rose and lychee, is the main component of rose and rose geranium. It is not only used to prepare flavors, such as rose, leaf, and other flower flavors, but it is also widely used in upscale cosmetics and the food industry. Surprisingly, many varieties of grapes also contain *Rose Oxide* broadly, which enriches their aroma features and makes wine purer and fresher [1]. As an oxide of monoterpenols, *Rose Oxide* has a strong volatility. It becomes a kind of recognizable aroma [2]; even its concentration reduced by three times during fermentation [3]. Therefore, *Rose Oxide* is an important sign for identifying the varieties, years, and origins of wine objectively [4]. Studies have shown that the existence of *Rose*

Oxide is highly significant correlated with whether grape and wine have a rose aroma [1]. Although *Rose Oxide* has a low concentration in wine, it also has a lower odor threshold with only 0.2 μg/L [4] and a higher odor activity value (OAV) of generally more than 100 [4,5]. This means that *Rose Oxide* could be easily perceived and contribute significantly to the overall aroma of wine. In addition, *Rose Oxide* could react with other terpene aroma substances and play a decisive role in the formation of aroma when other aroma substances hold a low concentration [2]. Many studies have revealed that *Rose Oxide* is the key compound for bringing out the floral, rose, and even lychee aromas of wine [6,7] and that it correlates with positive emotions and higher liking scores for wine consumers [8].

However, the quantification of *Rose Oxide* is complicated as a trace component. Gas chromatography (GC) coupled to at least one detector, such as a flame ionization detector (GC-FID) or a mass spectrometer (GC-MS), is the typical method for analyzing the *Rose Oxide* content in wine [9]. It is necessary to conduct a pre-treatment of the wine sample, such as extraction, heating, oscillation, analysis, as well as other steps, and finally use the instrument to conduct a qualitative and quantitative analysis. These methods are labor-intensive and costly, and they easily cause the loss of volatility of *Rose Oxide*. Therefore, a rapid, simple, and economical method for predicting the content of *Rose Oxide* as an alternative to the traditional analysis methods is required. Near-infrared (NIR) spectroscopy can address these limitations.

The NIR spectrum lies between the visible and IR regions of the electromagnetic spectrum in the wavelength range 780–2500 nm and involves the excitation of non-fundamental vibrations, overtones, and combination modes [10,11]. NIR spectroscopy mainly reflects the information of hydrogen-containing groups, including C–H (such as methyl, methylene, methoxy, carboxyl, and so on), O–H (hydroxyl), S–H (sulfhydryl), N–H (amino), and so on. There is also some other groups' information (such as C=C, C=O, and so on), but the intensity is weak. These groups are important components of organic compounds, which means that the structures and compositions of almost all organic compounds can be found in the near-infrared spectrum. The process of NIR generally includes spectrum pretreatment, outlier elimination, band screening, and quantitative model establishment. Band screening, also called spectrum allocation, aims to detect the feature wavebands of the chemical groups in targeting ingredients and ensuring the spectral fingerprint information of the ingredients. It is the basis for establishing the quantitative models and providing models with a theoretical note. NIR spectroscopy is a simple and non-destructive technique which generally does not require any sample pretreatment which may result in the loss of the substance under test. Therefore, it is widely used in agriculture, petroleum, chemical, tobacco, pharmaceutical, and food industries [12]. Several studies have used NIR spectroscopy for predicting compounds in wine, such as phenolic compounds [13], trace metal elements [14], and volatile compounds [9], as well as different terpenes in plants, such as α-pinene, β-pinene, myrcene, eugenol, cineole, and linalool [11,15]. To the best of our knowledge, no attempts have been made to establish an association between the NIR spectrum and terpene profiles of wines, let alone study the spectral characteristics of its molecular group and create a quick detection method in wines.

Here, we aimed to investigate the spectral properties of *Rose Oxide* and methods of quantitatively detecting it in wine. First, based on a single controlled environment (model wine), the NIR feature wavebands of *Rose Oxide* were screened using spectral preprocessing, outlier rejection, and synergy interval partial least squares (Si-PLS) methods. Second, a quantitative detection method for *Rose Oxide* was constructed based on these spectral wavebands using the partial least squares regression (PLSR) method in a relatively complex environment (de-aromatic wine). Third, wine samples containing *Rose Oxide* were used to effectively validate the accuracy and model transferability of the above method. In this study, NIR spectroscopy was used to analyze the typical terpene compounds in wine, establish a rapid detection method of *Rose Oxide*, and hopefully provide methodological support for the rapid and non-destructive detection of terpene compounds in wine.

2. Materials and Methods

2.1. Materials

Grape variety to be de-flavored: Cabernet Franc, collected from the Ningxia Helan Mountain's East Foothill Wine Experiment and Demonstration Station of Northwest A&F University in October 2021, which contains 222.5 g/L of residual sugar (expressed as glucose) and 4.6 g/L of acid (expressed as tartaric acid); bacterial strain: a strain of *S. cerevisiae* called ACTIFLORE F33 from Lafford Company in France; and the sample set of external verification: 21 Cabernet Franc dry red wines (produced from wineries at Ningxia Helan Mountain's East Foothill) with different concentrations of *Rose Oxide* were added to construct external verification wine samples.

2.2. Instruments and Reagents

The instruments and reagents used in the experiment include: HW.SY21-KP8 Electric Thermostatic Water Bath (Chengfeng Inc., Beijing, China); ME203E Electronic Balance (Mettler Toledo Inc., Shanghai, China); FE28pH meter (Mettler Toledo Inc., Shanghai, China); DW-YL270 Cryogenic Refrigerator (Zhongke Meiling Cryogenic Technology Inc., Hefei, Anhui, China); KH-500DE CNC ultrasonic cleaner (Hechuang Ultrasonic Instrument Inc., Kunshan, Jiangsu, China); Hei-VAP Table Rotary Evaporation Instrument (Hadolf Instrument Equipment Inc., Shanghai, China); GCMS-QP2020 Gas chromatography–mass spectrometry instrument (Shimazu Laboratory Equipment Inc., Shanghai, China); Bruker-TANGO-T Fourier Transform–near-infrared spectrometer (Brock Scientific Instruments Inc., Hong Kong, China), a built-in automatic background scanning program can timely eliminate the impact of environmental changes in detection results, equipped with Rock-SolidTM patent interferometer, multi-layer coating low OH quartz beam splitter, and InGAs digital detector.

The ultrapure water was obtained from the Milli-Q Pure Water Preparation System (Millipore Inc., Molsheim, France). The analytically pure-grade reagent, anhydrous ethanol, tartaric acid, and sodium hydroxide were purchased from Chemical Reagent Inc., Tianjin, China. The chromatographic grade reagents, 2-octanol (purity $\geq$ 99.0%) and (+)-*Rose Oxide* (purity $\geq$ 99.0%), were purchased from Sigma-Aldrich Corporation (Beijing, China).

2.3. Methods

2.3.1. Sample Preparation and Data Acquisition

Model wine preparation: In total, 120 mL of anhydrous ethanol, 880 mL of distilled water, and 5 g of tartaric acid were added to the blue silk-mouthed bottle and mixed evenly with ultrasonic waves; the pH was adjusted to between 3.2 and 3.4 using saturated NaOH. This was followed by the addition of *cis-Rose Oxide* into the configured model wine to the concentration of 0–40 μg/L (2 μg/L was a step), with a gradient that referred to the concentration range of *Rose Oxide* in real wine [16,17]. The sample was refrigerated at 4 °C after being prepared, sealed by the sealing film, and tested quickly and timely to reduce the volatilization of aroma substances.

De-aromatic wine solution: Ripened Cabernet Franc grapes were picked to make wine using 200 mg/L of Saccharomyces cerevisiae at 25–27 °C after hand-destemming and crushing. After the alcohol fermentation, the vacuum rotary evaporator was used to finish the deodorization procedure optimized based on Margaux Cameleyre [18]; the parameters of the rotary evaporator were set to 50 rpm and 30 °C, and two rounds of rotational evaporation were performed. For the first time, 500 mL of the original wine was spun in a water bath for 1.5 h. The matrix and ethanol water fractions of the wine sample were collected and blended to form 500 mL of initial de-aromatic wine, following which the second spinning was continued with identical parameters and was replenished with 12% ethanol–water to 500 mL again after 1.5 h. The matrix of de-aromatic wine and the distillate of anhydrous ethanol were collected and blended into the final de-aromatic wine. *Rose Oxide* was added to the de-aromatic wine, and the gradient settings and precautions were the same as those of the simulated wine samples.

Spectral data acquisition: The NIR spectra of different samples were recorded on a Fourier transform–near-infrared spectrometer equipped with an indium gallium arsenide detector from 11,500 to 4000 cm^{-1}. The temperature of the samples was equilibrated at 30 °C in the instrument. Each sample was scanned for 32 s with a spectral resolution of 8 cm^{-1}.

2.3.2. Data Pre-Processing and Outlier Rejection

The analysis process of spectral signals would be interfered with by the redundant information, spectral overlap, and baseline drift due to the complexity of the wine matrix. Min–max normalization (MMN) and vector normalization (VN) methods can effectively reduce redundant information and eliminate the effects of changes in the spectra, such as light range changes or sample dilution. First derivative (FD) and second derivative (SD) methods can eliminate the effects of baseline drift or smoothing background interference, distinguish overlapping peaks, and provide higher resolution and sharper spectral profile changes than the original spectra [19]. Thus, the above four methods were used to process the spectral signals of *Rose Oxide* in model wine and de-aromatic wine. For a spectral signal $x = (x_1, x_2, \ldots, x_n)$, the equations of MMN, VN, FD, and SD were as follows:

$$x_i^{\mathrm{MMN}} = \frac{x_i - x_{\min}}{x_{\max} - x_{\min}} \tag{1}$$

$$x_i^{\mathrm{VN}} = \frac{x_i}{\sqrt{\sum\limits_{i=1}^{n} x_i^2}} \tag{2}$$

$$x_i^{\mathrm{FD}} = \frac{x_{i+g} - x_i}{g} \tag{3}$$

$$x_i^{\mathrm{SD}} = \frac{x_{i+g} - 2x_i + x_{i-g}}{g^2} \tag{4}$$

where x_i was the i-th vector of x, $x_{\min}$ was the minimum vector of x, $x_{\max}$ was the maximum vector of x, and g was the window width.

During the acquisition of *Rose Oxide* spectral signals, human and instrumental errors may cause some signals to deviate severely from the true value, resulting in outliers. In this study, principal component analysis (PCA) was used to shine the simples upon the low-dimensional space, then the outliers would be found according to the Hotelling T^2 statistic under the coordinate of first and second principal components [20]. The confidence level of T^2 detection was calculated as follows:

$$x^T(t)P\Lambda^{-1}P^T x(t) \le \delta_{T^2} \tag{5}$$

$$\delta_{T^2} = \frac{(N-1)(N+1)}{N(N-K)}F \tag{6}$$

where δ_{T^2} was the confidence level of T^2 detection, $x(t)$ was the input matrix at the time t, Λ was the covariance matrix, N was the number of principal components, k was the kth principal component, and F was the F-distribution.

In this study, two T^2 confidence intervals with 99% and 95% were used to reject outliers. The specific screening criteria were as follows: samples located outside the 99% confidence interval were directly judged as outliers, and samples located between the two confidence intervals of 95% and 99% were judged as pending values and had to be validated to decide whether to reject them, while those located within the 95% confidence space were judged as excellent values and could be used for subsequent modeling.

2.3.3. Synergy Interval Partial Least Squares Regression (Si-PLSR)

The NIR spectrum contained abundant information regarding the molecular vibration absorption of hydrogen-containing groups, with most of them being redundant and unrelated. The removal of these extraneous spectral bands may significantly reduce the input variables and improve the accuracy of the prediction model. Therefore, synergy interval partial least square (Si-PLS) was used to screen the synergy intervals reflecting the *Rose Oxide* content [21,22]. Then, the rapid quantitative detection method of the *Rose Oxide* content in wine would be available based on the selected intervals. The specific calculation process was referred to [23]:

Step 1: Constructing PLSR models in the range of 11,500–4000 cm^{-1} for *Rose Oxide*. The root means square error of cross validation ($RMSE_{CV}$) was calculated as:

$$RMSE = \sqrt{\frac{1}{n}\sum_{i=1}^{n} (y_i - y')^2} \tag{7}$$

where y_i was the measured value of the i-th sample, y' was the predicted value of the i-th sample, and n was the number of samples.

Step 2: Dividing the spectral region of 11,600–4000 cm^{-1} into 19 equal-width subintervals into steps of 400 cm^{-1}. Establishing the regression model of each subinterval by the PLS correction analysis.

Step 3: Selecting the subintervals, of which the $RMSE_{CV}$ were smaller than the $RMSE_{CV}$ calculated by step 1.

Step 4: Establishing the new Si-PLSR model based on the selected subintervals and evaluating the performance of the model.

2.3.4. External Validation

Twenty-one commercial wines were used to verify the accuracy of rapid detection models. First, headspace solid-phase microextraction combined with gas chromatography–mass spectrometry (HS/SPME-GC-MS) was used for quantifying the measured value of *Rose Oxide* [24]. The details were as follows:

SPME sample processing: Volatiles were extracted using solid-phase microextraction using DVB/CAR/PDMS fiber (50/30 µm film thickness, 2 cm Stableflex) assembled with a 57330-U holder (Supelco, Bellefonte, PA, USA). A wine sample (8 Ml), 2.0 g of NaCl, 2-octanol (final concentration was 400 µg/L), and a magnetic stirring bar were mixed in a 20 Ml glass vial. The vial was incubated in a thermostatic water bath to equilibrate for 15 min at 40 °C, then the fiber was exposed for 30 min at 40 °C. This was immediately followed by thermos-desorption of the extraction fiber in the GC injector for 5 min at 280 °C prior to GC-MS analysis. The extraction operation was repeated twice for each wine sample.

GC-MS analysis: GC-MS-QP2020 equipped with a DB-WAX capillary column (60 mm × 0.25 mm × 0.25 µm; Agilent J & W, Santa Clara, CA, USA) was used. The carrier gas was high-purity helium (99.999%) without shunt, and the gas flow rate was 1.5 Ml/min. The temperature of the GC capillary column was maintained as follows: 40 °C for 3 min, increase to 160 °C at a rate of 4 °C/min, followed by an increase to 220 °C at the rate of 7 °C/min, and this temperature was maintained for 10 min. We set the temperature of the inlet as 250 °C, the ion source as 220 °C, and the connecting rod as 200 °C. We set the energy of the electron impact source as 70 Ev. Electron ionization mass spectrometric data were acquired within the mass range of 35–350 m/z at 0.2 s intervals combined with the selected ion monitoring mode for the quantitative analysis.

Qualitative and quantitative analysis: A calibration curve for the pure standard was established to analyze the *Rose Oxide* content in the simulated wine solution by the above HS/SPME-GC-MS method. *Rose Oxide* was identified by comparing the retention times, retention indexes, aroma characteristics, and mass spectra with those of the standards available in the NIST 17.0 mass spectral library. The concentration of *Rose Oxide* was quantitated by interpolating the relative area of the sample versus the area of the internal standard (2-octanol) using calibration curves previously established for pure standards. Then, the predicted value of *Rose Oxide* was obtained using the NIR model established in this study, and the validity and practical applicability of the model were judged by comparing the true values with the predicted values.

2.4. Data Analysis

Microsoft excel was used for processing and preliminary analysis of spectral data, MATLAB R 2021b (MathWorks Inc., Natick, MA, USA) was adopted for spectrum preprocessing, Si-PLS analysis and model establishment, and Unscrambler X 10.4 (Camo Inc., Oslo, Norway) was used to eliminate outliers. Origin 2019 (OriginLab Inc., Northampton, MA, USA) was used for data drawing.

3. Results

3.1. Original Spectral Analysis of Model and De-Aromatic Wine

The NIR spectral signals of the wine with added *Rose Oxide* are shown in Figure 1. The red curves expressed the model wine, and the blue curves expressed the de-aromatic wine. It could be observed that the different matrix backgrounds exerted a considerable effect on the height and width of the peaks in the curves, but they had little effect on the position of the peaks.

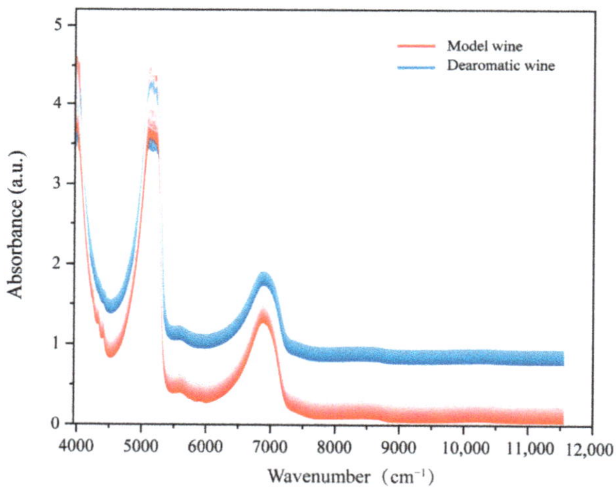

Figure 1. The raw spectra of different concentrations of *Rose Oxide* in simulated wine matrices and de-aromatic wine matrices.

The absorption peak of the de-aromatic wine was 0.30–1.00 (a.u.) higher than that of the model wine in 5500–11,600 cm^{-1} and 0.00–0.50 (a.u.) higher in 4000–5000 cm^{-1}. However, their original spectra overlapped at 5000–5500 cm^{-1}, probably because of the characteristic absorption of ethanol in this interval, while the ethanol content of both matrices was consistent. Although the matrix solution was different, the change in the spectral curves caused by a different content of *Rose Oxide* could be observed at 4500–5500 cm^{-1}, 5500–6000 cm^{-1}, 6000–7500 cm^{-1}, and 8000–9000 cm^{-1}. Therefore, we tentatively speculated that the characteristic waveband of *Rose Oxide* is mainly located in the wave number range of these four regions.

3.2. Spectral Pre-Processing and Outlier Rejection

The average data of all original spectral were pretreated using MMN, VN, FD, and SD. The spectral pre-processing results of *Rose Oxide* in model wine and de-aromatic wine are shown in Figure 2. The spectral curves were narrower and smoother after pre-processing by the MMN and VN algorithm, but there was no significant difference between the two curves. The FD and SD algorithms separated the overlapping regions in the different spectral curves and intensified the absorption peaks around 4500 cm^{-1}, 5500 cm^{-1}, and 7500 cm^{-1}. The *RMSE* and the determination coefficients (r^2) estimated the availability of four pre-processing methods as shown in Table 1. Among them, MMN had the highest r^2_{cv} value of 0.14 and the lowest $RMSE_{cv}$ value of 11.34 in de-aromatic wine, while VN showed the best performance, with an r^2_{cv} value of 0.31 and $RMSE_{cv}$ value of 10.10 in model wine. Considering MMN exhibits a better performance for both spectral curves, and the de-aromatic wine was more complex than the model wine, MMN was selected as the best pre-processing method and applied in subsequent data analysis and model construction, while VN was only selected as the pre-processing method to explore the theoretical spectral features of *Rose Oxide*, which was not used for the subsequent establishment of the *Rose Oxide* prediction model.

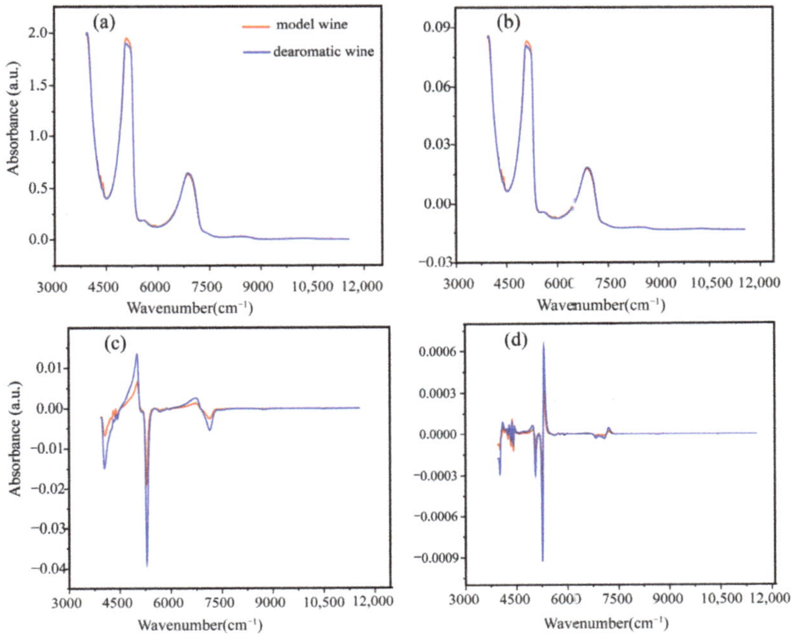

Figure 2. The spectral curves of *Rose Oxide* in model wine and de-aromatic wine based on (**a**) MMN; (**b**) VN; (**c**) FD; (**d**) SD.

Table 1. Results of spectral full-band modeling of model wine and de-aromatic wine by four pre-processing methods (before outlier rejection).

Pre-Processing Methods	Model Wine				De-Aromatic Wine			
	r^2_c	$RMSE_C$	r^2_{cv}	$RMSE_{CV}$	r^2_c	$RMSE_C$	r^2_{cv}	$RMSE_{CV}$
MMN	0.75	6.37	0.23	10.70	0.04	12.10	0.14	11.34
VN	0.78	5.97	0.31	10.10	0.06	11.90	0.03	12.30
FD	0.51	8.85	0.12	11.40	0.05	12.00	0.05	12.00
SD	0.14	11.34	0.09	11.80	0.06	11.90	0.09	11.80

Note: r^2_c, the correlation coefficient of calibration set (the closer to 1, the better); $RMSE_C$, the calibration set root mean square error; r^2_{cv}, the correlation coefficient of the cross-validation set $RMSE_{CV}$, root mean square error of calibration set (the smaller the better).

The outliers were found by the Hotelling T^2 statistics in the coordinate of PC1 and PC2, as shown in Figure 3. Out of a total of 133 model wine samples (No. 1 to 133), 3 samples (No. 90, 122, and 130) were excluded as outliers directly at a 99% confidence level. Sample No. 1 was distributed between the 95% and 99% confidence spaces and needed to be verified. As shown in Table 2, r_p^2 increased from 0.44 to 0.47, and $RMSE_P$ decreased from 9.03 to 8.67 after excluding the No. 1 sample, which indicated the it was an outlier. Out of a total of 133 de-aromatic wine samples (No. 1 to 133), 4 samples (No. 3, 63, 71, and 99) were outside the 99% confidence interval and were rejected directly. Additionally, five new samples (13, 31, 47, 109, and 128) were distributed between the 95% and 99% confidence spaces and needed to be verified. As shown in Table 2, the r_p^2 values increased, and the $RMSE_P$ values decreased by excluding samples No. 13, 109, and 128, while the model effect worsened after excluding samples No. 31 and 47. Therefore, samples No. 13, 109, and 128 were defined as outliers and samples No. 31 and 47 passed the verification. After removing the outliers, the *RMSE* values were reduced and the r^2 values were enhanced in most models (compared to the results in Table 1), suggesting that outlier rejection may improve the accuracy and stability of the prediction model. However, the model after pre-processing and outlier removal still did not work well because an *RPD* less than 1.5 meant a poor prediction performance. This may be caused by the interference of unrelated information in a full waveband. Thus, the feature wavebands of *Rose Oxide* in wines have to be screened out.

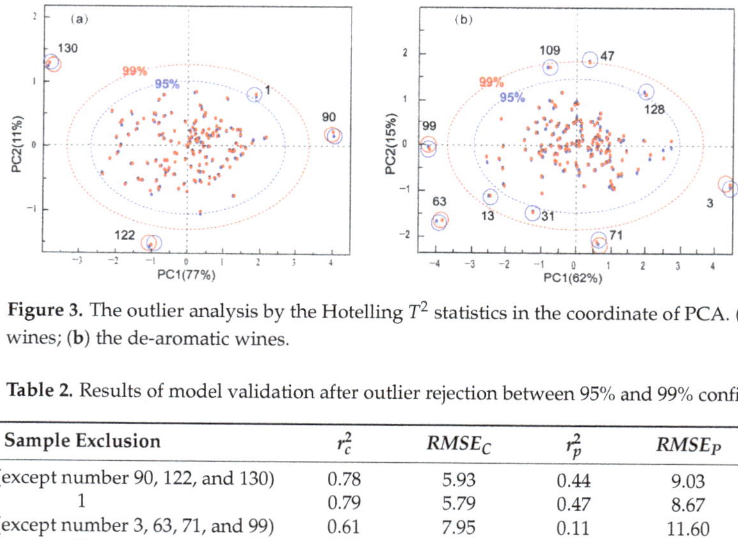

Figure 3. The outlier analysis by the Hotelling T^2 statistics in the coordinate of PCA. (**a**) The model wines; (**b**) the de-aromatic wines.

Table 2. Results of model validation after outlier rejection between 95% and 99% confidence space.

	Sample Exclusion	r_c^2	$RMSE_C$	r_p^2	$RMSE_P$	*RPD*
Model wine	All samples (except number 90, 122, and 130)	0.78	5.93	0.44	9.03	1.33
	1	0.79	5.79	0.47	8.67	1.38
De-aromatic wine	All samples (except number 3, 63, 71, and 99)	0.61	7.95	0.11	11.60	1.06
	13	0.70	7.06	0.14	11.30	1.08
	31	0.67	7.32	0.09	11.70	1.05
	47	0.57	8.37	0.08	11.80	1.04
	109	0.63	7.82	0.18	11.10	1.10
	128	0.71	6.86	0.22	10.80	1.13

Note: r_p^2, the correlation coefficient of the validation set; $RMSE_P$, validation set root mean square error; *RPD*, the ratio of prediction to deviation (*RPD* < 1.5: poor model. 1.5 ≤ *RPD* < 2.5: general model. 2.5 ≤ *RPD* < 5: good model. *RPD* > 5: excellent model).

3.3. Si-PLS Analysis

Rose Oxide, as a monoterpene cyclic ether compound with a tetrahydropyran ring, is attached to a methyl and an isobutylene group, and possesses C=C, C–H, C–O–C, $-CH_3$, and $-CH_2$ functional groups. In this study, the Si-PLS method was used to extract the feature bands. The full wavenumber (11,600–4000 cm^{-1}) range was divided into 19 equal subintervals, with each subinterval modeled separately. The experimental results were shown in Figure 4. The $RMSE_{CV}$ values for subintervals 8, 11, 12, 13, 14, 15, 18, and

19 were smaller than the $RMSE_{CV}$ values (10.7 and 12.4) modeled for the full band in model and de-aromatic wine, while the remaining subintervals had lager $RMSE_{CV}$ values, which suggested that the model built with these eight subintervals would be better. The eight subintervals of two different background matrices were identical, indicating that they contained the feature information of *Rose Oxide*. Indeed, the references regarding the function group and spectral structure of *Rose Oxide* were limited. However, Davis et al. [25] conducted a study on the same functional groups in other chemicals, such as alkanes, alkenes, ethers, and tetrahydropyran, as shown in Table 3. It can be shown that the overtones of CH_3 stretching and deformation modes were largely responsible for the strong absorption region of 5901–5909 cm^{-1} (the first overtone) and 8264–8696 cm^{-1} (the second overtone), while the combination C–H stretching vibration bands of the CH_3 group were at 4100, 4395, 4400, 4500–4545, 5520, 5814, 7355, and 7263 cm^{-1} in alkanes. The first overtone C–H stretching of the C–H=group bands were at 6100–6200 cm^{-1}, while the combination C=C stretching bands at 4482, near 4600, 4670–4780, and 6130 cm^{-1}, were found in alkenes. Moreover, the C–H stretching bands of CH and CH_2 functional groups in tetrahydropyran at 5565–6150 and 8040–9320 cm^{-1} were assigned to the first and second overtones, respectively, while bands at 3885–4795, 6500, and 7500 cm^{-1} were assigned to the combination regions. The C–O–C group was readily identified by the second overtone bands at 8300 and 8495 cm^{-1} and the combination bands at 6400–7515 cm^{-1} in ethers.

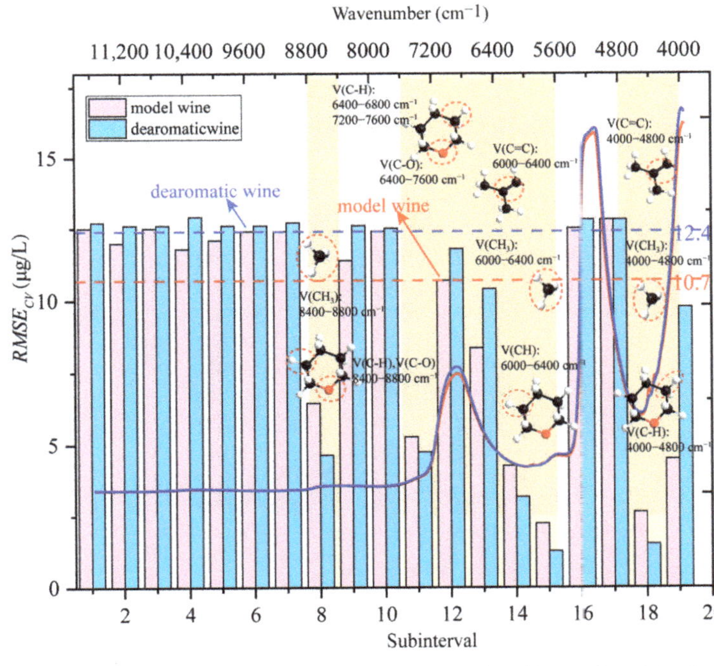

Figure 4. The $RMSE_{CV}$ values of Si-PLS based on each subinterval, and the feature wavebands analysis for the functional groups.

Table 3. Spectral mapping analysis of functional groups in similar substances of *Rose Oxide*.

Chemicals	Assignment Groups	Wave Numbers (cm^{-1})		
		First Overtone	Second Overtone	Combination Regions
Alkanes	V(C–H)	5555–5882	8264–8696	6666–7090, 4545, and 4500
	V(–CH_2–)	Near 6135	Near 8290	4545 and 4525
	V(–CH_3)	5901–5909	8264–8696	4500–4545, 4395, 4100, 4400, 5520, 5814, 7355, and 7263
Alkenes	V(C–H=)	6100–6200		
	V(=CH_2)		About 9260, 8787–9009, and 9091	
	V(C=C)			4482, near 4600, 4670–4780, and 6130
Tetrahydropyran	V(C–H)	5565–6150	8040–9320	3885–4795, 6500, and 7500
Ethers	V(C–H)			3800–4500 and 6400–7515
	V(–CH_2–)	5690 and 5790		
	V(–CH_3)	5898 and 5910		
	V(CH–O–)		8300	6400–7515
	V(CH_2–O–)		8495	

In this study, eight subintervals (4000–4400 cm^{-1}, 4400–4800 cm^{-1}, 5600–6000 cm^{-1}, 6000–6400 cm^{-1}, 6400–6800 cm^{-1}, 6800–7200 cm^{-1}, 7200–7600 cm^{-1}, 8400–8800 cm^{-1}) were recognized as the characteristic bands of *Rose Oxide* using the Si-PLS method. According to the assignments of the relevant groups mentioned in Table 3 and the eight characteristic intervals identified using the Si-PLS method, the group assignments of chemical structures in *Rose Oxide* were shown in Table 4. We observed that bands at the wave number region 5600–6000 cm^{-1} were due to the first overtone of C–H stretching in the tetrahydropyran ring and methyl group, as well as the combination of the C–H stretching of the CH_3 function groups. Furthermore, bands between 6000 cm^{-1} and 6400 cm^{-1} were assigned to the first overtone C–H stretching of the C–H= group and the combination C=C stretching in isobutyl and the wavenumber of 8400–8800 cm^{-1} belonged to the second overtone C–H stretching and C–O stretching in the tetrahydropyran ring, as well as the C–H stretching vibration in the methyl group. For the combination regions, 4000–4800 cm^{-1}, 6400–6800 cm^{-1}, and 7200–7600 cm^{-1} were assigned to the C–H stretching vibration, while 6400–7600 cm^{-1} was assigned to the C–O stretching vibration. These represented the spectral fingerprint information of *Rose Oxide* and are important for spectral identification and modeling applications.

Table 4. Group assignment of different chemical structures in *Rose Oxide*.

Chemical Structure	Assignment Group	Wave Numbers (cm^{-1})		
		First Overtone	Second Overtone	Combination Regions
Tetrahydropyran ring	V(C–H)	5600–6000	8400–8800	4000–4800, 6400–6800, and 7200–7600
	V(C–O)		8400–8800	6400–7600
Methyl	V(CH_3)	5600–6000	8400–8800	4000–4800, 5600–6000, 7200–7600
Isobutyl	V(C–H=)	6000–6400		
	V(C=C)			4400–4800, 6000–6400

Nevertheless, the individual intervals included limited information and did not completely reflect the absorption properties of the *Rose Oxide* spectra. Therefore, subintervals 8, 11, 12, 13, 14, 15, 18, and 19 (corresponding to wave numbers: 8800–8400 cm^{-1}, 7600–5600 cm^{-1}, and 4800–4000 cm^{-1}) were selected as joint intervals and re-modeled using the PLSR method. According to the results of joint interval modeling shown in

Table 5, the r_c^2 and $RMSE_c$ were 0.97 and 2.22 for the model wine and 0.97 and 2.36 for the de-aromatic wines, respectively; the r_{cv}^2 and $RMSE_{CV}$ were 0.96 and 2.55 for the model wine and 0.96 and 2.33 for the de-aromatic wines, which significantly improved the stability and predicted the accuracy of the model. The considerable improvement in the *RPD* value from 0.99 to 5.24 in de-aromatic wine indicated that a screened joint interval excluded a large amount of irrelevant information and condensed the spectral information of *Rose Oxide*, which laid the foundation for a further investigation of the spectral characteristics of its molecular groups and chemical bonds. The results of the best prediction model for the *Rose Oxide* were presented in Figure 5. It was apparent that the validation data (cross-validation method used in this study) were in good agreement with the resulting model. The correlation between the values tested by HS/SPME-GC-MS and the NIR calibration for the different wine substrates was good, and the models showed a satisfactory fitting result and predictive ability.

Table 5. Results of full-band and joint interval modeling of model wine and de-aromatic wine.

Interval Combinations	Model Wine					De-Aromatic Wine				
	r_c^2	$RMSE_C$	r_{cv}^2	$RMSE_{CV}$	*RPD*	r_c^2	$RMSE_C$	r_{cv}^2	$RMSE_{CV}$	*RPD*
Full waveband	0.75	6.37	0.23	10.70	1.19	0.04	12.10	0.14	11.34	0.99
Joint interval	0.97	2.22	0.96	2.55	4.78	0.97	2.36	0.96	2.33	5.24

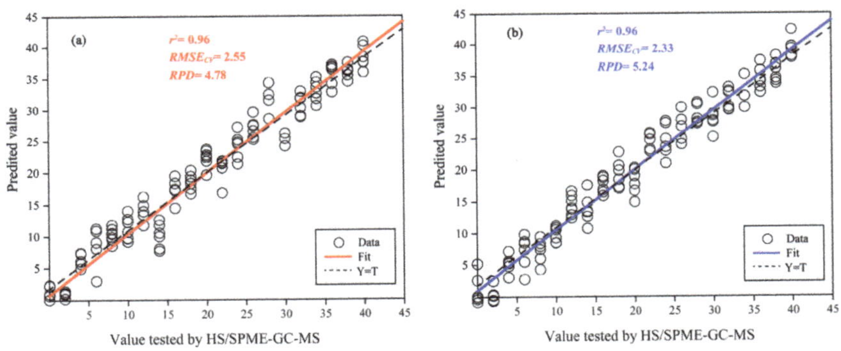

Figure 5. The results of the best prediction model for *Rose Oxide*. (**a**) The model wines; (**b**) the de-aromatic wines.

3.4. External Validation

The external validation aimed at estimating the predictive ability of the model based on a sample set that has not been included in the modeling process. In this study, the external validation of the PLS model for *Rose Oxide* in de-aromatic wine was conducted with the set of 21 samples, as shown in Table 6. The r_p^2 was 0.84 (higher than 0.80), indicating that the PLS models for *Rose Oxide* based on NIR spectra explained 84.00% of the variation in the data. The *RPD* value obtained for the *Rose Oxide* in the external validation was 2.36 (higher than 1.50) and the $RMSE_P$ value was 2.72, indicating the good prediction capacity of the NIR models for *Rose Oxide* in real wines. The regression equations are also presented in Table 6. It could be obtained that the *Rose Oxide* values tested by HS/SPME-GC-MS and the NIR calibration were similar. The results of external validation showed that this model could predict the *Rose Oxide* content of real wine to some extent, although the amount of information that can be explained was limited compared to the model built from the calibration set samples. On the one hand, the presence of other aromatic substances in real wine may affect the feature waveband of *Rose Oxide*, while on the other hand, it might be caused by the interference of other chemicals in the matrix. In conclusion, improving the accuracy of the external validation of the model is important for the transfer and application of the model, which still requires extensive research in the future.

Table 6. External validation of the established PLS models based on de-aromatic wine for *Rose Oxide* (μg/L) in real wines.

Spectral Number	External Validation			
	RMSE$_P$	*RPD*	r_p^2	Regression Equation
21	2.72	2.36	0.84	$y = 0.717x + 3.5288$

4. Discussion

4.1. Spectral Band Allocation of Rose Oxide

The NIR technology is also called "black box" technology. Most studies pay little attention to the connection between the chemical groups and spectral wavebands of substances in "black box". In fact, it is still beneficial to the analysis and application of near-infrared spectroscopy to master the distribution of the organic compounds near-infrared band. In this study, the near-infrared waveband allocation of *Rose Oxide* was analyzed based on the model wine substrate, and eight feature waveband subintervals were screened using the Si-PLS method for associating with the chemical groups of *Rose Oxide*. In fact, it is difficult to accurately attribute the near-infrared band because the near-infrared band may be the combination of several different fundamental frequency double and harmonic spectrum bands, and there is no sharp peak and baseline separation of spectral peaks, mostly overlapping peaks and wide peaks. In this study, the simulation of wine substrate and the Si-PLS method were adopted to avoid the above defects. On one hand, the model wine is a simple matrix with alcohol and pH values consistent with real wine, which reduces the interference of other chemical components while simulating the actual situation as much as possible. On the other hand, although it is hard to accurately locate the near-infrared band, the distribution range can be expanded by screening the feature bands in the form of a joint interval so that the broad peaks of near-infrared can be basically distributed in the sub-interval. In the process of chemical group allocation, the distribution of chemical bonds will always be the focus because the absorption of organic matter in the near-infrared band is generally caused by various chemical bond stretching vibrations. In this study, the frequency doubling and co-frequency absorption of tetrahydropyran rings, C–H bonds, C–O bonds, and C=C bonds at different positions on the methyl and isobutene groups of rose ether were assigned to eight selected characteristic bands. It can be observed that wavebands 5600–6000 cm^{-1} were related to the first overtone C–H stretching in the tetrahydropyran ring and methyl group, as well as the combination C–H stretching of the CH_3 function groups, which has been substantiated by the results of other studies. For example, Tosi and Pinto [26] found that bands near 5905 cm^{-1} in all the hydrocarbons can be attributed to the methyl group. Burns and Ciurczak [27] showed that 5700 cm^{-1}, 5810 cm^{-1}, and 5900 cm^{-1} were due to the 2v (C–H) vibration of the CH_2 functional group of cyclohexane and 2v (C–H) of the CH_3 groups of hydrocarbons with the methyl group. Bands between 6000 cm^{-1} and 6400 cm^{-1} were assigned to the first overtone C–H stretching of the C–H=group and the combination C=C stretching in isobutyl. As described by Gerasimov and Snavely [28], 6120, 6130, 6140, and 6200 cm^{-1} corresponded to the CH stretching bands of vinyl (CH_2=CH–) and vinylidene (CH_2=C<), which was consistent with the results obtained in this study. In addition, the distribution of other characteristic bands has been confirmed by relevant studies. There are also studies that used the fundamental frequency of chemical substances in the mid-infrared region to calculate their frequency doubling and frequency co-absorption bands in the near-infrared region, which can provide a new direction for the near-infrared spectral band attribution analysis of *Rose Oxide*.

4.2. Potential of Near-Infrared Spectroscopy Models of Rose Oxide

De-aromatic wine is often used to explore the perceptual interaction among aromas in wine because it is a modeling background substrate obtained from real wine based on a strict de-aromatic procedure. Except for the absence of aroma substances, the non-volatile substrate is consistent with real wine [18]. In this study, the prediction model of *Rose Oxide* was established based on the de-aromatic wine matrix. The experiment shows that the prediction model of *Rose Oxide* can explain 84% of the information with an *RPD* value greater than 1.5, indicating that the method had a certain feasibility. Indeed, many studies have taken the near-infrared detection of volatile aroma substances into consideration. For example, NIR technology combined with the PLS method was used to detect esters and higher alcohols in wine, and it achieved a good prediction [9]. In addition, NIR technology was used to detect volatile aroma substances such as esters and short-chain fatty acids in Riesling wine, and the model established based on PLS has also shown good prediction results [29]. A NIR correction model of oak volatiles was established in dry red wine using the PLS method, of which the r^2 was greater than 0.86 and the *RPD* was greater than 1.5 [30]. These results indicate that NIR spectroscopy can be used for the rapid detection of volatile aroma substances in wine. However, the majority of previous research has been aimed at the near-infrared analysis of volatile substances with a relatively rich content in wine, such as esters and higher alcohols. There are many compounds in wine with a low content, such as *Rose Oxide*, but with a great aroma contribution. Their rapid detection is also particularly important. Under the condition of modeling based on real wine substrate, the aroma substances of *Rose Oxide* are often ignored because the lengthy pre-treatment will lead to the loss of *Rose Oxide* in the wine sample, which would result in an error in the test results. In this study, a modeling sample set was constructed based on the standard for *Rose Oxide* in different concentrations and de-aromatic wine substrates. Under the background of a substrate simulating real wine samples to the greatest extent, the measured values of *Rose Oxide* could be accurately obtained, and a wide range of modeling concentrations could be obtained, which had a good universality for the rapid detection of the aroma substances of *Rose Oxide*. It provides methodological support for the near-infrared rapid nondestructive detection of similar aroma substances in wine.

5. Conclusions

In this study, NIR spectroscopy was used to address the waveband allocation and quantitative prediction analysis of *Rose Oxide*. First, MMN and VN were used to pro-process the spectral signal of the de-aromatic wine and the model wine, and PAC was used to find the outliers. Then, Si-PLS was used to select the related spectral subintervals of the *Rose Oxide*, which improved the accuracy of the prediction model. Finally, the prediction model of the *Rose Oxide* content in de-aromatic wine was established and verified in real wine. The prediction model with a high r^2 and low *RMSE* was effective for detecting the content of *Rose Oxide* in wines under certain conditions, and it provided methodology support for the quantitative analysis of other terpenes in wine using this method. Since the wine samples in this study were collected from the producing area in the wine region of the Ningxia Helan Mountain's East Foothill, whether the *Rose Oxide* model is applicable to the analysis of other producing areas or imported wine samples needs to be validated and optimized.

Author Contributions: Conceptualization, X.B.; methodology, X.B. and Y.X.; formal analysis, X.X.; resources, Y.X.; data curation, X.C. and B.D.; writing—original draft preparation, Y.X. and X.X.; writing—review and editing, X.B. and Y.T.; project administration, X.B. and Y.T.; funding acquisition, X.B. and Y.T. All authors have read and agreed to the published version of the manuscript.

Funding: This work was supported by the Natural Science Foundation of China (32202213), the Natural Science Foundation of China (31972199), the Shaanxi Science Fund for Distinguished Young Scholars (2020JC-22), and the Nature Science Foundation of Shaanxi (2022JQ-222).

Data Availability Statement: Not applicable.

Conflicts of Interest: The authors declare no conflict of interest.

References

1. Ruiz-García, L.; Hellín, P.; Flores, P.; Fenoll, J. Prediction of Muscat aroma in table grape by analysis of rose oxide. *Food Chem.* **2014**, *154*, 151–157. [CrossRef] [PubMed]
2. Chigo-Hernandez, M.M.; DuBois, A.; Tomasino, E. Aroma perception of rose oxide, linalool and α-terpineol combinations in Gewürztraminer wine. *Fermentation* **2022**, *8*, 30. [CrossRef]
3. Tang, Z.; Zeng, X.; Brennan, M.; Zhong, H.; Niu, D.; Huo, Y. Characterization of aroma profile and characteristic aromas during lychee wine fermentation. *J. Food Process. Pres.* **2019**, *43*, e14003. [CrossRef]
4. Bowen, A.J.; Reynolds, A.G. Odor potency of aroma compounds in Riesling and Vidal blanc Table Wines and Icewines by gas chromatography-olfactometry-mass spectrometry. *J. Agric. Food Chem.* **2012**, *60*, 2874–2883. [CrossRef] [PubMed]
5. Zhao, L.L.; Ruan, S.L.; Yang, X.K.; Chen, Q.L.; Shi, K.; Lu, K.; He, L.; Liu, S.W.; Song, Y.B. Characterization of volatile aroma compounds in litchi (Heiye) wine and distilled spirit. *Food Sci. Nutr.* **2021**, *9*, 5914–5927. [CrossRef] [PubMed]
6. Xu, Q.; Wang, H.; Lu, K.; Su, M.; Cai, J. Aroma Compounds Composition of "Rose Honey" Low-alcohol Rose Wine and the Influence of Commercial Yeast Strain. *China Fruit Veg.* **2020**, *40*, 49–55.
7. Maslov, L.; Tomaz, I.; Mihaljević Žulj, M.; Jeromel, A. Aroma characterization of predicate wines from Croatia. *Eur. Food Res. Technol.* **2016**, *243*, 263–274. [CrossRef]
8. Souza-Coutinho, M.; Brasil, R.; Souza, C.; Sousa, P.; Malfeito-Ferreira, M. Consumers Associate High-Quality (Fine) Wines with Complexity, Persistence, and Unpleasant Emotional Responses. *Foods* **2020**, *9*, 452. [CrossRef]
9. Genisheva, Z.; Quintelas, C.; Mesquita, D.P.; Ferreira, E.C.; Oliveira, J.M.; Amaral, A.L. New PLS analysis approach to wine volatile compounds characterization by near infrared spectroscopy (NIR). *Food Chem.* **2018**, *246*, 172–178. [CrossRef]
10. Bauer, R.; Nieuwoudt, H.; Bauer, F.F.; Kossmann, J.; Koch, K.R.; Esbensen, K.H. FTIR Spectroscopy for Grape and Wine Analysis. *Anal. Chem.* **2008**, *80*, 1371–1379. [CrossRef]
11. Birenboim, M.; Chalupowicz, D.; Barel, S.; Chen, Y.; Falik, E.; Kengisbuch, D.; Shimshoni, J.A. Optimization of sweet basil harvest time and cultivar characterization using near-infrared spectroscopy, liquid and gas chromatography, and chemometric statistical methods. *J. Sci. Food Agric.* **2021**, *102*, 3325–3335. [CrossRef] [PubMed]
12. Buratti, S.; Ballabio, D.; Giovanelli, G.; Dominguez, C.M.; Moles, A.; Benedetti, S.; Sinelli, N. Monitoring of alcoholic fermentation using near infrared and mid infrared spectroscopies combined with electronic nose and electronic tongue. *Anal. Chim. Acta* **2011**, *697*, 67–74. [CrossRef] [PubMed]
13. Silva, S.D.; Feliciano, R.P.; Boas, L.V.; Bronze, M.R. Application of FTIR-ATR to Moscatel dessert wines for prediction of total phenolic and flavonoid contents and antioxidant capacity. *Food Chem.* **2014**, *150*, 489–493. [CrossRef]
14. Cuq, S.; Lemetter, V.; Kleiber, D.; Levasseur-Garcia, C. Assessing macro- (P, K, Ca, Mg) and micronutrient (Mn, Fe, Cu, Zn, B) concentration in vine leaves and grape berries of vitis vinifera by using near-infrared spectroscopy and chemometrics. *Comput. Electron. Agric.* **2020**, *179*, 105841. [CrossRef]
15. Champagne, E.; Bonin, M.; Royo, A.A.; Tremblay, J.P.; Raymond, P. Predicting terpene content in dried conifer shoots using near infrared spectroscopy. *J. Near Infrared Spectrosc.* **2020**, *28*, 308–314. [CrossRef]
16. Yue, X.; Shi, P.; Tang, Y.; Zhang, H.; Ma, X.; Ju, Y.; Zhang, Z. Effects of methyl jasmonate on the monoterpenes of Muscat Hamburg grapes and wine. *J. Sci. Food Agric.* **2020**, *101*, 3665–3675. [CrossRef]
17. Zhang, B.; Shen, J.; Duan, C.; Yan, G. Use of Indigenous Hanseniaspora vineae and Metschnikowia pulcherrima Co-fermentation With Saccharomyces cerevisiae to Improve the Aroma Diversity of Vidal Blanc Icewine. *Front Microbiol.* **2018**, *9*, 2303. [CrossRef]
18. Cameleyre, M.; Lytra, G.; Tempere, S.; Barbe, J.C. Olfactory impact of higher alcohols on red wine fruity ester aroma expression in model solution. *J. Agric. Food Chem.* **2015**, *63*, 9777–9788. [CrossRef]
19. Diwu, P.Y.; Bian, X.H.; Wang, Z.F.; Liu, W. Study on the Selection of Spectral Preprocessing Methods. *Spectrosc. Spect. Anal.* **2019**, *39*, 2800–2806.
20. Bi, R.; Zhao, M. Online NIR modeling study based on PCA and Hotelling T^2 distribution. *Electron. Technol. Softw. Eng.* **2021**, *207*, 189–191.
21. Lu, B.; Liu, N.; Wang, X.; Hu, C.; Tang, X. A feasibility quantitative analysis of NIR spectroscopy coupled Si-PLS to predict coco-peat available nitrogen from rapid measurements. *Comput. Electron. Agric.* **2020**, *173*, 105400. [CrossRef]
22. Li, Y.; Li, F.; Yang, X.; Guo, L.; Huang, F.; Chen, Z.; Chen, X.; Zheng, S. Quantitative analysis of glycated albumin in serum based on ATR-FTIR spectrum combined with SiPLS and SVM. *Spectrochim. Acta A Mol. Biomol. Spectrosc.* **2018**, *201*, 249–257. [CrossRef] [PubMed]
23. Cai, J.; Wan, X.; Chen, Q. Feasibility study for the use of near-infrared spectroscopy in the quantitative analysis of TVB-N content in pork. *Acta Opt. Sin.* **2009**, *29*, 2808.
24. Zhu, Z.; Chen, S.; Su, J.; Tao, Y. Correlation Analysis Between Amino Acids and Fruity Esters During Spine Grape Fermentation. *Sci. Agric. Sin.* **2020**, *53*, 2272–2284.
25. Davis, J.C.; Koritzke, A.L.; Caravan, R.L.; Antonov, I.O.; Christianson, M.G.; Doner, A.C.; Osborn, D.L.; Sheps, L.; Taatjes, C.A.; Rotavera, B. Influence of the ether functional group on ketohydroperoxide formation in cyclic hydrocarbons: Tetrahydropyran and cyclohexane. *J. Phys. Chem. A* **2019**, *123*, 3634–3646. [CrossRef] [PubMed]

26. Tosi, C.; Pinto, A. Near-infrared spectroscopy of hydrocarbon functional groups. *Spectrochim. Acta A Mol. Biomol. Spectrosc.* **1972**, *28*, 585–597. [CrossRef]
27. Burns, D.A.; Ciurczak, E.W. *Handbook of Near-Infrared Analysis*; CRC Press: Beijing, China, 2007
28. Gerasimov, T.G.; Snavely, D.L. Vibrational photopolymerization of methyl methacrylate and quantitative analysis of polymerization results. *Macromolecules* **2002**, *35*, 5796–5800. [CrossRef]
29. Smyth, H.E.; Cozzolino, D.; Cynkar, W.U.; Dambergs, R.G.; Sefton, M.; Gishen, M. Near infrared spectroscopy as a rapid tool to measure volatile aroma compounds in Riesling wine: Possibilities and limits. *Anal. Bioanal. Chem.* **2008**, *390*, 1911–1916. [CrossRef] [PubMed]
30. Garde-Cerdán, T.; Lorenzo, C.; Alonso, G.L.; Salinas, M.R. Employment of near infrared spectroscopy to determine oak volatile compounds and ethylphenols in aged red wines. *Food Chem.* **2010**, *119*, 823–828. [CrossRef]

agronomy

Article

Construction and Test of Baler Feed Rate Detection Model Based on Power Monitoring

Huaiyu Liu [1,†], Ning Gao [1,†], Zhijun Meng [2], Anqi Zhang [1,*], Changkai Wen [1], Hanqing Li [1] and Jing Zhang [3]

1 Research Center of Intelligent Equipment, Beijing Academy of Agriculture and Forestry Sciences, Beijing 100097, China
2 Information Technology Research Center, Beijing Academy of Agriculture and Forestry Sciences, Beijing 100097, China
3 AgChip Science and Technology (Beijing) Co., Ltd., Beijing 100097, China
* Correspondence: zhangaq@nercita.org.cn
† These authors contributed equally to this work.

Abstract: The existing methods of measuring the baler feed rate seldom consider the influence of machine vibration on the sensor signal during field operation, which leads to the low detection accuracy and poor stability of feeding quantity detection. We established a feed rate detection model of a baler based on power monitoring of the pickup platform. Through the dynamic analysis of the pickup platform, the functional relationship between the working power of the pickup platform and the feed rate was constructed. A power monitoring system of the pickup platform was developed, and the model construction experiment of the working power and the feed rate was performed. The influence mechanism of different running speeds on the torque noise signal of the power input shaft of the pickup platform was explored. The frequency of the noise signal was mainly concentrated at 0.5–6 Hz and 9–13 Hz employing a fast Fourier transform, and the noise signal was eliminated by the frequency-domain-filtering method. The function model of working power and feed rate of the pickup platform was established based on signal processing, and the determination coefficient R^2 of the model was 0.9796. The field experiment results show that when the feed rate of the baler is between 1.6 and 4.88 kg/s, the determination coefficient R^2 and $RMSE$ between the actual and predicted feed rate are 0.989 and 0.2, respectively. The relative error range of feed-rate prediction is −9.37–8.77%, which indicates that the model has high detection accuracy and good stability and meets the requirements of feed-rate monitoring of a baler in field operation.

Keywords: baler; feed rate; pickup platform; working power; frequency domain filtering; model

Citation: Liu, H.; Gao, N.; Meng, Z.; Zhang, A.; Wen, C.; Li, H.; Zhang, J. Construction and Test of Baler Feed Rate Detection Model Based on Power Monitoring. *Agronomy* **2023**, *13*, 425. https://doi.org/10.3390/agronomy13020425

Academic Editor: William David Batchelor

Received: 17 December 2022
Revised: 19 January 2023
Accepted: 28 January 2023
Published: 31 January 2023

1. Introduction

Straw is a crucial biomass resource in agricultural production. The recycling of straw can not only promote the development of comprehensive straw utilization technology but also effectively improve the ecological environment [1–6]. To improve the collection efficiency of crop straw and reduce the transportation cost, balers were usually used to pick up the scattered straw in the field and press it into a high-density bundle structure [7,8]. The baler's feed rate is the critical index to evaluate the baler's functional performance and working state. If the feed rate is too low, the operation efficiency of the baler will be affected, and the operation cost will be increased. An excessive feeding amount will quickly lead to a blockage of the pickup platform, which will lead to operation failure of the transmission mechanism of the baler and seriously reduce the operation efficiency [9,10]. Therefore, obtaining the baler's feed rate in real time and keeping it in the best state can not only guide the driver to complete the straw-bundling operation efficiently but also have important practical significance for indirectly improving the comprehensive utilization rate of regional straw and improving the ecological environment.

Currently, the methods of crop feed-rate detection can be divided into extrusion force measurement, oil pressure measurement, image recognition, and torque and power measurement methods according to the principle [11–16]. Jie Zhan researched the relationship between feed rate, extrusion force and sensor signal based on the principle of extrusion pressure measurement [17]. Gomez-Gil developed a mathematical model of the grain-combine feed rate and running speed using GPS technology and particle weight sensors [18]. Sun Yifan developed a mathematical model of power-input-shaft torque, height, grain moisture content and feed rate of a combine harvester pickup platform based on the PSO-BP algorithm [19]. Wang Wei predicted the straw feed rate by monitoring the power of the screw conveyor and established a feeding amount detection model according to the grey theory [9]. Li Ping collected the pressure value of the header lifting cylinder and fitted the feed rate and pressure sensor signal by the least square method [20]. Liu Zhongpeng used an image segmentation algorithm to separate the wheat from the background and obtained the transformation relationship between the feed rate and image pixels by data fitting [15]. To sum up, limited by the working principle and mechanical structure of the baler's pickup platform, the feed-rate detection method based on extrusion force, oil pressure measurement and the image recognition principle was not suitable for the baler. Torque can better reflect the change in feeding amount. However, it is easily affected by uncontrollable variables, such as the angle of the pick-up mechanism in the detection process. As a linear quantity, power has fewer uncontrollable variables that affect its detection accuracy, which can more effectively reflect the change in machine feeding quantity [21]. In addition, the related research seldom considers the influence of machine vibration on sensor signals in actual operation, so it is not easy to ensure detection performance in a complex working environment.

Therefore, this study takes the square baler as the research object. Through an analysis of the dynamics of the pickup platform, the functional relationship between the working power of the pickup platform and the feed rate was established. A pick-up platform working power-monitoring system was built to obtain the original torque signal, and frequency domain analysis and filtering were performed. The specific function model of working power and feed rate of the pickup platform was obtained based on signal analysis and processing. Finally, field experiments tested the accuracy and stability of the feed-rate detection model. This study can provide a theoretical reference for the research of online detection technology of baler feed rate.

2. Materials and Methods

2.1. Dynamics Analysis of Pickup Platform

As shown in Figure 1, the overall structure of the pickup platform of the square baler is mainly composed of a power input shaft, picker shaft, pickup elastic teeth, screw conveyor and its shaft, which mainly completes the functions of straw pickup and centralizes backwards conveying. The working power of the pickup platform comes from the power input shaft. The power is transferred to the screw conveyor through the belt and drives the screw conveyor to rotate and then the power is transferred to the picker shaft through the chain. During continuous rotation, the picker shaft drives the elastic teeth to throw the ground straw to the screw conveyor. The screw conveyor squeezes the straw to the feeding mechanism during continuous rotation to complete the pick-up operation. According to the pitch diameter of the pulley and the transmission mode of each shaft, the transmission ratio among the power input shaft, screw conveyor shaft, and picker shaft is 1. Through the analysis of the operational characteristics of the pickup platform, it can be seen that the rotational speed of each shaft is consistent when the pickup platform works.

The power input shaft provides the working power for the square baler pickup platform, and the power input shaft power is the pickup platform power when the square baler is working. From the mathematical relationship of power, torque, and rotational speed, the total power of the pickup platform when the power input shaft is working is

$$P = Tn/9550 \quad (1)$$

where P is the total power of the power input shaft in kW; T is the total torque of the power input shaft in N·m; n is the rotational speed of each shaft of the pickup platform in r/min.

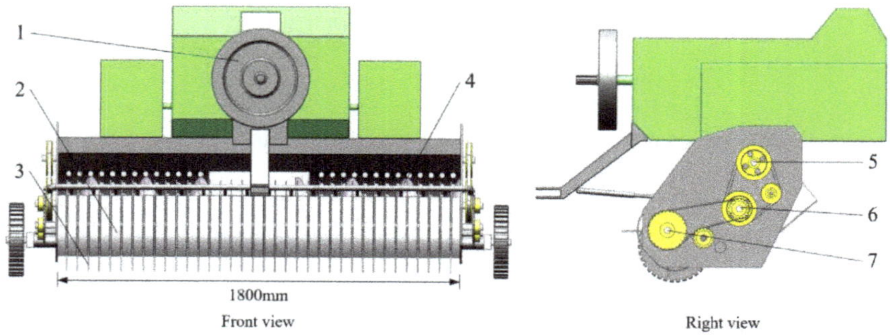

Figure 1. Structure diagram of square baler pickup platform: (**1**) flywheel; (**2**) pickup elastic teeth; (**3**) pickup plate; (**4**) screw conveyor; (**5**) power input shaft; (**6**) screw conveyor shaft; (**7**) picker shaft.

The pickup platform mainly consisted of a picker and a screw conveyor. According to the energy conservation theorem, the total power of the pickup platform can be decomposed into

$$P = P_0 + P_i + P_j \quad (2)$$

where P_0 is the total loss power of the transmission mechanism in kW, P_i is the total power of pickup in kW, and P_j is the total power of screw conveyor in kW.

(1) Torque analysis of picker shaft

The total torque T_i of the picker shaft can be decomposed into no-load torque T_{i1} and working torque T_{i2}, in which the no-load torque is related to its gravity. The working torque is mainly due to the torque produced by the gravity of the straw. Because the length of the elastic teeth extending out of the retaining ring is short, it can be considered that the straw is approximately concentrated at the end of the elastic teeth of the pickup, and thus

$$\begin{cases} T_{i1} = m_i g R \\ T_{i2} = \rho L h v_0 g R \end{cases} \quad (3)$$

$$q = \rho L h v_0 \quad (4)$$

$$T_i = (m_i + q) g R \quad (5)$$

where m_i is the weight of the picker in kg, ρ is the natural laying density of straw in kg/m^3, L is the pickup width in m, h is the natural laying thickness of straw in m, v_0 is the running speed of the baler in m/s, g is the acceleration of gravity in m/s^2, R is the radius of gyration of pickup elastic teeth in m, and q is the theoretical feed rate of straw in kg/s.

(2) Torque analysis of screw conveyor shaft

The total torque T_j of the screw conveyor shaft can be decomposed into no-load torque T_{j1} and working torque T_{j2}, and the no-load torque is related to its gravity. The working torque is mainly caused by the extrusion force F_r perpendicular to the surface of the spiral blade and the tangential friction resistance F_f along the spiral blade under the action of extrusion and the conveying of the straw by the spiral conveyor, which is not only affected by the extrusion force F_r perpendicular to the surface of the spiral blade. The friction resistance F_f is proportional to the extrusion force F_r of straw on spiral leaves. Assuming that the friction coefficient between the straw and screw conveyor is f, then

$$\begin{cases} T_{j1} = D(m_j g - F_r) \\ T_{j2} = Dqg + DF_f \end{cases} \tag{6}$$

$$F_f = f \cdot F_r \tag{7}$$

where D is the radius of gyration of the picker elastic teeth in m and m_j is the weight of the screw conveyor in kg.

According to [22–24], the extrusion force F_r is related to the degree of straw extrusion, and thus

$$F_r = K_p \cdot \left(\frac{C_{max}}{\delta}\right)^N \tag{8}$$

$$C_{max} = \frac{q\lambda}{[(1+\lambda)\rho V_s w]} \tag{9}$$

where K_p is the coefficient, C_{max} is the natural laying thickness of non-grain materials in mm, δ is the clearance between spiral blade and intaglio in mm, N is a real number, λ is the mass ratio of fruit and seedlings, V_s is the average linear velocity of straw in screw conveyor in m/s, and w is the width of the screw conveyor in m.

In this study, the experimental object is crop straw, and λ approximates it to infinity; thus, Formula (9) can be simplified as

$$C_{max} = \frac{q}{\rho V_s w} \tag{10}$$

From Equations (6)–(10), it can be seen that the total torque of the screw conveyor shaft is

$$T_j = Dqg + D[m_j g + K_p \left(\frac{1}{\rho V_s w \delta}\right)^N q^N (f-1)] \tag{11}$$

From Equations (2), (5), and (11), it can be seen that the total torque T of the power input shaft of the pickup platform is

$$T = T_0 + (m_i + q)gR + Dqg + D[m_j g + K_p \left(\frac{1}{\rho V_s w \delta}\right)^N q^N (f-1)] \tag{12}$$

(3) Power analysis of pickup platform

After excluding the torque caused by the gravity of the picker and screw conveyor, the working torque T' of the pickup platform is

$$T' = T_0 + qg(R+D) + DfK_p \left(\frac{1}{\rho V_s w \delta}\right)^N q^N \tag{13}$$

From Equations (1) and (13), it can be seen that the working power T' of the pickup platform is

$$P' = \frac{1}{9550} T_0 n + \frac{1}{9550}(R+D)gnq + \frac{1}{9550} DfnK_p \left(\frac{1}{\rho V_s w \delta}\right)^N q^N \tag{14}$$

It can be seen from the above formula that when other parameters are constant, and the baler is fed stably and evenly, the working power of the pickup platform is linearly related to the feed rate. The above formula can be simplified to

$$P' = K_0 + K_1 q + K_2 q^N \tag{15}$$

where K_0, K_1, and K_2 are all constant coefficients.

2.2. Field Test Method

2.2.1. Test Equipment and Materials

To explore the specific function model between the working power of the pickup platform and the feed rate, the power-monitoring system of the pickup platform was built, and the system structure is shown in Figure 2. The system consists of a sensor unit and a data-management unit. The sensor unit converts the torque and rotational speed signals of the power input shaft of the pickup platform into voltage signals and pulse signals, respectively. The built-in module of the acquisition instrument converts the two signals into torque and rotational speed data. Following this, it is transmitted to the vehicle data-management unit for display and storage through the RS485 bus. Finally, the power data of the pickup platform are calculated according to Equation (1).

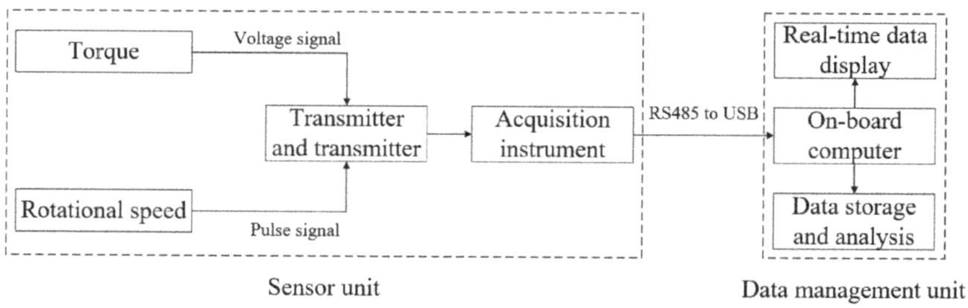

Figure 2. Structural block diagram of a monitoring system.

A DYN-200 torque and rotational speed sensor (Bengbu Dayang Sensing System Engineering Co., Ltd., Bengbu, China) is the test sensor. The torque and rotational speed data of the power input shaft of the pickup platform are collected synchronously through the supporting data management software. Its main parameters are as follows (Table 1):

Table 1. Main parameters of the sensor.

Parameters	Value
Range of torque	±500 N·M
Range of rotational speed	10,000 RPM
Error of measurement	<0.1%
Voltage	DC24 V

To accurately collect the torque and rotational speed data of the power input shaft, a separate power input shaft was designed. The sensor was installed between the power end and the load end of the power input shaft by two couplings and fixed on the rear protective plate of the pickup platform by an installation bracket. The installation diagram is shown in Figure 3.

In October 2022, the model construction test was conducted in the Xiaotangshan National Experiment Station for Precision Agriculture, Changping District, Beijing. The test equipment and materials are shown in Figure 4. Dry wheat straw is the test object. The torque and rotational speed sensor was installed on the power input shaft of the pickup platform of the 9YFQ-2.2 baler (Tianjin Xuanhe Agricultural Machinery Manufacturing Co., Ltd., Tianjin, China). The Lovol Oubao M800-D tractor (Weichai Lovol Heavy Industry Co., Ltd., Weifang, China) pulls the baler and provides operating power. In order to ensure the picking quality of the baler, AMG-300 tractor automatic driving navigation (AgChip Science and Technology (Beijing) Co., Ltd., Beijing, China) was adopted to control the tractor to drive automatically according to the planned path. In addition, other test materials include signs, stopwatches, tape measures, and electronic scales, which were used to divide the test site and accurately measure the working time of the baler.

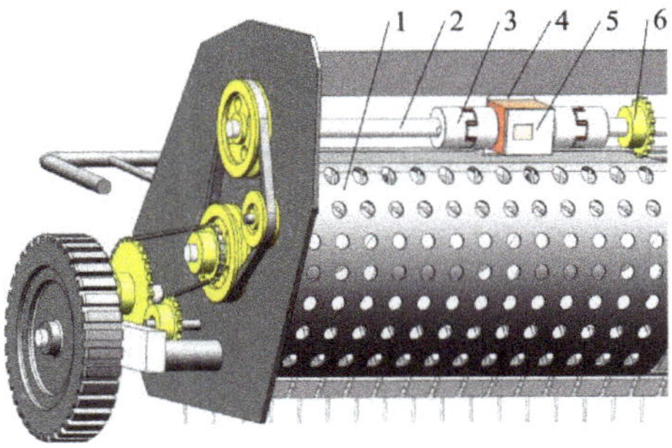

Figure 3. Sensor installation diagram: (**1**) rear protective plate of the pickup platform; (**2**) separate power input shaft; (**3**) coupling; (**4**) installation bracket; (**5**) torque and rotational speed sensor; (**6**) transmission gear.

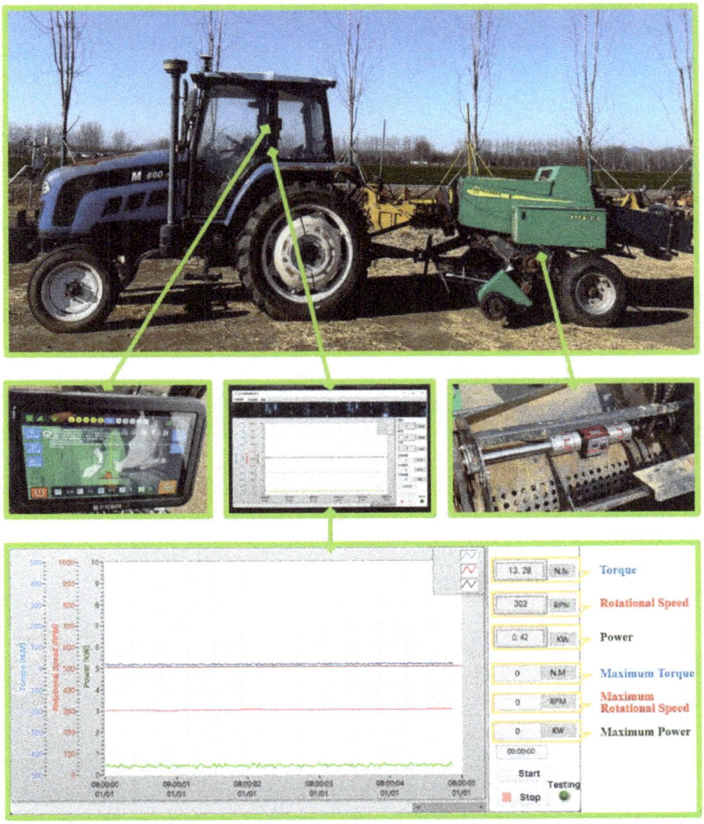

Figure 4. Test equipment.

2.2.2. Test Methods

In this study, the torque and rotational speed sensors were used to measure the baler's power input shaft's torque and rotational speed data at different running speeds under the field test conditions, and the working time and feed rate of the baler were obtained by manual measurement. To obtain sufficient experimental data and reduce the experimental cost, the size of the field experiment plot was determined to be 20 × 1.8 m. According to the actual yield and economic coefficient [25–28] of wheat in this experimental field, it can be calculated that 1.5 kg straw should be laid on each square meter of land. In order to ensure the uniform laying of straw, a plot with a length of 20 m was divided into ten sections. The straw was manually weighed and 5.4 kg straw was laid at each interval. The test site is shown in Figure 5.

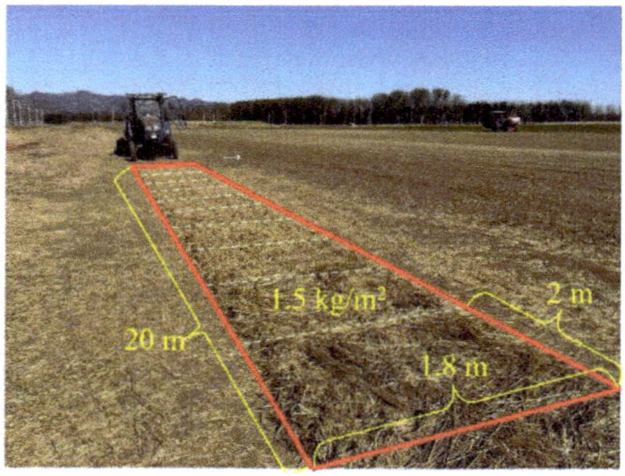

Figure 5. Test site.

Considering the influence of ground height on sensor data acquisition, this study determined through many no-load tests that the appropriate height for picking up elastic teeth from the ground of the baler is 50 mm. Before each test, the pickup elastic teeth were adjusted to the appropriate height, and the sensor was cleared. The tester used a stopwatch to measure the working time of the baler and inserted signs at the starting and ending positions of the test plot. The tester starts timing at the moment when the pickup platform of the baler touches the start sign and stops timing at the moment when the pickup platform touches the end sign. During data acquisition, a vehicle-mounted data-management software is used to measure the torque and rotational speed of the baler's power input shaft under no-load and load conditions. The difference between the average torque of the power input shaft under load and the average torque under no load is taken as the working torque of this test, and the rotational speed is the average value of the rotational speed data under load.

$$Q = \frac{M}{t} \tag{16}$$

where Q is the actual feed rate of the baler in kg/s, M is the total amount of straw harvested in kg, and t is the operation time in s.

2.3. *Signal Processing Method*

In the process of sensor signal acquisition, the vibration caused by the mechanical transmission of the baler and ground bumping greatly influences the sensor signal's stability [29]. In this study, the frequency domain information of the original torque signal of the power input shaft was obtained by Fourier transform, the concentrated frequency band of the noise signal was found, and the torque noise signal was eliminated by frequency domain filtering.

A discrete Fourier transform is a discrete form of Fourier transform in the time domain and frequency domain, which is the sampling of the Fourier transform of time domain signal in discrete time [30–32]. The main idea is to establish the function-mapping relationship between the signal's frequency spectrum with time as the independent variable and frequency as the dependent variable from the finite points of the Fourier transform in an ordered long sequence. The continuous Fourier transform formula of an analogue signal $x(t)$ is as follows

$$X(w) = \int_{-\infty}^{\infty} x(t)e^{-jwt}dt \tag{17}$$

$x(t)$ becomes $X(nT)$ after T sampling periods. If $X(nT)$ is an B point finite-length sequence, then DFT is:

$$X(k) = DFT[x(n)] = \sum_{n=0}^{N-1} x(n)W_N^{nk} = \sum_{n=0}^{N-1} x(n)e^{-j\frac{2\pi}{N}nk}, k = 0,1,\cdots,B-1 \tag{18}$$

Because the computational complexity of the DFT algorithm increases with the increase in ordered length signal, it does not meet the needs of practical engineering applications. The FFT algorithm transforms a long DFT operation into several short DFT operations employing periodicity $W_N^{nk} = W_N^{(n+N)k}$, symmetry $W_N^{n+N/2} = -W_N^n$, and reducibility $W_N^{nk} = W_{mN}^{mnk}$ of the DFT algorithm.

3. Results and Discussion

3.1. Signal Analysis and Processing Results

In this study, the test results of a baler working at 5.57 km/h are taken as an example. The time domain diagram of its original torque and rotational speed signal is shown in Figure 6. As seen from the figure, when the baler works, the mechanical vibration and ground turbulence significantly impact the torque sensor signal. The data variability and overall smoothness were poor. The rotational speed signal was relatively stable, and the overall distribution was about 278 r/min. The standard deviation was 6.18 r/min. In this study, we will analyze the causes of torque signal noise.

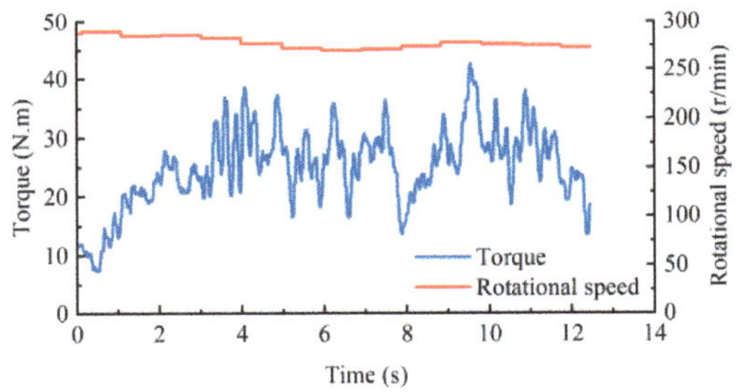

Figure 6. Time domain diagram of original torque and rotational speed signal.

The frequency domain information of torque signal under different running speeds was obtained by fast Fourier transform to explore the influence mechanism of running speed on the torque noise signal. Figure 7 is a spectrum diagram of the torque signal of the baler in the static and working state under different running speeds. As seen from Figure 7, under different running speeds, the amplitude of the torque noise signal when the baler works is more prominent than when it is at rest. With the baler's increased running speed, the noise signal's amplitude when the machine works gradually increases. The reason is that when the baler is running at a low speed, the vibration of the machine produced by the power system is the primary source of the noise signal, and the amplitude of the noise signal in the two states is equivalent. With the increased running speed of the baler, the bump caused by ground fluctuation becomes the primary source of the noise signal, and the amplitude of the noise signal increases with the increase in running speed.

As shown in Figure 7, the torque noise signals were mainly concentrated at 0.5–6 Hz and 9–13 Hz. In this study, the noise reduction of the original torque signal was conducted by the frequency-domain-filtering method. The torque signal collected by the baler at the running speed of 5.57 km/h was taken as the object. Figure 8 is a time domain diagram of working torque signals before and after processing. After filtering, the variability of torque data was reduced, and the overall smoothness of the data was improved. Data fluctuation was due to the change in torque caused by an uneven straw feed rate during baler operation.

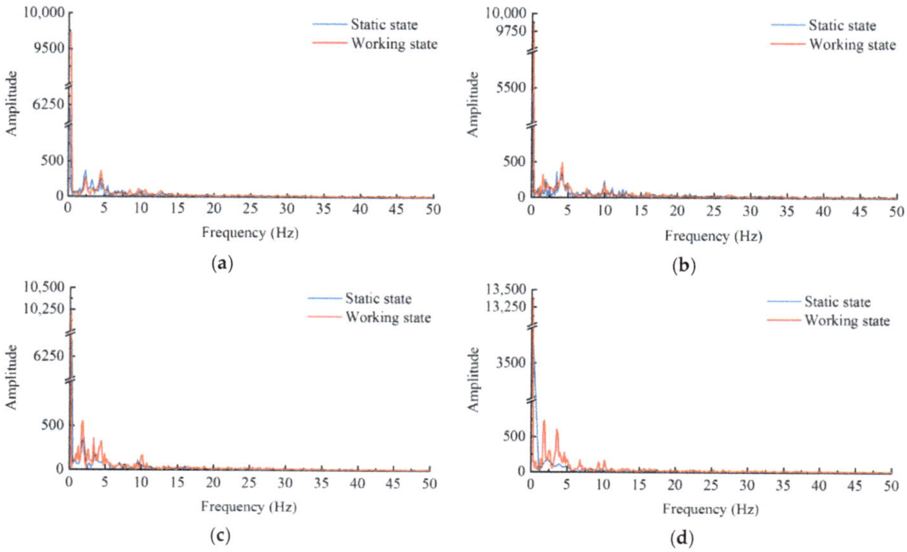

Figure 7. Spectrum diagram of torque signal at different running speeds: (**a**) 3.17 km/h; (**b**) 4.14 km/h; (**c**) 5.57 km/h; (**d**) 7.13 km/h.

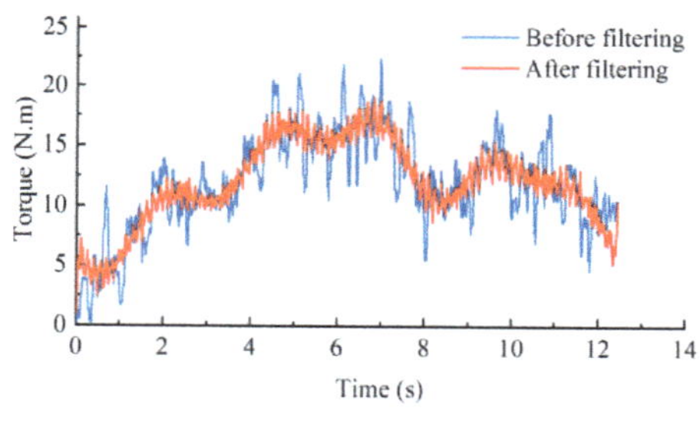

Figure 8. Time domain diagram before and after signal processing.

3.2. Construction Results of Feed-Rate Detection Model

In this study, we used the frequency-domain-filtering method to process all the original torque data in the above tests. According to Equation (1), the average working power of the pickup platform for each group of tests was calculated, and the test results are shown in Table 2.

In this study, the multivariate linear regression method was used to establish a specific functional model of working power and feed rate, and the performance of the model was evaluated by R^2 (efficient of determination) and *RMSE* (root mean square error). R^2 was used to evaluate the correlation degree between the actual value and the predicted value of the feed rate. The closer R^2 is to 1, the better the correlation degree between the actual value and the predicted value of the feed rate. *RMSE* was used to evaluate the prediction ability of the model. A smaller *RMSE* indicates a better generalization ability of the model.

In this study, the relationship curve between the working power of the pickup platform and the feed rate fitted by SPSS26 is shown in Figure 9.

Table 2. Test results.

Test	Running Speed (km/h)	Harvest (kg)	Working Torque (N·m)	Rotational Speed (r/min)	Working Power (kW)	Feed Rate (kg/s)
1	3.17	30.4	6.23	292.7	0.1909	1.34
2	4.24	32.28	8.66	393.3	0.3566	1.90
3	7.13	20.29	7.23	375	0.2839	2.01
4	5.57	30.61	11.97	345.8	0.4334	2.37
5	6.46	28.57	12.05	342.1	0.4317	2.56
6	5.78	32.81	15.6	365.5	0.5970	2.64
7	7.07	31.06	16.85	338.9	0.5980	3.05
8	8.41	29.63	17.95	391	0.7349	3.46
9	9.35	30.18	26.77	347.4	0.9738	3.92
10	10.64	32.02	33.79	384.3	1.3597	4.73
11	10.53	32.93	29.05	386.2	1.1748	4.81
12	10.48	36.13	40.38	379.9	1.6063	5.26

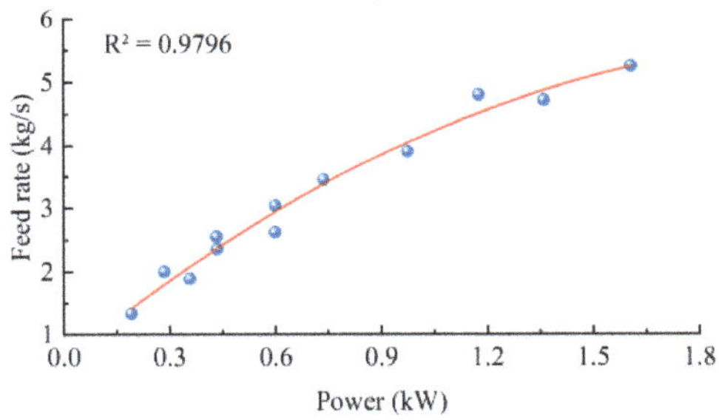

Figure 9. Fitting results of working power and feed rate.

As seen from Figure 9, the feed rate increases with the increase in power, and there is an excellent linear relationship between them. The determination coefficient R^2 was 0.9796, and the functional relationship between the working power of the pickup platform and the feed rate is shown in Equation (19).

$$Q = -1.0433P'^2 + 4.5791P' + 0.5859 \tag{19}$$

3.3. Field Verification Test Results

In order to verify the accuracy and effectiveness of the model in the field operation of the baler, 11 groups of field verification experiments were conducted according to the above test methods and signal processing methods. The test site is shown in Figure 10.

In the experiment, the running speed of the baler ranged from 4.14 km/h to 10.37 km/h, and the feed rate ranged from 1.62 kg/s to 4.88 kg/s. The feed-rate detection model calculated the predicted value of the feed rate of each group of tests. The relationship between the actual value and the predicted value of the feed rate in each group was compared and analyzed. Figure 11 shows the distribution of actual and predicted values of baler feed rate in each group of tests. According to the test results, the determination coefficient R^2 between the actual value and the predicted value of the baler feed rate is 0.989, the *RMSE* is 0.2, and the relative error range of the feed-rate prediction is −9.37–8.77%. Therefore, the results of field experiments prove that the model has high accuracy and good effectiveness, and meets the needs of monitoring the baler feed rate in the field operation.

Figure 10. Field experiment.

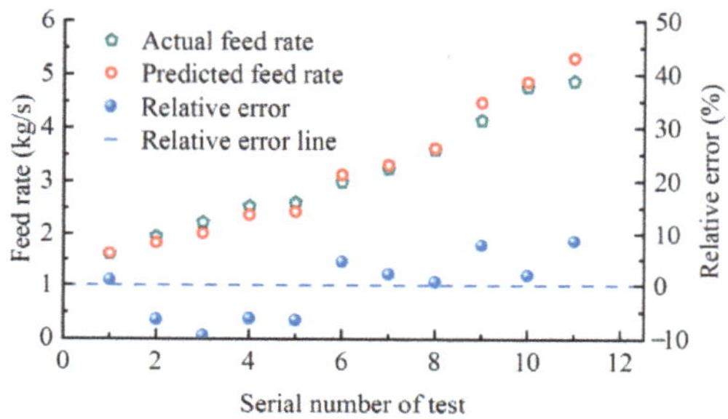

Figure 11. Field test results.

4. Conclusions

The existing methods of measuring the baler feed rate seldom consider the influence of machine vibration on the sensor signal during field operation, which leads to a low detection accuracy and poor stability in feeding-quantity detection. In this study, we established a model for measuring the baler feed rate based on the power monitoring of the pickup platform. We verified the accuracy and effectiveness of the model's predictions through field experiments. The conclusions of this study are as follows:

Through the dynamic analysis of the baler pickup platform, the functional relationship between the working power of the pickup platform and the feed rate was obtained. A power-monitoring system for the baler pickup platform was developed, which realizes the real-time collection of torque and rotational speed data of the power input shaft. The influence mechanism of the baler running speed on the torque noise signal was analyzed, the main frequency distribution range of the torque noise signal was determined, and the torque noise signal was eliminated by the frequency-domain-filtering method. Finally, a mathematical model of the working power and feed rate of the pickup platform was constructed, and a field experiment of model performance verification was performed.

Noise signal analysis results showed that when the sampling frequency of the system was 200 Hz, the frequency of torque noise signal was distributed in the ranges 0.5–6 Hz and 9–13 Hz. The results of the feed-rate detection model construction showed that the R^2 of the model was 0.9796 when the baler feed rate was between 1.34 and 5.26 kg/s. The results of field experiments showed that the R^2 between the actual and predicted values was 0.989, the $RMSE$ was 0.2, and the relative error range of model prediction was −9.37–8.77%. The feed-rate detection model has high accuracy and stability.

Currently, we have only established a model to detect the baler feed rate, but the model still needs to be integrated into the feed-rate monitoring device and has yet to be verified in field trials. In future research, we will develop a control system for the pickup operation of the baler based on feed-rate monitoring. The model in this study will be integrated into the control system. The real-time and accurate control of the baler feed rate in field operation may solve the problem that the control object (such as running speed and pickup height) does not match the feed rate due to the poor monitoring effect of the feed rate in the current baler feed-rate control system, which leads to a low operating efficiency of the baler and easy failure of the transmission mechanism. This study can provide a new technical scheme for the research and development of the feeding-monitoring system of the baler.

Author Contributions: Conceptualization, H.L. (Huaiyu Liu) and N.G.; methodology, Z.M. and A.Z.; software, C.W.; validation, H.L. (Hanqing Li); formal analysis, J.Z.; investigation, C.W.; resources, Z.M. and A.Z.; data curation, H.L. (Huaiyu Liu); writing—original draft preparation, H.L. (Huaiyu Liu) and N.G.; writing—review and editing, H.L. (Huaiyu Liu) and N.G.; visualization, H.L. (Huaiyu Liu); supervision, Z.M.; project administration, Z.M.; funding acquisition, A.Z. All authors have read and agreed to the published version of the manuscript.

Funding: This research was funded by the National Key Research and Development Plan of China (Grant No. 2021YFD200050302), the Innovation Capacity Building Project of Beijing Academy of Agriculture and Forestry Sciences (Grant No. KJCX20200416) and Focus on Research and Development Plan of Shandong Province of China (Major Scientific and Technological Innovation Project) (Grant No. 2022CXGC010608).

Data Availability Statement: All data are presented in this article in the form of figures or tables.

Acknowledgments: We would like to thank the "Research Center of Intelligent Equipment, Beijing Academy of Agriculture and Forestry Sciences", "Information Technology Research Center, Beijing Academy of Agriculture and Forestry Sciences" and "AgChip Science and Technology (Beijing) Co., Ltd.".

Conflicts of Interest: The authors declare no conflict of interest.

References

1. Li, S.; Ji, X.; Deng, K.; Zhu, J.; Li, C.; Jian, Y.; Peng, H. Analysis of Regional Distribution Patterns and Full Utilization Potential of Crop Straw Resources. *Trans. Chin. Soc. Agric. Eng.* **2020**, *36*, 221–228. [CrossRef]
2. Liu, S.; Li, D.; Huang, J.; Ma, C.; Wang, H.; Yu, Z.; Qu, X.; Zhang, L.; Han, T.; Du, J.; et al. Temporal and Spatial Distribution Characteristics of Rice Stalk Resources and its Potential of Synthetic Fertilizers Substitution Returning to Farmland in China from 1988 to 2018. *Trans. Chin. Soc. Agric. Eng.* **2021**, *37*, 151–161. [CrossRef]
3. Cong, H.; Meng, H.; Yu, J.; Ye, B.; Yao, Z.; Feng, J.; Yu, B.; Qin, C.; Huo, L.; Yuan, Y.; et al. Analysis of Long-term Mechanism for Development of Straw Industry in Northeast China under Guidance of "Green-concept". *Trans. Chin. Soc. Agric. Eng.* **2021**, *37*, 314–321. [CrossRef]
4. Huo, L.; Yao, Z.; Zhao, L.; Luo, J.; Zhang, P. Contribution and Potential of Comprehensive Utilization of Straw in GHG Emission Reduction and Carbon Sequestration. *Trans. Chin. Soc. Agric. Mach.* **2022**, *53*, 349–359. [CrossRef]
5. Meng, Z.; Liu, H.; An, X.; Yin, Y.; Jin, C.; Zhang, A. Prediction Model of Wheat Straw Moisture Content Based on SPA-SSA-BP. *Trans. Chin. Soc. Agric. Mach.* **2022**, *53*, 231–238, 245. [CrossRef]
6. Liu, C.; He, Z.; Lu, X. Optimization Analysis of Carbon Emission Reduction from Crop Straw Collection and Transportation under the Sustainable Development goals. *Trans. Chin. Soc. Agric. Eng.* **2022**, *38*, 239–248. [CrossRef]
7. Zhang, A.; Meng, Z.; Chen, L.; Wu, G.; Cong, Y.; An, X. Design and Experiment of Dynamic Weighing System for Small Square Baler. *Trans. Chin. Soc. Agric. Mach.* **2020**, *51*, 170–175. [CrossRef]
8. Liu, H.; Meng, Z.; Zhang, A.; Cong, Y.; An, X.; Fu, W.; Wu, G.; Yin, Y.; Jin, C. On-Line Detection Method and Device for Moisture Content Measurement of Bales in a Square Baler. *Agriculture* **2022**, *12*, 1183. [CrossRef]
9. Wang, W.; Gong, Y.; Bai, X.; Tan, R.; Huang, W. Investigation on Operating Speed Regulation System of Mobile Straw Granulator. *Trans. Chin. Soc. Agric. Mach.* **2021**, *52*, 186–195. [CrossRef]
10. Liu, J.; Guo, H.; Lv, Q.; Li, L. Design of Feeding Control System for Self-propelled Round Baler. *J. Chin. Agric. Mech.* **2021**, *42*, 13–18, 25. [CrossRef]
11. Jim, K. Computer control for the combine. *Agric. Eng.* **1983**, *64*, 7–9.
12. Liang, X.; Chen, Z.; Zhang, X.; Wei, L.; Li, W.; Che, Y. Design and Experiment of On-line Monitoring System for Feed Quantity of Combine Harvester. *Trans. Chin. Soc. Agric. Mach.* **2013**, *44* (Suppl. S2), 1–6. [CrossRef]

13. Li, Y.; Wang, J.; Xu, L.; Tang, Z.; Xu, Z.; Wang, K. Design and Experiment on Adjusting Mechanism of Concave Clearance of Combine Harvester Cylinder. *Trans. Chin. Soc. Agric. Mach.* **2018**, *49*, 68–75. [CrossRef]
14. Luo, S.; Jie, Z. Application of Wireless Dada Acquisition and Virtual Instrument Techniques to Measurement of Combine Feed Quantity. *J. Henan Univ. Sci. Technol. Nat. Sci.* **2010**, *31*, 72–75+111–112. [CrossRef]
15. Liu, Z.; Zhao, J.; Liu, L. Study on Wheat Harvester Crop Density Detection During Maturity Based on the Improved Algorithms. *J. Agric. Mech. Res.* **2015**, *37*, 49–52. [CrossRef]
16. John, S.; Michael, P.M.; Gary, W.K. Combine feed rate sensors. *Trans. ASAE* **1985**, *28*, 2–5. [CrossRef]
17. Jie, Z.; Chen, J.; Liu, H. Research on Fuzzy Control of Random Feed Quantity of GPS Combine. *Trans. Chin. Soc. Agric. Mach.* **2006**, *37*, 55–58.
18. Gomez Gil, J.; Lopez Lopez, L.J.; Navas Gracia, L.M. The Spatial Low-pass Filtering as an Alternative to Interpolation Methods in the Generation of Combine Harvester Yield Maps. *Appl. Eng. Agric.* **2011**, *27*, 1087–1097. [CrossRef]
19. Sun, Y.; Liu, R.; Zhang, M.; Li, M.; Zhang, Z.; Li, H. Design of Feed Rate Monitoring System and Estimation Method for Yield Distribution Information on Combine Harvester. In *Computers and Electronics in Agriculture*; Elsevier: Amsterdam, The Netherlands, 2022; Volume 201, p. 107322. [CrossRef]
20. Li, P.; Cui, J.; Feng, W.; Zhang, X.; Zhang, T.; Li, Y. Design and Test of Real-time Monitoring System for Feeding Quantity of Small Combine. *J. Agric. Mech. Res.* **2022**, *44*, 240–246. [CrossRef]
21. Sun, Y.; Liu, R.; Ou, H.; Zhang, Z.; Zhang, M.; Li, H. Analysis and Comparison of Feed Rate Detection Methods of Combine Harvester Based on Power Detection. *Trans. Chin. Soc. Agric. Mach.* **2020**, *51*, 118–123. [CrossRef]
22. Huynh, V.M.; Powell, T.; Siddall, J.N. Threshing and Separating Process-a Mathematical Model. *Trans. ASAE* **1982**, *20*, 65–73. [CrossRef]
23. Lu, W.; Zhang, D. Research on the Power Consumption Model of Threshing Cylinder. *J. Agric. Mech. Res.* **2012**, *34*, 44–47. [CrossRef]
24. Lu, W.; Deng, Z.; Zhang, D.; Gao, H.; You, H. Combine Feed Rate Modeling Research. *J. Agric. Mech. Res.* **2013**, *35*, 129–132. [CrossRef]
25. Yu, Q.; Sun, G. Study on the Change of Crop Straw Resources in China. *Mod. Agric.* **2018**, *09*, 13–15.
26. Wang, X. Spatial and Temporal Distribution of Crop Residue Resource in the East, Central-South, and Southwest Regions of China. Doctoral Dissertation, China Agricultural University, Beijing, China, 2014.
27. Wei, M.; Wang, X.; Xie, G. Field Residue of Field Crops and its Temporal Distribution among Thirty-one Provinces of China. *J. China Agric. Univ.* **2012**, *17*, 32–44.
28. Wang, X.; Xue, S.; Xie, G. Value-taking for Residue factor as a Parameter to Assess the Field Residue of Field Crops. *J. China Agric. Univ.* **2012**, *17*, 1–8.
29. Zhang, A.; Meng, Z.; Chen, L.; Wang, P.; Liu, H.; An, X. Signal Analysis and Processing of Dynamic Weighing System for Small Square Baler. *Trans. Chin. Soc. Agric. Mach.* **2020**, *51* (Suppl. S2), 243–248, 260. [CrossRef]
30. Yu, X. Application of Fast Fourier Transform in Signal Processing. *Inf. Rec. Mater.* **2021**, *22*, 184–186. [CrossRef]
31. Wu, J. Fast Fourier Spectrometer Signal Acquisition Technology Based on ZYNQ. Master's Thesis, University of China Academy of Sciences, Shanghai, China, 2021.
32. Wang, C. Research on Fast Fourier Transform Algorithm Based on Eisenstein. Master's Thesis, University of Science and Technology of China, Hefei, China, 2021.

agronomy

MDPI

Article

Design and Test of Obstacle Detection and Harvester Pre-Collision System Based on 2D Lidar

Yehua Shang [1,2], Hao Wang [2,3], Wuchang Qin [2], Qian Wang [2], Huaiyu Liu [4], Yanxin Yin [2,3], Zhenghe Song [1,*] and Zhijun Meng [2,*]

1 College of Engineering, China Agricultural University, Beijing 100083, China
2 Research Center of Intelligent Equipment, Beijing Academy of Agriculture and Forestry Sciences, Beijing 100097, China
3 State Key Laboratory of Intelligent Agricultural Power Equipment, Beijing 100097, China
4 AgChip Science and Technology (Beijing) Co., Ltd., Beijing 100097, China
* Correspondence: songzhenghe@cau.edu.cn (Z.S.); mengzj@nercita.org.cn (Z.M.)

Abstract: Aiming at the need to prevent agricultural machinery from colliding with obstacles in the operation of unmanned agricultural machinery, an obstacle detection algorithm using 2D lidar was proposed, and a pre-collision system was designed using this algorithm, which was tested on a harvester. The method uses the differences between lidar data frames to calculate the collision times between the farm machinery and the obstacles. The algorithm consists of the following steps: pre-processing to determine the region of interest, median filtering, and DBSCAN (density-based spatial clustering of applications with noise) to identify the obstacle and calculate of the collision time according to the 6σ principle. Based on this algorithm, a pre-collision system was developed and integrated with agricultural machinery navigation software. The harvester was refitted electronically, and the system was tested on a harvester. The results showed that the system had an average accuracy rate of 96.67% and an average recall rate of 97.14% for being able to stop safely for obstacles in the area of interest, with a summed average of 97% for both the accuracy and recall rates. The system can be used for an emergency stop when encountering obstacles in the automatic driving of agricultural machinery and provides a basis for the unmanned driving of agricultural machinery in more complex scenarios.

Keywords: lidar; obstacle detection; harvester; pre-collision system

Citation: Shang, Y.; Wang, H.; Qin, W.; Wang, Q.; Liu, H.; Yin, Y.; Song, Z.; Meng, Z. Design and Test of Obstacle Detection and Harvester Pre-Collision System Based on 2D Lidar. *Agronomy* **2023**, *13*, 388. https://doi.org/10.3390/agronomy13020388

Academic Editor: Shubo Wang

Received: 25 December 2022
Revised: 17 January 2023
Accepted: 26 January 2023
Published: 28 January 2023

1. Introduction

With the aging of the Chinese population and the continuous reduction in the farming labor force, who will farm and how will farming be conducted in the future are questions that urgently need to be considered [1]. Agricultural machinery can significantly reduce agricultural labor, and farming by machines replacing humans is widespread in China. Nevertheless, most agricultural machinery still needs to be driven by an operator. Some agricultural-machinery-assisted driving systems can already control the direction to go straight or turn without the operator's control [2]. However, when there are obstacles in the field, the assisted driving systems must be interrupted manually. Some unmanned driving is achieved by marking the coordinates of the obstacles and planning the path in a specific plot [3]. This method must obtain the plot coordinates and plan the path before each operation. It is not suitable for cross-regional work of agricultural machinery, which is very common in China. The detection of the farmland environment, especially the detection of farmland obstacles, is an essential part of realizing unmanned driving.

There have been many studies on the detection of obstacles in the field. Moreover, there are three main sensor technologies for detecting farmland obstacles: ultrasonic detection, machine vision detection, and lidar detection technology [4]. Each technical has advantages and disadvantages, and there are still some detection problems in unstructured field environments. Ultrasonic testing has a low cost and simple data processing. Dvorak tested an ultrasonic sensor's ability to detect several objects that are commonly encountered in outdoor agricultural or construction environments [5]. However, ultrasonic detection has a short measurement range and low ranging accuracy and is easily affected by temperature, environmental noise, and obstacle surface types. The application of visual methods for obstacle detection has been the subject of the most research. Many studies were based on binocular vision detection, and distance information could be obtained. Wei assumed that humans were the only potential obstacles in the field. By processing the obtained disparity map by setting thresholds, human-shaped obstacles and their motion states near the agricultural machinery could be detected [6]. Zhang proposed segmenting the obstacle area from the background by analyzing the brightness distribution on the scan line and performing fast stereo feature matching to obtain the spatial information of the human-shaped obstacle [7]. Aiming at visual obstacle detection for a combined harvester, Ding proposed a method that combined monocular color image segmentation and stereovision feature matching to determine the distances between obstacles and the harvester [8]. Yin denoised the depth image obtained by the 3D camera [9]. After noise removal, the pixels were filtered through coordinate conversion, and a height threshold was set. Then, the boundary of the obstacle is extracted using the four-connectivity method, and finally, information on the locations and sizes of obstacles was obtained. However, the method of visual inspection has certain limitations. For example, Wei and Yin's detection objects were only people in the field [6,9]. Zhang and Ding's experiments had strict requirements on the height and color differences between the obstacles and the surrounding environment [7,8]. The accuracy of the visual inspection scheme decreased as the detection distance increased and was greatly affected by ambient light, which cannot meet the needs of agricultural machinery for night operations, and there were certain limitations in the application of unstructured farmland inspection.

Research on the use of lidar for field obstacle detection is also an essential technical route. Brenneke provided a three-dimensional laser obstacle recognition algorithm that divides the three-dimensional point cloud into two categories [10]. The first category points are perpendicular to a line, such as tree trunks, walls, pits, and some artificial road signs; the second category points have no direct contact with the ground, such as tree branches and roofs. The identification of the two types of points and coloring them in the entire image can enable the recognition of obstacles. Jiménez provided an improved obstacle recognition algorithm, which solves the limitation of ordinary methods that only rely on the obstacle distance to recognize obstacles and calculates dynamic variables, such as speed, by extracting the characteristics of obstacles [11]. Asvadi designed a method to detect static and dynamic obstacles in the urban environment using point cloud data obtained using three-dimensional laser scanning and positioning data obtained using the inertial navigation system [12]. Zeng used three-dimensional lidar to collect point cloud data in an apple orchard and used MATLAB to develop an algorithm to segment trellis wires, support poles, and tree trunks in the point cloud images [13]. However, these studies all collected data and then performed offline processing, so real-time inspection and further agricultural machinery control could not be performed. In addition, due to the high price of 3D lidar, it is still difficult to apply it to agricultural machinery.

There is also some research on obstacle detection using two-dimensional lidar, and lidar and image fusion methods. Doerr used two-dimensional lidar to evaluate multiple feature recognition methods (average height, density, connectivity, and discontinuity methods) to identify three foreign objects placed in different environments under four crops [14]. Takahashi designed a LIDAR-based emergency obstacle avoidance module for obstacles on the sidewalk [15]. The module included an obliquely installed 2D lidar and an embedded microcontroller and used an autoregressive model to locate the obstacle's position. Peng designed an obstacle detection algorithm based on a two-dimensional lidar [16]. After denoising the laser point cloud, it was filtered, segmented, and clustered, and finally, the shape and position of the obstacle were output. However, it could only be used for stationary obstacles. Reina detected farmland obstacles and discerned traversable from non-traversable areas using stereovision, LIDAR, radar, and thermography fusion [17]. Kragh combined appearance- and geometry-based detection methods by probabilistically fusing lidar and camera sensing with semantic segmentation using a conditional random field [18]. Moreover, a small robot platform tested the algorithm in an orchard. Xue studied the fusion of 2D lidar data and image data to detect trunks [19]. These studies were based on the distances of obstacles and could not obtain obstacle size information or motion state information, such as speed, collision time, etc., and no two-dimensional lidar was used for multi-obstacle research. The above studies were tests in ideal environments indoors or outdoors, and there were few obstacles detected in actual farmland operation scenarios. In addition, most of these studies tested small robot platforms and have not been applied to real agricultural machinery.

Therefore, the objective of this research was to develop a real-time, low-cost, high-accuracy, pre-collision system as part of the autonomous driving of harvesters. Moreover, the detection algorithm proposed in this paper can detect an obstacle's distance and relative angle using lidar.

2. Materials and Methods

2.1. Equipment

The lidar used in this study was the PACECAT LDS-U50C-S two-dimensional lidar (Jinhua LANHAI photoelectricity Co., Ltd., Jinhua, China). The lidar is shown in Figure 1. This lidar is cost-effective and can be used in bright outdoor environments. The protection level can reach IP65, which is suitable for agricultural applications. The lidar parameters are shown in Table 1.

Figure 1. The lidar used in this research.

Table 1. Main parameters of the lidar.

Parameters	Value
Measuring range (m)	0.1–40
Ranging accuracy (cm)	±3
Field of view (°)	360
Angle resolution (°)	0.09–0.27
Rotating speed (r/min)	300–900

This study's laser scanning speed was 600 r/min, and the corresponding frame rate was ten frames per second. The angular resolution was 0.18 degrees, and each data frame had about 2000 points. The distance and angle data scanned by the lidar were output through the network port.

2.2. Lidar Processing Algorithm

In this study, the algorithm for lidar data processing was divided into three steps: pre-processing, clustering, and obstacle parameter calculation. Preprocessing mainly included three parts: the selection of the region of interest, coordinate transformation, and filtering to facilitate subsequent algorithm calculations. Clustering was mainly to detect and distinguish multiple obstacles. Obstacle parameter calculation was mainly to obtain the number of obstacles, the width of each obstacle, and the relative speed of each obstacle.

2.2.1. Pre-Processing

Pre-processing was carried out to first delineate the region of interest (ROI) according to the angle and distance and then only process the data in the region of interest, which could significantly reduce the processing time of the algorithm. Then, the data were converted from polar coordinates to rectangular coordinates to facilitate subsequent processing. The data were then filtered, mainly to remove noise due to dust or lidar instability.

The raw data output by the lidar were in polar coordinates. The angle was α, and the detected distance was d. The position of the 0-degree angle was directly in front of the lidar. The angle increased clockwise to 360 degrees. During pre-processing, first, the angular and distance ranges of the region of interest were determined according to the installation location and the application scenario. The left boundary angle of the ROI was α_L, and the right boundary angle was α_R. The data in the region of interest were extracted and converted into a rectangular coordinate system. A schematic diagram of the region of interest is shown in Figure 2. The conversion formula is shown in Formula 1.

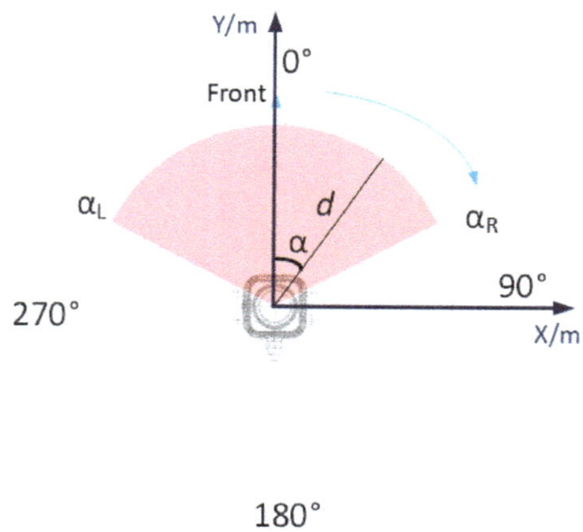

Figure 2. Region of interest and coordinate system.

$$\begin{cases} x = d \sin \alpha \\ y = d \cos \alpha \end{cases} \quad \alpha \in [\alpha_L, \alpha_R] \tag{1}$$

2.2.2. Filtering

Due to the special environment of agricultural machinery operation, some ambient light or dust interfered with the test data during the sensing process. This noise was an impulse noise. It was very similar to the salt and pepper noise in the field of image processing. This noise could be caused by sharp and sudden disturbances in the signal. An effective noise reduction method for this type of noise is a median filter or a morphological filter. To eliminate the influence of these noise points on obstacle detection, we performed median filtering on the collected data.

Median filtering is a nonlinear signal processing method to eliminate outlier noise. For each data point to be processed, the data in the left and right neighborhoods are selected for sorting according to the selected window length, and the value of the middle size is selected as the value of this point after filtering. Compared with the mean filter, the median filter could filter out the salt and pepper noise caused by the dust in the lidar perception. The window size of the median filter selected in this study was 5. The bubble sort method, which is a comparison sort, is named for the way the larger elements "bubble" up to the top of the list [20]. It was used to determine the value of the middle size in the window. An algorithm diagram of median filtering is shown in Figure 3.

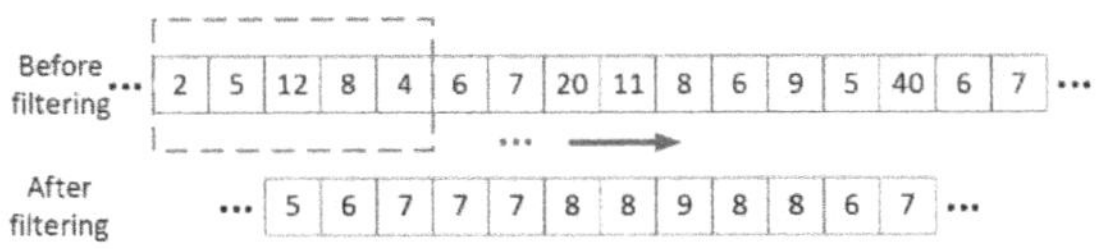

Figure 3. Schematic diagram of the median filter.

2.2.3. DBSCAN Clustering

Obstacles in the field include static obstacles and dynamic obstacles, and multiple obstacles may appear in the area of interest at the same time. In order to cope with such complex perception scenarios, multiple obstacles need to be distinguished to facilitate subsequent algorithms to calculate the size information and motion state information of each obstacle.

This study used a density-based spatial clustering of applications with noise (DBSCAN) method to distinguish multiple obstacles. The algorithm divided the points with sufficient density into clusters and found clusters of arbitrary shapes in the noisy data sequence. When applying this algorithm, we first determined the minimum number of points ($N_{\text{min\_}points}$) of each cluster and the maximum distance threshold (ε) between two adjacent points during clustering according to the sizes of common obstacles in the field and the resolution of the lidar. Each cluster was the largest collection of densely connected points. When clustering, an unclustered point was randomly selected as a seed point for each frame of data, and the distance (d_n) between the seed point and the point to be clustered was calculated.

$$d_n = \sqrt[2]{(x_{seed} - x_{cur})^2 + (y_{seed} - y_{cur})^2} \tag{2}$$

In the formula, x_{seed} and y_{seed} are the abscissa and the ordinate of the seed point, respectively, and x_{cur} and y_{cur} are the abscissa and ordinate of the current point, respectively. ε was the distance threshold of clustering. If $d_n \leq \varepsilon$, the current point was the direct density reachable point of the seed point, and the density reachable point of this point was calculated in turn. The set of these points was a cluster. Each point of the cluster was the center, and the cluster was extended with ε as the radius. If there were other unclustered points within this range, the points were expanded as points within the cluster. They expanded sequentially until the number of points in the cluster no longer expanded. If the number of points in the cluster was greater than $N_{\text{min\_}points}$, the cluster became an obstacle. Subsequently, the same calculation was performed on the other unclustered points in the region of interest until the number of clusters no longer increased. Then, the cluster was divided into multiple obstacles. The number of clusters is considered the number of obstacles. Each obstacle cluster had an ID value. A simple schematic diagram of the clustering process is shown in the figure below. In Figure 4, A is the core point, B and C are the boundary points of the cluster, and N is the outlier point of the cluster.

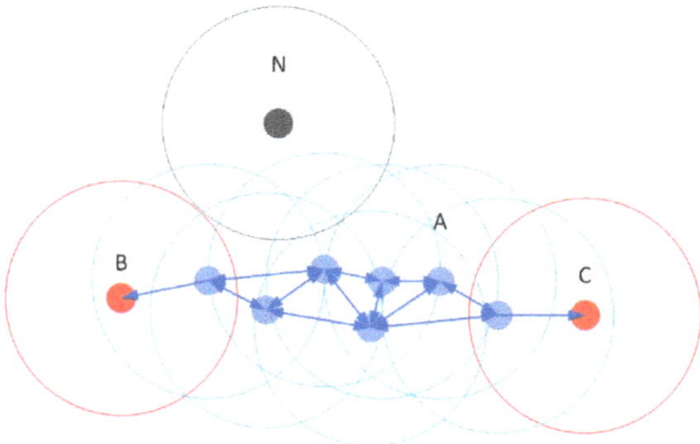

Figure 4. Schematic diagram of the DBSCAN algorithm.

2.2.4. Obstacle Information Calculation

After the above processing, each frame of data was clustered into one or more obstacles, and the number of obstacles was obtained from the number of clusters. Through the analysis of each obstacle cluster, the real-time width information of the obstacle could be obtained, and by analyzing the obstacle clusters between the two frames of data, the motion state information of the obstacles, such as the relative speed and time to collision (TTC), could be obtained.

The maximum and minimum values of the abscissa of each obstacle cluster point were calculated to obtain the horizontal boundary $[x_l, x_r]$ of the obstacle. Then, the real-time width (W) of the obstacle was

$$W = |x_r - x_l| \tag{3}$$

To obtain the motion state information of the obstacle, it was necessary to perform differential processing on the obstacle data in the two frames of data. In order to cope with a scene with multiple obstacles, it was first necessary to perform interframe matching on the obstacles between the two frames of data to ensure that the obstacle in the current frame and the obstacle in the previous frame represented the same obstacle. In this study, two criteria were used for interframe matching; one was the width of the obstacle, and the other was the point of the obstacle. For two obstacles to be the same obstacle in two frames of data, first, the difference between the widths of the two obstacles must be less than the threshold (ω), and second, in the current frame, the abscissa value (x_p) of a certain point in the obstacle cluster is within the range of the abscissa of an obstacle in the previous frame:

$$\left|W_{cur} - W_{pre}\right| \leq \omega \tag{4}$$

$$x_l \leq x_p \leq x_r \tag{5}$$

Each obstacle cluster in the previous frame and the current frame were traversed in turn, and the ID-matching pair of obstacles between the two frames was obtained. After obtaining the obstacle matching between the two frames, the relative speed and collision time of the lidar and the obstacle could be obtained by combining the distance difference of the obstacle between the two frames and the frame rate.

When calculating the distance between an obstacle and the lidar, if the intermediate point of the obstacle was directly calculated, there could have been deviations due to the appearance of individual deviation points in the data. In this study, the 6σ method was used to eliminate the deviation in the data, and the distance between the obstacle and the lidar was obtained using the average value of the qualified points. The 6σ method originated in quality management and is a statistical quality control method. In this study, it was used to deal with the longitudinal distance of lidar scanning obstacles. The ordinates of the cluster points of a certain obstacle in the current frame were $\{y_1, y_2 \cdots y_n\}$. Their average was calculated as $\overline{y}$, and the standard deviation was calculated as σ_y. The ordinates of the cluster points of the obstacle in the previous frame were $\{y'_1, y'_2 \cdots y'_n\}$. Their average value was calculated as $\overline{y}'$, and their standard deviation was calculated as $\sigma_y{}'$. The ordinate of any point of the obstacle in the current frame was y_i, and all ordinate points that satisfied the following formula were found:

$$|y_i - \overline{y}| < 3\sigma_y \tag{6}$$

The average value of all ordinates that met this requirement was calculated as the longitudinal distance (d_{cur}) of the obstacle from the lidar in the current frame. The ordinate of any point of the obstacle in the previous frame was y'_i, and the same method was used to calculate all ordinates in the previous frame that satisfied the following formula:

$$|y'_i - \overline{y}'| < 3\sigma_y{}' \tag{7}$$

The same method was used to calculate the average value of all ordinates that met this requirement as the longitudinal distance (d_{pre}) from the obstacle to the lidar in the previous frame. According to the distance change and the time interval between the two frames, the relative movement speed (v) was obtained.

$$v = \frac{d_{pre} - d_{cur}}{\Delta t} = \frac{d_{pre} - d_{cur}}{1/FPS} \tag{8}$$

In the formula, Δt is the time interval between the two frames of data and FPS is the frame rate of the lidar. According to the current distance between the obstacle and the lidar and the above-mentioned relative movement speed, the remaining time (t) before the obstacle would collide with the lidar could be obtained as:

$$t = \frac{d_{cur}}{v} = \frac{d_{cur}}{(d_{pre} - d_{cur}) * FPS} \tag{9}$$

Using the same method, the time for each remaining obstacle to collide with the lidar was calculated to provide parameter support for the subsequent development of the pre-collision system.

2.3. *Emergency Braking Strategy and Software Development*

2.3.1. Emergency Braking Strategy

After the above processing, parameters such as the number of obstacles, the width of each obstacle, the relative movement speed of the agricultural machine, and the time until collision with the agricultural machine were obtained. When these parameters were applied to the agricultural machinery pre-collision system, two main judgment strategies were used to decide whether to make an emergency stop: 1. A TTC safety threshold (T_ε) was set according to the speed of the vehicle and the movement speed of common obstacles in the work area. If there was any obstacle whose TTC is less than T_ε, the vehicle was stopped. 2. A dangerous area around the vehicle was set according to the operation type and the application scenario of the agricultural machinery. In this research, the area was symmetrical about the Y axis directly in front of the lidar, and the area could be set by the X and Y values. For each frame of obstacle data processed, it was judged whether the point closest to the lidar was in the dangerous area by setting the X and Y values. If there was any obstacle in the area, the vehicle stopped. The division of areas is shown in Figure 5.

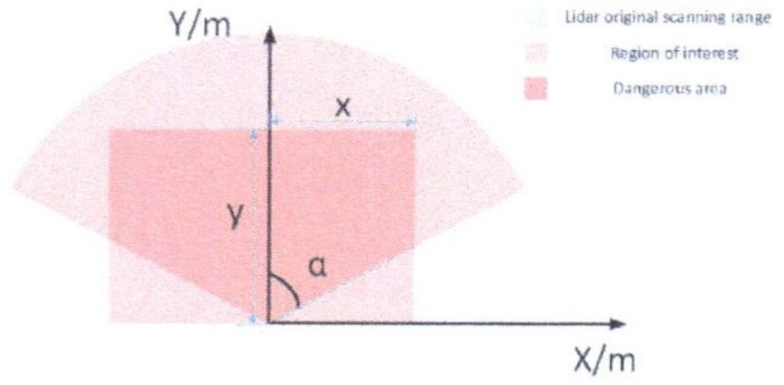

Figure 5. Data processing area comparison.

2.3.2. Software Development

To apply the algorithm to the actual agricultural machinery operation, we developed the program of the algorithm. The real-time pre-collision system used C++ programming and combines the MFC library [21] to compile the lidar communication and real-time detection display interface. The software could fill in the IP of the lidar so that the processor's network and the lidar were in the same network segment to establish communication. When the system was running, the black display area of the display interface would highlight the point of the obstacle. Moreover, if an obstacle was detected through algorithm processing that threatened driving safety and required a stop, the system would issue a stop command, and at the same time a prompt sentence would be output in the text box on the right side of the interface to allow the debugger or user to observe and confirm. A data processing flow diagram of the software is shown in Figure 6, and the software interface is shown in Figure 7.

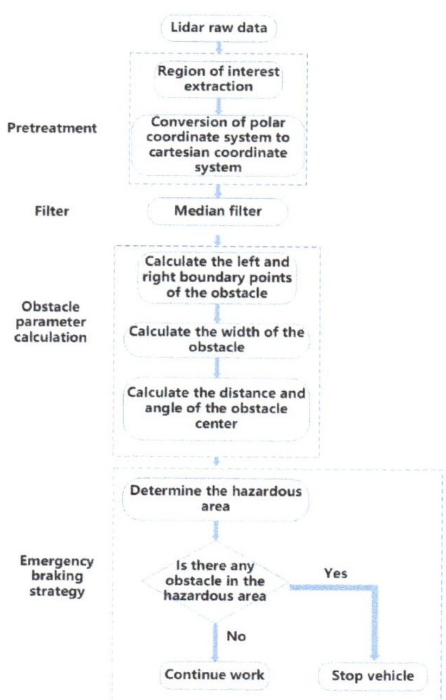

Figure 6. Schematic diagram of the data processing flow.

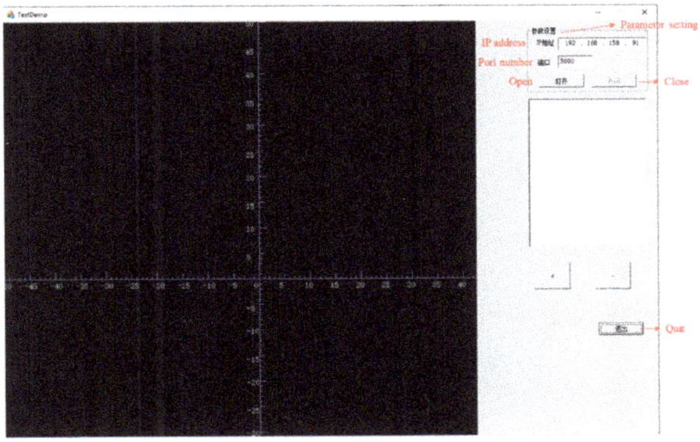

Figure 7. The software interface.

2.4. Design of a Harvester Pre-Collision System

To verify the effectiveness of the algorithm and apply it to actual agricultural machinery operations, we designed a harvester pre-collision system. The system directed the harvester to stop when encountering dangerous obstacles based on the obstacle information detected by the lidar. The system included a lidar, a display terminal, an automatic navigation controller, and actuators. A schematic diagram of the hardware design of the system is shown in Figure 8.

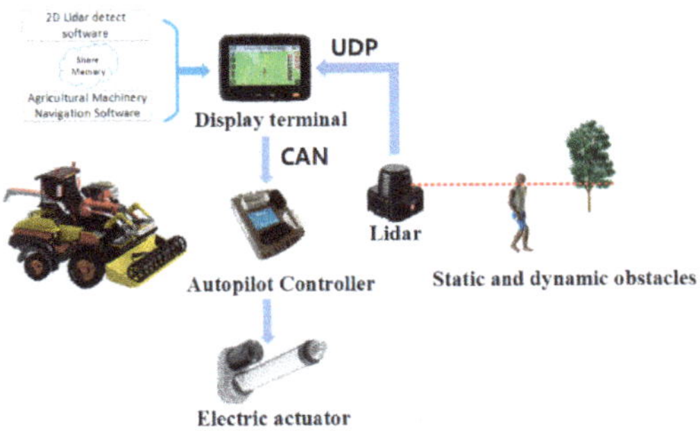

Figure 8. Design schematic diagram of the harvester pre-collision system.

The lidar sent the data to the display terminal through the network interface using the UDP protocol. The terminal had 2 G of memory, a 32 G solid-state hard drive, and a Windows 10 operating system. The agricultural machinery pre-collision system software developed in C++ ran on the terminal, and the software could use the algorithm mentioned above to process the lidar data. At the same time, to achieve an unmanned effect during the test, the system was integrated with the agricultural machinery automatic driving system, and the automatic driving software was also running on the terminal. The result, processed by the pre-collision system, was sent to the autonomous driving program on the terminal through shared memory. The terminal sent the instruction of whether to stop to the automatic driving controller through the CAN bus. Through the modification of the harvester, an electric push rod was used as an actuator to replace the gear in the cab to control the forward motion, stopping, and reversing of the vehicle body. The controller controlled the extension and contraction of the electric actuator through the IO port to control the driving and parking of the vehicle body.

The harvester used in this study was a Lovol Gushen GM-100 (Weifang City, Shandong Province, China)wheat combine harvester. Due to the particularity of the harvester, when selecting the installation position of the lidar, to avoid the false recognition of obstacles caused by lifting the header and reel, the lidar was installed above the cab in this study. Moreover, the height of the harvester was 3.2 m. To detect low obstacles while ensuring the detection range of the lidar and reducing blind spots, the scanning surface of the lidar was tilted down by 20°. During installation, the 0° direction of the lidar scan was the same as the front of the harvester, which was the positive direction of the Y axis, and the 90° direction was the positive direction of the X axis of the coordinate system. Lidar installation pictures are shown in Figure 9. To facilitate subsequent tests, the pre-collision system was integrated with the agricultural unmanned driving system. During the test, the display terminal and controller were installed in the cab to facilitate debugging and the viewing of obstacles. An installation picture is shown in Figure 10.

Figure 9. Lidar installation.

(**a**) (**b**)

Figure 10. Display terminal and controller pictures: (**a**) display terminal; (**b**) controller.

In order to allow the system to automatically stop when it detected an emergency, we carried out an electronic control modification to the harvester. An electric actuator replaced the function of shifting in the cab to realize a program-controlled stop. The electric actuator was connected with the hydraulic continuously variable transmission on the vehicle body through a linkage mechanism. An installation picture is shown in Figure 11. The body stopped when the electric push rod was in the neutral position. The contraction and extension of the electric push rod, respectively, controlled the forward and backward movement of the body. The position of the push rod was fed back to the controller through an angle sensor. The controller controlled the expansion and contraction of the push rod through the IO port to control the stopping and movement of the body.

Figure 11. Electric linear actuator installation.

2.5. Experiment

In order to test the detection error of the algorithm and evaluate the effectiveness of the system, we applied a system to the harvester and carried out field trials at the Beijing Xiaotangshan National Experiment Station for Precision Agriculture in June 2021. During the test, two main experiments were carried out; one was the multi-obstacle detection test, and the other was the pre-collision stopping vehicle test. The lidar was installed at the transverse center in front of the top of the harvester cab, with a downward tilt of about 17 degrees and an installation height of about 3.2 m. The lidar was fixed on the harvester with a bracket to ensure its relative position remained unchanged to avoid a change in position affecting the recognition results. Assuming that the height of wheat is 1 m, in order to avoid detecting wheat by mistake when detecting obstacles, the wheat part was not included when setting the region of interest. A schematic diagram is shown in Figure 12. The multi-obstacle detection test was carried out under an actual harvest scenario. During the test, several people were arranged to walk irregularly in front of the harvester to verify its detection effect. The pre-collision stopping vehicle test was a dynamic test. During the test, the harvester was driven into the wheat field for automatic driving operation, and the normal driving speed was 5 km/h. At the same time, people walked in front of the harvester as an obstacle to test whether the system detected and stopped effectively.

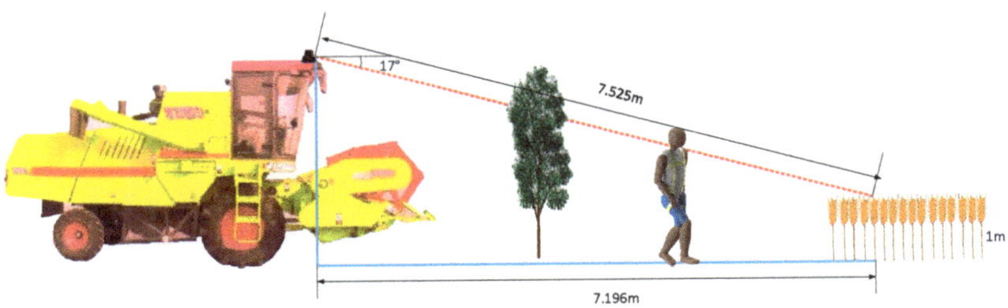

Figure 12. The installation position and obstacle detection of the lidar on the harvester.

In the experiment, the angular boundary range of the region of interest was set to [−70°, 70°]. Moreover, because of the harvester's particularity, the lidar's installation had an inclination angle. In order to avoid the false detection of wheat, according to the inclination angle and the height of the wheat (the average height in this experiment was 70 cm), the distance range was set to 7.3 m. The sliding window size of the median filter was 5. The minimum number of clustering points was 4, and the clustering distance threshold was 0.75 m. The TTC safety threshold was 30 s. According to the width of the harvester, the extended length of the header, and the speed of the vehicle, the x range of the dangerous area was set to [−3, 3], and the range of y was [0, 4]. A test picture is shown in Figure 13.

Figure 13. The experiment.

3. Results and Discussion

3.1. Multi-Obstacle Detection Results

In order to test the detection effect of the algorithm on multiple obstacles, during the test multiple people in the wheat field were tested as obstacles. The heights of the people in the experiment were 1.75–1.82 m. A test picture is shown in Figure 14. The lidar data were processed according to the above algorithm steps, and the data of each step were recorded for analysis. Each processing step is shown in Figure 15.

Figure 14. Multi-obstacle detection.

Figure 15 shows the processing effect of the algorithm at each step. Figure 15a is the original lidar data in the polar coordinate system. It can be seen in the figure that due to the installation position and angle of the lidar and because its left and right sides were blocked, most of the returned lidar data were in the front. The data from the region of interest that were extracted and converted to the rectangular coordinate system are shown in Figure 15b. It can be seen in the figure that there was some reduction in the data after the region of interest was extracted, and there were some outliers on the right side of the figure. After filtering, the partial separation group points disappeared, as shown in Figure 15c. The result of clustering the processed data is shown in Figure 15d. There are six clusters in the figure. Compared with the previous figure, two points are not shown in the figure because they did not meet the clustering conditions. Each different cluster is represented by a different color.

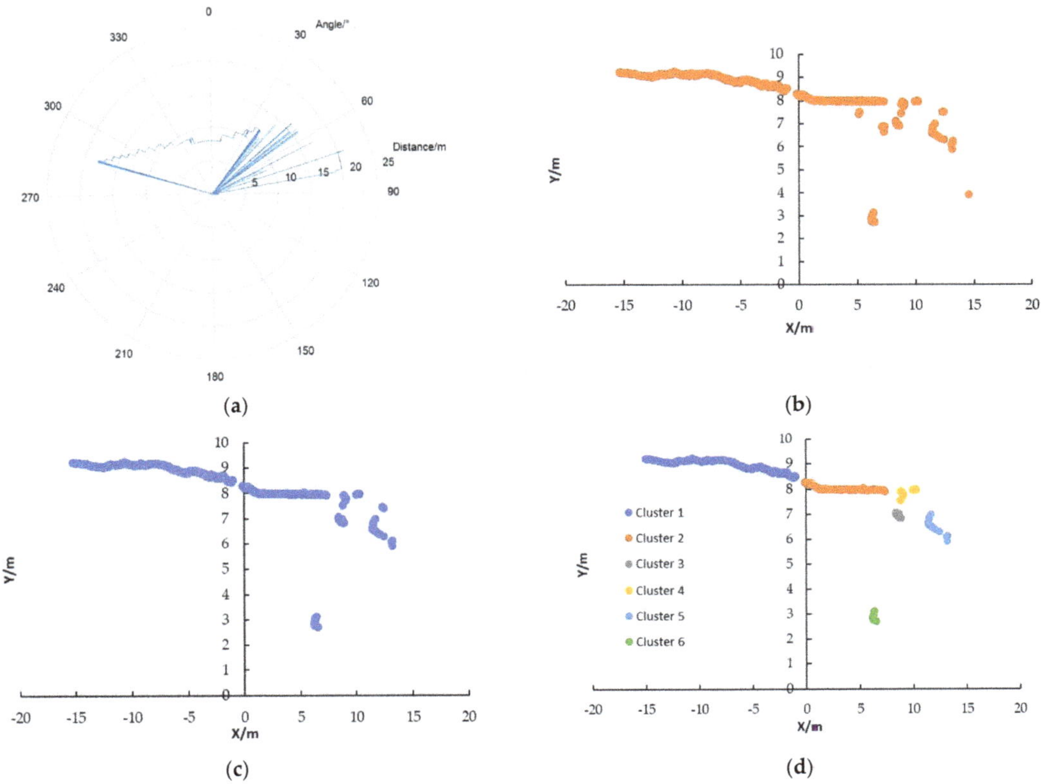

Figure 15. Data processing diagram of each step: (**a**) raw data in a polar coordinate system; (**b**) data in the ROI in a Cartesian coordinate system; (**c**) filtered data; (**d**) clustered data.

Figure 15 was compared with Figure 14 to analyze the obstacle information. Cluster 1 in blue was the harvested ground, with few ups and downs. Cluster 2 in orange was an unharvested wheat field with a relatively stable height. The gap between the two clusters was the harvest boundary. Since the forward center of the installed lidar was close to the left side of the vehicle body and the harvester was closer to the harvesting boundary on the left side, the lateral center of the collected lidar data was near the harvesting boundary. Because the lidar had an installation inclination angle, combined with the wheat's height and the stubble's height after harvest, the longitudinal distance of the detected wheat and the ground was also consistent with the actual situation. Cluster 3 in gray, cluster 4 in yellow, and cluster 5 in light blue in the upper right of the graph correspond to the four people in the front right and far away from the harvester in the experimental graph. Since the lidar installation had a certain scanning angle and these people were relatively close, two people were clustered as one obstacle during the clustering. However, it did not affect the overall pre-collision effect. Cluster 6 in green at the bottom right of the graph corresponds to the white-clothed person near the harvester in the experimental picture. It can be seen from the experiment that the application of this algorithm can realize low-cost two-dimensional lidar harvester multi-obstacle detection.

A total of 223 frames of data were collected and clustered during the multi-obstacle detection, of which 212 frames were correctly clustered. The accuracy rate of the multi-obstacle detection was 95.06%.

3.2. Pre-Collision Results

In order to test whether the pre-collision system could stop autonomously after detecting dangerous obstacles, we conducted a test. In order to test the effect of the pre-crash system, experiments were carried out in this research. During the test, humans acted as obstacles within the detection range of the lidar to test whether the system could stop autonomously after detecting dangerous obstacles. During the test, the number of obstacles that occurred in front of the harvester and the number of vehicle stops were recorded. The test results' precision rate and recall rate were calculated to analyze the system's abnormal situation.

$$P = \frac{T_P}{T_P + F_P} \tag{10}$$

$$R = \frac{T_P}{T_P + F_N} \tag{11}$$

where P is the precision rate and R is the recall rate; T_P is the number of times an obstacles occurred and the vehicle stopped correctly; F_P is the number of times an obstacles occurred and the vehicle did not stop; and F_N is the number of times the vehicle stopped without obstacles. The harmonic mean value (F) of the precision rate and recall rate is

$$F = \frac{\left(\alpha^2 + 1\right) \times P \times R}{\alpha^2 (P + R)} \tag{12}$$

In the formula, α is the value of the harmonic parameter. Under normal circumstances, the value of α is 1. In this case, F is

$$F_1 = \frac{2 \times P \times R}{P + R} \tag{13}$$

Five groups of experiments were conducted in this study. Each group included 4–6 anti-collision tests. The number of trials and the number of successful trials in each group were recorded. The recorded data analysis is shown in Table 2.

Table 2. Analysis of pre-collision system test.

Test Group	Number of Frames	The Number of Obstacles That Occurred	The Number of Vehicle Stops	Precision Rate	Recall Rate	Harmonic Mean Value
1	791	5	5	100%	100%	100%
2	849	4	4	100%	100%	100%
3	610	5	4	80%	100%	88.89%
4	1274	6	5	83.33%	100%	90.91%
5	341	5	5	100%	100%	100%

It can be seen in Table 2 that in the five groups of experiments that were carried out, tests 1, 2, and 5 could all stop correctly. However, in test groups 3 and 4, the number of vehicle stops was less than the number of obstacles that occurred; that is, there was a situation where an obstacle appeared, but the vehicle did not stop. This led to a drop in the recall rate. The reason may be that the obstacle was not within the effective detection range of the lidar, causing the failure to stop the vehicle. Alternatively, the obstacle was not in the danger zone, and the time to collision had not reached the set threshold, so the vehicle did not stop. In this situation, the harmonic mean values of the third and fourth groups were 88.89% and 90.91%, respectively. The average precision rate was 92.67%. The average harmonic mean value was 95.96%.

4. Conclusions

This research used low-cost two-dimensional lidar to detect obstacles in farmland. An algorithm for identifying multiple obstacles and calculating each obstacle's size and movement status was proposed. The algorithm performed pre-processing to delineate the region of interest, perform coordinate conversion, and filter the data. Then, it used the density-based clustering method to obtain the number and width of each obstacle. Finally, the relative motion speed and the time to the collision for each obstacle and the vehicle body were obtained based on the interframe difference algorithm. This algorithm was used to realize the software development of agricultural machinery pre-collision systems, using Windows as the system platform and C++ as the programming language. Multi-obstacle tests and pre-collision tests were carried out on the system. The test results showed that the system could effectively detect multiple obstacles. The accuracy rate of multi-obstacle detection was 95.06%. The accuracy rate of stopping when dangerous obstacles were detected was 92.67%. The harmonized average of the accuracy and recall rates was 95.96%. This research provides a foundation for the safe autonomous driving of agricultural machinery. In the future, more complex and efficient obstacle circumvention strategies can be studied based on the number, width, speed, and other parameters of the detected obstacles.

Author Contributions: Conceptualization, Y.S., Z.S. and Z.M.; methodology, Y.S., Z.S. and Z.M.; software, H.W.; validation, W.Q.; formal analysis, Q.W.; investigation, H.L.; resources, Z.S. and Z.M.; data curation, Y.Y.; writing—original draft preparation, Y.S.; writing—review and editing, Y.S.; visualization, H.W.; supervision, Z.S.; project administration, Z.M.; funding acquisition, Z.S. and Z.M. All authors have read and agreed to the published version of the manuscript.

Funding: This research was funded by the National Key Research and Development Program of China (grant number: 2022YFD2001603) and the Jiangsu Provincial Agricultural Science and Technology Independent Innovation Fund Project (grant number: CX (20)1007).

Data Availability Statement: All data are presented in this article in the form of figures or tables.

Acknowledgments: We would like to thank the "College of Engineering, China Agricultural University", "Research Center of Intelligent Equipment, Beijing Academy of Agriculture and Forestry Sciences", and "AgChip Science and Technology (Beijing) Co., Ltd.".

Conflicts of Interest: The authors declare no conflict of interest.

References

1. Zhang, S.; Tian, X.; Wang, S. Impact of aging agricultural labor on mechanization and technical efficiency in wheat production: A perspective analysis based on landforms. *J. China Agric. Univ.* **2018**, *23*, 174–182. [CrossRef]
2. Zhang, M.; Ji, Y.; Li, S.; Cao, R.; Xu, H.; Zhang, Z. Research progress of agricultural machinery navigation technology. *Trans. Chin. Soc. Agric. Mach.* **2020**, *51*, 1–18. [CrossRef]
3. Rahman, M.M.; Ishii, K.; Noguchi, N. Optimum harvesting area of convex and concave polygon field for path planning of robot combine harvester. *Intel. Serv. Robot.* **2019**, *12*, 167–179. [CrossRef]
4. He, Y.; Jiang, H.; Fang, H.; Wang, Y.; Liu, Y. Research progress of intelligent obstacle detection methods of vehicles and their application on agriculture. *Trans. Chin. Soc. Agric. Eng.* **2018**, *34*, 21–32. [CrossRef]
5. Dvorak, J.S.; Stone, M.L.; Self, K.P. Object detection for agricultural and construction environments using an ultrasonic sensor. *J. Agric. Saf. Health* **2016**, *22*, 107–119. [CrossRef] [PubMed]
6. Wei, J.; Han, S.; Reid, J.F.; Han, S. Obstacle detection using stereo vision to enhance safety of autonomous machines. *Trans. ASABE* **2005**, *48*, 2389–2397. [CrossRef]
7. Zhang, L.; Wang, S.; Chen, B.; Liu, Z. Detection of obstacles in farmland based on binocular vision. *J. China Agric. Univ.* **2007**, *12*, 70–74.
8. Ding, Y.; Wang, S.; Chen, H. Obstacle detection in the working area of agricultural vehicle based on machine vision. *Trans. Chin. Soc. Agric. Mach.* **2009**, *40* (Suppl. 1), 23–27.
9. Yin, X.; Noguchi, N.; Ishi, K. Development of an obstacle avoidance system for a field robot using a 3D camera. *Eng. Agric. Environ. Food* **2013**, *6*, 41–47. [CrossRef]

10. Brenneke, C.; Wagner, B. A Scan Based Navigation System for Autonomous Operation of Mobile Robots in Man-Made Environments. In Proceedings of the International Conference of Systems Engineering, Institute for System Engineering, Hannover, Germany, 1 January 2003.
11. Jiménez, F.; Naranjo, J.E. Improving the obstacle detection and identification algorithms of a laser scanner-based collision avoidance system. *Transp. Res. Part C Emerg. Technol.* **2011**, *19*, 658–672. [CrossRef]
12. Asvadi, A.; Premebida, C.; Peixoto, P.; Nunes, U. 3D Lidar-based static and moving obstacle detection in driving environments. *Robot. Auton. Syst.* **2016**, *83*, 299–311. [CrossRef]
13. Zeng, L.; Feng, J.; He, L. Semantic segmentation of sparse 3D point cloud based on geometrical features for trellis-structured apple orchard. *Biosyst. Eng.* **2020**, *196*, 46–55. [CrossRef]
14. Doerr, Z.; Mohsenimanesh, A.; Lagu, C.; Mclaughlin, N. Application of the LIDAR technology for obstacle detection during the operation of agricultural vehicles. *Can. Biosyst. Eng. Le Genie Des Biosyst. Au Can.* **2013**, *55*, 2.9–2.16. [CrossRef]
15. Takahashi, M.; Kobayashi, K.; Watanabe, K.; Kinoshita, T. Development of prediction based emergency obstacle avoidance module by using LIDAR for mobile robot. *IEEE Int. Symp. Soft Comput. Intell. Syst.* **2015**, *2014*, 561–564. [CrossRef]
16. Peng, Y.; Qu, D.; Zhong, Y.; Xie, S.; Luo, J.; Gu, J. The Obstacle Detection and Obstacle Avoidance Algorithm Based on 2-D Lidar. In Proceedings of the IEEE International Conference on Information and Automation, Lijiang, China, 10 August 2015. [CrossRef]
17. Reina, G.; Milella, A.; Rouveure, R.; Nielsen, M.; Worst, R.; Blas, M.R. Ambient awareness for agricultural robotic vehicles. *Biosyst. Eng.* **2016**, *146*, 114–132. [CrossRef]
18. Kragh, M.; Underwood, J. Multi-Modal Obstacle Detection in Unstructured Environments with Conditional Random Fields. *J. Field Robot.* **2017**, *37*, 53–72. [CrossRef]
19. Xue, J.; Fan, B.; Yan, J.; Dong, S. Trunk detection based on laser radar and vision data fusion. *Int. J. Agric. Biol. Eng.* **2018**, *11*, 20–26. [CrossRef]
20. Cormen, T.H.; Leiserson, C.E.; Rivest, R.L.; Stein, C. *Introduction to Algorithms*, 2nd ed.; MIT Press: Cambridge, MA, USA; McGraw-Hill: New York, NY, USA, 2001; Problem 2–2; p. 40, ISBN 0-262-03293-7.
21. Visual Studio 2019 Release Notes. Available online: Msdn.microsoft.com (accessed on 24 July 2019).

agronomy

MDPI

Article

Factors Affecting Droplet Loss behind Canopies with Air-Assisted Sprayers Used for Fruit Trees

Shijie Jiang, Wenwei Li, Shenghui Yang, Yongjun Zheng *, Yu Tan and Jiawei Xu

College of Engineering, China Agricultural University, Beijing 100083, China
* Correspondence: zyj@cau.edu.cn

Abstract: Air-assisted sprayers are widely employed in orchards, but inappropriate spray parameters can lead to large droplet losses, pesticide waste, and environmental pollution. To investigate the factors affecting the droplet loss of an air-assisted sprayer behind canopies, a two-factor, five-level full experiment was conducted in an actual orchard, where the two factors were the power gradient and foliage area volume density (FAVD). In addition, the location of the sampling point was also considered in the data analysis, including horizontal distance, forward distance, and height. The results show that all factors significantly affected droplet coverage (p-value < 0.01). The droplet coverage showed an increase and then a decrease with an increasing power gradient, and the maximum coverage was measured at power gradient P3 (forward speed: 0.49 m/s, spray pressure: 0.30 MPa, and spray flow rate: 7.13 L/min) or P4 (forward speed: 0.58 m/s, spray pressure: 0.35 MPa, and spray flow rate: 8.44 L/min). The effect of FAVD on droplet coverage had obvious regularity, and this regularity did not change with the power gradient. At different positions behind canopies, the droplet coverage had great differences. The droplet coverage gradually decreases with increasing horizontal distance and height, while increasing with forward distance. This study provides a reference for the air-assisted sprayers to reduce droplet loss, and data support for subsequent research on precision spraying based on FAVD.

Keywords: air-assisted sprayer; behind canopies; droplet loss; foliage area volume density (FAVD); environmental pollution; orchard

Citation: Jiang, S.; Li, W.; Yang, S.; Zheng, Y.; Tan, Y.; Xu, J. Factors Affecting Droplet Loss behind Canopies with Air-Assisted Sprayers Used for Fruit Trees. *Agronomy* **2023**, *13*, 375. https://doi.org/10.3390/agronomy13020375

Academic Editor: Yanbo Huang

Received: 25 December 2022
Revised: 25 January 2023
Accepted: 25 January 2023
Published: 27 January 2023

1. Introduction

Air-assisted sprayers are widely employed in orchards to improve spray efficiency [1,2]. Droplets rely on the air-assisted sprayer's fan to obtain a greater initial velocity, while the airflow generated by the fan enables leaves to turn over. The flow enhances the droplet deposition within canopies. However, if the operating parameters are not properly set, droplets will be carried away by the airflow, resulting in droplet loss, which mainly contains the deposition loss to the ground and the diffusion loss to the air [3–5]. Off-target losses lead to the pollution of the orchard environment. Especially, the loss behind canopies facilitates pesticide residues in neighboring crops and soils [6–8], which is the main form of pesticide pollution to the environment [9,10].

Current research on anti-drift air-assisted sprayers focuses on developing anti-drift technologies to enhance droplet deposition in canopies to reduce drift [11–13]. Computational Fluid Dynamics (CFD) simulation is a common method to study droplet loss and drift. Pascuzzi et al. performed a numerical simulation to analyze the drift patterns of the droplets with different diameters influenced by airflow directions [14]. Hong et al. evaluated the droplet deposition and drift loss from an air-assisted sprayer in the form of CFD simulation [15]. Duga et al. developed a CFD model for drift from an air-assisted sprayer, and analyzed the effect of nozzle position, number of nozzles, and fan speed on droplet drift based on this model [16]. Although CFD simulations predict the deposition and drift of droplets, there are differences between the simulated and actual environments.

The simulation results can only serve as a reference and still need to be verified by actual operation. In terms of anti-drift technology and mechanism improvement, conical airflow anti-drift devices [17,18], anti-drift nozzles [19,20], and other mechanisms were designed [21]. Rathnayake et al. evaluated the downwind drift characteristics of air-assisted sprayers in apple orchards and found that the drift distance of droplets in the downwind direction would exceed 183 m [22]. In addition, applicable methods for the evaluation of pesticide droplet drift or loss were developed [23,24]. These studies provided useful implications for the improvement of air-assisted spraying technology and the selection of application parameters. The canopy characteristics had a great influence on droplet drift in actual operations [25–27]. However, these characteristics were not considered in the studies. Therefore, the factors affecting droplet drift or loss in practice are not well-understood, and further research is needed.

In this paper, an experiment was conducted in an orchard to investigate the factors influencing the droplet loss of air-assisted sprayers behind canopies. The purpose is to clarify the effects of different factors on droplet loss behind canopies and quantify each factor's impact. This study provides a reference for air-assisted sprayers to reduce droplet loss in orchard spraying.

2. Materials and Methods

2.1. Air-Assisted Sprayer

A tower-type air-assisted sprayer, G6S, was selected (Figure 1A), which was produced by Shandong Guohaha Agricultural Machinery Co., Ltd. (Linyi, China). The main technical parameters are: the overall dimensions of the G6S air-assisted sprayer are 2500 × 1050 × 1500 mm, the maximum capacity of the tank is 350 L, the maximum spray pressure is 2.0 MPa, the pump is a three-cylinder plunger pump, and the maximum flow rate of the pump is 48 L/min. The nozzle is a small bee nozzle (Shandong Guohaha Agricultural Machinery Co., Ltd., Linyi, China), which is a two-way nozzle that combines a fan-shaped and a conical nozzle (Figure 1B). Spraying was performed with the fan-shaped nozzle during the experiment, and the spray angle of the fan-shaped nozzle is 65°. All 18 nozzles are symmetrically distributed on both sides of the tower. The forward speed, fan speed, and pressure of the sprayer are directly controlled by the throttle, which is commonly used for air-assisted sprayers in orchards in China at present.

Figure 1. G6S tower air-assisted sprayer.

2.2. Experimental Scheme

The experiment was conducted from 8 to 25 April 2020 in Tianyang District, Baise City, Guangxi Zhuang Autonomous Region, China. The fruit tree is mango with large canopies. Planting parameters such as row spacing, plant spacing, and plant height were determined in the experimental area by randomly measuring 30 times with a tape measure. Row spacing, plant spacing, and plant height were on average 4.5, 3.5, and 4.5 m, respectively.

A two-factor, five-level full experiment was conducted. The two factors were the power gradient and the foliage area volume density (FAVD) of the fruit tree canopy.

FAVD is the sum of the leaf area per unit volume at a certain height (m^2/m^3). The FAVD at different positions of the canopy is different. Therefore, the average FAVD at different positions of fruit tree canopies was used to represent the FAVD of the whole tree. Five fruit trees were randomly selected as the five levels of the FAVD, labeled Tree1, Tree2, Tree3, Tree4, and Tree5. The measurement method is described as follows:

(a) A 50 × 50 × 50 cm cubical frame was selected to randomly frame the canopy 5 times.
(b) The number of leaves in the frame was counted, and the average value was calculated.
(c) The average area of individual leaves of fruit trees was calculated using the following image processing method [28]:
   - Pre-processing the acquired image.
   - Segmenting the labeled images to obtain target images.
   - Calculating the actual area of the leaf by the reference method.

(d) The FAVD of each fruit tree was calculated based on Equation (1):

$$FAVD = \frac{N \times S_{leaf}}{V} \quad (1)$$

where S_{leaf} is the area of the leaf, N is the average number of leaves in the cubical frame, and V is the volume of the cubical frame.

Spray experiments were conducted at five power gradient levels. These levels were the five positions of the air-assisted sprayer throttle, marked P1, P2, P3, P4, and P5. The five power gradients were calibrated before the start of the experiment. Per power gradient, the average forward speed was determined by recording the time to cross a distance of 30 m, 5 times. The time of each case was recorded, and then the ratio of distance to time was calculated to obtain the average speed. The fan speed was measured five times by a UNI-T UT370 photoelectric tachometer (UNI-Trend Technology (China) Co., Ltd., Dongguan, China) for each power gradient and then averaged. The airflow rate at each nozzle outlet was measured by a Testo 405i thermal wireless anemometer (Testo AG, Titisee-Neustadt, Germany), so there were 18 anemometers in total. Per power gradient, data were continuously collected for 30 s (about 15 data points). The average of these 15 data points was selected as the outlet airflow rate of the corresponding nozzle. The Testo 405i thermal wireless anemometer was connected to an android phone via Bluetooth to save the data in real time.

Spray pressure and spray flow rate can be calculated according to Equations (2) and (3):

$$P_{pressure} = \frac{V_{test}}{V_{max}} \times P_{max} \quad (2)$$

$$F_{flow} = \frac{V_{test}}{V_{max}} \times F_{max} \quad (3)$$

where $P_{pressure}$ is the spray pressure, V_{test} is the forward speed calibration value, V_{max} is the maximum forward speed, P_{max} is the maximum spray pressure, F_{flow} is the spray flow rate, and F_{max} is the maximum spray flow rate.

The experimental scheme is shown in Figure 2. First, 3 × 3 × 3 sampling points (27 sampling points) were arranged behind the canopies of 5 fruit trees and divided into 3 layers: top, middle, and bottom. Second, the sampling points of each layer were labeled in the order of 1–9. To express each location more clearly, measurement locations were defined. For example, the location of sampling point 1 at the top layer was described as '1 top', and that at the bottom layer was defined as '1 bottom'. The horizontal distance from the central line of the sprayer to the tree center was 1.3 m. The horizontal distance from the center of the tree to the first sampling point was 1.8 m. In each layer, the distance between sampling points was 1.2 m, the distance between each layer was 0.8 m, and the distance from the bottom layer to the ground was 0.8 m. Therefore, the sampling distances from the center of the tree were 1.8, 3.0, and 4.2 m, respectively. The heights were 0.8, 1.6, and 2.4 m, respectively. The distances along the trees were 0, 1.2, and 2.4 m, respectively. Water-sensitive paper (WSP) (76 × 26 mm) was used to collect droplets, which was fixed at each sampling point with paper clips. For safety reasons and the impact on fruit trees, water was chosen as the solution during the experiment.

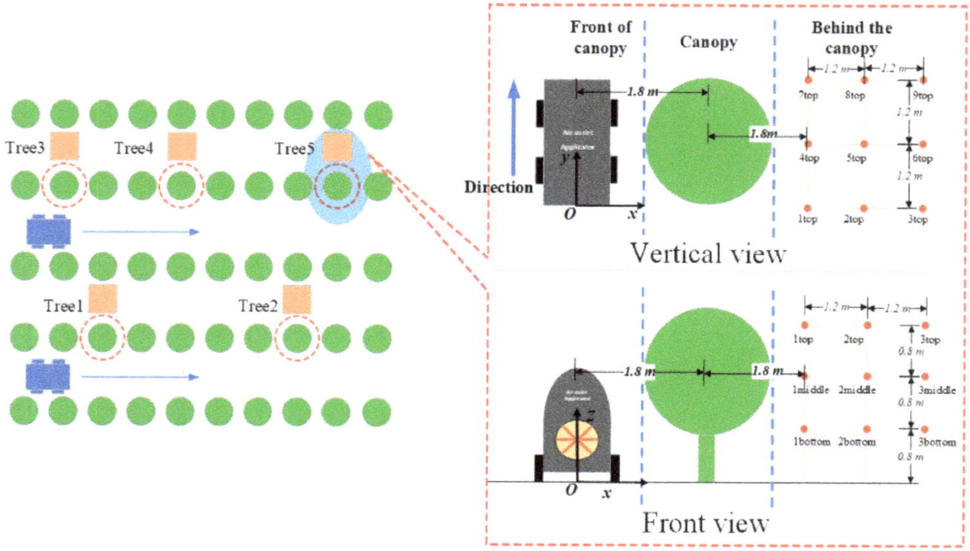

Figure 2. Schematic of the experimental scheme.

Before each experiment, the WSPs were arranged on the sampling points. Then, the air-assisted sprayer was started, the power gradient was varied, the nozzle was turned on, and the sprayer was steadily run to pass by the test fruit trees at a uniform speed along the centerline between the rows. After traveling to a distance of 20 m from the test fruit trees, the sprayer was turned off. After drying, the WSPs were collected and scanned. The above experimental process was repeated in the order of Tree1 to Tree5 and power gradient P1 to P5, with a minimum interval of 30 min between each experimental group. In total, 675 samples were thus collected.

To facilitate the analysis of each sampling point's location, a three-dimensional coordinate system (O-XYZ) was constructed. The intersection of the plane containing the sampling points 1, 2, and 3 was taken as well as the sprayer central axis and the ground as the coordinate origin O. The direction vertical to the forward orientation was the X-positive direction, the direction of forward orientation was the Y-positive direction, and the vertical direction upward from the ground was the Z-positive direction.

X represents the horizontal diffusion direction of droplets, which was used for the subsequent analysis of the correlation between droplet coverage and horizontal distance. Y represents the forward diffusion direction of droplets with the sprayer, which was used for the analysis of the correlation between droplet coverage and forward distance. Z represents the vertical diffusion direction of droplets, which was used for the analysis of the correlation between droplet coverage and height.

2.3. Data Analysis Methods

2.3.1. Selection of Dependent Variables

The WSPs collected by the experiment were processed using DepositScan™ software (version 1.2). The parameters used to represent the spray effect commonly include droplet coverage and droplet quantities. The data of the two parameters were read and saved in an Excel file. In this study, the F-test was used to obtain the consistency between the two parameters, and the appropriate parameter was selected as the dependent variable for the subsequent analysis.

As the units of the two parameters were not the same, the values were first normalized separately according to Equation (4) before the analysis:

$$x^* = \frac{x_i - x_{min}}{x_{max} - x_{min}} \quad (4)$$

where x^* is the normalized result, x_i is the original data, x_{min} is the minimum value of the original data, x_{max} is the maximum value of the original data, and i is the ordinal number.

2.3.2. Variables' Significance Based on Analysis of Variance (ANOVA)

Five variables (FAVD, power gradient, horizontal distance (X), forward distance (Y), and height (Z)) were selected as factors during the experiment. The droplet coverage at the sampling site locations was selected as the dependent variable in a multi-way ANOVA, with the magnitude of the p-value indicating the significant factors.

Based on the ANOVA, the effect of these five variables on droplet coverage was separately analyzed and drawn by Origin 2019 for visualization.

For analyzing the effect of the power gradient, the power gradient was utilized as the horizontal axis, and droplet coverage was used as the vertical axis.

For analyzing the effect of FAVD, FAVD was utilized as the horizontal axis, and droplet coverage was used as the vertical axis. The droplet coverage was averaged for all sampling point locations behind canopies. The calculation method is shown in Equation (5):

$$\overline{C} = \frac{\sum_{j=1}^{n} C_j}{n} \quad (5)$$

where $\overline{C}$ is the average droplet coverage, C_j is the droplet coverage of the j-th sampling point, n is the number of sampling points, $n = 27$, and j is the ordinal number.

For analyzing the droplet coverage difference of the position behind canopies, the three sections in the Y direction (Y = 0, Y = 1.2 m, and Y = 2.4 m) were selected to examine the variation of droplet coverage with horizontal distance and forward distance.

3. Results

3.1. Results of FAVD Calculation and Power Gradient Calibration

FAVD calculation results are shown in Table 1. The power gradient calibration results are shown in Table 2 and Figure 3.

Table 1. FAVD calculation results.

Tree Number	FAVD/($m^2 \cdot m^{-3}$)
Tree1	2.72
Tree2	2.32
Tree3	1.50
Tree4	1.92
Tree5	2.22

Table 2. Power gradient calibration results.

Power Gradient	Forward Speed ($m \cdot s^{-1}$)	Fan Speed (RPM)	Spray Pressure (MPa)	Spray Flow Rate ($L \cdot min^{-1}$)
P1	0.25	9850	0.15	3.63
P2	0.31	10,789	0.19	4.51
P3	0.49	18,820	0.30	7.13
P4	0.58	20,930	0.35	8.44
P5	0.65	22,985	0.39	9.45

Note: All calculated values in the table are average values.

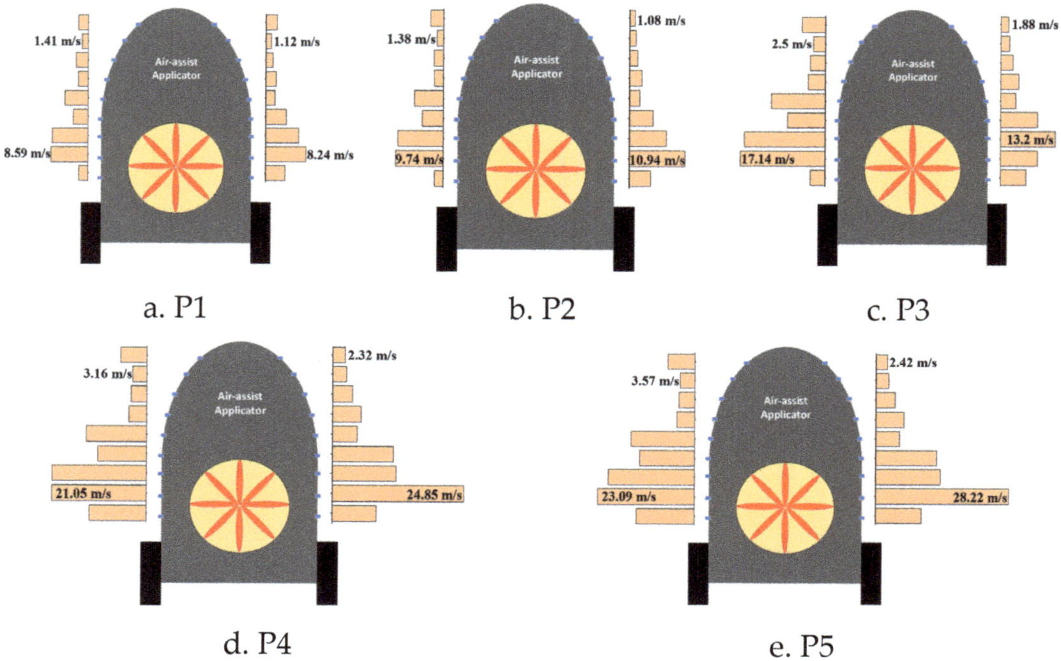

Figure 3. Minimum and maximum airflow rates (m/s) at the nozzle outlets on both sides of the air-assisted sprayer for each power gradient.

3.2. Result of Dependent Variable Selection

An F-test was conducted to observe the consistency between droplet coverage and droplet quantities, and the results are shown in Table 3.

Table 3. F-test results for droplet coverage and droplet quantities.

Tree Number	*p*-Value
Tree1	0.07
Tree2	0.11
Tree3	0.13
Tree4	0.57
Tree5	0.25

According to the results of the F-test, the *p*-values of all five trees were greater than 0.05, indicating no significant difference between droplet coverage and droplet quantities.

Therefore, droplet coverage was selected as the dependent variable for subsequent analysis in this study.

3.3. Results of Variables' Significance

The ANOVA results are shown in Table 4.

Table 4. Results of ANOVA on factors influencing droplet coverage behind canopies.

Factor	Spray Variable	F Value	p-Value	Significance
FAVD	Droplet coverage	7.74	5.56×10^{-3}	**
Power gradient		116.63	$<2.2 \times 10^{-16}$	***
Horizontal distance (X)		23.17	1.83×10^{-6}	***
Forward distance (Y)		6.75	9.55×10^{-3}	**
Height (Z)		80.40	$<2.2 \times 10^{-16}$	***

Note: Significance: *** 0.001, ** 0.01.

As can be seen from Table 4:

(1) For droplet coverage, all five factors were significant. Among them, power gradient, horizontal distance, and height were the most significant, followed by FAVD and forward distance. Although the five factors showed different levels of significance, the p-value of each factor was less than 0.01.

(2) The actual spray needs to combine two factors, FAVD and power gradient, to reduce the droplet coverage behind the canopy. For the actual spray, FAVD is the fruit tree variable and power gradient is the sprayer variable, both of which are decisive for droplet coverage, so it is important to focus on these two variables to reduce the droplet coverage of a non-target (non-canopy).

3.4. Effect of the Power Gradient on Droplet Coverage behind Canopies

The variation of droplet coverage with power gradient is shown in Figure 4.

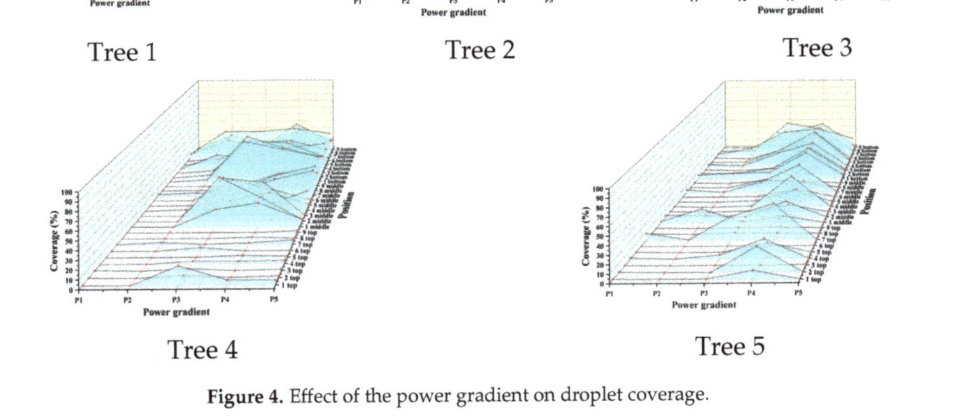

Figure 4. Effect of the power gradient on droplet coverage.

From Figure 4, it can be seen that:

(1) The power gradients of P1 and P2 may be more suitable for fruit tree spraying. For the five fruit trees, the loss behind the canopy, as expressed as droplet coverage on WSPs, was lower at P1 and P2. According to the authors' previous research [29], the droplet coverage within the canopy could meet the spray requirements when power gradients were P1 and P2. Therefore, P1 or P2 could be preferred for fruit tree spraying to reduce the droplet coverage behind canopies.

(2) The use of power gradients P3 and P4 could be best avoided when spraying mango trees. The droplet coverage increased first and then decreased with the power gradient. In Figure 4, P3, P4, and P5 reached the highest losses, quantified by droplet coverage. For Tree2 and Tree3, the average droplet coverage at P3 (Tree2: 56.97%, Tree3: 47.22%) was larger than at P5 (Tree2: 54.39%, Tree3: 39.10%), so P3 or P4 resulted in the highest losses, and the actual spraying should pay attention to the serious loss formed by these two power gradients.

3.5. Effect of FAVD on Droplet Coverage behind Canopies

Figure 5 shows the variation of droplet coverage with FAVD. It could be indicated that:

(1) With the increase of the FAVD, the average coverage of droplets behind canopies under different power gradients showed a decrease, increase, and then a decrease. Overall, the droplet coverage of Tree4 and Tree5 was smaller than that of Tree1, Tree2, and Tree3.

(2) The power gradient did not change the relationship between FAVD and droplet coverage, but changed the maximum value of average droplet coverage, e.g., P1 maximum value was 12.73% (Tree3), while P5 maximum value was 56.97% (Tree2).

(3) The effect of FAVD on droplet coverage had obvious regularity, and this regularity did not change with the power gradient. Therefore, in actual spraying, the effect of FAVD on droplet coverage should be considered.

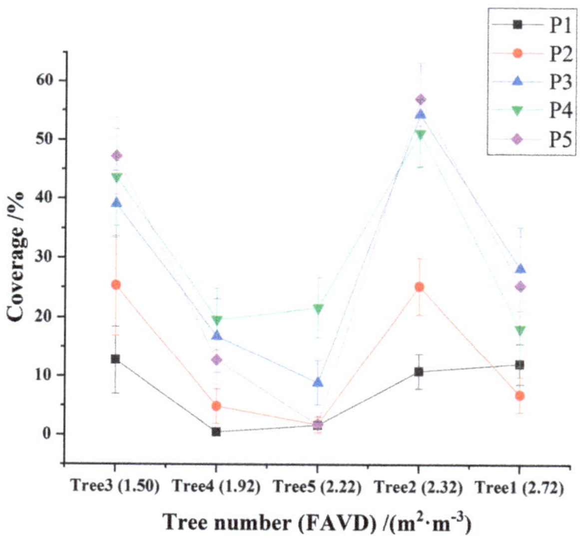

Figure 5. Mean droplet coverage (%) for different FAVD (m^2/m^3) over different power gradients.

3.6. Droplet Coverage Difference of the Position behind Canopies

The variation of droplet coverage in the horizontal distance (X) and height (Z) at Y = 0, Y = 1.2 m, and Y = 2.4 m was separately analyzed. Some of the results are shown in Figures 6–8. All the results are shown in Appendices A–C.

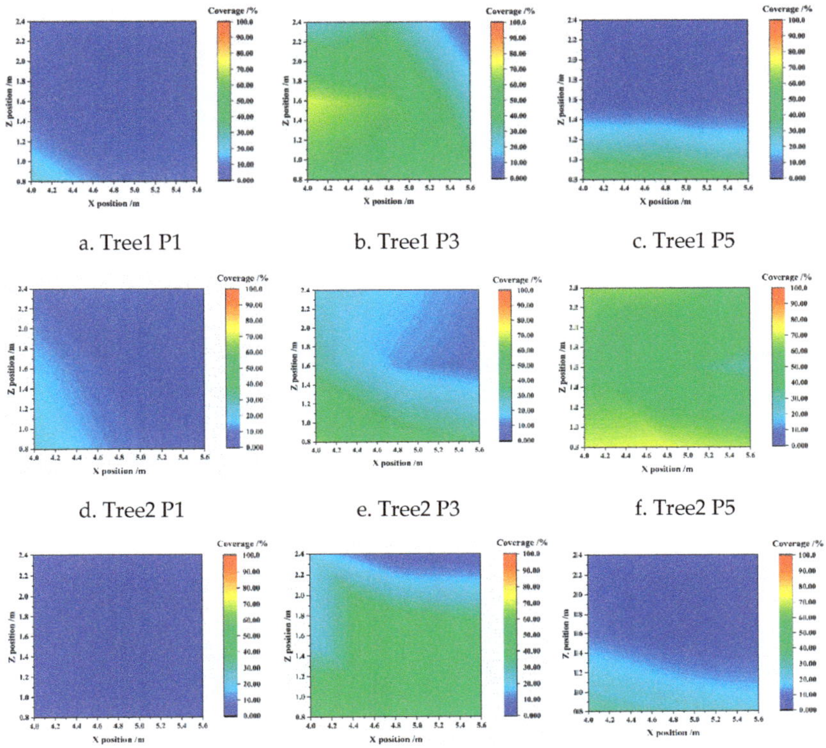

a. Tree1 P1 b. Tree1 P3 c. Tree1 P5

d. Tree2 P1 e. Tree2 P3 f. Tree2 P5

g. Tree4 P1 h. Tree4 P3 i. Tree4 P5

Figure 6. Droplet coverage difference at the position Y = 0.

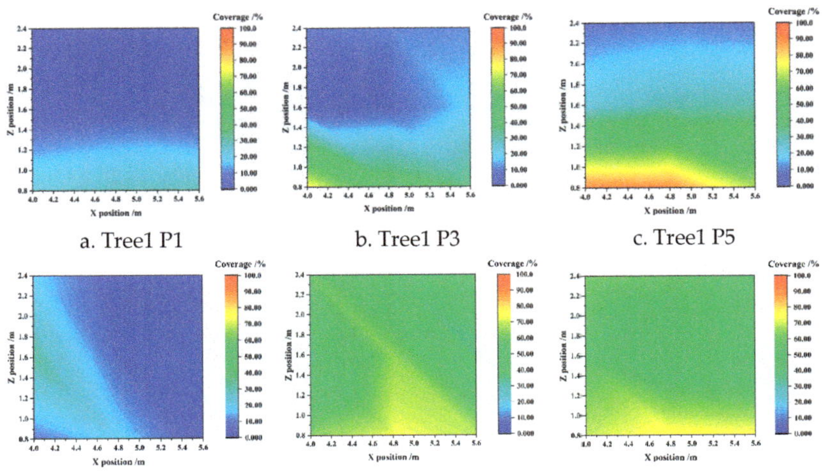

a. Tree1 P1 b. Tree1 P3 c. Tree1 P5

d. Tree2 P1 e. Tree2 P3 f. Tree2 P5

Figure 7. *Cont.*

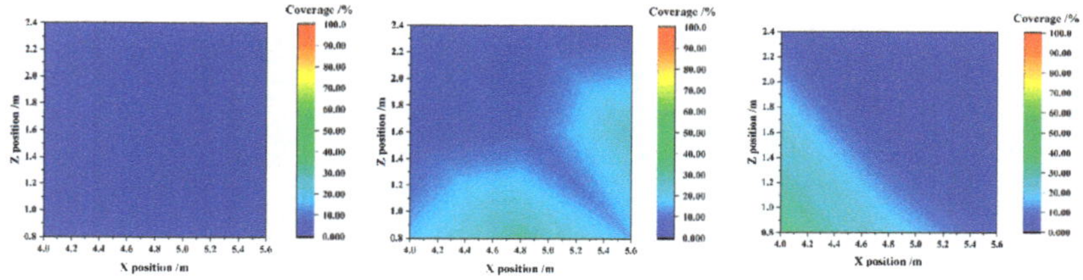

g. Tree4 P1 h. Tree4 P3 i. Tree4 P5

Figure 7. Droplet coverage difference at the position Y = 1.2 m.

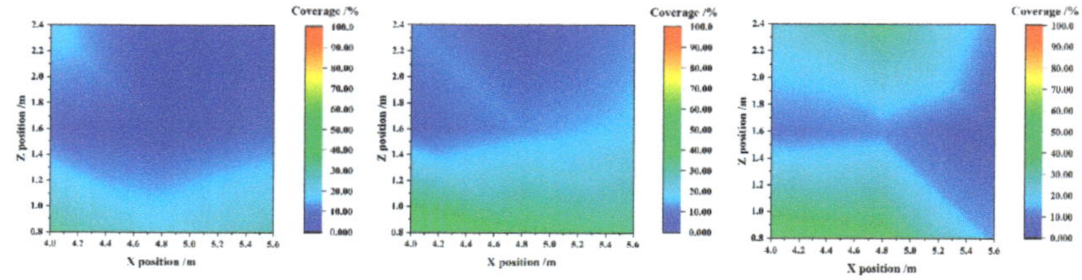

a. Tree1 P1 b. Tree1 P3 c. Tree1 P5

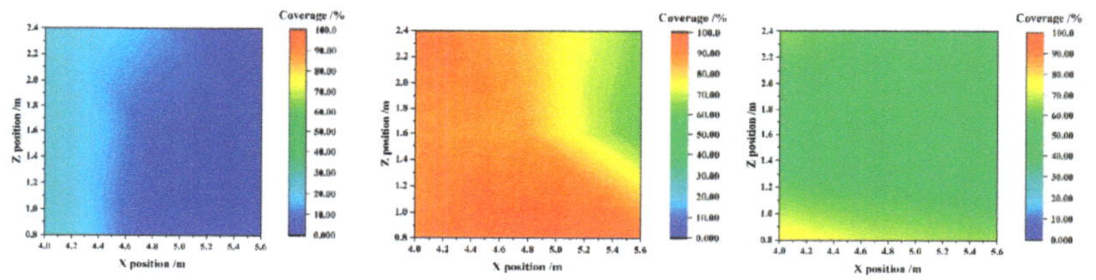

d. Tree2 P1 e. Tree2 P3 f. Tree2 P5

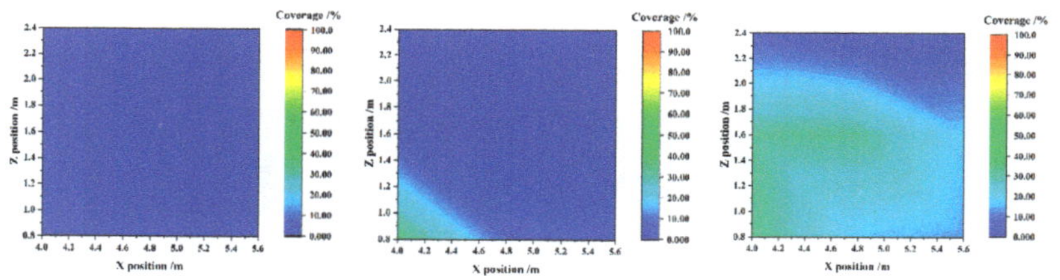

g. Tree4 P1 h. Tree4 P3 i. Tree4 P5

Figure 8. Droplet coverage difference at the position Y = 2.4 m.

From Figures 6–8, it can be seen that:

(1) As the horizontal distance (X) and height (Z) increased, the droplet coverage gradually decreased. During actual spraying, the spray parameters can be appropriately adjusted to improve the droplet coverage of the target and reduce the droplet loss.

(2) The droplet coverage gradually increased in the Y direction (forward direction). When the FAVD and power gradient were constant, the droplet coverage gradually increased along the Y direction. Although the trends under some conditions were different (such as Tree4 under the P3 power gradient), it did not affect the overall regularity. This may be due to experimental errors caused by environmental factors.

(3) The possible reasons for the increase of droplet coverage along the Y direction were: the forward speed of the sprayer in the Y direction makes the droplets have an initial velocity, and the droplet coverage in the current section was formed by the

superposition of the current droplets and the previous droplets. Thus, the droplet coverage gradually increased along the Y direction.

(4) For different power gradients, the variation regularity of droplet coverage along horizontal distance and height did not change, but the value of droplet coverage changed.

4. Discussion

In this study, to explore the loss of droplets behind canopies during spraying of fruit trees by air-assisted sprayers, a two-factor, five-level full experiment was conducted. The FAVD of fruit trees was tested as a factor. To examine the effect of FAVD on droplet loss, the experiment results showed that the FAVD significantly affected the droplet coverage (Table 4 and Figure 5). The FAVD was rarely involved in other studies on the effects of spraying in orchards. It is usually focused on the influence of spray pressure, airflow rate, and other parameters on droplet loss [10,30–32]. Although Sun et al. considered FAVD when studying the droplet penetration model of fruit trees, they did not clarify the effect of FAVD on droplet loss [33].

The application of air-assisted sprayers in orchards has significantly improved spraying efficiency, but it has also caused many problems, such as excessive spraying, pesticide residues, and environmental pollution. Many studies have proven that the adjustment of the pesticide application rate according to the characteristics of the fruit tree canopy is an effective means to reduce pesticide residues and environmental pollution [34–38]. However, the acquisition of canopy features was generally based on the 'sensor scanning/acquisition–solving geometric equations–fitting canopy profile' method [39,40], which simplified the asymmetric and porous structure of the canopy. Canopy characteristics mainly include canopy volume, leaf wall area, FAVD, etc. FAVD is the parameter that best reflects the characteristics of the canopy. Based on this study, studying the detection method of FAVD, and adjusting the spraying parameters based on this, will significantly reduce the droplet loss.

This study provides data support for subsequent research on precision spraying based on FAVD. However, there are still some shortcomings:

(1) In this paper, only one air-assisted sprayer was used as the experimental equipment, and a variety of sprayers can be used for comparative experiments in the future.

(2) The G6S air-assisted sprayer uses the throttle (power gradient) to simultaneously adjust parameters such as forward speed, fan speed, spray pressure, and spray flow rate. These parameters cannot be separately adjusted, so experiments with different parameter combinations were not conducted.

(3) In this paper, mango trees with large canopies were used as the experimental trees. Later, experiments can be carried out on different types of fruit trees to explore the regularity of droplet loss in different types of fruit trees.

5. Conclusions

A two-factor, five-level full experiment was conducted in an actual orchard. The effects of the power gradient and FAVD on droplet coverage behind canopies were analyzed, and the droplet coverage difference behind the canopies was clarified. Based on a detailed analysis of the experimental data, the following conclusions were obtained:

(1) Droplet coverage and droplet quantities were consistent, and five factors (power gradient, FAVD, horizontal distance (X), Forward distance (Y), and height (Z)) were significantly affected by droplet coverage behind the canopies (p-value < 0.01).

(2) In this study, power gradients P1 (forward speed: 0.25 m/s, spray pressure: 0.15 MPa, and spray flow rate: 3.63 L/min) and P2 (forward speed: 0.31 m/s, spray pressure: 0.19 MPa, and spray flow rate: 4.51 L/min) resulted in the lowest losses behind the canopy, as quantified by droplet coverage on WSPs. Based on the results found in this study, the use of power gradient P3 (forward speed: 0.49 m/s, spray pressure: 0.30 MPa, and spray flow rate: 7.13 L/min) or P4 (forward speed: 0.58 m/s, spray pressure: 0.35 MPa, and spray flow rate: 8.44 L/min) should be avoided.

(3) The effect of FAVD on droplet coverage had obvious regularity. With the increase of FAVD, droplet coverage first decreased, then increased, and finally, decreased again. This regularity did not change with the power gradient. At different positions behind canopies, the droplet coverage considerably varied. As the horizontal distance (X) and height (Z) increased, the droplet coverage gradually decreased. However, the droplet coverage gradually increased in the Y direction (forward direction).

This study revealed the loss regularity of droplets behind the fruit tree canopy, which can provide reference for the selection of operation parameters. The effect of FAVD on droplet loss was clarified, and data support was provided for the subsequent research on variable spraying based on FAVD.

Author Contributions: S.J., data curation, tests, and writing—original draft; W.L., visualization and tests; S.Y., methodology and writing—review and editing; Y.Z., supervision, methodology, and resources; Y.T., methodology and resources; J.X., tests. All authors have read and agreed to the published version of the manuscript.

Funding: This study was supported and funded by the National Natural Science Foundation of China (NSFC) (32171901), and the YANTAI Locality and University Cooperation Development Project (2021XDRHXMPT29).

Institutional Review Board Statement: Not applicable.

Informed Consent Statement: Not applicable.

Data Availability Statement: Not applicable.

Conflicts of Interest: The authors declare no conflict of interest.

Appendix A

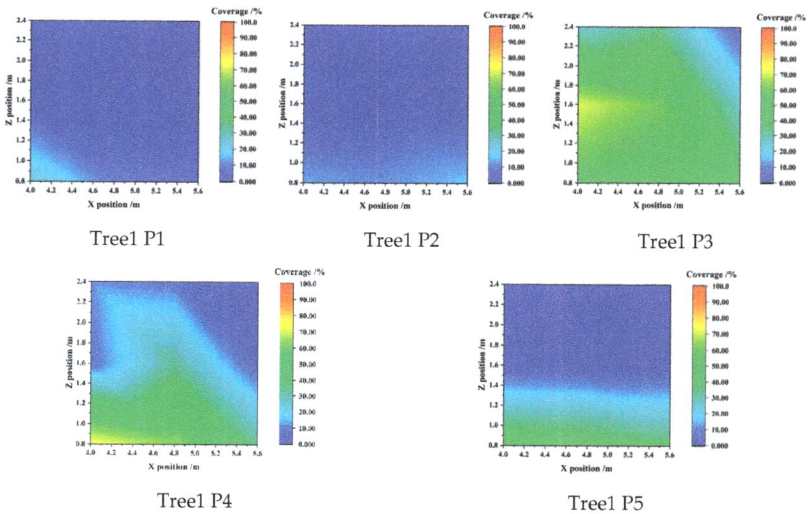

Figure A1. *Cont.*

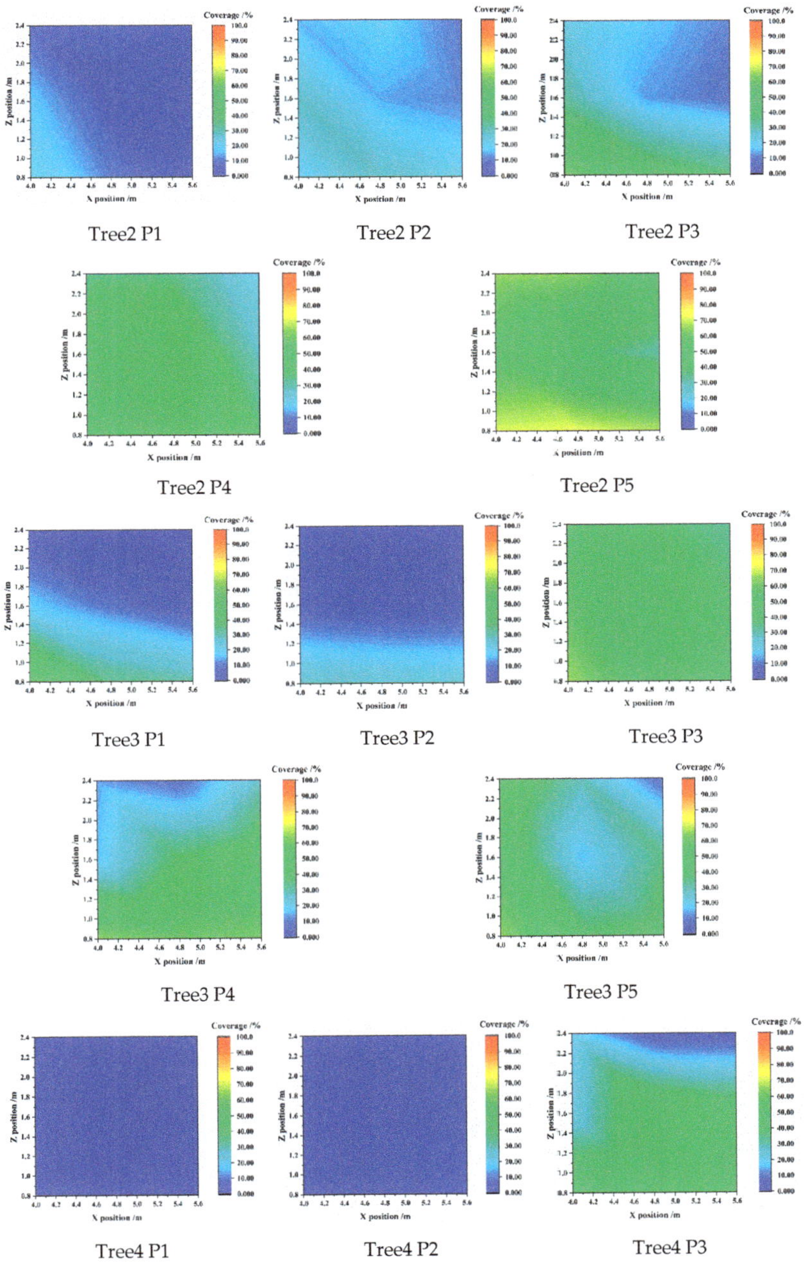

Figure A1. *Cont.*

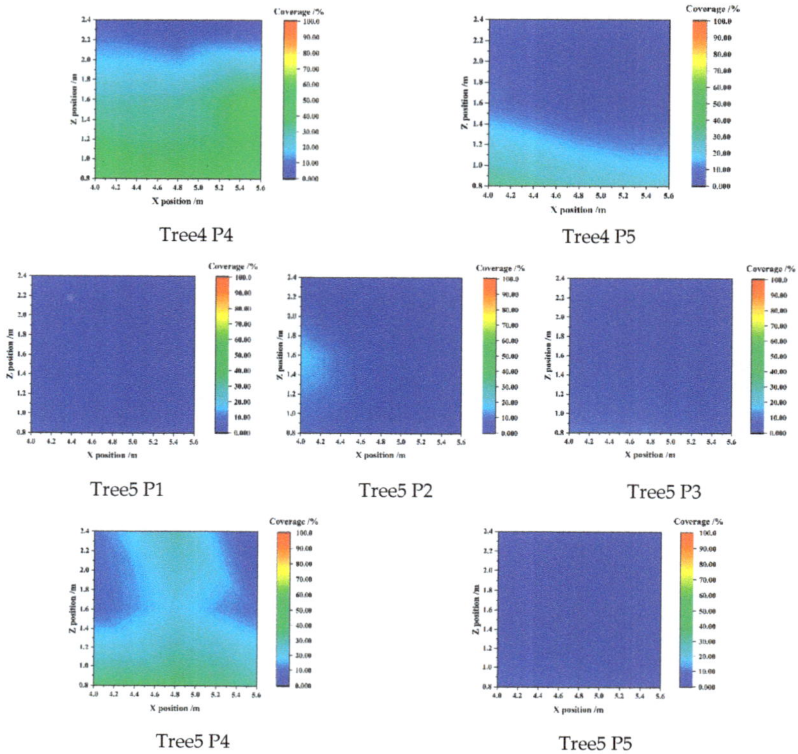

Figure A1. Droplet Coverage Difference at the Position Y = 0.

Appendix B

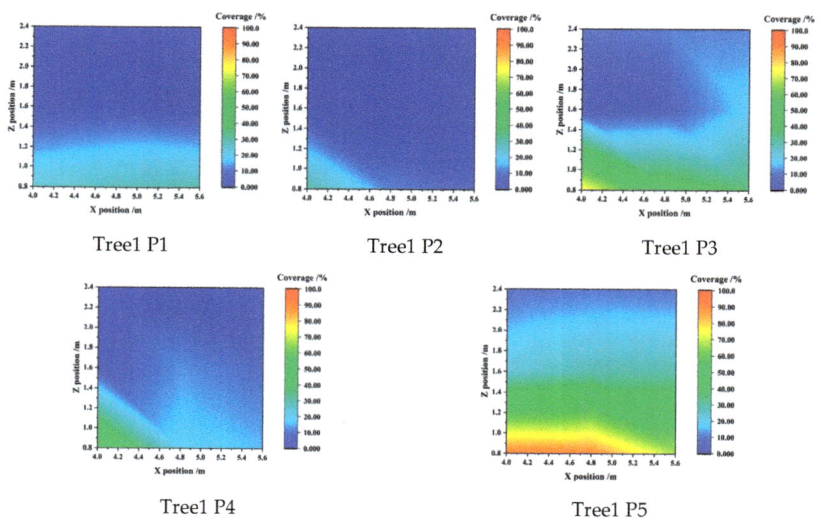

Figure A2. *Cont.*

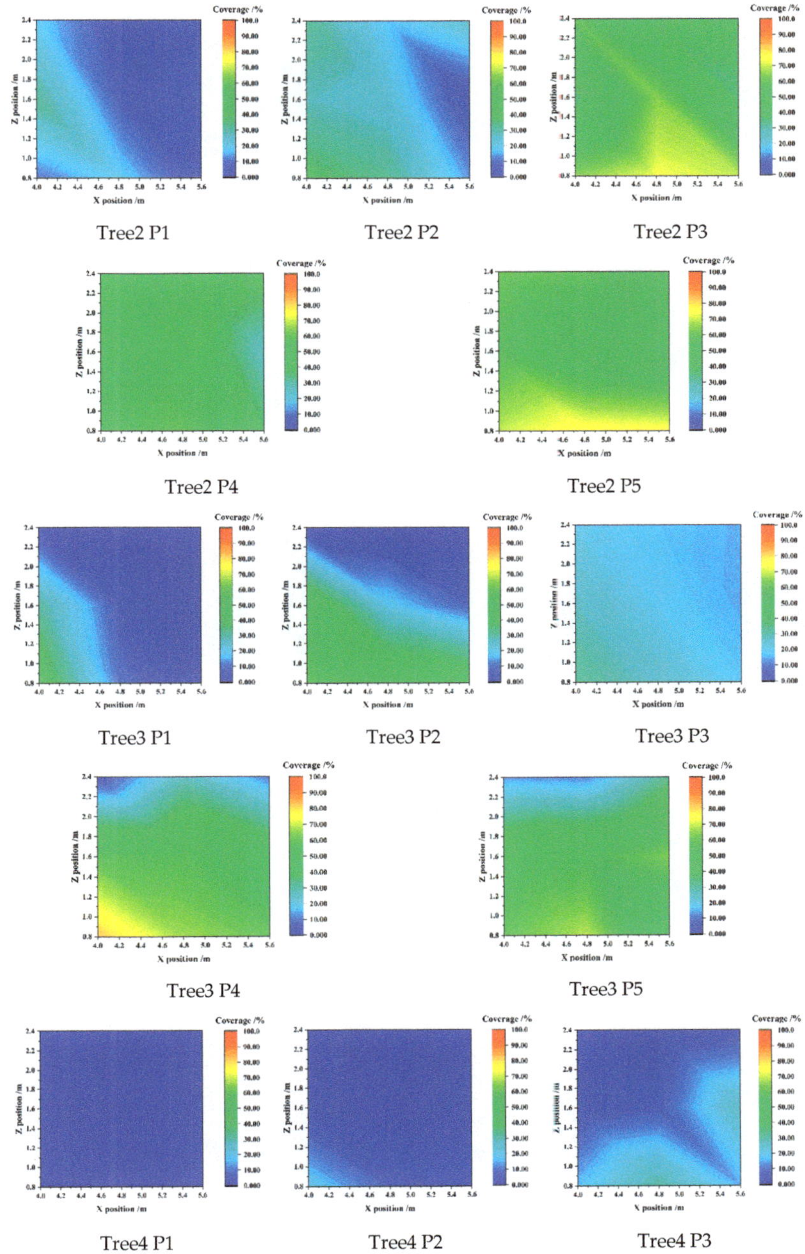

Figure A2. *Cont.*

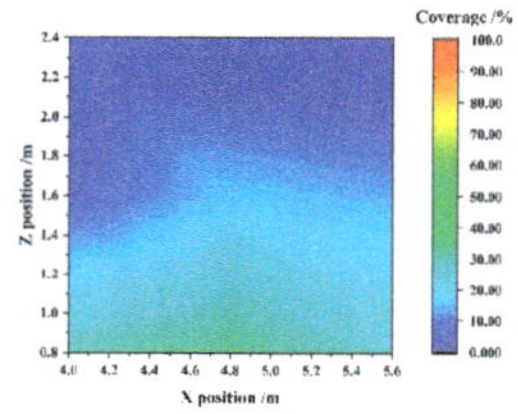

Tree4 P4

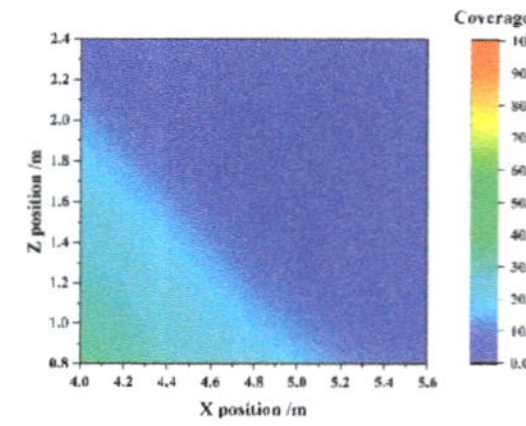

Tree4 P5

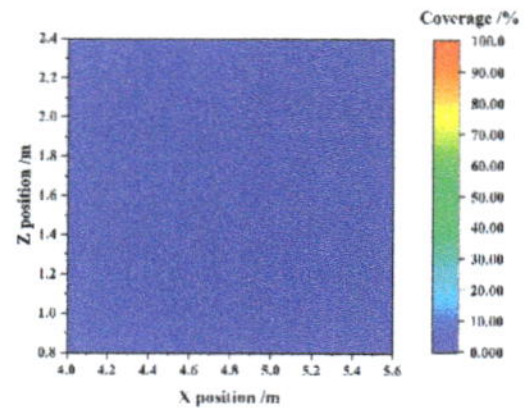

Tree5 P1

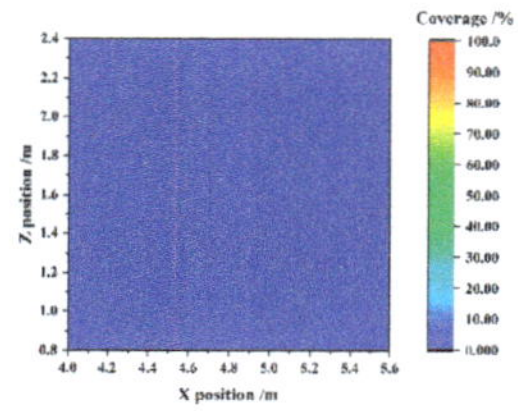

Tree5 P2

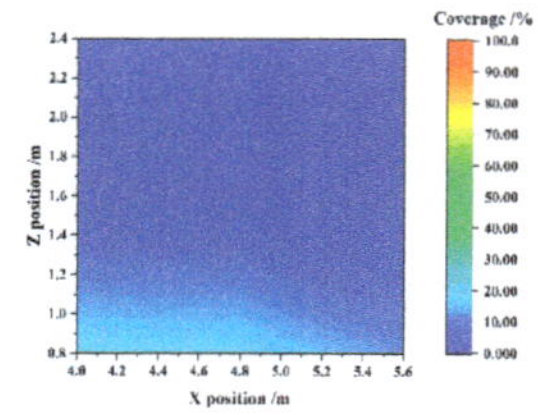

Tree5 P3

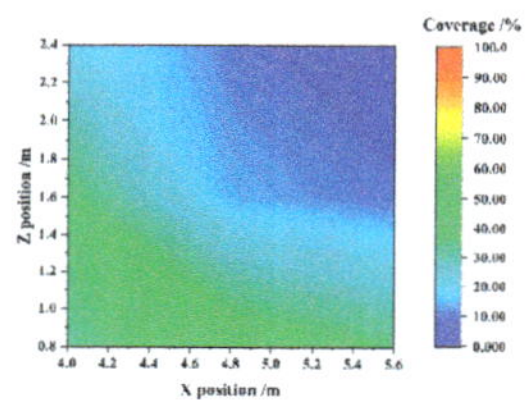

Tree5 P4

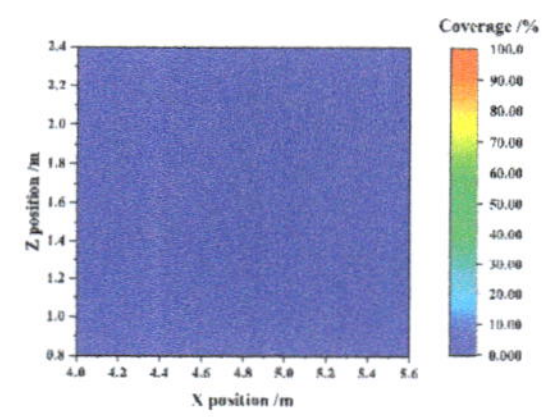

Tree5 P5

Figure A2. Droplet Coverage Difference at the Position Y = 1.2 m.

Appendix C

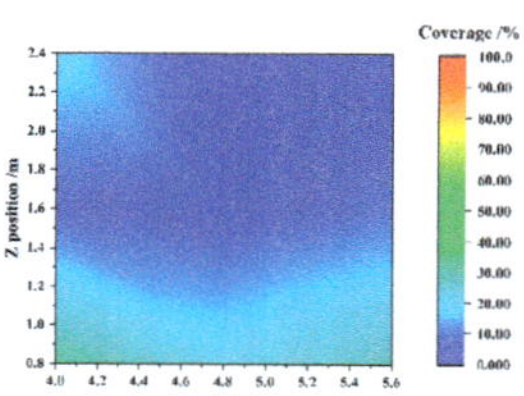

Tree1 P1

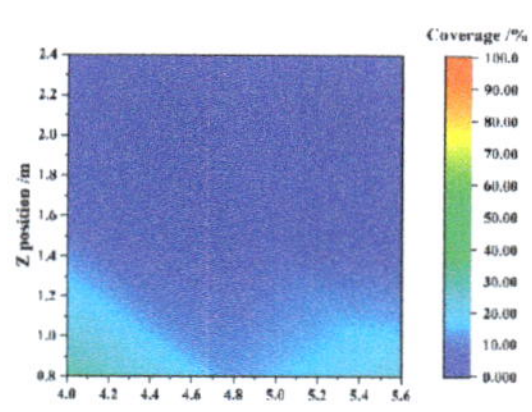

Tree1 P2

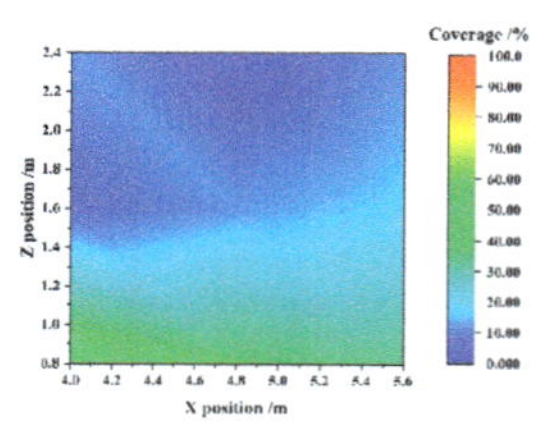

Tree1 P3

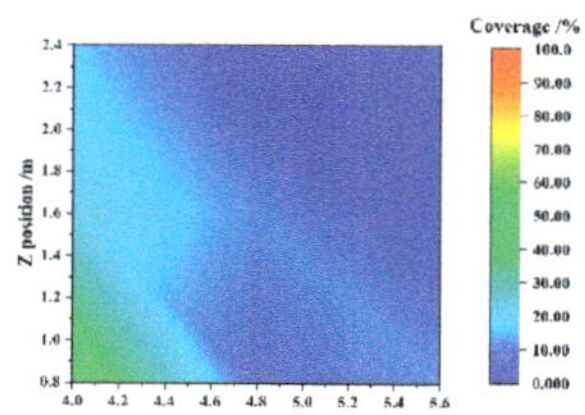

Tree1 P4

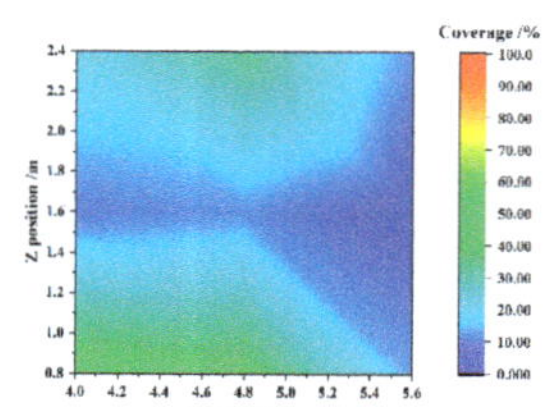

Tree1 P5

Figure A3. *Cont.*

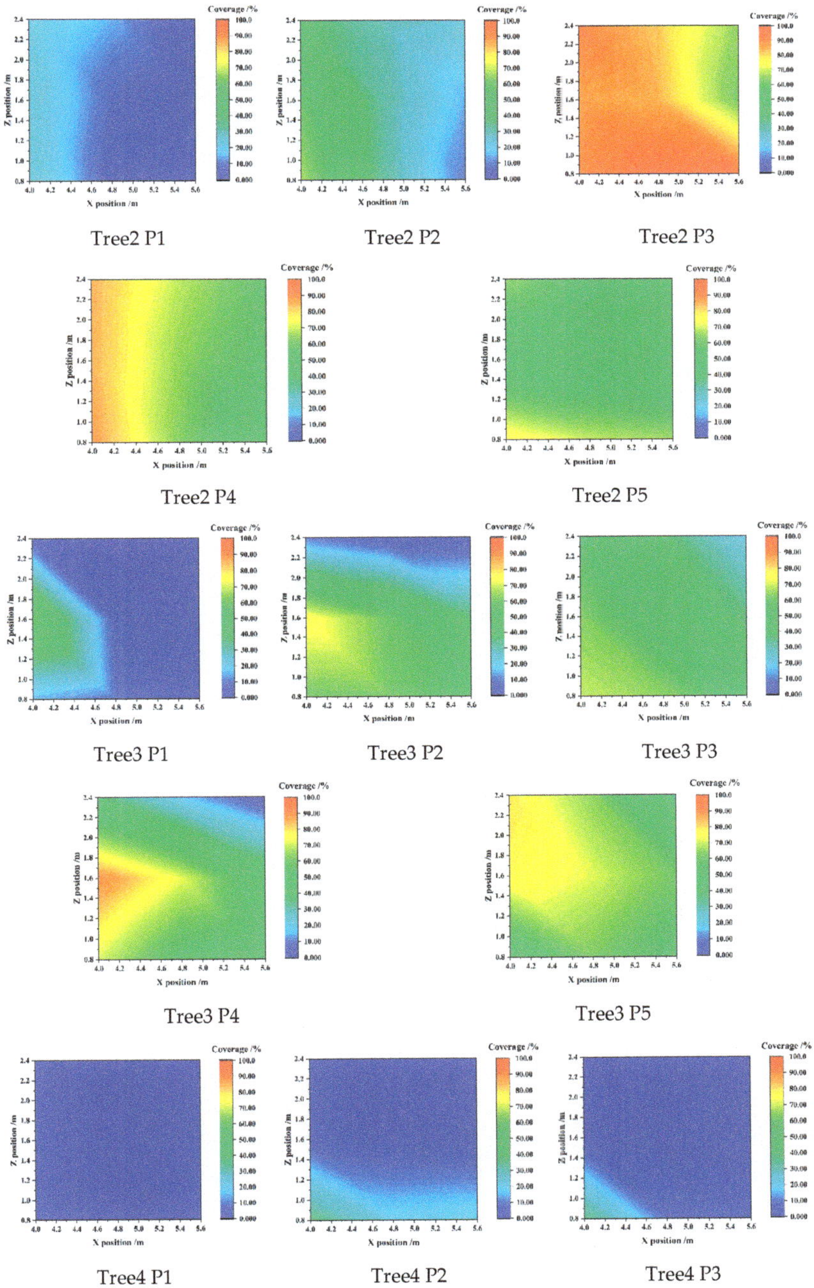

Figure A3. *Cont.*

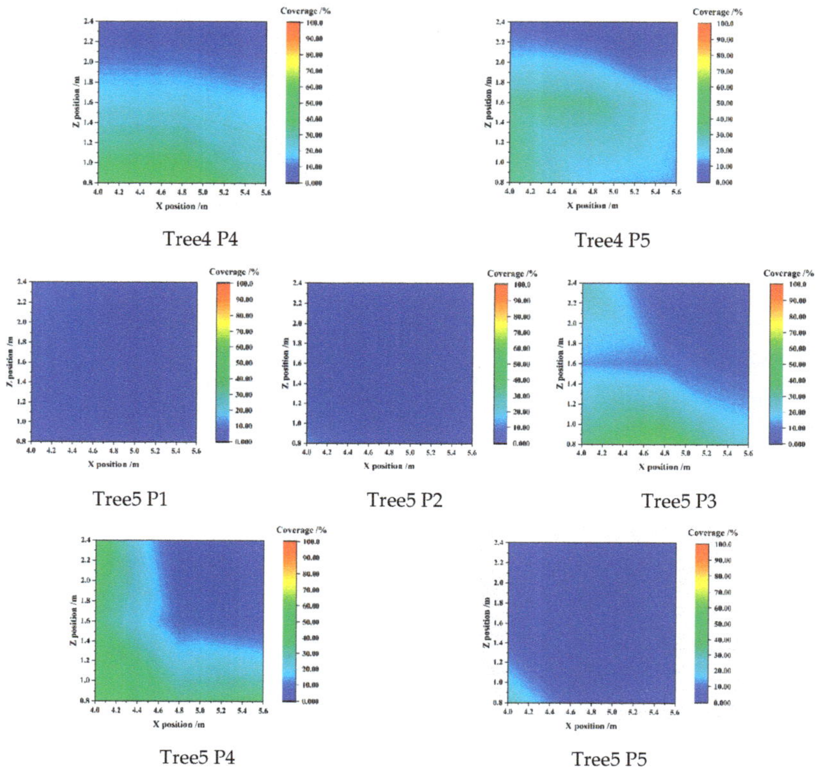

Figure A3. Droplet Coverage Difference at the Position Y = 2.4 m.

References

1. Zheng, Y.; Chen, B.; Lv, H.; Kang, F.; Jiang, S. Research progress of orchard plant protection mechanization technology and equipment in China. *Trans. CSAE* **2020**, *36*, 110–124.
2. Zheng, Y.; Jiang, S.; Chen, B.; Lv, H.; Wan, C.; Kang, F. Review on technology and equipment of mechanization in hilly orchard. *Trans. Chin. Soc. Agric. Mach.* **2020**, *51*, 1–20.
3. Wang, X. Study on Spray Drift and Anti-Drift Method. Ph.D. Thesis, China Agricultural University, Beijing, China, 2017.
4. Liu, X.; Yuan, L.; Shi, X.; Du, Y.; Yang, D.; Yuan, H.; Yan, X. Research progress on spray drift of droplets of plant protection machinery. *Chin. J. Pestic. Sci.* **2022**, *24*, 232–247.
5. Gil, Y.; Sinfort, C. Emission of pesticides to the air during sprayer application: A bibliographic review. *Atmos. Environ.* **2005**, *39*, 5183–5193. [CrossRef]
6. Zhang, H.; Zheng, J.; Zhou, H.; DORR, G. Droplet Deposition Distribution and Off-target Drift during Pesticide Spraying Operation. *Trans. Chin. Soc. Agric. Mach.* **2017**, *48*, 114–122.
7. Soheilifard, F.; Marzban, A.; Raini, M.G.; Taki, M.; van Zelm, R. Chemical footprint of pesticides used in citrus orchards based on canopy deposition and off-target losses. *Sci. Total Environ.* **2020**, *732*, 139118. [CrossRef] [PubMed]
8. Salcedo, R.; Vallet, A.; Granell, R.; Garcerá, C.; Moltó, E.; Chueca, P. Eulerian–Lagrangian model of the behavior of droplets produced by an air-assisted sprayer in a citrus orchard. *Biosyst. Eng.* **2017**, *154*, 76–91. [CrossRef]
9. Arvidsson, T.; Bergström, L.; Kreuger, J. Spray drift as influenced by meteorological and technical factors. *Pest Manag. Sci.* **2011**, *67*, 586–598. [CrossRef]
10. Grella, M.; Gallart, M.; Marucco, P.; Balsari, P.; Gil, E. Ground Deposition and Airborne Spray Drift Assessment in Vineyard and Orchard: The Influence of Environmental Variables and Sprayer Settings. *Sustainability* **2017**, *9*, 728. [CrossRef]

11. Fornasiero, D.; Mori, N.; Tirello, P.; Pozzebon, A.; Duso, C.; Tescari, E.; Bradascio, R.; Otto, S. Effect of spray drift reduction techniques on pests and predatory mites in orchards and vineyards. *Crop Prot.* **2017**, *98*, 283–292. [CrossRef]
12. Ru, Y.; Chen, X.; Liu, B.; Wang, S.; Lin, M. Optimized design and performance test of axial flow orchard sprayer air delivery system. *Trans. Chin. Soc. Agric. Mach.* **2022**, *53*, 147–157.
13. Miranda-Fuentes, A.; Rodríguez-Lizana, A.; Cuenca, A.; González-Sánchez, E.; Blanco-Roldán, G.; Gil-Ribes, J. Improving plant protection product applications in traditional and intensive olive orchards through the development of new prototype air-assisted sprayers. *Crop Prot.* **2017**, *94*, 44–58. [CrossRef]
14. Pascuzzi, S.; Bulgakov, V.; Santoro, F.; Anifantis, A.S.; Ivanovs, S.; Holovach, I. A Study on the Drift of Spray Droplets Dipped in Airflows with Different Directions. *Sustainability* **2020**, *12*, 4644. [CrossRef]
15. Hong, S.-W.; Zhao, L.; Zhu, H. CFD simulation of pesticide spray from air-assisted sprayers in an apple orchard: Tree deposition and off-target losses. *Atmos. Environ.* **2018**, *175*, 109–119. [CrossRef]
16. Duga, A.T.; Delele, M.A.; Ruysen, K.; Dekeyser, D.; Nuyttens, D.; Bylemans, D.; Nicolai, B.M.; Verboven, P. Development and validation of a 3D CFD model of drift and its application to air-assisted orchard sprayers. *Biosyst. Eng.* **2017**, *154*, 62–75. [CrossRef]
17. Liu, C.; Hu, J.; Li, Y.; Zhao, S.; Zhang, W.; Li, Q. Optimization of the inner flow channel of conical wind field anti-drift spray device and anti-drift characteristics. *Trans. CSAE* **2021**, *37*, 11–20.
18. Hu, J.; Liu, C.; Chu, X.; Li, Y.; Sun, S.; Zhang, W. Droplet deposition characteristics of conical wind field anti-drift device. *Trans. Chin. Soc. Agric. Mach.* **2020**, *51*, 142–149, 174.
19. Sun, D.; Zhan, X.; Liu, W.; Xue, X.; Xie, J.; Li, Z.; Song, S.; Wang, W. Compensation of spray angle to droplet drift under crosswind. *Trans. CSAE* **2021**, *37*, 80–89.
20. Grella, M.; Miranda-Fuentes, A.; Marucco, P.; Balsari, P.; Gioelli, F. Development of Drift-Reducing Spouts For Vineyard Pneumatic Sprayers: Measurement of Droplet Size Spectra Generated and Their Classification. *Appl. Sci.* **2020**, *10*, 7826. [CrossRef]
21. Grella, M.; Miranda-Fuentes, A.; Marucco, P.; Balsari, P. Field assessment of a newly-designed pneumatic spout to contain spray drift in vineyards: Evaluation of canopy distribution and off-target losses. *Pest Manag. Sci.* **2020**, *76*, 4173–4191. [CrossRef] [PubMed]
22. Rathnayake, A.P.; Khot, L.R.; Hoheisel, G.A.; Thistle, H.W.; Teske, M.E.; Willett, M.J. Downwind Spray Drift Assessment for Airblast Sprayer Applications in a Modern Apple Orchard System. *Trans. ASABE* **2021**, *64*, 601–613. [CrossRef]
23. Wang, Z.; He, X.; Li, T.; Huang, M.; Zhang, Y.; Xu, L.; Deng, X. Evaluation method of pesticide droplet drift based on laser imaging. *Trans. CSAE* **2019**, *35*, 73–79.
24. Grella, M.; Marucco, P.; Balsari, P. Toward a new method to classify the airblast sprayers according to their potential drift reduction: Comparison of direct and new indirect measurement methods. *Pest Manag. Sci.* **2019**, *75*, 2219–2235. [CrossRef]
25. Fessler, L.; Fulcher, A.; Lockwood, D.; Wright, W.; Zhu, H. Advancing Sustainability in Tree Crop Pest Management: Refining Spray Application Rate with a Laser-guided Variable-rate Sprayer in Apple Orchards. *Hortscience* **2020**, *55*, 1522–1530. [CrossRef]
26. García-Ramos, F.J.; Serreta, A.; Boné, A.; Vidal, M. Applicability of a 3D Laser Scanner for Characterizing the Spray Distribution Pattern of an Air-Assisted Sprayer. *J. Sens.* **2018**, *2018*, 1–7. [CrossRef]
27. Chen, L.; Wallhead, M.; Reding, M.; Horst, L.; Zhu, H. Control of Insect Pests and Diseases in an Ohio Fruit Farm with a Laser-guided Intelligent Sprayer. *Horttechnology* **2020**, *30*, 168–175. [CrossRef]
28. Ren, D.; Feng, T.; Li, Y.; Ren, H. Research and implementation of living leaf area measurement based on plant image. *Intell. Comput. Appl.* **2019**, *9*, 173–176.
29. Jiang, S.; Yang, S.; Xu, J.; Li, W.; Zheng, Y.; Liu, X.; Tan, Y. Wind field and droplet coverage characteristics of air-assisted sprayer in mango-tree canopies. *Pest Manag. Sci.* **2022**, *78*, 4892–4904. [CrossRef]
30. Ding, S.; Xue, X.; Dong, X.; Gu, W.; Zhou, Q. Effects of spraying parameters on droplet deposition performance. *Trans. Chin. Soc. for Agric. Mach.* **2020**, *51*, 308–315.
31. Qiu, W.; Zhao, S.; Ding, W.; Sun, C.; Lu, J.; Li, Y.; Gu, J. Effects of fan speed on spray deposition and drift for targeting air-assisted sprayer in pear orchard. *Int. J. Agric. Biol. Eng.* **2016**, *9*, 53–62.
32. Li, T.; Qi, P.; Wang, Z.; Xu, S.; Huang, Z.; Han, L.; He, X. Evaluation of the Effects of Airflow Distribution Patterns on Deposit Coverage and Spray Penetration in Multi-Unit Air-Assisted Sprayer. *Agronomy* **2022**, *12*, 944. [CrossRef]
33. Sun, C.; Liu, C. Construction and application of droplet canopy penetration model for air-assisted spraying pattern. *Trans. CSAE* **2019**, *35*, 25–32.
34. Salcedo, R.; Zhu, H.; Zhang, Z.; Wei, Z.; Chen, L.; Ozkan, E.; Falchieri, D. Foliar deposition and coverage on young apple trees with PWM-controlled spray systems. *Comput. Electron. Agric.* **2020**, *178*, 105794. [CrossRef]
35. Manandhar, A.; Zhu, H.; Ozkan, E.; Shah, A. Techno-economic impacts of using a laser-guided variable-rate spraying system to retrofit conventional constant-rate sprayers. *Precis. Agric.* **2020**, *21*, 1156–1171. [CrossRef]
36. Cai, J.; Wang, X.; Gao, Y.; Yang, S.; Zhao, C. Design and performance evaluation of a variable-rate orchard sprayer based on a laser-scanning sensor. *Int. J. Agric. Biol. Eng.* **2019**, *12*, 51–57. [CrossRef]

37. Li, L.; He, X.; Song, J.; Liu, Y.; Zeng, A.; Liu, Y.; Liu, C.; Liu, Z. Design and experiment of variable rate orchard sprayer based on laser scanning sensor. *Int. J. Agric. Biol. Eng.* **2018**, *11*, 101–108. [CrossRef]
38. Xiao, K.; Ma, Y.; Gao, G. An intelligent precision orchard pesticide spray technique based on the depth-of-field extraction algorithm. *Comput. Electron. Agric.* **2017**, *133*, 30–36. [CrossRef]
39. Zeng, L.; Feng, J.; He, L. Semantic segmentation of sparse 3D point cloud based on geometrical features for trellis-structured apple orchard. *Biosyst. Eng.* **2020**, *196*, 46–55. [CrossRef]
40. Yandún, N.; Salvo, D.; Prieto, P.; Torres-Torriti, M.; Auat, F. LiDAR and thermal images fusion for ground-based 3D characterisation of fruit trees. *Biosyst. Eng.* **2016**, *151*, 479–494. [CrossRef]

agronomy

MDPI

Article

Research on the Slip Rate Control of a Power Shift Tractor Based on Wheel Speed and Tillage Depth Adjustment

Changhai Luo [1,2], Changkai Wen [2], Zhijun Meng [2], Huaiyu Liu [3], Guoqiang Li [3], Weiqiang Fu [2,*] and Chunjiang Zhao [4,*]

1 College of Mechanical and Electrical Engineering, Shihezi University, Shihezi 832003, China
2 Research Center of Intelligent Equipment, Beijing Academy of Agriculture and Forestry Sciences, Beijing 100097, China
3 AgChip Science and Technology (Beijing) Co., Ltd., Beijing 100097, China
4 National Engineering Research Center of Intelligent Equipment for Agriculture, Beijing 100097, China
* Correspondence: fuwq@nercita.org.cn (W.F.); zhaocj@nercita.org.cn (C.Z.)

Abstract: The existing control methods for the slip rate of the driving wheel of a test prototype have limitations that cause low-quality tillage and finishing operations. We propose a slip rate control method based on the dual factor adjustment of wheel speed and tillage depth, taking the power shift tractor New Holland T1404 as an example to verify the algorithm. This method employs the wheel speed control principle based on the power transmission ratio calculation, throttle adjustment, and wheel speed control methods, as well as the slip rate control method, with wheel speed–slip rate control as the main factor and tillage depth–slip rate control as the secondary factor. A tractor test prototype was built to validate the method. The wheel speed control method enabled the tractor to accurately control the wheel speed under three working conditions: no load on a cemented ground, no load in a field, and subsoiling operation. For the subsoiling operation, the slip rate control method gradually reduced the tractor wheel speed when the slip rate of the tractor's drive wheel was too high until it met the requirements. When the wheel speed was adjusted to the lower limit, suspension control was performed to reduce the tillage depth and improve vehicle trafficability. In the 130 s validation test, it took 14.1 s for the tractor with the slip rate control function to have a wheel slip rate exceeding 20%, which was 25.4% lower than that of the tractor without this function. The proposed method controls the slip rate within the optimal range while ensuring maximum operation quality (tillage depth).

Keywords: high-power tractors; subsoiling operation; multiple factors; joint-control method

Citation: Luo, C.; Wen, C.; Meng, Z.; Liu, H.; Li, G.; Fu, W.; Zhao, C. Research on the Slip Rate Control of a Power Shift Tractor Based on Wheel Speed and Tillage Depth Adjustment. *Agronomy* **2023**, *13*, 281. https://doi.org/10.3390/agronomy13020281

Academic Editor: Shubo Wang

Received: 15 December 2022
Revised: 15 January 2023
Accepted: 16 January 2023
Published: 17 January 2023

1. Introduction

As the most widely used selfpropelled power machinery in field and management operations, tractors can complete the traction and driving operations of agricultural equipment [1,2]. In the process of land preparation, the operation type (such as ploughing, subsoiling, and rotary tillage), operation parameters (such as the number of ploughs, depth, and width), and the tractor driving parameters (including speed and wheel slip rate) affect the operational performance of the tractor directly [3,4]. Owing to the relatively complex field operation environment and fluctuations in the operating load, the tractor in the operational process easily produces an excessive drive-wheel slip rate, which reduces the operating efficiency and damages the soil environment [5–7].

Approximately 20–55% of the available tractor energy is wasted at the tire-soil interface [8]. A high slip rate leads to wasted tractor power and accelerates the wear of the tires [9,10]. Authoritative research has shown that a tractor wheel slip rate of 10–20% is optimal [11–13]. Therefore, with the integration of agronomy and agricultural machinery for soil protection requirements and the increasing degree of intelligence of medium and large tractors [14], drive antiskid operations have gradually become a research hotspot in the field of intelligent operation control of tractor units.

In the 1980s, Chancellor and Zhang discovered that controlling the slip rate could save energy and fuel and become efficient in tractor operations [15]. Mirzaeinejad and Mirzaei investigated the effect of the ratio of front and rear axle loads on the slip rate of the drive wheels in a tractor. They found that with a front axle load ratio of 45–55%, the tractor has a low slip rate regarding the drive wheels and good traction performance [16]. Pranav designed an automatic slip rate control system and analysed the effectiveness of slip rate control in terms of fuel consumption and traction capacity [17]. Pasillas-Lépine studied an automatic control method for tractor slip rate with the slip rate and acceleration calibration as the control objectives, but the control accuracy was poor [18].

After years of research, there have been more studies on slip-rate control. Most slip-rate control studies are based on automatic tillage depth control and algorithms. Gupta designed and developed a slip-rate control system based on an Arduino Mega 2560 microcontroller for a two-wheel-drive tractor. The control strategy is to drive the motor to reduce the tillage depth when the slip rate exceeds the upper limit of the set range and increase the tillage depth if the slip rate is less than the lower limit [14]. Zhang et al. proposed a nonlinear slip-rate control method for tractor-drive wheels based on sliding mode variable-structure control. The method uses traction as the primary regulation target and slip rate as the secondary regulation target. The drive-wheel slip rate of the tractor was effectively controlled at approximately 20% of the optimal value [19]. Based on the fuzzy control theory, Soylu developed an automatic slip-rate control system for agricultural tractors. He established a control relationship between the slip rate and tillage depth using the drive-wheel slip rate as a separate input. He determined the variation in tillage depth in the fuzzy controller [20].

In summary, to reduce the slip rate of the driving wheels, most existing slip-rate control studies reduce the traction resistance of the tractor by adjusting the tillage depth. These methods are straightforward and effective; however, the reduced tillage depth reduces the working quality. Therefore, more factors should be incorporated into the slip-rate control method, such as the tractor travel speed, fuel consumption, tractive effort, and engine load.

In this study, we aimed to develop a tractor slip-rate control system and propose a method to control the slip rate of the drive wheel by adjusting the tractor operating wheel speed and the three-point hitch mechanism height. This method mainly includes the wheel speed control principle based on the methods of the power transmission ratio calculation, throttle adjustment, and wheel speed control, and the slip rate control method, with wheel speed–slip rate control as the main factor and tillage depth–slip rate control as the secondary factor. A test platform was built based on a power shift tractor. The typical working conditions are cement ground with no load, field flat ground with no load, and subsoiling. Field experiments were conducted to verify that the proposed method can effectively control the tractor wheel speed and reduce the time when the wheel slip rate exceeds 20% during the subsoiling operation of the tractor. A block diagram of the study design is shown in Figure 1.

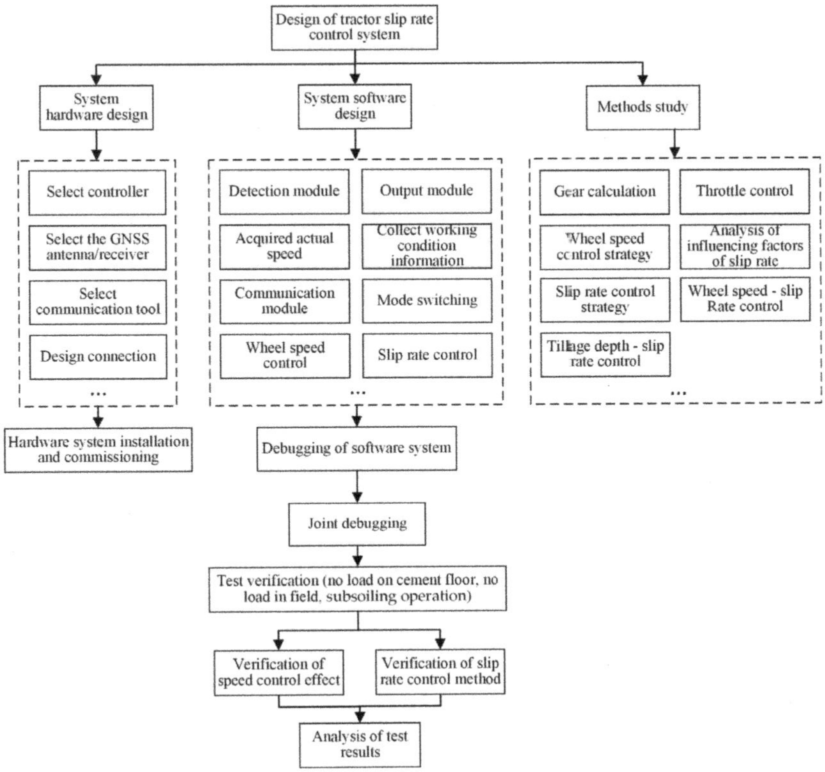

Figure 1. Block diagram of the study design.

2. Materials and Methods

2.1. Control Method

A method was proposed to control the slip rate of the drive wheel by adjusting the tractor operating wheel speed and three-point hitch mechanism height. This method mainly included the wheel speed control principle based on the methods of the power transmission ratio calculation, throttle adjustment, and wheel speed control, and the slip-rate control method with wheel speed-slip rate control as the main factor and tillage depth-slip rate control as the secondary factor.

2.1.1. Wheel Speed Control Principle

In this study, we aimed to control tractor wheel speed precisely. The wheel speed control method can calculate the current gear based on the target speed and target engine speed, fine-tune the real-time engine speed, and achieve a target speed to achieve accurate control of the wheel speed.

The engine speed and transmission ratio of the transmission system jointly determine tractor wheel speed. By referring to the New Holland T1404 tractor operation manual, a wheel speed for the tractor of 2200 r/min rated engine speed for each gear can be obtained. A comparison of gears and speeds is presented in Table 1.

From this, the relationship between the engine speed in each gear and tractor wheel speed can be calculated, as shown in Figure 2.

Table 1. Comparison of gears and speeds.

Gears	Speed (km·h^{-1})
A1	2.4
A2	2.9
A3	3.4
A4	4.1
A5	4.9
A6	5.8
B1	5.5
B2	6.6
B3	8.0
B4	9.6
B5	11.3
B6	13.6
C1	15.7
C2	18.8
C3	22.6
C4	27.2
C5	32.0
C6	38.5

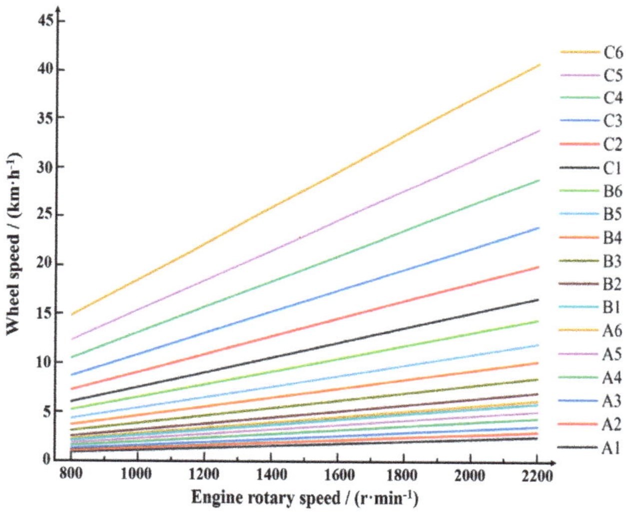

Figure 2. Relationship between tractor wheel speed and engine speed in each gear.

Various combinations of engine speed and gearbox power transmission ratios exist to meet the requirements at the same wheel speed, which must be selected according to the actual operational conditions. The fuel economy was relatively high when the engine was operated at an economical speed. Because wheel speed was the control target, yet the gearbox was stepped, it was often impossible to meet the engine speed and wheel speed requirements. We used the power transmission ratio calculation, throttle adjustment, and wheel speed control methods.

- Gear Calculation Method

The wheel speed and engine speed commands were obtained; the wheel speed–engine speed relationship in the ideal gear is indicated by the dashed line G′ in Figure 3. Due to the limitation of the gearbox structure, only the neighbouring gears, G1 or G2, could be selected. The target speed was adjusted to n1 or n2 (the one with a smaller gap to the target speed is preferred) to achieve the wheel speed as the control target; the engine speed was close to the requirement. The speed controls covered in this study were all wheel-based speed controls.

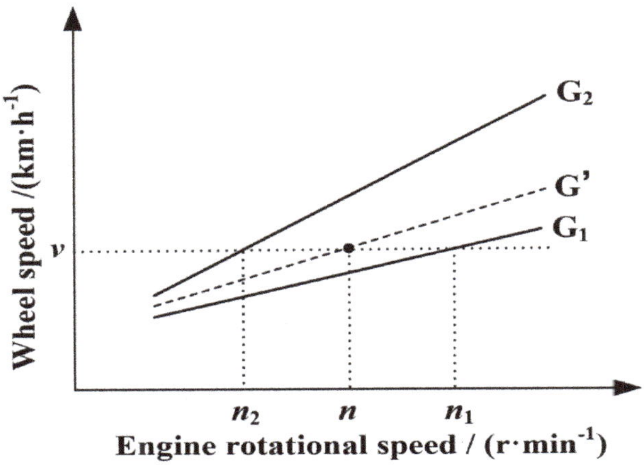

Figure 3. Gear calculation principle.

- Throttle-Adjustment Method

When the tractor gear remains unchanged, the tractor wheel speed is only related to engine speed. Therefore, the control system controls the engine speed after gear changes by adjusting the throttle to control the tractor drive wheel speed. This calibration test used a voltmeter to collect the voltage analogue output values of the maximum and minimum throttle openings, set as 0% and 100%, respectively. A voltage analogue input was used to control the throttle opening. Then, the neutral gear was maintained, and the throttle opening was increased by 5% from 5%. The engine speed at each throttle opening was acquired using the CAN information acquisition equipment. Finally, the engine speed and throttle opening were fitted. The fitting curve is shown in Figure 4.

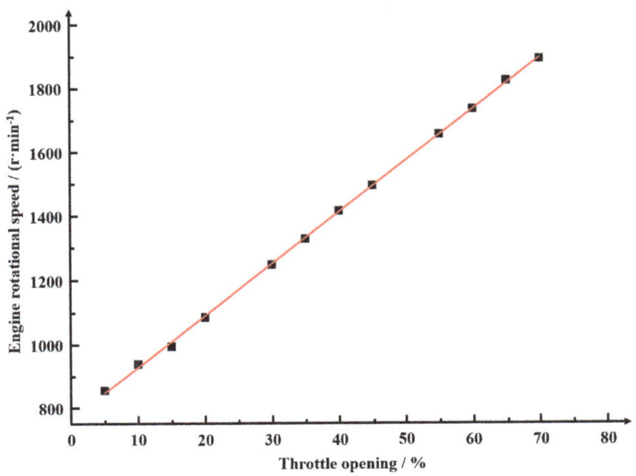

Figure 4. The fitting curve between throttle opening and engine speed.

The correspondence between engine speed and throttle opening was obtained from the fitting formula in Equation (1).

$$n = 16.12\alpha + 767.5 \tag{1}$$

where n is the engine speed in r/min, and α is the throttle opening in the range of 0–100%.

When the tractor gearbox gear remains unchanged, the engine speed determines the tractor wheel speed. Therefore, after the gear shift, the throttle adjustment program controlled the throttle opening to adjust the tractor wheel speed to the target value. The tractor throttle adjustment mechanism was generally a foot pedal and hand throttle pusher. The two control mechanisms work simultaneously, and the control mechanism with a more significant signal is adequate. We considered the hand throttle pusher as the control object.

The difference between the current and target wheel speeds was calculated by adjusting the throttle to control the tractor wheel speed. The single-wheel speed adjustment was determined according to the wheel speed control algorithm. The engine speed adjustment amount was calculated using the relationship between the tractor wheel speed of the current gear and engine speed. The relationship between throttle and engine speed was simplified to a proportional function with a slope of 1. The throttle-opening adjustment amount could be calculated based on the rotating speed and converted into the output value of the digital-to-analogue converter (DAC). The throttle-adjustment process is shown in Figure 5.

- Wheel Speed Control Method

After receiving the wheel speed control instruction from the upper computer, the control system must implement the corresponding control method according to the actual state of the tractor. First, according to the instructions and the current tractor wheel speed, it can be divided into three situations: speed increase, speed decrease, and stop. Speed increase should be divided into two situations: starting and driving speed increase.

When speeding up, to prevent the engine load from increasing or even stalling due to the decrease in transmission ratio after upshifts, the throttle was adjusted to increase the engine speed to near the target speed, and then the gear was adjusted. After a successful shift, the PID controller adjusted the throttle to control the tractor wheel speed.

When the tractor starts, it must select the forward gear and delay it for some time after shifting to approach the target wheel speed gradually from the static state. If the delay is too short, the PID controller will work ahead of time, leading to an excessive overshoot or unstable wheel speed control.

When the tractor slowed down, the shift operation was performed first, then the speed was reduced to near the target speed, and finally, the PID controller adjusted the wheel speed. This is because the tractor has a heavy load when pulling the unit, and a sudden speed reduction may lead to an engine flameout.

When the target wheel speed is 0, the control system executes the stop program, puts the gearbox in the neutral state, adjusts the throttle opening to the lowest value, turns off the PID controller, and clears the flag position.

The wheel speed control process is shown in Figure 6.

2.1.2. Slip Rate Control Method

The slip rate control method used was based on the slip rate control system of the tractor built in this study. A tractor wheel speed influencing factor was added to the control system. A control strategy was proposed with wheel speed–slip rate control as the primary control method and tillage depth–slip rate control as the second control method. The control system is described in the next section.

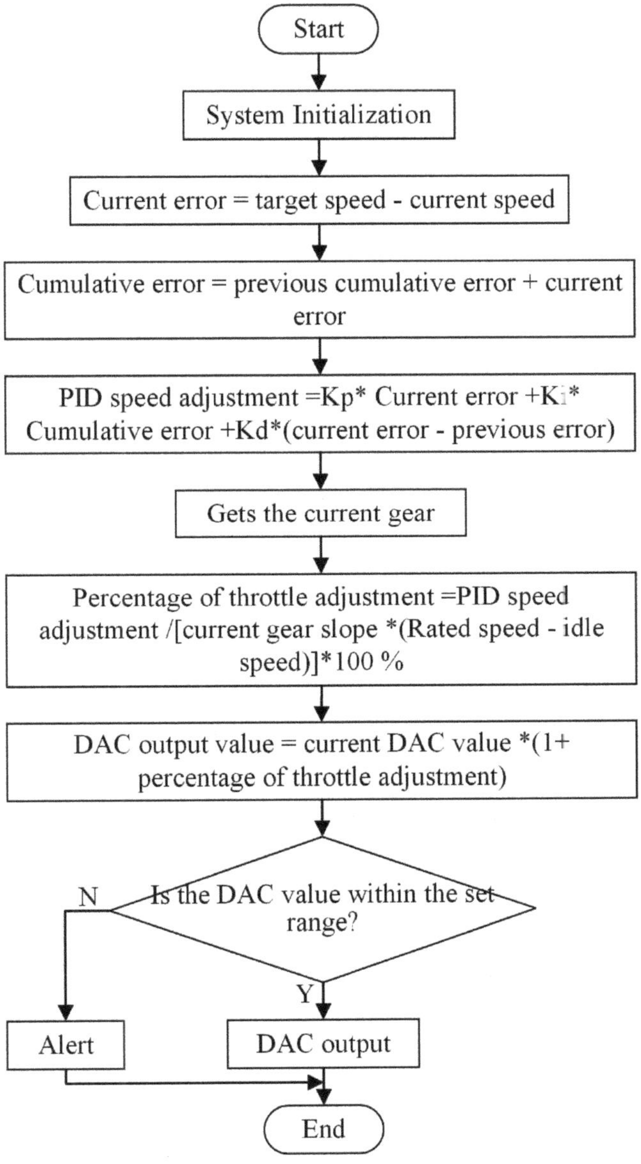

Figure 5. Flow chart of throttle adjustment procedure.

The slip rate of the tractor driving wheels is calculated from the tractor wheel speed and ground speed, according to Equation (2).

$$S_r = \frac{u_t - u_a}{u_t} \times 100\% \quad (2)$$

where u_a is the ground speed of the tractor, and u_t is the wheel speed of the tractor. We used the real-time kinematic (RTK) carrier phase difference technology to obtain the tractor ground speed.

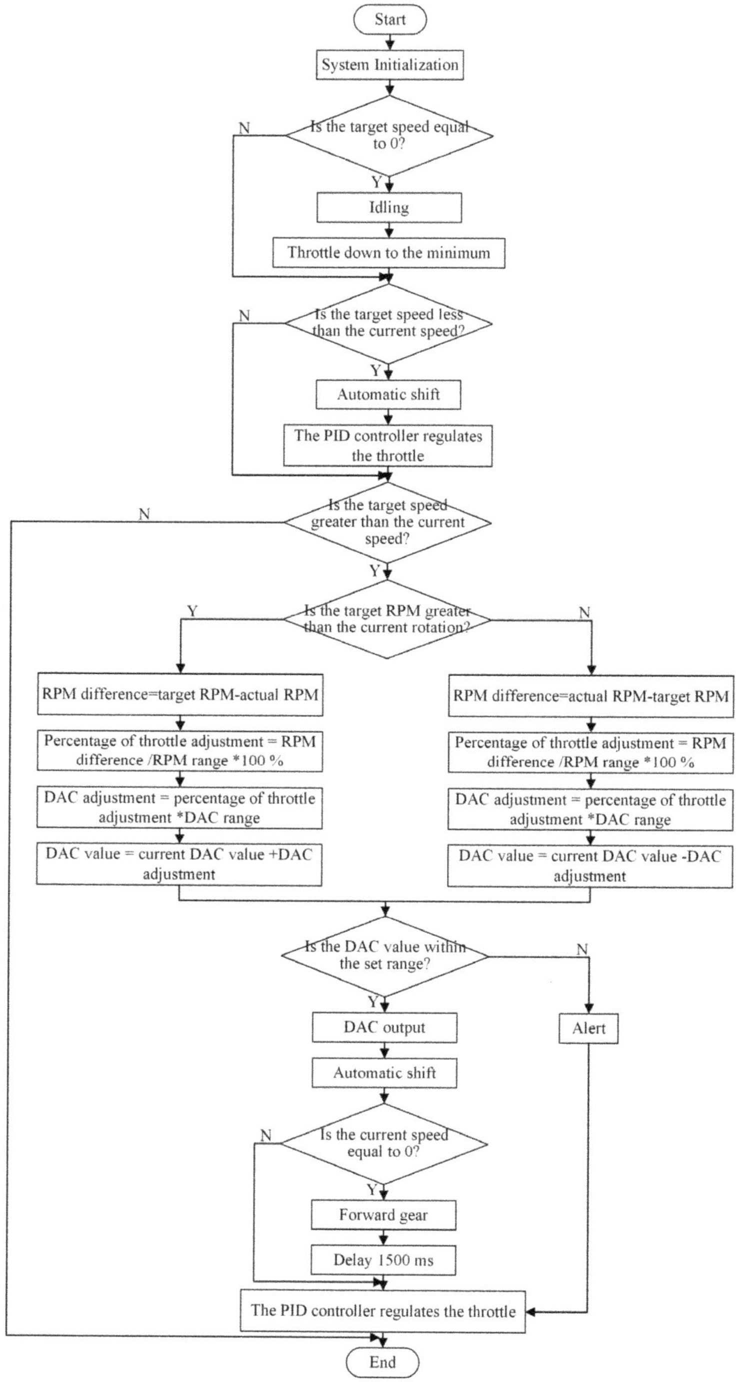

Figure 6. Flow chart of the wheel speed control program.

The slip rate control strategy proposed in this study considers the slip rate of the tractor-driving wheels as the control target and controls two factors: tractor wheel speed (based on wheel speed) and tillage depth. The tractor's engine, gearbox, and hitch were controlled by taking the slip rate range as the input and the target wheel speed and tillage depth as the output. This strategy uses wheel speed adjustment as the primary control mode. The target wheel speed was first adjusted when the slip rate exceeded the threshold. When the wheel speed adjustment reached the lower threshold, the tillage depth was taken as the control object. These two adjustment methods control the slip rate by changing the traction resistance. This is a two-factor, single-objective control strategy. When compared to the slip-rate control strategy based on automatic control of tillage depth (single-factor and single-objective), it is closer to the driver's operation and can ensure operation quality (tillage depth) to the greatest extent, as well as improving the energy efficiency of agricultural machinery. Therefore, a three-stage control strategy was adopted to control the slip rate within the best 10–20% and ensure operation quality (tillage depth).

In the first stage, the tillage depth remained unchanged, and the wheel speed–slip rate was used to control it. When the slip rate exceeded 15%, the target wheel speed gradually reduced, and the original target wheel speed was gradually restored after the slip rate was restored to 10%. The step speed regulation was performed with a step size of 0.5 km/h, and the adjustment frequency was 1 Hz. Because the traction efficiency of the tractor decreases when the speed decreases and the operating speed range of the subsoiler is 5–10 km/h, the lower limit of the target wheel speed in this stage is set to 3 km/h.

In the second stage, when the slip rate was still higher than 20% after reaching the lowest set wheel speed in the previous stage, it was considered that the working conditions in this area were poor, which reduced the working quality and improved the passability of the tractor. Therefore, tillage depth–slip rate control was adopted, and there have been many studies on tillage depth–slip rate control. The tillage depth was controlled only by the switching value. The hitch is lifted when the slip rate exceeds 20%, and the machine is dropped after returning to 10%.

In the third stage, if the wheel slip rate was too high after the hitch system was lifted to 0, it was judged that the soil condition was terrible and could not pass, stop, or give an alarm to the upper computer.

When the slip rate exceeded the upper limit, the three stages were executed sequentially. The three stages were executed in reverse order when the slip rate was restored. In other words, when the slip rate increases, the wheel speed is first reduced, and if the wheel speed is adjusted sufficiently to make the slip rate return to the normal range, the second stage is not started. When the wheel speed adjustment made it difficult to restore the slip rate, the tillage depth was reduced. After the slip rate was restored, the tillage depth was restored first, and then the wheel speed was restored. A flow diagram of the slip-rate control strategy is shown in Figure 7.

2.2. Control System Implementation

The overall structure of the control system (VS-SMHCS) is shown in Figure 8. The software component of the system includes both manual and automatic control modes. The manual mode is the original operation function of the tractor, which includes reversing the handle, gear button, handle throttle push rod, and three-point hitch mechanism lift button. The automatic control mode includes a communication module, work information collection module, vehicle speed collection module, throttle control module, gear control module, and hitch-mechanism control module. In the automatic control mode, the communication module is mainly used for data and operation log transmissions between the onboard computer and controller. The working information collection module collects tractor speed, wheel speed, gearbox gear, and hitch mechanism height lifting data. It transmits the collected data to the controller through the controller area network (CAN) bus or serial communication. The automatic control mode is mainly used to control the tractor engine, gearbox, and hitch mechanism through high-level and low-level signals.

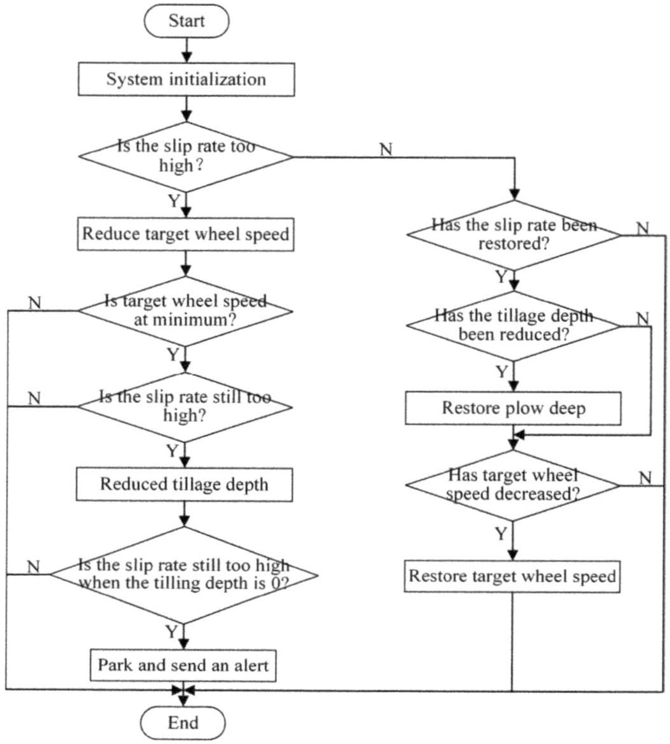

Figure 7. The slip rate control strategy flow chart.

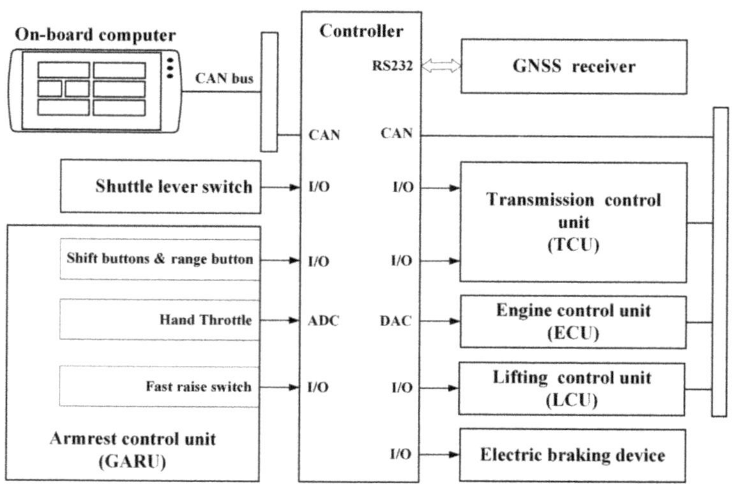

Figure 8. The overall structure of the control system.

The subsoiling unit is illustrated in Figure 9, which includes the tractor, subsoiler, controller, onboard computer, GNSS antenna, receiver, and sensors. A New Holland T1404 tractor equipped with a power-shift gearbox with 18 forward gears and six reverse gears was used in the test. The forward gears are divided into sections A, B, and C. Electronic control technology is widely used. Reversing, shifting, and hitch are switch controls, and the throttle is an analogue control with a rated power of 104 kW and a rated speed of 2200 r/min. The tractor was equipped with a speed sensor, hitch position sensor, and hitch force sensor. The CAN supports the CAN bus protocol ISO 11783 for agricultural vehicles, where the CAN obtains real-time working information. Additionally, the controller was an MC1206 controller from Beijing Xinhai Rongke Technology Co., Ltd. (Beijing, China). The receiver was an AMG_PFZ202 GNSS receiver (Nongxin Technology Co. Ltd., Beijing, China).

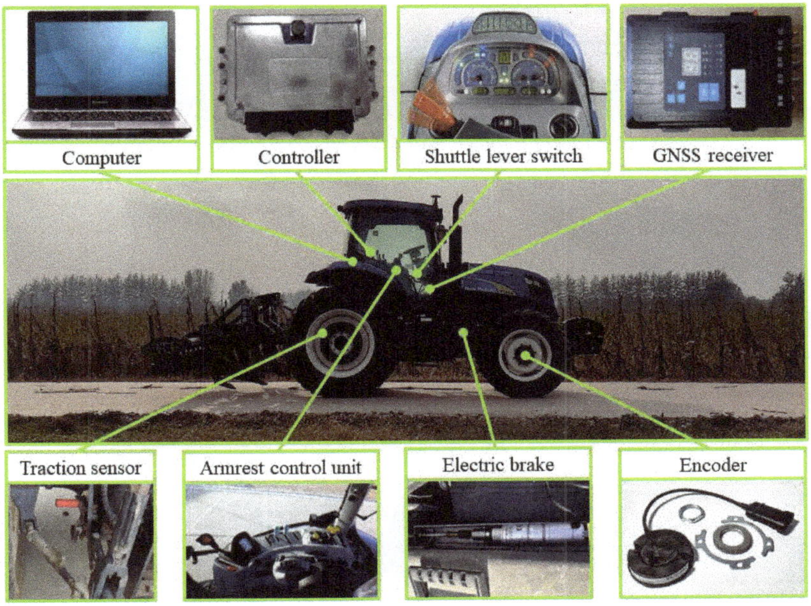

Figure 9. Subsoiling unit.

The control program is divided into the following parts: an initialisation module, a timer module, a digital-to-analogue conversion module, an algorithm access module, a serial communication module, and a CAN bus communication module. After the control system is started, the program is initialised, and then the working status information of the tractor is obtained through the CAN bus/serial communication module. When the control system receives the vehicle speed command, it first calculates the transmission gear and engine speed and determines whether to accelerate or decelerate. If acceleration is required, the system adjusts the throttle so that the engine speed reaches the target speed and then adjusts the engine gear to the target speed. If the vehicle speed is reduced, the gear should be changed directly to the target speed.

The control system adopted a proportional-integral-derivative (PID) algorithm to control the throttle of the tractor to accurately and stably adjust the target speed. After the vehicle speed is adjusted and stabilised, the control system judges the running status according to tractor information, such as vehicle speed and the horizontal traction of the hitch. If the wheel slip ratio of the tractor is above the set threshold value of 15%, the system activates the slip-ratio control module. When the control system detects that the tractor load is greater than the set threshold of 6500 N, the system determines that the traction resistance of the hitch mechanism is excessive and lifts the hitch mechanism to reduce the working depth of the subsoiler, protecting the tractor. A flowchart of the system and software is shown in Figure 10.

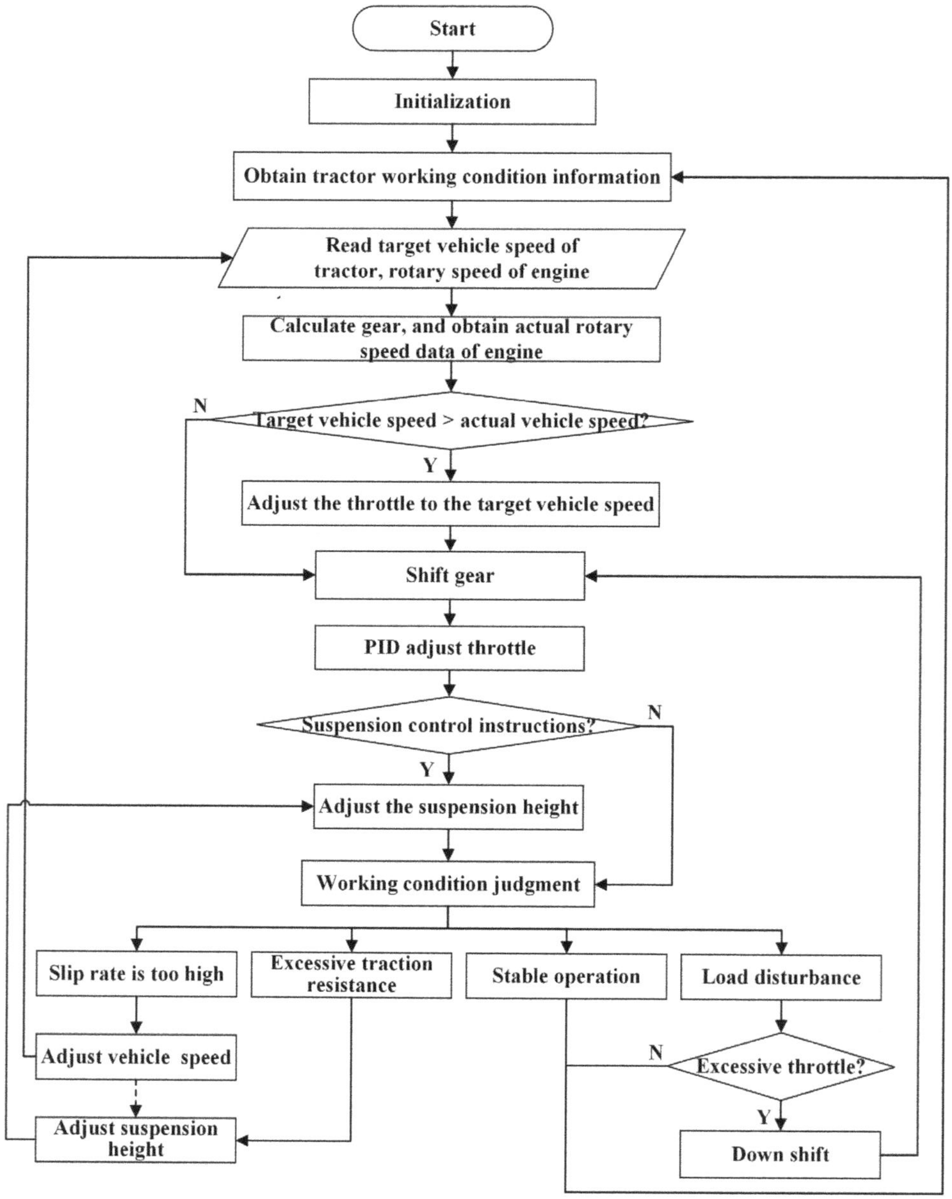

Figure 10. Software flowchart.

2.3. *Field Test*

2.3.1. Preparation for Test

- Wheel Speed Calibration Based on the CAN Bus

The proposed slip-rate control strategy considers the wheel speed as the primary control factor, which is read from the tractor CAN bus. If the error is significant, then the control effect and test results are significantly affected. Therefore, when the tractor runs at a constant speed, it records the time the tractor passes through the specified distance, calculates the actual wheel speed, and compares it with the wheel speed read from the CAN Bus.

The test site was a straight and level concrete road running from south to north, with a street light pole at each end. The road had good traction conditions, and the tractor wheels could be seen to be in a pure rolling motion. When the test was carried out, the tractor started timing when it passed the first pole and ended when it reached the second pole.

- Hitch Position-Tillage Depth Calibration

A calibration test was conducted in the field to determine the relationship between the tractor hitch position and tillage depth. The extension length of the three-point hitch piston rod of the tractor used for lifting the hydraulic cylinder corresponds to the rotation angle of the lower pull rod. The lower pull rod's rotation angle determined the tillage depth of the subsoiler. The rotation angle of the lower pull rod CAN was read by the CAN Bus tractor. The data range was 0–100%, which corresponded to the lowest position to the highest position of the hitch.

- Slip Rate Characteristics of the Subsoiling Unit under Different Wheel Speeds

This section verifies the influence of wheel speed on traction resistance, the relationship between wheel speed and slip rate under no-load conditions, and the influence of wheel speed on the slip rate during the subsoiling operation. This study explored the steady-state characteristics of the subsoiling units at different wheel speeds through field experiments. The experiment was divided into three parts:

1. Test of the Relationship between Wheel Speed and Traction Resistance

The depth of subsoiling was set to 30 cm, the target wheel speed was set to 3, 5, and 7 km/h in the same field with uniform soil conditions, and the data collected by hanging horizontal force sensors were recorded.

2. Test of the Relationship between Wheel Speed and Slip Rate under the No-Load Condition

Under the no-load condition, the tractor travelled in the same direction on ordinary land (average soil moisture content of 22.2%) and soft land with high soil moisture content (average soil moisture content of 30.5%). The target wheel speeds were set at 3, 5, and 7 km/h, respectively, and the slip rate was recorded.

3. Test of the Relationship between Wheel Speed and Slip Rate under Subsoiling Conditions.

The depth of subsoiling was set to 30 cm, the target wheel speed was set to 3, 5, and 7 km/h along the same direction in the same field with variable soil conditions, and the slip rate was recorded.

2.3.2. Field Experiments

Field experiments were conducted to verify the effectiveness of the strategy of controlling the slip rate by adjusting the wheel speed and tillage depth. The specific test was divided into two parts: the speed regulation effect verification test of the wheel speed control system and the slip rate control strategy verification test.

The field test was conducted at the Xiaotangshan National Experiment Station for Precision Agriculture, Changping District, Beijing, from 14 to 20 August 2021. The test sites were a 6 m × 480 m cement road and a 110 m × 320 m flat field. Before the experiment, the local area experienced a rainy season lasting for 1 month, the soil moisture content was relatively high, and the difference in terrain led to uneven soil moisture content. The slip rate of the tractor was high during the test, and the slip rate fluctuated significantly when the tractor passed through the field with high moisture content. There were a few weeds, some corn stalks had no set ears because of the weather, and the height of the stalks was approximately 1 m.

First, a verification test of the tractor wheel speed control method was conducted. To verify the primary function of the wheel speed control method, we performed an automatic gear shift according to the instructions and adjusted the speed regulation accuracy and efficiency of the throttle. The experiments were conducted under three conditions: no load in the cement ground, no load in the field, and subsoiling operation.

- Zero Vehicle Speed Starting Test

The tractor was set to the rear-wheel-drive mode, and the test conditions were selected for no load cement ground, no load field flat ground, and subsoiling operation. No implements were installed in the no-load test of the cement ground or the no-load test of the field flat ground. When the target speed of the engine was set to 1400 r/min, and the target wheel speed was set to 3, 5, and 7 km/h, the controller controlled the tractor to adjust from a static state to the target wheel speed and finally stabilised within a 5% error range of the target wheel speed.

- Target Wheel Speed Switching Test

Test conditions were selected for no load cement ground, no load field flat ground, and subsoiling operation. No implements were installed for the no-load test of the cement ground or the no-load test of the field flat ground. When the controller controls the tractor to run steadily at a target wheel speed of 3, 6, and 7 km/h, the vehicle-mounted computer sends a new target wheel speed of 5 km/h to the controller via a serial port. Then, the controller adjusts the tractor wheel speed to the new target value.

Conversely, a tractor slip rate control strategy verification test based on wheel speed and tillage depth was carried out. In the same field with unstable soil conditions, the subsoiling depth was set to 30 cm, and the target wheel speed was set to 5 km/h. The tractor unit performed subsoiling operations and recorded the relevant data in the same direction under the conditions of manual operation (constant wheel speed and manual hitch control) and automatic slip rate control (automatic wheel speed and tillage depth control).

3. Results and Discussion

3.1. Results of the Preparation Tests

3.1.1. Wheel Speed Calibration Results

The distance between the two lampposts was 89.7 m, the measured time was 81.53 s when the tractor travelled at 3.96 km/h (CAN bus data), the actual wheel speed was 3.96075 km/h, and the error was 0.00075 km/h. When the tractor moved at 5.04 km/h (CAN bus data), the measured time was 64.16 s, the actual wheel speed was 5.03304 km/h, and the error was 0.00696 km/h. The error was small, so the CAN bus wheel speed could be regarded as the actual wheel speed.

3.1.2. Three-Point Hitch Position–Tillage Depth Calibration Results

The left-lower, right-lower, and upper pull rod length was 98 cm, 100.5 cm, and 76 cm, respectively. The left and right lifting rod length was 85.5 cm and 87 cm, respectively. When the hanging pull rod was in different positions, the vertical distance between the subsoiling shovel tip and the soil surface was taken as the tillage depth, and the curve was plotted (Figure 11).

The corresponding relationship between engine speed and throttle opening was obtained from the fitting formula, as shown in Equation (3):

$$h = 0.00344x^2 - 0.75216x + 31.61615 \tag{3}$$

where h is the tillage depth in cm, and x is the hitch position in the range of 0–100%.

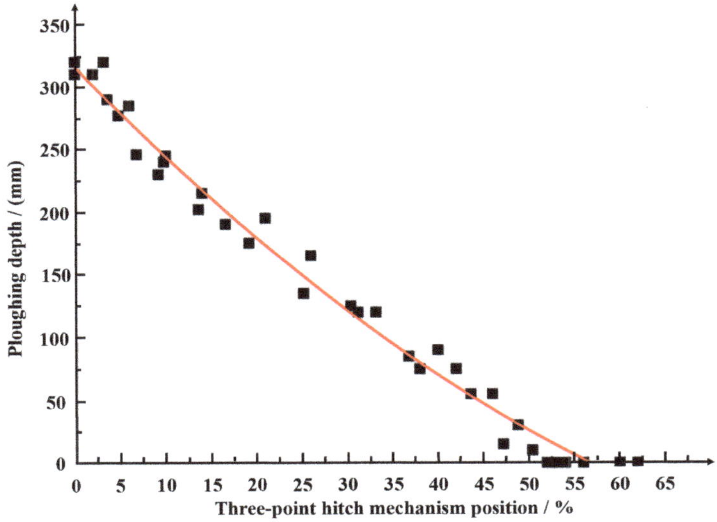

Figure 11. A fitting curve of hitch position–tillage depth.

3.1.3. Slip Rate Characteristics of the Subsoiling Unit under Different Wheel Speeds

The Results of the Wheel Speed-Traction Resistance Relationship

The traction resistance of the tractor subsoiling unit at different working speeds (wheel speeds) is shown in Figure 12. The data analysis results are presented in Table 2.

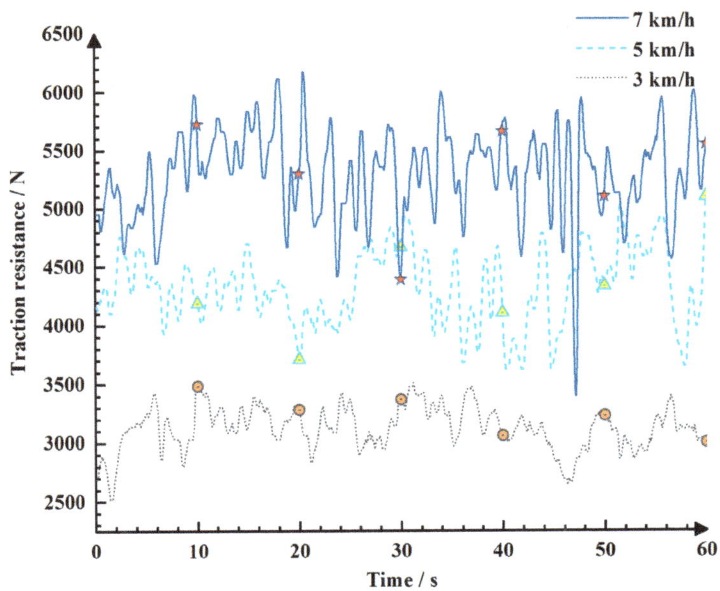

Figure 12. Traction resistance at different wheel speeds during subsoiling.

Figure 12 shows that the subsoiling resistance is stable in a specific range when the wheel speed is constant, and the subsoiling resistance increases with an increase in wheel speed. Table 1 shows that, during subsoiling operation, when compared with the 3 km/h operation, the average traction resistance of the tractor working at 5 km/h and 7 km/h increased by 1196 N and 2169 N, respectively, which increased by 38% and 69%, respectively. The maximum traction resistance increased by 1587 N and 2663 N, respectively, which increased by 45% and 76%, respectively.

Table 2. Traction resistance data analysis.

Wheel Speed (km·h^{-1})	3	5	7
Average traction resistance (N)	3125	4321	5294
Maximum traction resistance (N)	3510	5097	6173

- Relationship between Wheel Speed and Slip Rate under the No-Load Condition

The slip rates of the tractor subsoiling unit under different working speeds (wheel speeds) and different soils are shown in Figure 13. The data analysis results are presented in Table 3.

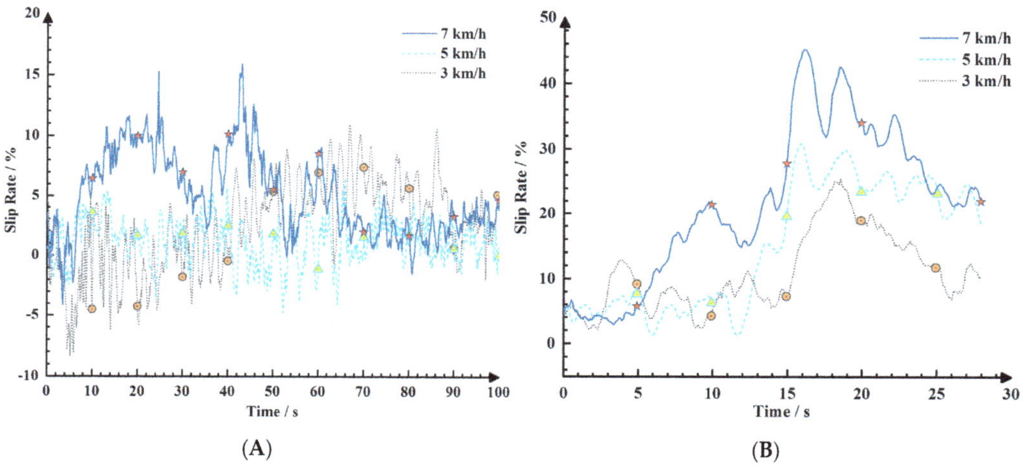

Figure 13. The slip rates of the tractor subsoiling unit at different wheel speeds during subsoiling: (**A**) ordinary field; (**B**) soft field.

Table 3. Slip rate at different wheel speeds without load.

Wheel Speed (km·h^{-1})		3	5	7
General road condition	Average slip rate (%)	3.28	3.52	5.39
	Maximum slip rate (%)	9.08	6.83	9.26
Soft road condition	Average slip rate (%)	10.99	15.19	21.60
	Maximum slip rate (%)	25.67	30.94	45.22

As shown in Figure 13A,B, the relationship between wheel speed and slip rate is more evident on soft road surfaces. In Figure 13B, from 13–25 s, the slip rate for the 7 km/h drive is >the slip rate of the 5 km/h drive, which is >the slip rate of the 3 km/h drive. When compared with the drive at 3 km/h, the average slip rate of the tractors driving at 5 km/h and 7 km/h on ordinary land increased by 0.24% and 2.11%, respectively, and the maximum value increased by 2.25% and 0.18%, respectively (Table 2). In the soft fields, the average slip rate increased by 4.2% and 10.61%, respectively, and the maximum slip rate increased by 5.27% and 19.55%, respectively.

Test Results of Wheel Speed–Slip Rate under Subsoiling Conditions

The slip rate of the tractor under different working speeds (wheel speeds) during the subsoiling operation is shown in Figure 14. The data analysis results are presented in Table 4.

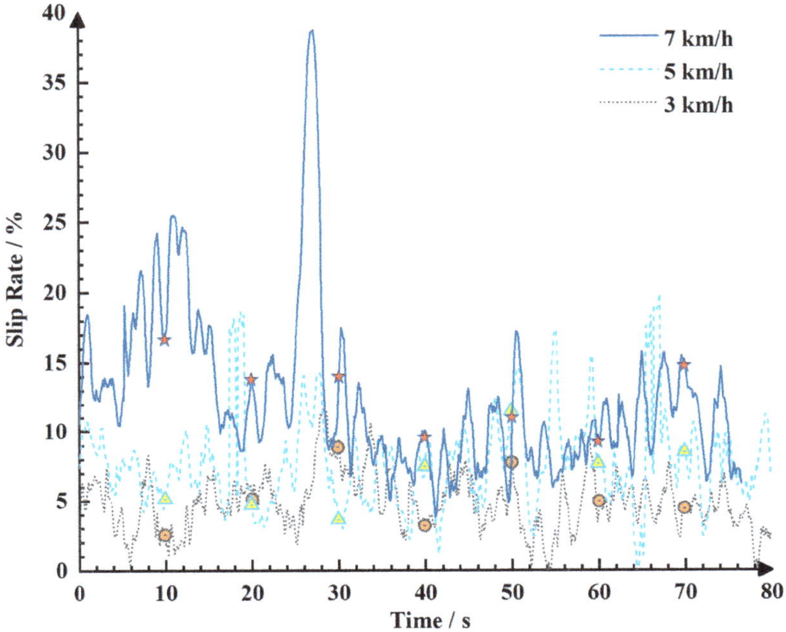

Figure 14. Slip rate at different wheel speeds during subsoiling operation.

Table 4. Slip rate data at different wheel speeds during subsoiling operation.

Wheel Speed (km·h^{-1})	3	5	7
Average slip rate (%)	4.85	7.86	12.41
Maximum slip rate (%)	11.72	19.92	38.76

The slip rate decreased at 7 km/h, 5 km/h, and 3 km/h (Figure 14). In the subsoiling operation, when compared with the 3 km/h operation, the average slip rate of the drive wheels operating at 5 km/h and 7 km/h wheel speeds increased by 3.01% and 7.56%, respectively. The maximum slip rate increased by 8.2% and 27.04%, respectively (Table 4).

In summary, when the depth of subsoiling is constant, the traction resistance of the hitch and slip rate of the driving wheel increase with an increase in the wheel speed. There is a directly proportional relationship between the wheel speed and slip rate in a no-load tractor. This relationship was evident when the ground adhesion conditions were poor but not when they were good.

3.2. Results of Wheel Speed Control Tests

3.2.1. Zero Vehicle Speed Starting Test

When the target engine speed was 1400 r/min, the tractor wheel speeds reached the target wheel speeds of 3 km/h, 5 km/h, and 7 km/h (Figure 15).

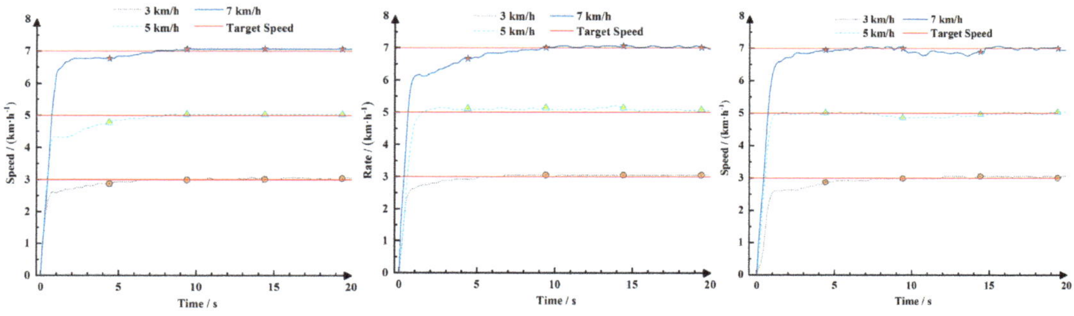

(A) (B) (C)

Figure 15. Parking start test under three working conditions: (**A**) no-load test on a cement floor; (**B**) no-load test in flat fields; (**C**) subsoiling operation.

The wheel speed increased rapidly during the clutch engagement stage (Figure 15). The throttle was adjusted after the clutch was fully engaged so that the wheel speed gradually approached the target wheel speed and stabilised at the target value. The absolute error was 0.23 km/h, the relative error was 3.3%, and the maximum adjustment time (the sum of the clutch engagement time and throttle adjustment time) was 4.4 s under the three working conditions (Table 5). The shift time was not shown in the chart; the average single shift time was 1.5 s, according to the data analysis.

Table 5. Analysis of experimental data of 0-speed starting tests.

Wheel Speed (km·h^{-1})	Working Condition	Absolute Error (km·h^{-1})	Relative Error (%)	Accommodation Time (s)
3	No-load test on a cement floor	0.06	2.0	3.7
	No-load test in flat fields	0.06	2.0	2.7
	Subsoiling operation	0.07	2.3	4.4
5	No-load test on a cement floor	0.04	0.8	4.3
	No-load test in flat fields	0.15	3.0	1.2
	Subsoiling operation	0.14	2.8	1.1
7	No-load test on a cement floor	0.07	1.0	1.7
	No-load test in flat fields	0.13	1.9	4.2
	Subsoiling operation	0.23	3.3	1.8

3.2.2. Target Wheel Speed Switching Test

The adjustment of the tractor wheel speed when the target wheel speed was switched from 3 km/h, 6 km/h, and 7 km/h to 5 km/h is shown in Figure 16.

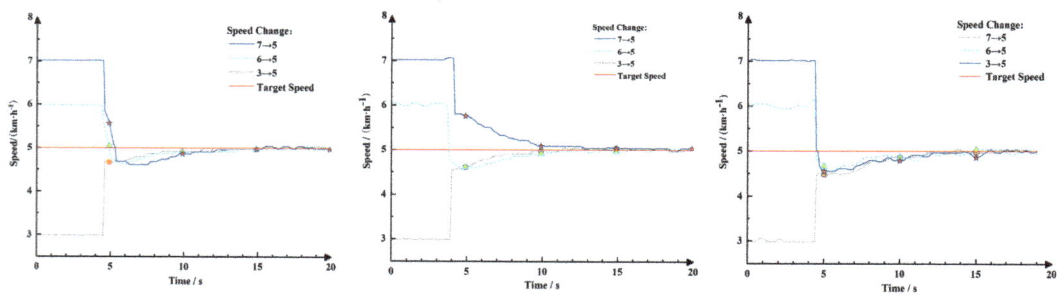

(A) (B) (C)

Figure 16. Wheel speed switching experiment under three working conditions: (**A**) no-load test on a cement floor; (**B**) no-load test in flat fields; (**C**) subsoiling operation.

The wheel speed was adjusted for 4.5 s and stabilised at the new target value at approximately 10 s (Figure 16). The adjustment curves of 3–5 km/h, 6–5 km/h, and 7–5 km/h under the three working conditions were roughly the same. The absolute error was 0.14 km/h, the relative error was 3%, and the maximum adjustment time (the sum of the shift time and throttle adjustment time) was 4.4 s after switching the new target wheel speed to a steady state under the three working conditions (Table 6). The test results show that the speed regulation accuracy of the wheel speed control method was 0.23 km/h, the relative error was 3.3%, and the maximum regulation time was 4.4 s, which meets the production requirements and shows high stability under various working conditions.

Table 6. Analysis of speed switching experimental data.

Wheel Speed (km·h^{-1})	Working Condition	Absolute Error (km·h^{-1})	Relative Error (%)	Accommodation Time (s)
3	No-load test on a cement floor	0.09	3.0	2.0
	No-load test in flat fields	0.04	1.3	2.0
	Subsoiling operation	0.07	2.3	4.3
6	No-load test on a cement floor	0.04	0.8	2.6
	No-load test in flat fields	0.09	1.8	3.4
	Subsoiling operation	0.09	1.8	2.8
7	No-load test on a cement floor	0.04	0.6	4.4
	No-load test in flat fields	0.06	0.9	4.0
	Subsoiling operation	0.23	3.3	1.8

3.3. Results of Slip Rate Control Tests

The verification test results of the slip rate control method are shown in Figure 17.

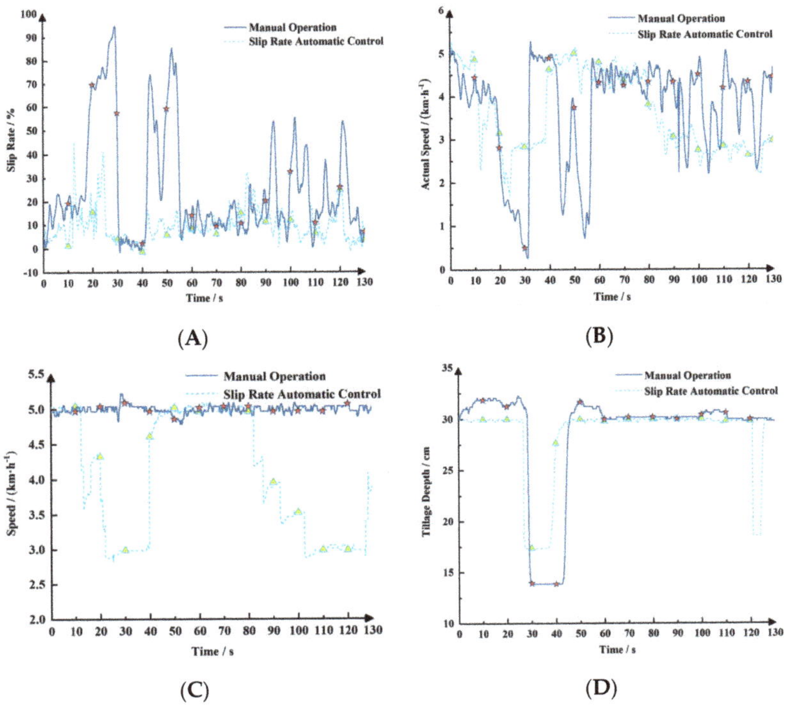

Figure 17. Comparison test between manual operation and automatic control of the slip rate: (**A**) comparison of slip rate test data; (**B**) comparison of ground velocity test data; (**C**) comparison of wheel speed test data; (**D**) comparison of tillage depth test data.

When the slip rate was automatically controlled, it increased to more than 15% in 11.1 s (Figure 17). When the speed regulation strategy was started, the target wheel speed gradually decreased, and the ground speed and slip rate decreased. The target wheel speed also dropped to the lower limit of 3 km/h in 22.0 s. At 23.3 s, the slip rate increased again and exceeded the threshold value, and an automatic ploughing depth control strategy was activated. At 26.8 s, the hitch was lifted, and the slip rate was reduced. The hitch was lowered to 36.8 s. At 43.6 s, the target wheel speed recovered to 5 km/h. At 43.6–82.0 s, the tractor operated normally at a wheel speed of 5 km/h. The automatic slip rate control strategy was repeated at 82.0–130.0 s. In Figure 17B, the average ground speed under manual driving is shown; the slip rate automatic control modes were 3.74 km/h and 3.61 km/h, respectively, and the actual operating efficiency was reduced by only 3.4%. The slip rate control test data are listed in Table 7.

Table 7. Slip rate control test data.

Data	Manual Operation	Automatic Slip Rate Control
Average slip rate (%)	25.9	9.8
Maximum slip rate (%)	94.8	45.4
Time when slip rate exceeds 20%	47.0	14.1

The average and maximum slip rates under manual driving and automatic slip rate control were 25.9% and 94.8%, and 9.8% and 45.4%, respectively (Table 7). During a sampling time of 130 s, the time at which the slip rate under manual driving and automatic control exceeded 20% was 47.0 s and 14.1 s, respectively, which account for 36.2% and 10.8%, respectively. Under the combined control of wheel speed and tillage depth, the average slip rate decreased by 16.1%, the maximum slip rate decreased by 49.4%, and the time when the slip rate was higher than 20% decreased by 25.4%.

In summary, the test results show that when the slip rate exceeds the threshold value, the controller can implement an automatic control strategy for the slip rate and, in turn, adjust the wheel speed and hitch height. When compared with manual operation, the time when the slip rate exceeded 20% under a sampling time of 130 s was reduced by 25.4%. The effectiveness of the combined control strategy of wheel speed and tillage depth was demonstrated.

4. Conclusions

Currently, the primary control mode for tractor slip rate is the automatic control of tillage depth, which uses an electro-hydraulic hitch as the control object and only controls the slip rate by adjusting the hitch height. These methods do not consider the influence of operation speed on the slip rate, leading to varying tillage depths and reducing the quality of tillage and subsoiling operations. Therefore, we proposed a slip rate control method based on wheel speed and tillage depth and used the New Holland T1404 power shift tractor to verify the algorithm. When compared to single-factor adjustment (automatic control of tillage depth), this method can ensure the greatest extent of operational quality (tillage depth).

Typical working conditions include cement ground with no load, flat field ground with no load, and subsoiling. Field experiments were conducted to verify that the wheel speed control method could ensure that the tractor accurately controlled the wheel speed under three working conditions: no load on the cement ground, no load in the field, and subsoiling operation. For the subsoiling operation, the slip rate control method could gradually reduce the tractor wheel speed when the slip rate of the tractor's drive wheel was too high until it met the requirements. When wheel speed was adjusted to the lower limit, suspension control was performed to reduce the tillage depth and improve vehicle trafficability. In the 130 s validation test, the time for the tractor with the slip rate control function to have a wheel slip rate exceeding 20% was only 14.1 s, which was 25.4% lower than that of the tractor without this function. In summary, the control method proposed in this study controls slip rate within the optimal range while ensuring operation quality (tillage depth) to the greatest extent.

In addition, the strategy of controlling slip rate by adjusting the two factors proposed in this paper verifies that wheel speed can be used as an essential factor in slip rate control and is not organically combined with the regulation of tillage depth. Therefore, in future research, we will determine the combination mode and prioritise tillage depth and wheel speed regulation by identifying changes in slip rate. Moreover, we will further optimise the control system of the power-shift tractor designed in this study.

Author Contributions: Conceptualization, C.L. and C.Z.; methodology, C.L. and C.Z.; software, Z.M. and W.F.; validation, C.W.; resources, Z.M. and C.Z.; data curation, G.L.; writing—original draft preparation, C.L. and C.W.; writing—review and editing, C.L., C.W. and H.L.; supervision, W.F.; project administration, W.F.; funding acquisition, Z.M. All authors have read and agreed to the published version of the manuscript.

Funding: This research was funded by the National Key Research and Development Plan of China (Grant No. 2022YFD2002401), China Postdoctoral Science Foundation (Grant No. 2022M720490) and the Major Science and Technology Projects of Inner Mongolia Autonomous Region of China (Grant No. 2021ZD0003).

Data Availability Statement: All data are presented in this article in the form of figures or tables.

Acknowledgments: We would like to thank the "College of Mechanical and Electrical Engineering, Shihezi University", "Research Center of Intelligent Equipment, Beijing Academy of Agriculture and Forestry Sciences", "AgChip Science and Technology (Beijing) Co., Ltd." and "National Engineering Research Center of Intelligent Equipment for Agriculture".

Conflicts of Interest: The authors declare no conflict of interest.

References

1. Wen, C.; Xie, B.; Li, Z.; Yin, Y.; Zhao, X.; Song, Z. Power density based fatigue load spectrum editing for accelerated durability testing for tractor front axles. *Biosyst. Eng.* **2020**, *200*, 73–88. [CrossRef]
2. Wen, C.; Xie, B.; Song, Z.; Yang, Z.; Dong, N.; Han, J.; Yang, Q.; Liu, J. Methodology for designing tractor accelerated structure tests for an indoor drum-type test bench. *Biosyst. Eng.* **2021**, *205*, 1–26. [CrossRef]
3. Matache, M.; Voicu, G.; Cardei, P.; Vladut, V.; Persu, C.; Voicea, I. Accelerated test of MAS 65 deep soil loosening machine frame. In Proceedings of the 43rd International Symposium on Agricultural Engineering, Opatija, Croatia, 24–27 February 2015; pp. 131–139. [CrossRef]
4. Croitoru, S.; Vladut, V.; Marin, E.; Matache, M.; Dumitru, I. Determination of subsoiler traction force influenced by different working depth and velocity. In Proceedings of the 15th International Scientific Conference "Engineering for Rural Development", Latvia, Jelgava, 25–27 May 2016.
5. Wu, Z.; Xie, B.; Li, Z.; Chi, R.; Ren, Z.; Du, Y.; Inoue, E.; Mitsuoka, M.; Okayasu, T.; Hirai, Y. Modelling and verification of driving torque management for electric tractor: Dual mode driving intention interpretation with torque demand restriction. *Biosyst. Eng.* **2019**, *182*, 65–83. [CrossRef]
6. Lee, J.H.; Gard, K. Vehicle–soil interaction: Testing, modeling, calibration and validation. *J. Terramech.* **2014**, *52*, 9–21. [CrossRef]
7. Cuong, D.M.; Zhu, S.; Ngoc, N.T. Study on the variation characteristics of vertical equivalent damping ratio of tire–soil system using semi-empirical model. *J. Terramech.* **2014**, *51*, 67–80. [CrossRef]
8. Burt, E.C.; Bailley, A.C. Load and Inflation Pressure Effects on Tires. *Trans. ASABE* **1982**, *25*, 881–884. [CrossRef]
9. Simikić, M.; Dedović, N.; Savin, L. Power delivery efficiency of a wheeled tractor at oblique drawbar force. *Soil Tillage Res.* **2014**, *141*, 32–43. [CrossRef]
10. Porteš, P.; Bauer, F.; Čupera, J. Laboratory-experimental verification of calculation of force effects intractor's three-point hitch acting on driving wheels. *Soil Tillage Res.* **2013**, *128*, 81–90. [CrossRef]
11. Raheman, H.; Jha, S.K. Wheel slip measurement in 2WD tractor. *J. Terramech.* **2007**, *44*, 89–94. [CrossRef]
12. Pranav, P.K.; Pandey, K.P.; Tewari, V.K. Digital wheel slip meter for agricultural 2WD tractors. *Comput. Electron. Agric.* **2010**, *73*, 188–193. [CrossRef]
13. Lyne, P.W.; Meiring, P.A. Wheel Slip Meter for Traction Studies. *Trans. ASABE* **1977**, *20*, 238–242. [CrossRef]
14. Gupta, C.; Tewari, V.K.; Ashok Kumar, A.; Shrivastava, P. Automatic tractor slip-draft embedded control system. *Comput. Electron. Agric.* **2019**, *165*, 104947. [CrossRef]
15. Chancellor, W.; Zhang, N. Automatic Wheel-Slip Control for Tractors. *Trans. ASABE* **1987**, *32*, 17–22. [CrossRef]
16. Mirzaeinejad, H.; Mirzaei, M. A novel method for nonlinear control of wheel slip in anti-lock braking systems. *Control Eng. Pract.* **2010**, *18*, 918–926. [CrossRef]

17. Pranav, P.K.; Tewari, V.K.; Pandey, K.P. Automatic wheel slip control system in field operations for 2WD tractors. *Comput. Electron. Agric.* **2012**, *84*, 1–6. [CrossRef]
18. Pasillas-Lépine, W.; Loría, A.; Gerard, M. Design and experimental validation of a nonlinear wheel slip control algorithm. *Automatica* **2012**, *48*, 1852–1859. [CrossRef]
19. Zhang, S.; Du, Y.; Zhu, Z.; Mao, E.; Liu, J.; Shi, J. Integrated control method of traction & slip ratio for rear-driving high-power tractors. *Trans. CSAE* **2016**, *32*, 47–53. [CrossRef]
20. Soylu, S.; Carman, K. Fuzzy logic based automatic slip control system for agricultural tractors. *J. Terramech.* **2021**, *95*, 25–32. [CrossRef]

agronomy

MDPI

Article

Dual-Manipulator Optimal Design for Apple Robotic Harvesting

Zicong Xiong [1], Qingchun Feng [2,3,*], Tao Li [2,3], Feng Xie [2], Cheng Liu [2], Le Liu [4], Xin Guo [2,3] and Chunjiang Zhao [1,2,3]

[1] School of Mechanical Engineering, Guangxi University, Nanning 530004, China
[2] Intelligent Equipment Research Center, Beijing Academy of Agriculture and Forestry Sciences, Beijing 100097, China
[3] National Research Center of Intelligent Equipment for Agriculture, Beijing 100097, China
[4] Fruit Industry Service Center, Pinggu District Agriculture Bureau, Beijing 101200, China
* Correspondence: fengqc@nercita.org.cn; Tel.: +86-10-51503192

Abstract: In order to ensure canopy area coverage with the most compact mechanical configuration possible, this paper proposes a configuration optimization design method of dual-manipulator to meet the research and development needs of an apple-efficient harvesting robot using the typical tree shape of a "high spindle" in China as the object. A Cartesian coordinate dual-manipulator with two groups of vertically synchronous operations and a three-degree range of motion based on the features of the spatial distribution of fruits under a typical canopy of dwarf and close planting was designed. Two-stage telescoping components that can be driven by both gas and electricity are employed to ensure the picking robotic arm's quick response and accessibility to the tree crown. Based on the quantitative description of the working space and configuration parameters of the dual-manipulator, a multi-objective optimization model of the major configuration parameters is constructed. A comprehensive evaluation method of the dual-manipulator configuration based on the CRITIC–TOPSIS combined method is proposed. The optimal solutions of the lengths and elevations of upper and lower telescopic parts of the dual-manipulator and the distance from the mounting base of the outer frame of the dual-manipulator to the center of the tree trunk are determined, which are 1119.3 mm and 39.4°, 898.7 mm and 26°, 755.3 mm, respectively. The interaction between the configuration parameters of the dual-manipulator and its working area is then simulated and examined in order to verify the rationality of the optimum configuration settings. The results show that the optimal configuration of the dual-manipulator can fully cover the target working space, and the redundancy rate is 16.62%. The results of this study can be utilized to advance robotic fruit-picking research and development.

Keywords: picking robots; robotic configurations; dual-manipulator; design optimization

Citation: Xiong, Z.; Feng, Q.; Li, T.; Xie, F.; Liu, C.; Liu, L.; Guo, X.; Zhao, C. Dual-Manipulator Optimal Design for Apple Robotic Harvesting. *Agronomy* **2022**, *12*, 3128. https://doi.org/10.3390/agronomy12123128

Academic Editor: Shubo Wang

Received: 17 November 2022
Accepted: 8 December 2022
Published: 9 December 2022

1. Introduction

China leads the world in apple planting area and production, generating 55% of all apples [1], which is a crucial assurance for the secure supply of fruits for humans and a rise in farmers' incomes. However, in recent years, with the decrease in the agricultural labor force, the labor cost of apple planting and production has been rising, reaching about 66% of the total production cost [2]. In orchard production, the mechanization rate of fruit harvesting is the lowest, less than 3%, especially in the harvest season of short-term explosive employment demand, and the problem of "difficult and expensive employment" is particularly prominent. Because fresh apples need to have good appearance quality, developing robots with selective and accurate harvesting ability is expected to solve the problems faced by apple production at present and achieve the goal of alleviating labor intensity and improving harvesting efficiency. This research direction has been a wide concern in the world in recent years [3,4].

A picking manipulator is the key component of a picking robot to locate the position and posture of the picking gripper according to the growth posture of fruit [5], which directly determines the working space and efficiency of the picking robot. To guarantee proper functioning of the robot, it is crucial to combine the planting mode and production efficiency needs of a particular picking object, choose the configuration of the picking robotic manipulator and optimize the design of its structural parameters.

According to the number of manipulators that can be driven, there are currently two types of fresh-fruit-picking robots: single-manipulator configuration and multi-manipulator configuration. When choosing goods with a relatively narrow distribution area in greenhouses, such as strawberries, tomatoes, sweet peppers, etc., the single-manipulator picking robot's configuration is frequently utilized. The single-manipulator configuration is mainly articulated to adapt to the narrow operating space in the greenhouse and meet the harvesting requirements of mature fruits in specific areas. The "RUBION" automated damage-free strawberry-picking robot was created by Preter et al. [6] and has a customized five-DoF (degree of freedom) articulated mechanical arm. Its prototype can pick a strawberry in 4 s and is mostly employed in tiny working spaces, such as greenhouse scenes. Arad et al. [7] developed a sweet pepper harvesting robot named "SWEEPER", and its prototype includes a six-DoF articulated industrial manipulator model Fanuc LR Mate 200iD. The robotic manipulator of "SWEEPER" is installed on a movable trolley with a scissor lift mechanism and harvests sweet peppers within a 200–290 mm reachable three-dimensional area in front of the robot body. The average cycle time for harvesting a single fruit (including fruit localization, obstacle localization, visual serving, detaching fruit) is 24 s. The cherry tomato picking robot developed by Feng et al. [8] used a DENSO VS-6556 six-DoF articulated robotic manipulator to pick tomatoes in the 600–1200 mm area with a distance of 550 mm and put the fruits into the basket after picking. Its picking efficiency is 8 s in each cycle. The apple harvesting robot by Abundant Robotics [9] adopts a parallel mechanical arm and negative pressure air suction picking end, and the average picking efficiency is one apple per second, which greatly reduces manpower work.

For the fruits distributed in a tall tree canopy, the picking robot needs to have a larger working area and picking efficiency. The standard mechanical manipulator configuration used by the greenhouse fruit and vegetable picking robot is difficult to meet the needs, and the multi-manipulator configuration with parallel operation capability [10–14] has been continuously applied to picking robots [15].

Xiong et al. [16] developed an automatic strawberry-picking robot. The overall picking manipulator is composed of two three-DoF Cartesian coordinate picking robotic arms running on a single track and a self-developed and designed end. The machine vision system is used to detect the area of 1200 mm × 500 mm within 500 mm of the robot body for positioning and picking ripe strawberries. Working with both arms at the same time can improve the work efficiency by half, and the average picking time of a strawberry is 4.6 s. Williams et al. [17] developed a kiwifruit selective intelligent harvesting robot. The robot has four separate platforms, and each platform is installed with a three-DoF articulated harvesting robotic arm. Each robot arm independently picks kiwifruit cultivated in the pergola-style frame within a range of about 1.7–2.0 m from the ground and about 3 m wide. Compared with the single-arm picking cycle of 3.36 s per piece, the picking robotic arms on the four platforms work at the same time, and the average harvest time of kiwifruit is only 0.84 s per piece. Israeli company FFRobotics (*Gesher HaEts*, Israel) has developed an apple harvesting robot for apple orchards that adopts fence cultivation mode [18]. Three parallel tracks are arranged on each side of the prototype, and two picking robotic arms are installed on each track, a total of 12 three-DoF robots. The Cartesian coordinate picking robotic manipulator can simultaneously pick multiple apple tree crowns in the fruit rows on both sides of the fruit road. FFRobotics company claims that the picking efficiency of high-quality apples is 10 times higher than that of manual work. AGROBOT Robotics of Spain (Huelva, Spain) has developed a picking robot for strawberry high-ridge cultivation and frame cultivation [19]. Six independent parallel tracks are set on the main body, and each track runs four three-DoF Cartesian coordinate picking robotic arms (a total of 24 independent arms) working together on four rows of strawberries at the same time, which greatly improves the efficiency of strawberry picking. In view of this, a brief summary of the above-mentioned manipulator configuration is carried out, as shown in Table 1. By selecting representative manipulator configurations and products and analyzing their shortcomings, some ideas are provided for the configuration of this paper.

Table 1. A table with shortcomings in representative manipulators and products.

Classification of Existing Manipulators	Picking Object	Large-Scale Operation Area	Compact Structure	Representative Product	Shortcoming
Articulated manipulator	Greenhouse strawberry, sweet pepper	No	Yes	RUBION [7], SWEEPER [8]	Used for greenhouse fruits, which cannot cover a large working area
Parallel manipulator	Apple	Yes	Yes	Abundant Robotics [10]	The manipulator has complex configuration and difficult to design multiple parallel manipulators
Multi-Cartesian manipulator	Apple	Yes	Yes	FFRobotics [18]	Mainly aimed at the American vertical trellis cultivation mode

The optimization and determination of its structural characteristics are also a prerequisite for the design of the manipulator in terms of workspace coverage, dynamic control precision and cloud route planning depending on the requirements of various jobs. According to the research and development needs of the greenhouse cucumber picking robot, Feng Qingchun [20] et al. established the quantitative relationship of each arm length, installation position and picking space of a five-DoF articulated picking arm and selected the optimal arm length parameter, which effectively covered over 90.5% of the fruit distribution area. Using finite element virtual simulation technology, Sun Feng et al. [21] constructed a multi-objective optimization model according to the dynamic characteristics of six-DoF articulated manipulator joints and solved the optimal combination of manipulator structural parameters, which reduced the average angular velocity of each manipulator's joint to varying degrees. The maximum descending range can reach 9.3654% compared with that before optimization, which effectively improves the stability of the end position and posture of the manipulator. Zhao Jiangbo [22] et al. established a multi-objective optimization model with a linear weighting of three factors in order to comprehensively optimize the cooperative workspace, load capacity and end motion accuracy of the manipulator with two six-DoF joints and obtained the optimal structural parameters of the manipulator by using the particle swarm optimization algorithm. The cooperative workspace of the two arms increased by 9.5%, the end motion speed increased by 7.8%, and the load capacity of the mechanical arm increased by 6.1%.

For the apple-picking robot, the effective coverage of the working area of the tall tree canopy by the robot working space is a necessary condition for the design of the picking robotic manipulator. In this paper, the high-spindle-shaped apple tree widely planted in China is taken as the object, and the vertical parallel operation of the dual-manipulator picking mechanism is studied according to the spatial shape of the tree canopy. According to the relationship between the structural parameters of the manipulator and its working space, a multi-objective optimization model of the parameters of its key components is established. By solving the optimal configuration parameter combination, the purpose of covering the fruit picking area with a compact machine is achieved. On this basis, the optimal configuration is analyzed and verified by simulation experiments. This research can provide a design basis for research and development of apple-picking robots.

2. Design Specifications for Apple Harvesting Robot

2.1. Standard Spindle-Shaped Tree

In recent years, application of dwarf and close planting cultivation technology in standard apple orchards has become a hot trend in different apple-producing areas in China [23], which also laid a good foundation for intelligent and mechanized operation of orchard production. To improve the efficiency of orchard production management and facilitate mechanized operations, the spindle-shaped standard fruit tree shape is widely used in China [24], which has the characteristics of early fruiting, high land utilization and large yield. As shown in Figure 1a, the fruit branches are pruned in a standardized manner, and there are support rods behind the fruit trees, which are fixed and constrained by vertical equal-spaced steel wire ropes installed along the tree row so that about 25 branches of each fruit tree grow along the steel wire ropes. The crown is a high spindle shape. The row spacing of fruit trees is 3500 mm, the plant spacing is 1000 mm and the tree height is 3000–3500 mm. Fruit trees have a crown width between 1000 and 1500 mm during the fruiting season and a side thickness between 400 and 500 mm. A vertical fruit wall is created when the crowns of fruit trees next to one another touch and cross over one another. To collect fruit from either side of the crown, the robot alternately goes back and forth between the rows of fruit trees. (For an illustration, see Figure 1a,b).

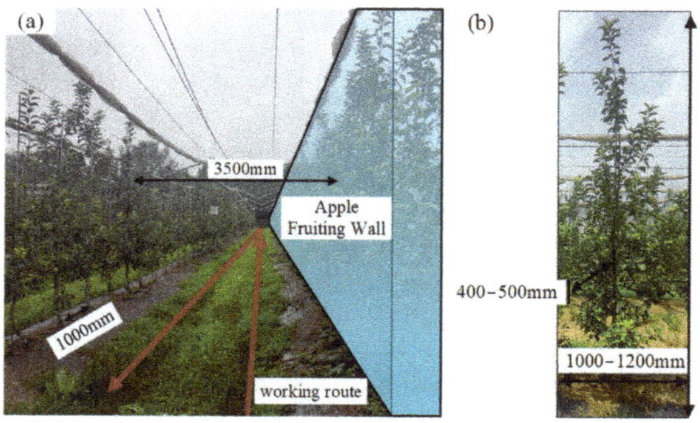

Figure 1. (**a**) Apple orchard with standard cultivation; (**b**) spindle-shaped canopy.

2.2. Definition of Harvesting Workspace

Constrained by the shape of spindle-shaped branches, the spatial distribution densities of fruits in the canopy vary (as shown in Figure 2a,b). The canopy of 3-year-old Fuji apple trees in northern China was divided into sampling blocks with length, width and height of 50 cm × 20 cm × 50 cm in different layers and orientations, and the sampling blocks with height less than or equal to 1000 mm from the ground belonged to the bottom of the canopy, 1000–2000 mm belonged to the middle of the canopy and more than or equal to 2000 mm belonged to the top of the canopy. Each layer is divided into the inner chamber (<200 mm away from the trunk) and outer chamber (>200 mm away from the trunk), with the trunk as the central axis. The number of fruits in each sampling block and canopy level was counted separately. It is obvious that the fruit density of different canopy layers of the apple high-spindle canopy gradually increases from top to bottom. The fruit density of the upper layer was lower, 5 fruits/m^2, and the density of the middle and lower layers was higher, 12 fruits/m^2 and 13 fruits/m^2. In general, the fruits in the canopy are mainly distributed at 1000 mm–2500 mm from the ground, accounting for about 95% of the total number of fruits.

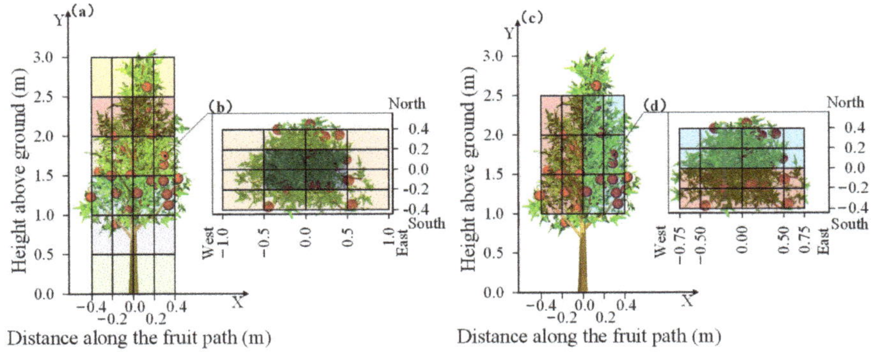

Figure 2. Fruit distribution statistics. (**a**) Front view of the canopy space; (**b**) top view of a middle-level canopy space; (**c**) front view of the harvesting workspace: the red and blue areas indicate that the canopy is harvested on both sides; (**d**) top view of a middle-level harvesting workspace.

Because of the low density of fruit distribution at the edge of the canopy, combined with the compact design requirements of the robot picking manipulator, it is considered that the middle and lower areas of the canopy with high fruit density are the picking operation space, so the target space of picking operation is determined as follows: the trunk of the fruit tree as the center, a space with a length of 1500 mm, a width of 400 mm and a height of 1000–2500 mm from the ground (as shown in Figure 2c,d).

3. Dual-Manipulator Prototypes for Apple Harvesting

3.1. Mechanical Configuration Design

The design requirement of the picking robotic arm is to drive the picking claws to move between the tree crown and the robot body according to the spatial position of the fruit in the robot's field of view to separate the fruit from the branches and put it into the fruit collection basket. For the standard apple tree shape as described above, the apple fruits are concentrated in the area of a tall and large cuboid. The Cartesian coordinate manipulator combined with the linear motion mechanism has a good match between the working space and the fruit growing area, so it has good applicability. In addition, unlike the manipulator with a tandem joint coordinate configuration, the Cartesian coordinate manipulator has no singularity problem. The current picking scene does not require the rotation of the manipulator, and the Cartesian coordinate manipulator can have a larger picking work area. To improve the working space and efficiency of the traditional Cartesian coordinate manipulator, this paper designs a dual-manipulator configuration arranged vertically in parallel to meet the needs of picking robots for fruit picking in tall tree canopies.

The dual-manipulator prototype (as shown in Figure 3a) primarily consists of two sets of picking mechanical arms aligned longitudinally and their respective independent driving components. Three degrees of freedom—horizontal, vertical and telescopic—are available for each picking arm. The mounting base of the outer frame is set as the origin O, the horizontal movement is parallel to the X-axis direction and the vertical movement is parallel to the Z-axis direction. Two sets of horizontal driving components travel up and down along the vertical driving components in turn, and its vertical motion driving mechanism is fixed to the picking robot body. The translating picking arms move along the corresponding horizontal drive members. The picking arm's telescoping movement is a two-stage driving system that consists of a pneumatic telescopic component and an electric drive servo component. The pneumatic telescopic part is driven by a cylinder, which is used to drive the picking gripper to move back and forth quickly in the non-picking area between the robot and the crown, improving the positioning efficiency of the robot to the picking end gripper. The electric drive servo component is driven by a servo motor, and the motion displacement and speed are precisely controllable (as shown in Figure 3b).

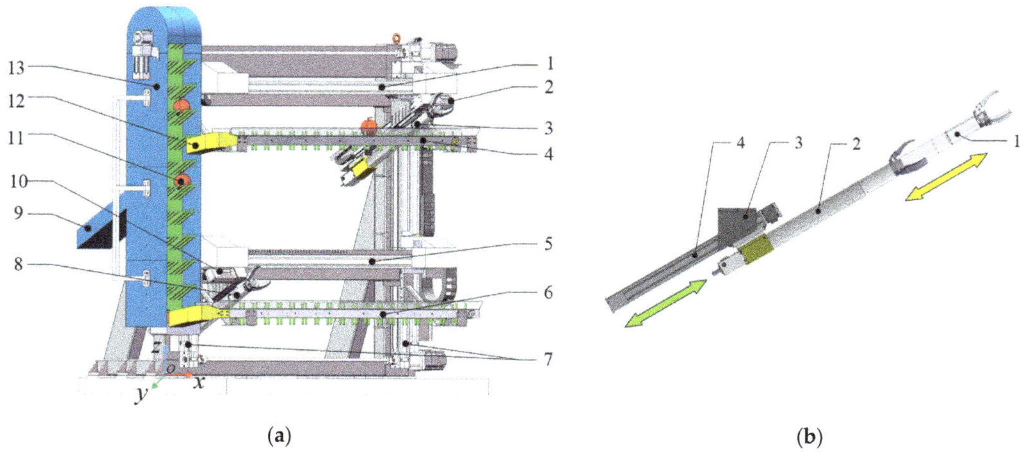

Figure 3. (**a**) Dual-manipulator configuration assembly. (1). Upper horizontal track (2). Vision sensor (upper) (3). Upper picking arm (4). Upper transverse conveying device (5). Lower horizontal track (6). Lower transverse conveying device (7). Vertical tracks (8). Lower picking arm (9). Back outlet of the vertical conveying device (10). Vision sensor (lower) (11). Harvested fruit (12). Transposed elbow (13). Vertical conveying device. (**b**) Two-stage telescopic wireframe diagram of the robotic arm (1). Picking gripper (2). Pneumatic telescopic parts (3). Adjustable angle connecting parts (4). Electric servo telescopic parts.

The picking robot controller receives the three-dimensional position data of the fruit in the canopy obtained by the vision sensor and divides the picking area of the dual-manipulator in real time. According to the close-up cameras installed on the telescopic parts, the two sets of robotic arms pick and collect the fruits in the picking area in turn. The picking gripper that performs the picking task has two degrees of freedom: rotation and opening and closing, which can "screw" the target fruit. The main material of the claws is made of aluminum alloy, which can ensure rigidity and light weight. The contact part of the fruit is made of flexible material, and it is coated on the aluminum alloy frame of the claw, which can well protect the surface of the picked fruit. Conveying devices for fruit collection tasks are divided into horizontal and vertical. Horizontal conveying comprises upper and lower transverse conveying devices, wherein the device is provided with a conveying channel, flexible baffles are arranged on both sides of the conveying direction and transposition elbows are installed at the conveying endpoint, which, respectively, follows the upper and lower horizontal track for synchronous lifting, collect fruits harvested by picking grippers and transport them to the vertical conveying device through the transposition elbows. The vertical conveying device is designed with two conveying passages, which, respectively, convey the fruits delivered by the upper and lower horizontal transposed elbows. The fruits flow out from the back outlet of the vertical conveying device and are collected by the universal fruit transfer frame.

3.2. Definition of Dual-Manipulator's Workspace

For the picking operation of spherical fruits, the picking robotic manipulator drives and positions the picking grippers with the three-dimensional coordinates of the center of the fruit as the target position. The working space of the picking manipulator refers to the set of spatial points that its end can reach due to its configuration and structure [25]. Obtaining the maximum coverage of the target working space with a compact configuration size is an important index to evaluate the configuration design of the manipulator. Therefore, establishing the relationship between the configuration parameters of the picking manipulator and its working space is the premise of the configuration optimization design. Especially, for the approximately rectangular canopy area, the accessibility of the picking robotic manipulator in the canopy height and depth directions is a quantitative representation of the picking robot's working space.

The picking manipulator and the canopy target working space are parameterized (as shown in Figure 4). The height of the mounting base of the picking robot arm from the ground (denoted by B) is 300 mm; taking the mounting base of the robotic arm as the origin O, the picking direction of the picking arm is the X-axis, the height direction of the robot body frame is the Z-axis and, on the XOZ plane, create a coordinate system. Assume that the depth of the target picking canopy is represented by W, the height of the upper limit position of the target picking canopy from the ground is represented by T and the height of the lower limit position of the target picking canopy from the mobile platform is represented by K.

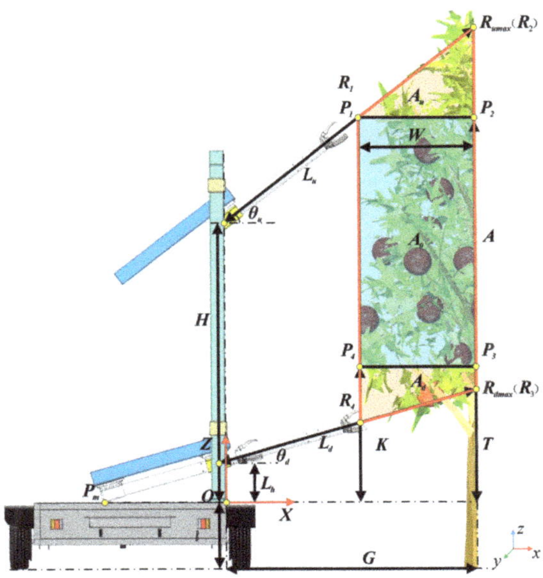

Figure 4. Dual-manipulator simplified configuration parameterization.

The quadrilateral region produced by extending the edge of the canopy to the center of the canopy at each elevation angle is designated as A on the XOZ plane when the double picking robotic arms are at the upper and lower height limit locations. On the XOZ plane, the projected area of the chosen target working space is A_0. The double-arm invalid picking space A_1 is the remaining space, and the predicted invalid picking areas for the robotic upper and lower picking arms are A_u and A_d, respectively. The center distance between the outer frame installation base and the trunk of the tree is G, and the maximum reachable height of the upper arm is H. Set the length of the upper picking manipulator arm as L_u and the inclination angle with the horizontal as θ_u. The length of the lower picking manipulator arm is L_d, and the inclination angle to the horizontal is θ_d. The total structure length L_m is the sum of H, L_u, L_d. When the lower picking arm is at the lower limit position, the height from the bottom of the mobile platform is L_h. The length β of the telescopic module in the rear section of each picking arm is set to 3/4 of the length of its associated picking arm (when the picking task is completed, the picking arm can be retracted into the frame as much as possible to improve the utilization rate of the internal space of the frame).

Taking the trunk of the tree as the center, the fruits on one side of the tree are distributed in the blue rectangular area. Four boundary points P_i ($i = 1, 2, 3, 4$) are the initial conditions, and the reachable space at the end of the picking robot arm is the red area R_j ($j = 1, 2, 3, 4$), the upper picking limit point at the end of the upper picking robot arm is $R_{umax}(R_2)$ and the lower picking limit point at the end of the lower picking robot arm is $R_{dmax}(R_3)$.

4. Optimization and Evaluation of Key Configuration Parameters

4.1. Two-Objective Optimization Model

Considering that the size of the component configuration directly determines the overall volume, load and stiffness of the manipulator, it is necessary to use the compact picking manipulator structure to cover the picking target area. In view of the fact that the working space of the picking manipulator is mainly determined by the length of the upper and lower picking arms and their elevation angle and is directly related to the overall size of the picking manipulator, this paper takes the configuration parameters as the parameters that need to be optimized.

As shown in Figure 4, when the upper and lower picking arms are at their respective extreme positions, this makes the R_{umax} point on the extension line of the P_2P_3 line R_{umax} coincide with point P_3 or be below point P_3, and the lower picking arm is at the bottom. When the limit position is retracted, the lower endpoint P_m of the rear section of the manipulator may interfere with the moving platform. In order to minimize the redundancy ratio γ of the workspace (i.e., the invalid workspace A_1) of the picking robot, an objective optimization function of "avoiding invalid workspace" should be established, which accords with the structural parameters of each picking arm and the conditions of three limit feature points: R_{umax}, R_{dmax} and P_m. Specifically:

$$\begin{cases} \delta(L_u, \theta_u, L_d, \theta_u, G) = A_u + A_d \\ A_u = \frac{1}{2}W \times [G \times \tan\theta_u + H - T] \\ A_d = \frac{1}{2}W \times [2K + (W - 2G) \times \tan\theta_d - 2L_h] \end{cases}$$

where $H = T - (G - W) \times \tan\theta_u$, $L_h = \beta \times L_d \times \sin\theta_d$, $\gamma = \frac{A_1}{A}$;

At the same time, to make the picking robot move more flexibly when performing the picking task, the configuration should be compact on the premise of meeting the requirements of the target picking space task of the dual-arm picking robot. Therefore, an objective optimization function is established to make the total structure length L_m small and achieve a "compact structure".

The three characteristic limit points R_{umax}, R_{dmax}, P_m and the proposed objective optimization function are used to limit the optimization conditions and set the parameter change interval of the distance between the outer frame mounting base and the center of the tree trunk as G as $[G_{io}, G_{ic}]$; the parameter variation interval of the upper picking arm length L_u and the lower picking arm length L_d is $[L_{io}, L_{ic}]$; the parameter variation intervals of the inclination angle θ_u between the upper arm and the horizontal and the inclination angle θ_d between the lower arm and the horizontal are $[\theta_{io}, \theta_{ic}]$. Reasonably limit the range of these five design variables so that the two objectives of "avoiding invalid workspace" $\delta(L_u, \theta_u, L_d, \theta_u, G)$ and "compact structure" $\varphi L_u, \theta_u, L_d, \theta_d, G$ can be minimized when meeting the picking requirements, namely:

$$\min\begin{cases} \delta(L_u, \theta_u, L_d, \theta_d, G) \\ \varphi(L_u, \theta_u, L_d, \theta_d, G) \end{cases}$$

$$s.t.\begin{Bmatrix} L_{io} \le L_u, L_d \le L_{ic}; \theta_{io} \le \theta_u, \theta_d \le \theta_{ic}; \\ G_{io} \le G \le G_{ic}; L_u, L_d, \theta_u, \theta_d, G \in 0.1 \times N; \end{Bmatrix}$$

The multi-objective optimization equation in the above formula is solved by the non-dominated sorting genetic algorithms (NSGA-II) [26,27]. NSGA-II is a genetic algorithm method that uses fast non-dominated sorting and elite mechanism to solve multi-objective optimization problems. The implementation steps are as follows:

(1) Taking $a_w = \{1,\ 1,\ 1,\ 1\}$ as the corresponding fusion weighted initial values of images in different wavebands, adding random numbers $r \in (-1, 1)$ to each coefficient to generate an initial population of fusion coefficients and performing fast non-dominant sorting on the initial population; (2) carrying out the binary selection, crossover and variation on the initial population to obtain a new population; (3) merging the new population with the initial population into a new population, performing non-dominant ranking on all individuals and calculating the crowding degree in the non-dominant set; (4) when generating a proper number of first-generation populations, the individuals with low non-dominant level are preferred, and, when they are at the same level, the individuals with high crowding distance are preferred to ensure individual diversity; (5) if the evolutionary algebra reaches 1000, the algorithm is stopped and the contemporary population is taken as the optimal Pareto solution set. Otherwise, go to step (2) to continue the loop. In order to better understand the NSGA-II algorithm, its flowchart is as follows (Figure 5).

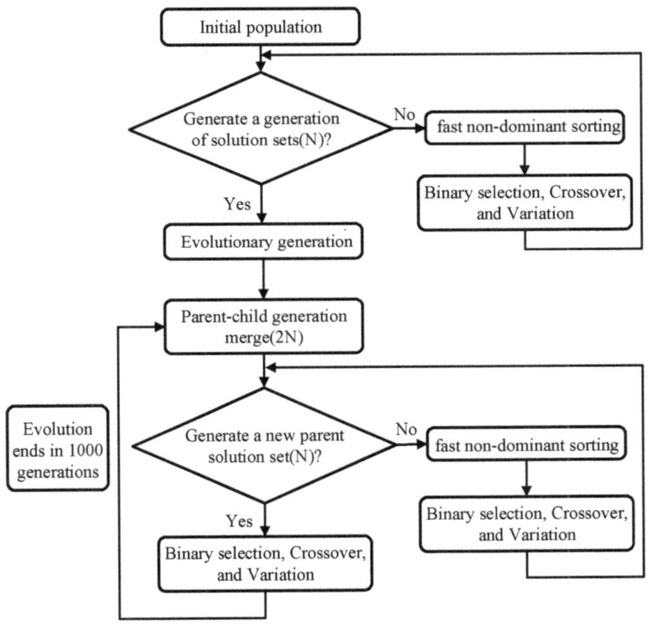

Figure 5. NSGA-II algorithm flowchart.

4.2. Optimal Solution Selection Based on CRITIC–TOPSIS

Because multi-objective optimization problems often obtain a solution set consisting of many non-inferior solutions (effective solutions), it is necessary to combine the actual needs to determine the optimal objective solution. The TOPSIS method [28] (technique for order preference by similarity to an ideal solution method) is an effective method for multi-objective decision analysis of finite programs. It obtains a comprehensive evaluation of the individuals in the target solution set by calculating the distance between the evaluation-specific solution and the corresponding positive and negative ideal targets. In particular, considering the difference in volatility and correlation of $\delta(L_u, \theta_u, L_d, \theta_d, G)$, $\varphi(L_u, \theta_u, L_d, \theta_d, G)$ solution sets (hereinafter referred to as A_1, L_m), their contributions to the optimal results are different. This paper introduces the CRITIC weight method [29], which normalizes the solution sets of the two optimization objectives and then weighs and evaluates them. The calculation steps are:

(1) Pareto solution sets have different schemes. The solution sets of A_1, L_m are composed of decision matrix S:

$$S = \left(S_{ij}\right)_{n\times 2}\ (i = 1,\ 2, \cdots, n; j = 1,\ 2)$$

(2) To eliminate the influence of dimensions on the evaluation results, each element in the decision matrix is standardized and both objectives are reverse indicators:

$$\beta^*_{ij} = \frac{S_{jmax} - S_{ij}}{S_{jmax} - S_{jmin}}$$

where S_{jmax}, S_{jmin} are the largest and smallest elements in column j.

(3) Calculate the standard deviation of each optimization objective in the decision matrix as follows:

$$D_j = \sqrt{\frac{\sum_{i=1}^{n}\left(S_{ij} - X_j\right)^2}{n-1}}$$

where X_j is the average value of the elements in column j, and D_j is the standard deviation of the indicators in column j, which is used to express the difference and fluctuation of the values within each indicator. If the standard deviation of A_1 solution set is larger, it reflects that the numerical difference of A_1 solution set is larger, more information can be reflected and the evaluation intensity of index A_1 is stronger. When assigning weights to A_1, more weights will be distributed, which also means that this goal has a greater influence on the selection of the whole optimal goal solution.

(4) Calculate the correlation and conflict quantitative index values among optimization objectives:

$$r_{ij} = \frac{\sum_{i}^{n}\left(S_{ij} - X_j\right)\left(S_{ik} - X_k\right)}{\sqrt{\sum_{i=1}^{n}\left(S_{ij} - X_j\right)^2 \sum_{i=1}^{n}\left(S_{ik} - X_k\right)^2}}$$

$$R_j = \sum_{j=1}^{2}\left(1 - r_{ij}\right)\ (i = 1, 2, \cdots, n; j = 1; k = 2)$$

where r_{ij} is the correlation coefficient between the i-th index and the j-th index, and j is the conflict between the j-th decision index and other indexes. The smaller the value of R_j is, the smaller the conflict is, which means that the information of the optimization goal is repeated too much in the solution set and the smaller the contribution to determining the optimal solution.

(5) The weight of each optimization target is quantified after comprehensively measuring the contrast intensity and conflict among indexes, and its comprehensive information amount C_j is calculated as follows:

$$C_j = D_j \sum_{j=1}^{2}\left(1 - r_{ij}\right) = D_j \times R_j$$

where C_j is the amount of information contained in the solution set of the j-th optimization objective. The greater the value of C_j, the greater the contribution of the j-th objective to the determination of the optimal solution and the greater its corresponding weight.

(6) The weights ω_j^* of the two optimization objectives are calculated by using the obtained information C_j:

$$\omega_j^* = \frac{C_j}{\sum_{j=1}^{2} C_j}$$

(7) Each element in the decision matrix is weighted to obtain a weighted decision matrix T:

$$T = (t_{ij})_{n\times 2}\ (i = 1,\ 2,\ \ldots, n; j = 1,\ 2)$$

where $t_{ij} = \beta^*_{ij} \times \omega^*_j$.

(8) The ideal optimal solution $V_j^+ = \underset{1\le i\le n}{max}\ T_{ij}$ and the worst solution $V_j^- = \underset{1\le i\le n}{min}\ T_{ij}$ are obtained according to the weighting matrix T, and the distances P_i^+, P_i^- of each element t_{ij} of the weighting matrix T to V_j^+, V_j^- are calculated:

$$P_i^+ = \sqrt{\sum_{j=1}^{2} (t_{ij} - V_j^+)^2}\ P_i^- = \sqrt{\sum_{j=1}^{2} (t_{ij} - V_j^-)^2}\ (i = 1, 2, \cdots, n)$$

The approximation degree between the i-th solution and the ideal optimal solution is expressed as K_i, and the value of K_i is in the interval of (0, 1). The more K_i tends to 1, the closer the i-th scheme is to the optimal scheme:

$$K_i = \frac{P_i^+}{P_i^+ + P_i^-}$$

5. Test and Results

5.1. Test

According to the final solution of the configuration parameters of the dwarf and close planting apple orchard picking robot and the structural parameters of the manipulator, combined with the picking target workspace, the simulation verification of the picking working space of the robot is carried out.

The parameters of the picking operation area in the canopy of a typical fruit tree in Section 2.2 are substituted into the multi-objective optimization model. Input the standard orchard environmental working parameters and the actual task requirements of the dual-manipulator picking robot and set the numerical variation range of the relevant variables. Considering the movement limit range of the manipulator extending and retracting into the frame, and referring to the average thickness of the fruit tree crown on one side of 400–500 mm, the parameter interval of the length L_u of the upper arm and the length of the lower arm is set as 700–1500 mm. In particular, since the average height of the canopy from the ground is 1000–1200 mm, the lower arm must meet the premise of accommodating the target workspace. The parameter variation range of the inclination angle θ_d between the lower arm and the horizontal is 20–40°. The difference between the horizontal inclination angle of the upper arm and the horizontal inclination angle of the lower arm should not be too large, which will affect the lifting stroke of the horizontal linear tracks (components 1 and 5 in Figure 3a) and the picking of fruits in the middle area of the lifting stroke. Therefore, the parameter variation range of the inclination angle θ_u between the picking manipulator and the horizontal is set to be 20–50°. The row spacing of the ... is about 3500 mm. Due to the limitation of the width of the bottom of the car body ... position of the picking robot between the fruit rows when working, the parameter ... variation range of the distance G between the outer frame mounting base and the center of the tree trunk is 450–1000 mm.

$$s.t.\begin{cases} 700 \le L_u, L_d \le 1500; 20 \le \theta_u \le 50; 20 \le \theta_d \le \ldots \\ 450 \le G \le 1000; L_u, L_d, \theta_u, \theta_d, G \in 0.1 \times \ldots \end{cases}$$

In the parameter setting of the NSGA-II genetic algorithm ... performance of crossover probability and mutation probability affects the behavior and ... performance of the whole algorithm and even directly affects the convergence of the algorithm. The crossover probability is generally selected in the range of 0.9– ... the greater the crossover probability, the faster the new individuals will be produced ... when the crossover probability is too high, then there is a greater possibility that the genetic model will be destroyed, which will make the individual structure with high fitness quickly destroyed. However, if the crossover probability is too small, the whole algorithm will slow down or even stagnate. As far as the variation probability is concerned, the general value range is 0.01–0.1, and, if the value is too small, it is difficult to produce a new individual structure. If the value is too large, the genetic algorithm loses its constraint and becomes a random search algorithm.

Therefore, in this solution, the initial population size of the NSGA-II genetic algorithm is set as 50, the crossover probability is 0.9, the crossover distribution index is 20, the mutation probability is 0.04 and the maximum iterative generation is 1000 generations (it is recommended to set it to 20 times the number of design variables, and the number of variables in this paper is five). The objective optimization function and constraints are compiled and the Pareto solution set and the changes in various parameters are obtained as shown in Figure 6.

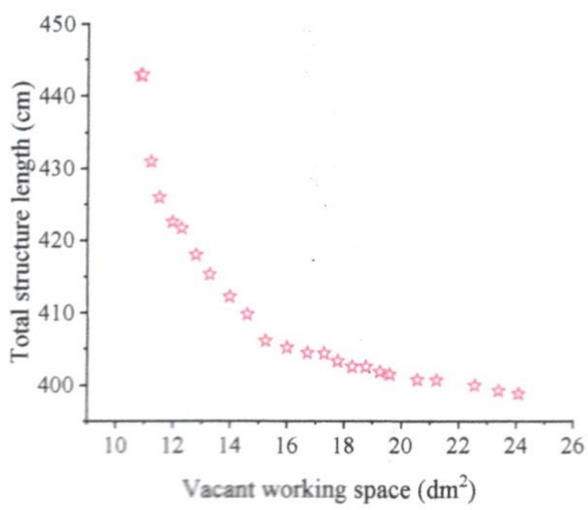

Figure 6. Pareto optimal solution set.

The weight of total structure length L_m is 23.4%, and the weight of dual-manipulator invalid picking space A_1 is 76.6% by objective weight calculation with the CRITIC method. The top 10 solutions in TOPSIS's sorting of Pareto solution sets are shown in the following Table 2. Finally, the first structural parameter is determined as the final optimal structural parameters, which are bold in Table 2.

Table 2. Top 10 Pareto optimal solutions after TOPSIS ranking.

Order ber	Design Variables				Design Objective			Combined Score Index (K_i)
	L_u (mm)	θ_u (°)	L_d (mm)	θ_d (°)	G (mm)	A_1 (dm²)	L_m (cm)	
	1119.3	**39.4**	**898.7**	**26.0**	**755.3**	**11.96**	**422.6**	**0.784**
	1119.1	42.2	881.7	26.1	753.1	12.78	418.1	0.783
	1132.3	40.5	887.1	26.1	753.6	12.28	421.7	0.780
	1115.7	42.5	858.9	26.0	750.3	13.26	415.4	0.775
	1137.2	38.3	902.4	26.1	754.9	11.49	426.0	0.773
	1129.1	44.4	838.5	26.1	751.6	13.97	412.3	0.751
	1137.2	37.2	942.4	25.6	755.7	11.20	431.0	0.747
9	1116.1	45.6	831.9	26.1	741.9	14.59	409.9	0.722
10	1111.4	48.2	832.1	26.3	740.3	15.23	406.2	0.693
	1155.9	36.5	1040.7	24.4	760.6	10.84	442.9	0.688

The solution with the highest comprehensive score index (Figure 7a) is the optimal configuration parameter for this optimization. The position of this solution in the Pareto optimal solution set is shown in Figure 7b.

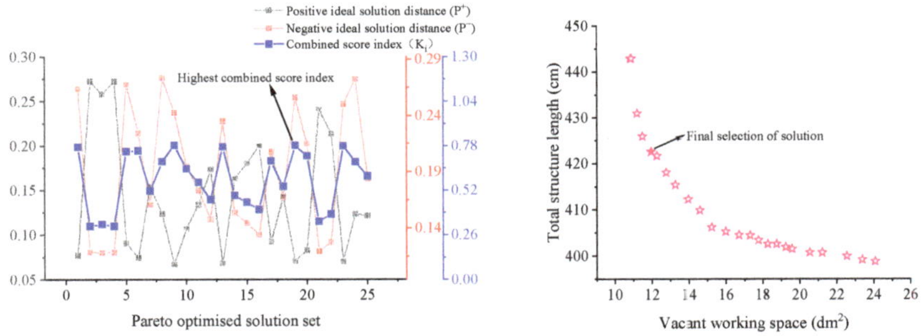

(a) (b)

Figure 7. (**a**) Composite score index (Ki) calculated by the Pareto optimal solution set by the weighted TOPSIS method; (**b**) the position of the solution obtained by the final decision in the Pareto optimal solution set.

By investigating the actual orchard working parameters, the orchard 3D simulation environment was drawn and the working space and target picking space of the picking robot were analyzed (as shown in Figure 8). The main section of the accessible picking working area formed by the upper limit position and the lower limit position of the picking robot arm can completely contain the target picking area, which can not only meet the requirements of the best recovery ratio of the picking robot but also reach a larger accessible area with a smaller structural length of the robot arm. The overall configuration is compact, which is suitable for working in an unstructured orchard environment.

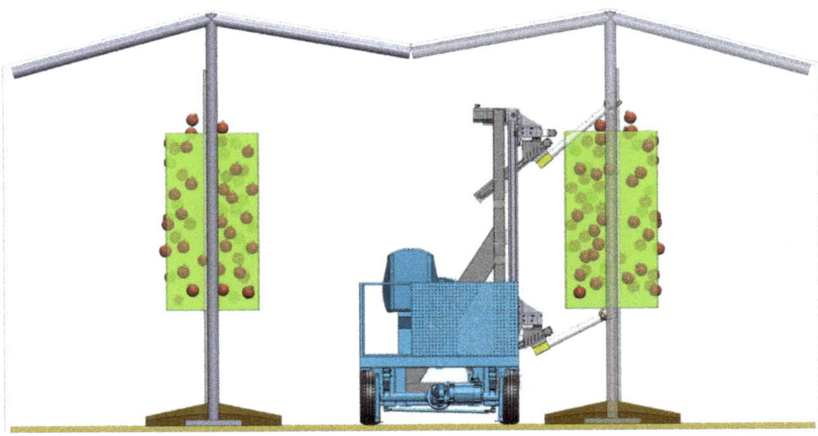

Figure 8. Optimization results of picking arm structure of dual-manipulator picking robot in orchard simulation environment.

5.2. Analysis of the Validity of the Optimal Parameters

In order to verify the unique validity of the optimal parameter solution, it is necessary to clarify the relationship between the optimization objective and each parameter. On the premise that the two-arm working space covers the target space, the optimal results of $\varphi(L_u, \theta_u, L_d, \theta_d, G)$ (i.e., L_m) and $\delta(L_u, \theta_u, L_d, \theta_d, G)$ (i.e., A_1) are set as fixed constant values C_{L_m} and C_{A_1}, respectively, to verify whether another optimization objective has a unique optimal solution within the numerical range of each configuration parameter, that is, to transform a multi-objective optimization into a single-objective optimization problem for comparison and verification. The curve of the result value of the optimization objective function $\min\delta(L_u, \theta_u, L_d, \theta_d, G)$ varying with each structural parameter is shown in Figure 9a–e.

$$\begin{cases} \min\delta(L_u, \theta_u, L_d, \theta_d, G) \\ \varphi(L_u, \theta_u, L_d, \theta_d, G) = C_{L_m} \end{cases} \quad \begin{cases} \delta(L_u, \theta_u, L_d, \theta_d, G) = C_{A_1} \\ \min\varphi(L_u, \theta_u, L_d, \theta_d, G) \end{cases}$$

$$s.t. \begin{Bmatrix} 700 \le L_u, L_d \le 1500; 20 \le \theta_u \le 50; 20 \le \theta_d \le 40; \\ 450 \le G \le 1000; L_u, L_d, \theta_u, \theta_d, G \in 0.1 * N; \end{Bmatrix}$$

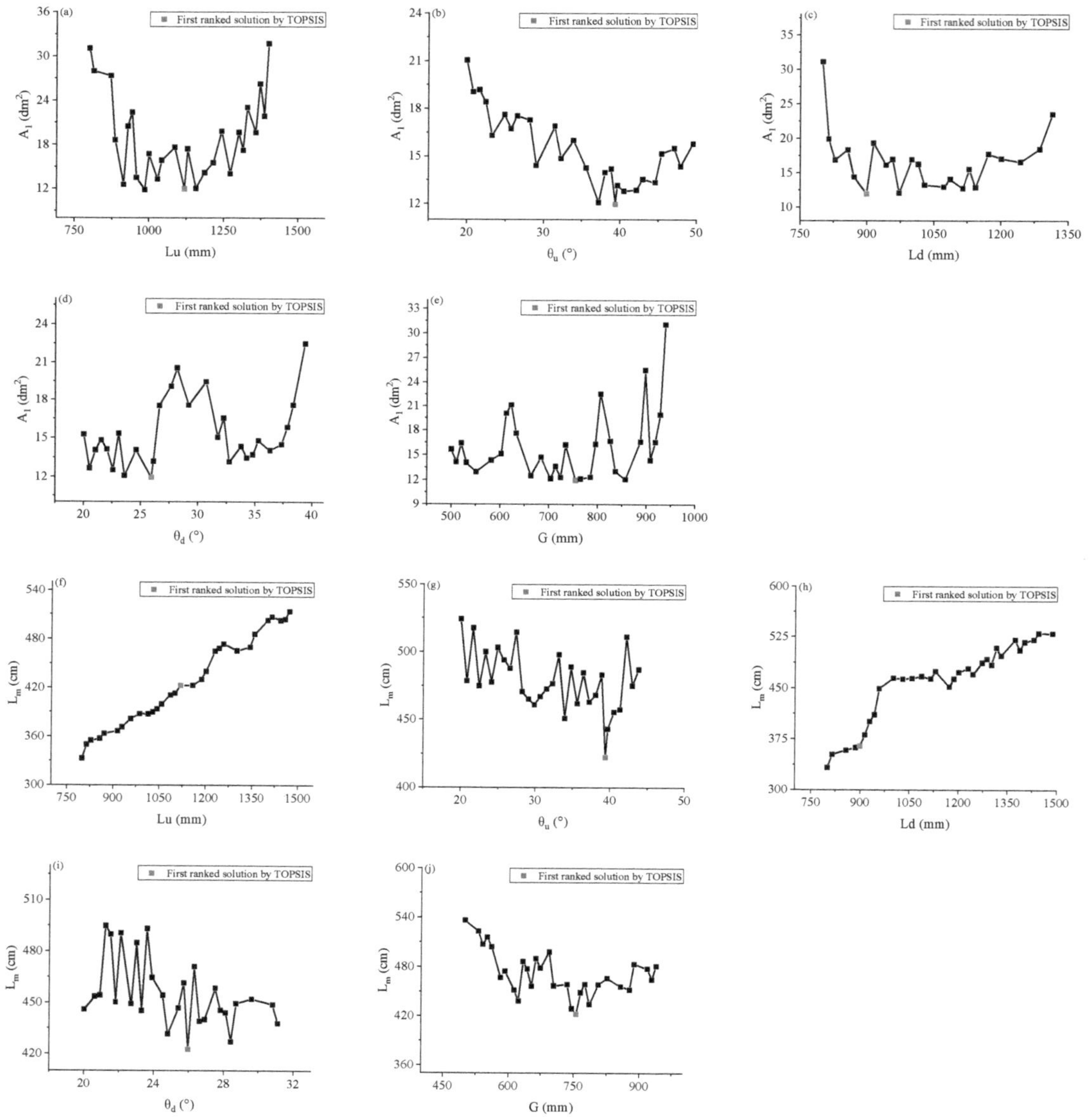

Figure 9. (**a–e**) The relationship between the structural parameters of the dual-manipulator and the optimization target A_1. (**f–j**) The relationship between the structural parameters of the dual-manipulator and the optimization target L_m.

It can be seen in Figure 9a that the projected area A_1 of the robot's invalid picking space increases with the length of the upper arm L_u, which presents an approximate "U"-shaped curve state. L_u is in the range of 1000–1200 mm, and there is a low point of the downward trend of A_1, which means that the optimal L_u is generated in the range of 1000–1200 mm. Different from Figure 9a, it can be seen in Figure 9b that, when the inclination angle θ_u between the upper arm and the horizontal increases, the projected area A_1 of the invalid picking space can be changed to about 40° at θ_u when there is a trend change inflection point that first descends and then rises, and the overall curve is similar to the letter "V" shape, which means that the optimal θ_u is generated within a small area of about 40°. Compared with the curve change in Figure 9a, the curve of Figure 9c oscillates less, and the projected area A_1 of the invalid picking space has several lower values of A_1 in the range of 900–1100 mm of L_d, and the range of variation interval is large, indicating that the optimal L_d has little influence on the result value of A_1. It can be seen in Figure 9d that the corresponding change curve of the inclination angle θ_d between the lower arm and the horizontal and the projected area A_1 of the invalid picking space is similar to the letter "W" as a whole, at 25° and 35°. There is an inflection point of the changing trend of A_1 value on the left and right of the angle, and, after 35°, the upward amplitude of A_1 increases sharply, which means that, when the inclination angle θ_d between the lower arm and the horizontal is greater than 35°, it is difficult to obtain the optimal θ_d value. It can be seen from Figure 9e that, when the center distance G is in the range of 650–800 mm, the change range of the ineffective picking space projected area A_1 is small and the low value of A_1 appears intensively; especially in the range of 750–800 mm, the trend is more obvious.

Similarly, the optimization objective function $\min\varphi(L_u, \vartheta_u, L_d, \theta_d, G)$ shows the curve of the result value changing with each structural parameter, as shown in Figure 9f–j.

It can be seen in Figure 9f that the optimal L_u value is located in the region of 1050–1200 mm in the middle of the variation scheme. Although the total structure length L_m corresponding to this range shows an upward trend with the increase in L_u, the change range is small. It can be seen in Figure 9g that, when the inclination angle θ_u between the upper picking manipulator and the horizontal increases, the total structure length L_m will first decrease and then increase when θ_u changes to about 40°. The trend change inflection point, the optimal θ_u is at this angle. The unique solution produced around 40° is characteristically distinct. It can be seen in Figure 9h that the total structure length L_m increases correspondingly with the increase in L_d, but the increased range is different. The optimal L_d is in the range of 800–900 mm, and the value of the total structure length L_m remains in this range, at lower levels, and the increase is small. It can be seen in Figure 9i that, when the inclination angle θ_d between the lower arm and the level changes from 20° to 24°, the total structure length L_m is greatly disturbed, and the obtained L_m value is at a high level. In contrast, in the 24–28° region, the variation in L_m is small and there are many low points of L_m in this region, and the optimal θ_d is more likely to be generated in this region. It is not difficult to see from Figure 9j that, when the center distance G varies from 700 mm to 800 mm, the total structure length L_m presents a "U"-shaped curve, the low value of L_m frequently appears around 750 mm and the optimal G has obvious characteristics around 750 mm.

To sum up, the results of the two optimization objectives are fixed by constants, and whether the other optimization objective has a unique optimal solution within the numerical range of each configuration parameter is explored. By comparing the result intervals of each optimal configuration parameter after analysis, it is found that the intersection part of the intervals is the result of the final selected optimization parameter scheme, which can verify the effectiveness of the bi-objective optimization. When L_u is 1119.3 mm, θ_u is 39.4, L_d is 898.7 mm, θ_d is 26 and G is 755.3 mm, the redundant working space of the upper and lower picking manipulators can be minimized, and the structure length of the dual-arm picking robot is small and the overall configuration is compact.

6. Conclusions

This paper investigates the picking conditions of the dwarf and close planting apple orchard in the Beijing area and obtains the distribution law of mature fruit in the apple canopy by using the crown three-dimensional zoning method. The main configuration of the apple picking robot was determined through analysis of the orchard target workspace, and the structure of the picking robot arm driven by electric and pneumatic telescope was designed by using three-dimensional mechanical design software. Parametric analysis of the relationship between the configuration of the apple-picking robot, the structure parameters of the dual-manipulator and the target workspace is established, and a multi-objective optimization function is established with "avoiding invalid workspace" and "compact structure" as performance indexes.

The CRITIC–TOPSIS combined method is used to evaluate and screen the multi-objective optimal solution. The final optimization result is that the length of the upper picking arm is 1119.3 mm, the horizontal elevation angle is 39.4°, the length of the lower picking arm is 898.7 mm, the horizontal elevation angle is 26° and the outer frame installation base is 755.3 mm from the center of the tree trunk. The working space of the dual-manipulator completely covers the target working space, and the redundancy rate of the working space is 16.62%.

The experimental results show that the sensitivity of the two optimization objectives to the change in configuration parameters is different, and the optimal solution has unique validity within the numerical range of composition parameters, thus achieving the design goal of minimizing the invalid working space and compacting the structure of the dual-manipulator picking robot. The optimization method in this paper can be used for a standardized scene, similar to the standardized dwarf densely planted orchard. At the same time of research and development of picking robots, agronomic requirements need to be completed simultaneously, and the research and development of robots and agronomy complement each other. The dual-manipulator apple picking robot proposed in this paper can also be iterated into a four-manipulator apple picking robot; that is, two groups of arms are combined into two groups of four-arms, and the upper and lower group arms have the same structural parameters, so this iteration has certain practical application significance.

Author Contributions: Conceptualization, Z.X. and Q.F.; methodology, Z.X. and Q.F.; formal analysis, Q.F., T.L. and F.X.; investigation, Z.X., Q.F. and X.G.; resources, Z.X. and Q.F.; data curation, Z.X., Q.F., L.L. and X.G.; writing—original draft preparation, Z.X. and Q.F.; writing—review and editing, Z.X., T.L., F.X., C.L., L.L. and X.G.; project administration, C.Z.; funding acquisition, C.Z. All authors have read and agreed to the published version of the manuscript.

Funding: This research was funded by the Beijing Science and Technology Plan Project (grant number Z201100008020009), the Beijing Nova Program (grant number 20220484023) and the BAAFS Innovation Capacity Building Project (grant number KJCX20210414).

Institutional Review Board Statement: Not applicable.

Data Availability Statement: Not applicable.

Conflicts of Interest: The authors declare no conflict of interest.

References

1. Chen, H.; Wang, Q.; Gao, Q. Analysis of apple industry development and its influencing factors in China-based on panel data of seven major producing provinces. *China Fruits* **2019**, *1*, 92–95. [CrossRef]
2. Zhu, Q.; Lu, R.; Li, F. Research status and development trend of apple picking machinery. *For. Mach. Woodwork. Equip.* **2021**, *49*, 4–9, 15. [CrossRef]
3. Israeli Robotics Firm FFRobotics Aims to Release Robotic Apple Picker by Next Year. Available online: https://www.therobotreport.com/israeli-robotics-firm-FFRobotics-aims-release-robotic-apple-picker-next-year (accessed on 16 November 2022).
4. Li, T.; Qiu, Q.; Zhao, C.; Xie, F. Task planning of multi-arm harvesting robots for high-density dwarf orchards. *Trans. CSAE* **2021**, *37*, 1–10. [CrossRef]
5. Davidson, J.; Bhusal, S.; Mo, C.; Karkee, M.; Zhang, Q. Robotic manipulation for specialty crop harvesting: A review of manipulator and end-effector technologies. *GJAAS* **2020**, *2*, 25–41. [CrossRef]

6. De Preter, A.; Anthonis, J.; De Baerdemaeker, J. Development of a robot for harvesting strawberries. *IFAC-PapersOnLine* **2018**, *51*, 14–19. [CrossRef]
7. Arad, B.; Kurtser, P.; Barnea, E.; Harel, B.; Edan, Y.; Ben-Shahar, O. Controlled Lighting and Illumination-Independent Target Detection for Real-Time Cost-Efficient Applications. The Case Study of Sweet Pepper Robotic Harvesting. *Sensors* **2019**, *19*, 1390. [CrossRef]
8. Feng, Q.; Zou, W.; Fan, P.; Zhang, C.; Wang, X. Design and test of robotic harvesting system for cherry tomato. *Int. J. Agr. Biol. Eng.* **2018**, *11*, 96–100. [CrossRef]
9. Thorne, J. Apple-Picking Robots Gear Up for U.S. Debut in Washington State. Available online: https://www.geekwire.com/2019/apple-picking-robots-gear-u-s-debut-washington-state/ (accessed on 16 November 2022).
10. Li, G.; Ji, C.; Gu, B.; Xu, W.; Dong, M. Kinematics analysis and experiment of apple harvesting robot manipulator with multiple end-effectors. *Trans. Chin. Soc. Agric. Mach.* **2016**, *47*, 14–21, 29. [CrossRef]
11. Mann, M.; Zion, B.; Shmulevich, I.; Rubinstein, D.; Linker, R. Combinatorial optimization and performance analysis of a multi-arm cartesian robotic fruit harvester-extensions of graph coloring. *J. Intell. Robot. Syst. Theory Appl.* **2016**, *82*, 399–411. [CrossRef]
12. Zion, B.; Mann, M.; Levin, D.; Shilo, A.; Rubinstein, D.; Shmulevich, I. Harvest-order planning for a multiarm robotic harvester. *Comput. Electron. Agric.* **2014**, *103*, 75–81. [CrossRef]
13. Zhao, Y.; Gong, L.; Liu, C.; Huang, Y. Dual-arm robot design and testing for harvesting tomato in greenhouse. *IFAC-PapersOnLine* **2016**, *49*, 161–165. [CrossRef]
14. Saunders, S. The Robots that Can Pick Kiwi-Fruit. Available online: https://www.bbc.com/future/bespoke/follow-the-food/the-robots-that-can-pick-kiwifruit.html (accessed on 20 April 2022).
15. Yuan, J. Research progress analysis of robotics selective harvesting technologies. *Trans. Chin. Soc. Agric. Mach.* **2020**, *9*, 1–17. [CrossRef]
16. Xiong, Y.; Ge, Y.; Grimstad, L.; From, P.J. An autonomous strawberry-harvesting robot: Design, development, integration, and field evaluation. *J. Field Robot.* **2019**, *37*, 202–224. [CrossRef]
17. Williams, H.; Ting, C.; Nejati, M.; Jones, M.H.; Penhall, N.; Lim, J.; Seabright, M.; Bell, J.; Ahn, H.S.; Scarfe, A.; et al. Improvements to and large-scale evaluation of a robotic kiwifruit harvester. *J. Field Robot.* **2020**, *37*, 187–201. [CrossRef]
18. The Latest on FF Robotics' Machine Harvester. Available online: https://basinbusinessjournal.com/news/2021/apr/12/machine-picked-apples/ (accessed on 16 November 2022).
19. Zitter, L. Berry Picking at Its Best with AGROBOT Technology. Available online: https://www.foodandfarmingtechnology.com/news/harvesting-technology/berry-picking-at-its-best-with-agrobot-technology.html (accessed on 16 November 2022).
20. Feng, Q.; Ji, C.; Zhang, J.; Li, W. Optimization Design and Kinematic Analysis of Cucumber-harvesting-robot Manipulator. *Trans. Chin. Soc. Agric. Mach.* **2010**, *41*, 244–248.
21. Sun, F.; Yin, X. The optimization of structural parameters of robot arm for hotel delivery. *Mach. Des. Manuf. Eng.* **2019**, *11*, 23–27. [CrossRef]
22. Zhao, J.; Xiu, B.; Wang, J.; Zhang, X. Structural Parameters Design of Rescue Manipulator Based on Multi-Objective Optimization. *Trans. Beijing Inst. Technol.* **2022**, *5*, 493–501. [CrossRef]
23. Zhang, Q.; Wei, Q.; Shang, Z. Analysis of high-quality and abundant tree structure and light condition of dwarf anvil apple orchards in Beijing. *J. Fruit Sci.* **2013**, *30*, 586–590. [CrossRef]
24. Dong, R.; An, G.; Zhao, Z.; Mei, L.; Li, M. Comparison of intra-crown illumination and its growth and yield of dwarf root-root anvil apples in different tree shapes. *Sci. Agric. Sin.* **2013**, *46*, 1867–1873. [CrossRef]
25. Tian, H.; Ma, H.; Wei, J. Research on working space and structural parameters of tandem robot manipulator. *Trans. Chin. Soc. Agric. Mach.* **2013**, *44*, 196–201.
26. Deb, K.; Jain, H. An evolutionary many-objective optimization algorithm using reference-point-based nondominated sorting approach, part I: Solving problems with box constraints. *IEEE Trans. Evol. Comput. A Publ. IEEE Neural Netw. Counc.* **2014**, *18*, 577–601. [CrossRef]
27. Chang, H.; Li, W.; Dong, F.; Guo, X. Research on multi-objective optimization of cold chain logistics distribution path based on NSGA-II. *Technol. Econ. Areas Commun.* **2022**, *2*, 8–17.
28. Shih, H.; Shyur, H.; Lee, E. An extension of TOPSIS for group decision making. *Math. Comput. Model.* **2007**, *45*, 801–813. [CrossRef]
29. Diakoulaki, D.; Mavrotas, G.; Papayannakis, L. Determining objective weights in multiple criteria problems: The critic method. *Comput. Oper. Res.* **1995**, *22*, 763–770. [CrossRef]

MDPI AG
Grosspeteranlage 5
4052 Basel
Switzerland
Tel.: +41 61 683 77 34

Agronomy Editorial Office
E-mail: agronomy@mdpi.com
www.mdpi.com/journal/agronomy

www.ingramcontent.com/pod-product-compliance
Lightning Source LLC
Chambersburg PA
CBHW061554220926
51850CB00005B/335
9783725816200